Visual Basic 5

Environment, Programming, & Applications

Alan L. Eliason
Ryan Malarkey

201 West 103rd Street
Indianapolis, IN 46290

Visual Basic 5: Environment, Programming, & Applications

For information, address Que Education & Training, Macmillan Computer Publishing, 201 W. 103rd St., Indianapolis, IN 46290.

International Standard Book Number:
1-57576-867-4

Library of Congress Catalog Card Number:
97-67453

00 99 98 4 3 2 1

Interpretation of the printing code: the rightmost double-digit number is the year of the book's printing; the rightmost single-digit, the number of the book's printing. For example, a printing code of 98-1 shows that the first printing of the book occurred in 1998.

Trademark Acknowledgments

Screens reproduced in this book were created using Collage Plus from Inner Media, Inc., Hollis, NH.

Composed in *AGaramond* and *MCPdigital* by Que® Education & Training

Publisher
Robert Linsky

Executive Editor
Randy Haubner

Managing Editor
Nancy E. Sixsmith

Product Marketing Manager
Susan L. Kindel

Development Editor
Songlin Qiu

Production Editor
Thomas Cirtin

Technical Editor
Alfonso Hermida

Copy Editors
Gina Brown
Susan Hobbs
Brian Sweany
Leah Williams

Operations Manager
Elizabeth Brown

Book Designer
Louisa Klucznik

Production
Kim Cofer
Cindy Fields
Daniela Raderstorf
Laure Robinson

Indexer
Sandra Henselmeir

Contents at a Glance

Table of Contents

About the Authors

Alan L. Eliason is an associate professor of Computing and Information Science and Applied Information Management at the University of Oregon in Eugene. He currently is a visiting professor at Willamette University. His areas of specialization include visual programming languages, business computing applications, systems analysis, and database design. He, together with Ryan Malarkey, conducts numerous workshops on Visual Basic and database design.

Ryan Malarkey is the president of Malarkey and Associates, a software development and corporate training firm located in Portland, Oregon. He specializes in visual programming languages and client/server design. He conducts workshops on Visual Basic, Java, and database design.

Acknowledgments

Many people were important in helping us write and polish this book. We would like to thank the many students who attended our Visual Basic workshops, giving us comments and suggestions on improving the chapter materials. They helped us select topics, which they considered to be essential in their learning of the language.

We also must mention the marvelous development team at Que Education & Training. Special thanks is given to Randy Haubner, executive editor, for his enthusiastic backing of the work; Songlin Qiu, development editor, for her careful eye in reviewing the material and keeping us on schedule; Tom Cirtin, production editor, for his careful review of all chapter material; and Alfonso Hermida, technical editor, for checking all program submissions and catching those errors and omissions that are so difficult to spot. Finally, we want to thank the production team at Que E & T for making the book so pleasing to follow and read.

Introduction

Visual Basic is the most popular programming language for the world's most popular operating system. By encapsulating the complexities of the Windows application program interface (API) into easily manipulated objects, Visual Basic is the first language people consider when they want rapid application development for the Windows environment. The capability of custom controls to easily extend the language has made Visual Basic a popular choice for an amazingly wide variety of programming tasks.

However, the easy accessibility of the language and its enormous breadth pose challenges to both the students and the instructor. Students approach the language from a wide variety of backgrounds and abilities: Some are new to programming; some have extensive programming experience in other languages (often character-based procedural languages); some want to learn the language to accomplish a very specific task.

Frequently, instructors are challenged by the variety of students who come together in a course. A senior engineer from an aerospace company sits right next to a programming neophyte. Some students are comfortable working with a visual programming environment, while others find the design paradigm quite difficult. Consequently, teaching materials must be flexible enough to accommodate a broad range of backgrounds.

Key Features of This Book

Because knowledge is cumulative, each chapter of this book builds on the previous one, as the concepts and exercises become progressively more challenging. The material begins with easy-to-learn design environments and computer programs that require no previous knowledge of programming. This book, therefore, works well for the beginning programmer.

All of the chapters contain essentially the same elements, which are designed to explain the concepts involved in developing Visual Basic programs and guide the students through putting these concepts into practice. The chapter elements are as follows:

- Each chapter begins with a list of tasks that students will be able to perform after reading the chapter material and working through the exercises.
- Each chapter is organized around objectives that distill the tasks into meaningful units of explanatory text and exercises. Visual Basic concepts are introduced and then put into a practical context. Our belief is that by working through a number of exercises, students will quickly learn the language.
- Each chapter ends with a summary of the key concepts and a set of skill-building exercises that call upon knowledge acquired in the chapter. These final exercises provide opportunities for students to wrestle with programming problems on their own so that the concepts and techniques presented in the chapter become solid skills.

New Features

We provided a more comprehensive examination of object types, variables, and collections in this book than in the previous edition. We added a complete chapter on designing class modules (Chapter 13). We modified the applications part of the book to provide special emphasis on modular design (Chapter 22), ActiveX DLLs (Chapter 23), working with Objects from the Microsoft Office suite (Chapter 26), and using ActiveX documents for creating Internet applications (Chapter 28).

Prerequisites

A working knowledge of Microsoft Windows 95 is assumed, as is familiarity with the use of a mouse, basic file-handling skills, and basic editing skills. For example, students should know how to open a file and save it to a disk, and know how to cut, copy, and paste text. Students should also be conversant with basic computer terminology, such as *icon, window,* and *menu bar.*

Using This Book

We have discovered that the best way to learn the language is by working through a large number of exercises. To this end, each chapter contains several exercises for the students to study and complete.

The Student Disk provided with the book contains portions of most of the exercises, but not all. For example, after students learn to design a user interface, we provide the interface. What remains is to complete the programming. For large programs, most of the code is also provided, and the students complete the programming. The Student Disk contains no help for the skill-building exercises at the end of each chapter: The students are expected to be able to complete those on their own.

The Organization of This Book

To simplify the learning process required in mastering Visual Basic, this book follows a three-part sequence of general topics: environment, programming, and applications.

Part I: Environment

The first part of the book addresses the Visual Basic design environment, teaching some programming, but very little. We stress the importance of design in visual programming and the event-driven nature of the language. In this way, we can introduce the standard Visual Basic controls and reveal the nature of an event-oriented language to those students who come to Visual Basic from other programming backgrounds.

Part II: Programming

Part II is more challenging and contains some complex, yet important, subjects. We introduce core programming concepts, some of which will be new to most students—including those with previous programming experience. These include standard Visual Basic data types, conditional tests, variable and procedure scoping, loops, object types, class modules, user events, single and multidimensional arrays, numeric and string functions, and file processing.

Contrary to our own expectations that this portion of the text would prove the least adaptable to the needs of a wide variety of students, the opposite has proven to be the case. The topics in this part of the book often lead to our most successful classroom experiences.

Part III: Applications

Material from Part I and Part II is combined in Part III for the building of actual Visual Basic applications. We introduce application development with a heavy emphasis on application design, Visual Basic's extensive record-handling routines, and Visual Basic's built-in Jet database engine. Some books slight the record-handling to focus more attention on databases; this one gives appropriate attention to both.

This part of the book becomes challenging when Data Access Objects are introduced and used in the exercises (important in database applications). This final part also examines working with objects from Microsoft Office, creating ActiveX documents, and building Internet applications with these documents.

Additional Resources

The appendices in this book provide a handy reference for two important topics: the entire ANSI code as supported by Microsoft Windows, which is frequently referred to in this text, and the collections, properties, and methods of various Data Access Objects, which are vital to the chapters in Part III.

Resources of particular importance are the Student Disk, which is bundled with this book, and the Instructor's Resource Disk, which is provided by Que Education & Training free of charge upon request.

The Student Disk contains a README file and a program that contains compressed files for the chapters. The README file provides step-by-step instructions for uncompressing the self-extracting file stored on this disk. The resulting files can be viewed using Microsoft Explorer.

Also on the uncompressed disk are chapter folders (for example, Ch06), which hold the chapter exercises. When we placed an exercise on the disk, part of the design might have been completed or part of the program might have been written for the students, but not all of it. The students must complete the exercise. Further, the databases that are used throughout the book are included on the Student Disk.

The Instructor's Resource Disk contains lecture notes, answers to the end-of-chapter exercises, test bank questions and answers, as well as instructional resources.

Part I Environment

1

Running a Visual Basic Program

Welcome to the world of Visual Basic. Released by Microsoft in 1991, Visual Basic was designed to be a visually oriented programming language in contrast to the popular languages of that time (Pascal, C, COBOL, and FORTRAN). Although Visual Basic is similar to QBasic—the *procedural* language supplied with every version of MS-DOS beginning with version 5.0—it contains important extensions that make it more of an *object-oriented* language.

The newest version of Visual Basic is more object-oriented than ever. It is capable of handling software development projects of enormous scope and depth. Visual Basic is now one of the most flexible and powerful visual object-oriented computer languages available, and it remains the most popular language for the world's most popular operating system.

One way of describing Visual Basic's nature is to say that when the computer programmer develops programs in Visual Basic, data is more often than not approached as an *object* rather than just numeric or text information. Data objects, like real-world objects, such as desks and chairs, have properties. Desks and chairs could be said to have a "LegCount" property, which describes the number of legs for that object, whether it is a three-legged stool or a four-legged desk. Similarly, a data object that held information about a store's customers might have a `CustomerCount` property. Unlike the "LegCount" physical property of a real-world chair, you can easily change the `CustomerCount` property of a data object.

Objects in Visual Basic have many features that are introduced in this chapter and explored throughout this book. When you complete this chapter, you will

- comprehend the basic concepts of object-oriented programming;
- understand the Visual Basic design environment;
- understand the essential programming components of Visual Basic;
- understand the three key object features: properties, methods, and events;
- be able to launch Visual Basic;

- be able to use Object Browser to examine Visual Basic objects;
- be able to run Visual Basic programs.

Chapter Objectives

Every chapter of this book contains a set of objectives. Read through them prior to reading each chapter, and review them at the end of the chapter to determine whether you have realized the objectives. In this chapter, you will not write a Visual Basic program. Instead, you will explore the Visual Basic program design environment and run several sample Visual Basic programs.

Before you turn on your computer and start working, you should consider briefly the Visual Basic programming language to understand how it differs from other types of programming languages. Consider the Visual Basic architecture and learn a number of terms, including forms, controls, properties, methods, events, statements, and system objects. The objectives of this chapter are as follows:

1. Learning how Visual Basic differs from other programming languages
2. Understanding the relationship of Visual Basic to Windows
3. Understanding objects as instances of classes
4. Running sample Visual Basic programs

Objective 1.1 Learning How Visual Basic Differs from Other Programming Languages

Of the 1,000 or so computer languages that have been developed, each language can be categorized based upon the following criteria:

- Low-level or high-level
- Procedure-oriented or event-oriented
- High-visual or low-visual
- Interpreted or compiled

Imagine: 1,000 or so languages! Where does Visual Basic fit? This section helps to explain not only where Visual Basic fits, but why it is different from many other programming languages.

1.1.1 Low-Level and High-Level Languages

A computer language can be described as a low-level or high-level language based on how close the language is to machine language (which depicts a low-level language) or to English (which depicts a high-level language). Let's look at what we mean by this difference in language.

Low-Level Languages

Low-level languages are machine oriented. These languages work close to machine language, which is limited to 0s and 1s. Why only 0s and 1s? This is the language the computer understands. By making a language closer to 0s and 1s, the speed at which the computer processes data improves.

An example of a low-level language is *assembly* language code. With this language, such instructions as the following tell the computer to save, move, add, and store the results of processing:

```
PUSH BX
PUSH AX
MOV  AX, @A
MOV  BX, @B
ADD  BX, AX
MOV  @B, BX
POP  BX
POP  AX
```

Assembly language code uses *mnemonics* (memory aids), for which such words as `ADD` and `POP` make it easier to remember what an assembly instruction does. While not quite 0s and 1s, you deal with the exact steps the computer must take in processing data when programming in assembly language.

Low-level languages are said to have a one-to-one relationship with the computer: A programmer must write an explicit instruction for every operation of the machine. A low-level language is precise. Programmers use low-level language code for such tasks as writing operating systems (the software that enables your computer to operate).

High-Level Languages

High-level languages are more people-oriented. These languages have a one-to-many relationship, in which one instruction leads to a series of machine-level instructions. These languages feature more English and English-like words. In Visual Basic, examples of these English-like words are `If`, `Else`, `Dim` (for *dimension*), and `OpenDatabase`.

Because of this one-to-many relationship, high-level languages are easier to learn, use, and understand. However, they do require more machine time to translate a single instruction into a set of machine-level instructions.

Programmers (those who write computer programs) use high-level languages in writing application programs. For example, a high-level language, such as Visual Basic, would be used to write the instructions for processing a company payroll. Application programs define the ways by which users are able to use the computer.

1.1.2 Procedure-Oriented and Event-Oriented Languages

Besides low-level and high-level languages, computer languages can be classified as procedure-oriented or event-oriented. Procedure-oriented languages tend to run without human interference or the taking of some action by the user: A computer program is executed by a simple *run* instruction, and usually runs from top to bottom, with all the code executed until the program ends. Event-oriented languages are different in that they depend on the user: They wait for the user to take some action before they execute. As you will soon discover, the program waits for an event (or happening) to occur before beginning a program execution.

Procedure-Oriented Languages

Prior to 1990, most commercial high-level languages were procedure-oriented languages. The emphasis in writing a computer program was to identify a set of processing tasks and to describe the steps important to each task. Collectively, the set of tasks represented a procedure: a listing of a set of tasks required to perform an activity. As an example, consider the procedure required in processing an employee paycheck, in which the steps of the procedure are expressed as tasks:

1. Get employee name.
2. Get hours worked.
3. Get hourly wage.
4. Multiply hours worked by the hourly wage to compute gross pay.
5. Compute taxes based on gross pay.
6. Subtract taxes and other deductions (such as union dues) from gross pay to compute net pay.
7. Print the employee's check.

This procedure could be used in writing a QBasic program. The following code listing shows a partial QBasic program written to print an employee's check. Even though you may not know the QBasic language, you should be able to understand, step by step, how the computer processes an employee paycheck.

```
'Compute and print a payroll check
'Initialize variables
emp.name$ = "Roger Rabbit"
pay.date$ = "06/15/99"
hours.worked = 40           'Total hours worked
rate = 7.50                 'Pay per hour
tax.rate = 0.25             'Tax percentage

'Compute gross and net pay
gross.pay = hours.worked * rate
taxes = gross.pay * tax.rate
net.pay = gross.pay - taxes

'Display the results
LPRINT TAB(40); "Date: "; pay.date$
LPRINT
LPRINT "Pay to the order of: "; emp.name$
LPRINT
LPRINT "Pay the full amount of: "; gross.pay
LPRINT TAB(25); "---"
```

```
LPRINT
LPRINT TAB(40); "-----------------"
LPRINT TAB(40); "W. Pinchpenney, Treasurer"
```

Event-Oriented Languages

Event-oriented languages became possible with the advent of the Macintosh operating system for Apple Macintosh computers and Microsoft Windows for MS-DOS computer systems. Both environments were designed to bring hardware and software together into a standard user interface by employing a *graphical user interface,* or GUI (pronounced goo-ey). A GUI simplifies learning: Once you learn how to work with one application using the interface, it is easy to learn another application because the interface remains the same.

An event-oriented language implies that an application (the computer program) waits for an *event* to occur before taking any action. What is an event? It might be the press of a key on the keyboard or the click of a mouse button. With these events (there are many types), the computer waits for a key press or a mouse click (pushing a button on a hand-held mouse).

1.1.3 Low-Visual and High-Visual Languages

Computer languages can be described as having a low-visual or a high-visual orientation. Prior to 1990, most languages were low-visual languages, so programmers had considerable difficulty designing the computer forms and reports, data entry screens, and navigation tools by which to move from one area of a computer program to another. High-visual languages greatly simplify the tasks of designing forms, screens, and navigation tools. For this and other reasons, Visual Basic is known as a *rapid prototyping language.* Its design tools let you quickly design a version or prototype of a computer application.

Low-Visual Languages

Let's consider how a low-visual language differs from a high-visual language. Low-visual languages are not supported by a GUI. Instead, the programmer usually works with a blank terminal screen, adding line after line of instruction to that screen. After the instructions are entered, the programmer issues a `Run` or `Execute` command to execute the instructions. Only at this time do visual images appear on the screen. Those of you who use common DOS-level commands, such as `Copy`, `Del`, or `CD`, use a low-visual language.

High-Visual Languages

High-visual languages are supported by a GUI design environment. Figure 1.1 shows the Visual Basic design environment after startup. The appearance of the design environment is conceived to improve speed in program design and can even be customized to suit your needs. At the top of the screen is a menu bar, which contains the File, Edit, View, Project, Format, Debug, Run, Tools, Add-Ins, Window, and Help menus. Below the menu bar is a toolbar (see Figure 1.2).

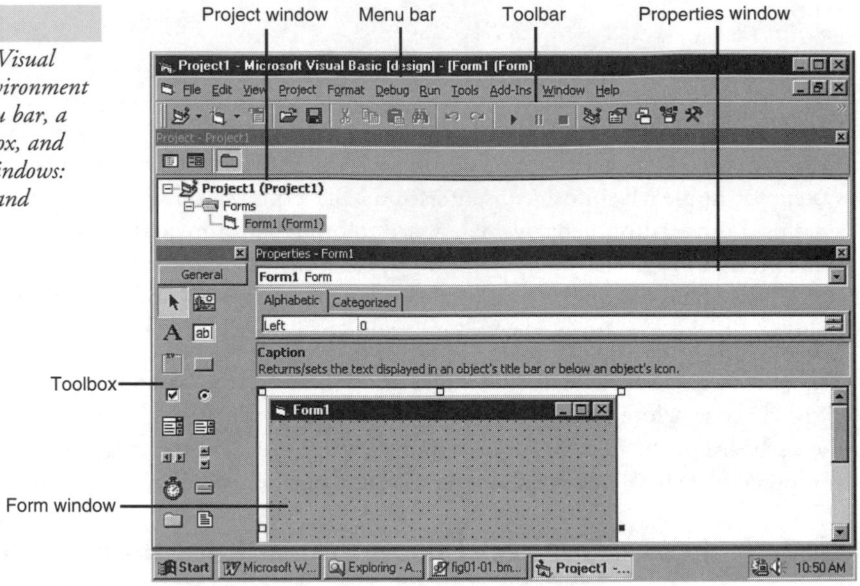

Figure 1.1

The integrated Visual Basic design environment contains a menu bar, a toolbar, a toolbox, and three types of windows: Form, Project, and Properties.

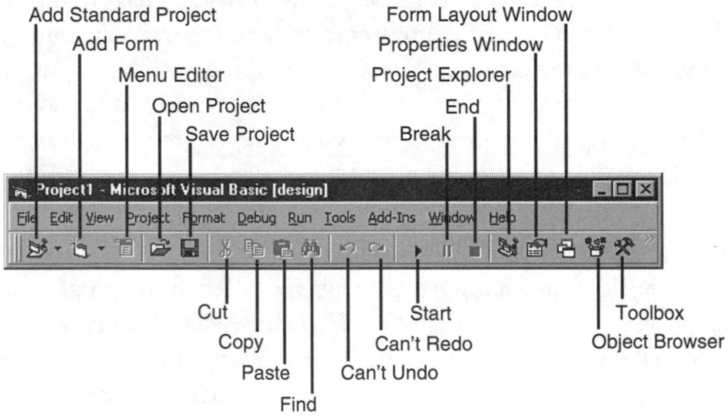

Figure 1.2

The Visual Basic toolbar contains commonly used buttons designed to speed your program development.

The buttons appearing on the toolbar allow quick access to the most commonly used commands. Figure 1.3 shows the toolbox, which you can drag around the screen and place where you want. As you work your way through the exercises contained in this book, you will become familiar with the use of each tool in this toolbox. Finally, the Visual Basic startup screen often contains a Form window, a Project (Explorer) window, and a Properties window. Figure 1.1 shows all three of these windows. Chapter 2, "Writing and Running Your First Visual Basic Program," asks you to use each in constructing a Visual Basic program.

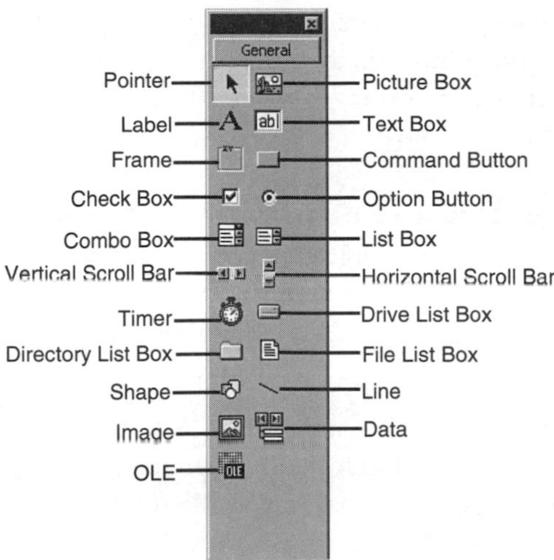

Figure 1.3
The Visual Basic standard toolbox contains the standard controls for constructing a visual application.

Pause and Breathe for a Moment!

Looking at the toolbar and toolbox the first time might fill your heart with fear. You might exclaim, I can't remember what all those icons mean! With Visual Basic, help is readily on hand. To quickly identify a control, just let your mouse pointer linger over an item in the toolbox. A ToolTip appears by your pointer and identifies the control. The same applies for buttons on the toolbar. You can also click the button you want to inspect and press the Help key: the F1 function key on the keyboard. The Visual Basic Help screen appears with a description of that button.

1.1.4 Interpreted or Compiled Languages

Computer languages can be interpreted or compiled. An interpreter executes a program, while a compiled language uses a compiler to translate the high-level language into machine language. Visual Basic is now both an interpreted language and a compiled language. The developer has a choice when creating a Visual Basic executable whether to make a compiled program or an interpreted program. You will create both types of programs in an exercise in Chapter 2.

The main advantage of an interpreted language is immediate response. Program development often goes faster because the code instructions can be easily modified and immediately tested without being compiled (or translated) first. This saves you considerable time in writing and testing a program. The main disadvantage of an interpreted program is speed of execution—especially in processor-intensive instructions. An interpreted program must translate instructions each time a program is run. This is not required of a compiled program.

The advent of compilation in Visual Basic 5 has dramatically sped up numeric calculations and most other computations. The speed by which forms now load into memory in Visual Basic 5 is one of the language's more dramatic speed improvements. But compilation alone does not guarantee speed. Design of a program is one of the more important determinants of performance. Visual Basic programs, whether they are interpreted or compiled, rely upon other runtime libraries of functions in order to execute. These libraries are probably the largest factor in Visual Basic performance.

1.1.5 The Recent Evolution of Visual Basic

In the last several years, Visual Basic has evolved rapidly. Until recently, Visual Basic was a proprietary language used only by Microsoft products. Microsoft now licenses Visual Basic for Applications to those software developers who want to add programmability to their applications. This will increase Visual Basic's prevalence.

Visual Basic is now the universal macro language for the Microsoft Office suite of applications. The newest version of Word, for instance, has migrated to Visual Basic for Applications. Other large programmable applications, such as Microsoft Project, also use Visual Basic for Applications.

Another important development of the Visual Basic language is the advent of the Visual Basic Scripting Edition (also known as VBScript) for developing Internet-enabled applications. The capability of Visual Basic to create downloadable ActiveX components and the capability of VBScript to manipulate Internet browsers and Internet documents (HTML documents) suggests Visual Basic will play a major role in the explosion of Internet and *intranet* applications.

This text assumes you have the Standard, Professional, or Enterprise Editions of Visual Basic. All chapters other than the database chapters can be completed using the Standard Edition of Visual Basic, but you will need the Professional Edition to work with Data Access Objects in Chapter 25, "Using Data Access Objects." Microsoft Office will also need to be installed on your computer to take advantage of Chapter 26, "Working with Objects from Microsoft Office," which introduces working with Microsoft Office applications from Visual Basic.

Objective 1.2 Understanding the Relationship of Visual Basic to Windows

All application programs (the kind of programs developed with such languages as Visual Basic) must be developed to work with a specific operating system. Application programs interact with operating systems by using operating system functions that applications developers can exploit. These functions are usually referred to as *application program interface* (API) functions. API functions allow programmers to add such functionality to their programs as opening and saving files, checking system hardware, using the computer's internal clock, and, in the Microsoft Windows family of operating systems, creating, modifying, and communicating with Windows.

The Windows API is enormously complex. It has over 800 functions, some of which can perform a wide variety of tasks. The value of Visual Basic is that it greatly simplifies

Windows programming. Visual Basic *encapsulates* the Windows API into objects that Visual Basic programmers can manipulate. For example, when you use the C programming language, creating a program that greets the user with the message "Welcome to Visual Basic" can take several lines of instructions. In Visual Basic, however, the same program requires only a single line.

1.2.1 Forms and Controls

What, exactly, does the phrase "encapsulates the Windows API" mean? Encapsulation is a key concept for any object-oriented programming language. It refers to the capability of an object to hide its internal workings from other objects. In turn, the object allows itself to be manipulated by the programmer and other objects through the use of three key object features: *properties*, *methods*, and *events*.

Let's consider two kinds of objects that you have seen in the Visual Basic startup screen: *forms* and *controls*. Figure 1.1 shows a Window that contains a form named `Form1`. This form acts as a container for controls. You could think of `Form1` as a blank piece of paper that you must fill in. You use this blank form to design the user interface for a Visual Basic application. When you begin the design of a Visual Basic program, you start with a blank form and, with a set of tools, begin to design the user interface.

The tools you use to construct the interface are called controls, which are attached to forms. Examples of controls are shown on the toolbox (see Figure 1.3) and include a text box (a box the user types text into), a command button (an on-screen button that the user can click), and a label (an area on the form to place a text heading). If the toolbox is not visible, select View, Toolbox on the menu bar, or click the Toolbox button on the toolbar to display it.

Because forms and controls are objects, they can be manipulated by properties, methods, and events. Let's consider each of these object features in turn.

Object Properties

A *property* is a named attribute of an object. Properties are used to change the appearance and behavior of objects, such as forms and controls. These properties include attributes affecting the color, size, and name of the object.

Let's take an everyday object, such as a shirt. Suppose you write the following:

```
Shirt.color = "Blue"
```

In Visual Basic, this would tell you that you have an object named `Shirt`. One of the properties of the shirt is `color`. You can change the color by assigning a new one, such as `"Blue"`. You might want to set the `Shirt` property as

```
Shirt.launder = "Clean"
```

as opposed to

```
Shirt.launder = "Dirty"
```

In the example, the condition of the shirt—clean or dirty—can be set directly. Wouldn't it be great if your shirts were always clean?

For a Visual Basic object, such as a form, you can write instructions to alter it. The following instruction changes the color of a form named `Form1` to blue:

```
Form1.BackColor = vbBlue
```

Object Methods

The concept of an object method is more difficult than the concept of a property. *Methods* are computer instructions (code) held by an object that operate on the object's data. Methods are written in a different way than properties. The *syntax* (required way) for describing the application of a method to an object is written as follows:

```
Object.Method
```

Perhaps an analogy will make methods more understandable. Suppose you had a very compliant dog named Sparky. By invoking the following methods of Sparky, you would get the behavior from Sparky that you desire:

```
Sparky.eat
Sparky.bark
Sparky.scat
```

Object methods in Visual Basic are similar. One method for a List Box control (a control that displays a list of different text items) is the `Clear` method. To clear the contents of the list box, you would write the following instruction:

```
List1.Clear
```

When working with object properties in Visual Basic, you are either assigning them a new value (for example, `Form1.BackColor = vbBlue`), or you are reading their values (for example, `mycolor = Form1.BackColor`). Object methods, by contrast, are invoked with the `Object.Method` syntax (for example, `List1.Clear`). The distinctions between object methods and object properties will become familiar to you as you work more with Visual Basic objects.

Object Events

While the value of a property can be set and a method can be invoked, an object *event* is an action taken by the object when notified by a message. An object event might be a mouse click. The handling of object events is an operating system responsibility. The Windows operating system sends messages to windows (objects) for different events.

Remember the words *message* and *action.* If the user clicks a mouse button on a Visual Basic form, the operating system sends a message to the form. The Visual Basic designer can choose to execute code (the action) in response to the message or ignore it. For example, an image on a form might be programmed to take some action when it is clicked. It might take some other action when it is dragged across the form. The Visual Basic programmer writes coded instructions for the events that should be assigned to the form or control. Visual Basic code, for instance, executes instructions by invoking a form's `Print` method when the user clicks the form:

```
Private Sub Form_Click
    Form1.Print "You clicked the form!"
End Sub
```

1.2.2 A Closer Look at Encapsulation

Object properties, methods, and events make up what is known as the objects *interface.* This interface may be considered the *public view* of the object, and it describes how that object can be manipulated. Objects also have a *private view:* the data and code that the object does not expose to public view, but retains in order to execute its functions. When you invoke the `Clear` method of a List Box control, for example, you do not know how the list box clears its contents. It just does it, and you know that it works. To illustrate encapsulation, let's conduct a brief and simple exercise.

Exercise 1.1 Illustrating Encapsulation

In this exercise, you don't have to write any code. You simply run an executable that is included on the companion disk.

You must have Visual Basic installed on your computer; otherwise, the executable will not run. Using the Windows Explorer, you can run the demo1.exe program from the companion disk (in the ch01 folder) or copy it to your computer's hard disk and run it from there.

1 To run the program, click it. You should see the interface shown in Figure 1.4. The program consists of only two Text Box controls placed on a form. No Visual Basic code was written at all.

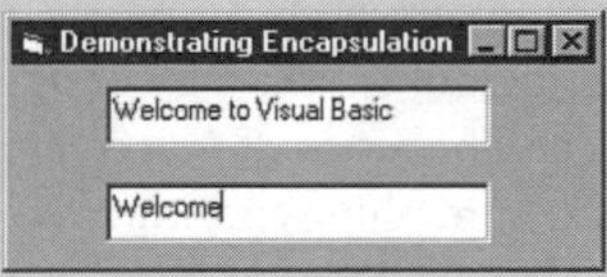

Figure 1.4

This application places two text boxes on a form, but contains no Visual Basic code.

2 Set the insertion point in the upper text box, and type the text **Welcome to Visual Basic**.

continues

3 Select the text you have written using your mouse (or hold the Shift key down and use the directional keys).
4 Copy the selected text onto the Windows Clipboard (Ctrl+C).
5 Using the Tab key, shift to the lower text box and paste the text (Ctrl+V) from the Clipboard into the lower text box.
6 Using the Backspace key, delete part of the text that you pasted.
7 Exit the application by clicking the Close button in the upper-right corner of the form or using the System menu (the choices presented when you click the small button in the upper-left corner of the form).

Consider all the functionality contained within this application without a single line of Visual Basic code being written. Three windows, a form, and two text boxes, were created. One of these windows, the form, uses the default colors specified in the Windows Control Panel. This form can be resized, minimized into an icon, or maximized to fill the entire screen. It can be moved around the screen with the mouse.

The two text boxes allow the user to erase existing text and type text. Text could be copied and pasted. The text controls even indicate the user's location by displaying a cursor.

How does the Text Box control manage the insertion point? How does it "know" to erase text when the Backspace key is pressed? You don't know. That's the *private view* of the Text Box object. You don't really have to know the inner workings of the Text Box object. It works, and you can use it to build your own applications. That's the power of encapsulation!

1.2.3 Visual Basic Statements and Instructions

The discussion of objects and encapsulation in section 1.2.1, "Forms and Controls," uses several lines of Visual Basic code, including examples of object properties being assigned values and methods being invoked. The terms *code* and *instructions* can be used interchangeably, but the term *statement* means something specific in Visual Basic.

Visual Basic Statements

A *statement* differs from the assignment of a property or invocation of a method. Consider the following definition: A statement is a *reserved word* that causes a computer program or the operating system to do something. Statements are instructions (code) that are built in to the Visual Basic language. Methods are code held by an object.

If you look at the Visual Basic Help, you discover that some *keywords* are statements. For example, the keyword `End` is a statement that ends (stops) the execution of a Visual Basic program. The keyword `Kill` (sorry, this is in the language) is a statement that deletes a file from a disk.

As another example, the statement to force explicit declaration of all variables is written as follows:

```
Option Explicit
```

In this example, no reference to an object is required because the statement is built in to the language.

Visual Basic Program Instructions

Methods, statements, and Visual Basic instructions are different. An instruction is simply a line of code in a Visual Basic application. Instructions contain the names of constants, variables, functions, properties, methods, and built-in Visual Basic statements. The rule to remember regarding instructions is that all instructions begin with a statement or the invocation of an object method. For example, an instruction beginning with a statement is as follows:

```
End
```

The following is an instruction beginning with a method:

```
List1.AddItem Entry
```

The method is `AddItem` and the object is `List1`.

Finally, some instructions beginning with methods or statements also contain what are known as statement or method *arguments*. Consider the following:

```
Form1.Move Left + (Width \ 10), Top + (Height \ 10)
```

In this instruction, `Move` is a method and `Left`, `Width`, `Top`, and `Height` are properties of the object, `Form1`, that are used as method arguments. An example of a built-in Visual Basic statement that takes an argument is the `AppActivate` statement:

```
AppActivate "Calculator"
```

In this case, the statement shifts focus to a running instance of the Calculator utility bundled with Windows.

This all may seem a bit confusing right now, but you will get used to working with objects, methods, and statements once you begin writing Visual Basic programs.

Objective 1.3 Understanding Objects as Instances of Classes

In Exercise 1.1, the two text boxes and the form are referred to as *windows* as well as objects. In a Windows program, just about everything that you see on the screen is a window, and many items that you don't see are windows.

In a standard Windows word processing program, such as Word or WordPerfect, every button on the toolbars is a separate window. Every open document is a separate window, as are the dialog boxes that permit you to select different font properties, and the rulers that allow you to set tabs and adjust page width. These windows are *objects,* which have their own properties, methods, and events. The Windows operating system sends messages to the windows to respond to user events.

1.3.1 Object Classes

How many different kinds of windows can there be? As many as a programmer can imagine and create! The design of the Windows operating system provides several standard window types called *classes.* There can be many classes of objects in addition to the standard types. Whether you know it or not, you probably are already familiar with several standard Windows classes and their behaviors, including the following:

- `Button`
- `ComboBox`
- `Edit`
- `ListBox`
- `Scrollbar`

As you might suspect by now, some of the controls in the Visual Basic toolbox are related to these standard Windows classes. For example, a Text Box control is closely related to the base Windows Edit class.

A *class* is like a cookie cutter: It defines how an object should look and behave when it is first created. To work with Visual Basic objects, you first use this cookie cutter to cut out the uniform shape of your object. After you have this basic shape (called an instance of the class), you can manipulate the object by altering its properties or by calling one of the object's methods to make the object behave in a certain way.

Often Visual Basic creates an instance of a class for you. When you add a Text Box control to a form, Visual Basic creates an instance of the Visual Basic Text Box control class. After it is created, you can manipulate it. To alter text displayed in a text box object named `txtDisplay`, for instance, you would execute the following instruction:

```
txtDisplay.Text = "Welcome to Visual Basic"
```

In other circumstances, you will be required to create an instance of the class (an object) yourself. In this book, you will even define your own classes and then create multiple instances (objects) from the class you have defined.

1.3.2 Invisible Objects in Visual Basic

Not all objects in Visual Basic are visible and displayed to the user. Neither are all object windows. Visual Basic contains system objects that are available to use, although they are not displayed in the toolbox, including the following:

- `App`—This object stores information about the application, including its name, path (where it is stored), and the files it requires. This object can also activate an application.
- `Clipboard`—This object is a temporary storage location. It allows text, graphics, and computer code to be transferred from one part of an application to another part, or from one application to another.
- `Screen`—This object references the entire video display screen. It manipulates the size and placement of forms on the screen.
- `Debug`—This object permits the display of information about a program while it is being developed.
- `Printer`—This object determines how text and graphics are printed on a page and how output is sent to the printer.

These are system objects because the Windows operating system keeps information about each one. For example, when you set up Windows, you must designate the computer printer to be used to send output to the printer. When a specific printer is selected, Windows stores information about its makeup, such as the paper width and length. In Visual Basic, you never have to use your printer's name in a program. Instead, you can instruct your printer to print directly by using the `Printer` object.

System objects will become important later. For now, you should simply be aware that they exist and of what they do.

1.3.3 Software Components

After all of this discussion of classes and objects, you might be asking yourself, Why all the fuss about objects? What are the advantages of using objects?

The real benefit of objects is that, developed properly, they provide software that is reusable and robust. The continuing dilemma of software development has been to create reliable code much faster and at lower cost. Some claim that it is an impossible task. More than a few software developers have a large sign posted above their doors: "Software Developed: Fast, Cheap, Good. Choose Two."

Although object-oriented design has certainly not proved to be a magic bullet to shoot the beast of software development costs, it has helped create reliable, reusable software *components*. Visual Basic components include the following:

- Visible controls (the tools represented in the Visual Basic toolbox)
- Software libraries that can be reliably used by many applications (ActiveX DLLs)
- Executable programs that can be used by other programs (ActiveX EXEs)

These components make single or multiple objects available to the developer.

Different types of software components are introduced and used in this book. You will build applications using software components, and in the case of ActiveX DLLs, build a software component yourself. Software components make possible rapid development of applications by acting as the building blocks of an application. Each software component provides a particular service to the application. One way to think about software components is to see them as *servers* to *client* applications.

How do you know which software components are available to you? Visual Basic provides the Object Browser utility to let you examine your choices.

Exercise 1.2 Launching Visual Basic and the Object Browser

In this exercise, you will start Visual Basic and then use Object Browser to examine Visual Basic objects. You are provided step-by-step instructions. Don't be alarmed if you see unfamiliar windows. The Visual Basic environment is explained in greater detail in Chapter 2, "Writing and Running Your First Visual Basic Program," and Chapter 3, "Adding Controls and Event Procedures to Form Modules." This exercise assumes Visual Basic has been installed on your computer.

1 Start Visual Basic from the Windows Start menu.

2 When Visual Basic launches, a dialog box similar to that shown in Figure 1.5 appears. If this dialog box does not appear, select Project, New from the Visual basic menu.

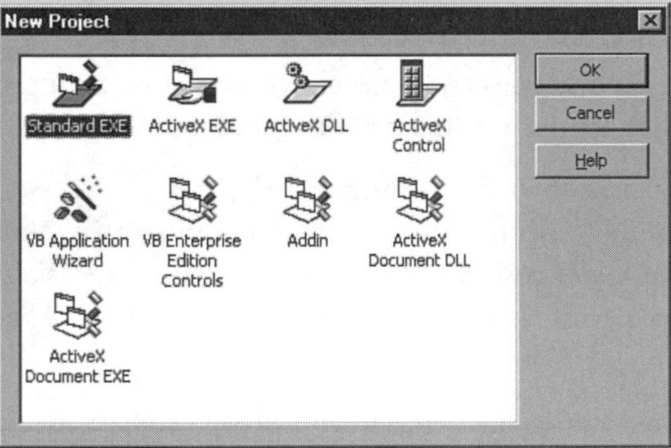

Figure 1.5

The Visual Basic startup dialog box contains several types of new projects, including the Standard EXE project.

3 Click the Open button for a new Standard EXE project (or click OK if you opened the dialog box from the Visual Basic menu).

The Visual Basic design environment will appear. It should look somewhat similar to the design environment you see illustrated in Figure 1.1. It will appear *similar*, but not identical. As you will discover in the next section of this chapter, the Visual Basic design environment is almost infinitely configurable by you. There really is no standard Visual Basic design environment.

4 Press the F2 key or select View, Object Browser from the Visual Basic menu. Object Browser will appear as shown in Figure 1.6.

5 Select the VB library from the combo box in the upper-left corner of Object Browser, as shown in Figure 1.6. (You have used this kind of Windows object before, but you may not have known that was called a combo box.)

Figure 1.6

The VB Object Browser lists the software classes available to you from the Visual Basic class library.

6 The classes through which you can scroll are listed on the left-hand side of Object Browser. Select the `TextBox` class of the VB library and then examine the properties, methods, and events of the `TextBox` class in the right-hand list box.

Notice that the Object Browser icons indicate whether the class element (called a class member) is a property (the icon depicting a small hand holding a piece of paper), a method (the icon depicting a small green rectangle), or an event (the icon depicting a small thunderbolt). Figure 1.7 shows these three icons.

continues

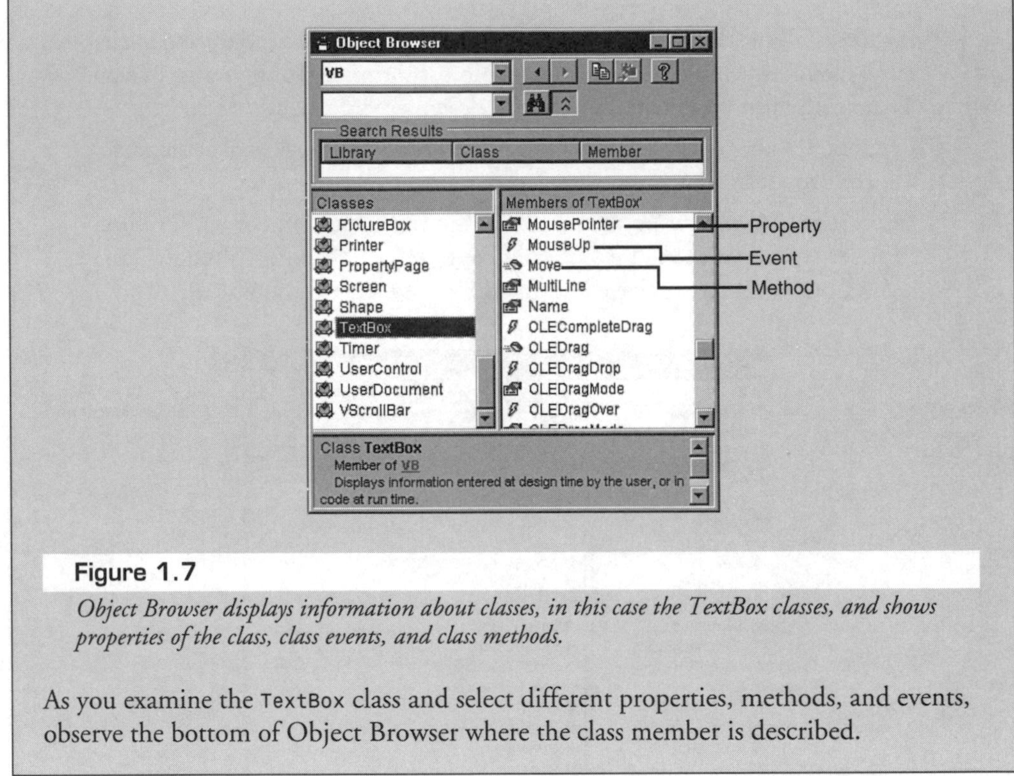

Figure 1.7

Object Browser displays information about classes, in this case the TextBox classes, and shows properties of the class, class events, and class methods.

As you examine the `TextBox` class and select different properties, methods, and events, observe the bottom of Object Browser where the class member is described.

Objective 1.4 Running Sample Visual Basic Programs

Given this brief introduction to Visual Basic, it is time for you to become more familiar with the Visual Basic programming environment. Still, you are not expected to write a Visual Basic program at this time; right now, you'll start Visual Basic and become a bit more familiar with the way in which the environment can be configured. Then, you will run three sample programs stored on disk.

1.4.1 A Quick Tour of the Visual Basic Design Environment

The Visual Basic design environment is almost infinitely configurable. This makes it difficult to describe a standard Visual Basic user interface. It depends upon the way the design environment was last configured.

There are two overall design environment choices: MDI or SDI. The abbreviation MDI stands for *multiple-document interface,* and it is the default Visual Basic setting. It features an overall container window and movable and *dockable* windows inside the container. It is the environment displayed in Figure 1.1. The abbreviation SDI stands for *single-document interface*. It features free-floating windows on the Windows desktop, which is the environment that would be most familiar to you if you have worked with previous versions of Visual Basic.

Whether the environment is MDI or SDI, the Visual Basic startup environment frequently displays the windows and utilities listed in Table 1.1. These windows and utilities are described in detail in Chapter 2 and Chapter 3.

Table 1.1 Visual Basic Startup Windows

Project Explorer window	This window is frequently positioned on the right of the Visual Basic container. It lists the modules that make up a Visual Basic project.
Form module window	This window displays a form if a form is selected for viewing in the Project Explorer. This form is the "canvas" upon which controls are placed.
Properties window	This window displays the properties for whatever object has focus—whether a module or control.
Toolbox window	This window displays available controls for placing on forms.

The MDI Design Environment: Using Dockable Windows

The multiple-document interface for Visual Basic 5.0 is new. It allows you to configure the Visual Basic design environment in many ways, but it can be confusing. In order to use it effectively, you must have a large screen set to a fairly high resolution; otherwise, you will quickly suffer from screen clutter. It also requires that you know how to work with *dockable* windows, an interface feature designed for those who prefer to use a mouse, rather than the keyboard, as a principal input device.

If you have not used dockable windows inside a container before, you might find them frustrating at first. Dockable windows allow you to take any window inside a container and *dock* it on the perimeter of the container. As you move a window inside a container by keeping your mouse button depressed on the window title bar while moving it, an outline of the window is displayed. The width of this band gives you a visual cue as to what you can do with the window. A thick band outline indicates it will become *free-floating* if released by the mouse button. Figure 1.8 shows the toolbox docked to the right-hand side of a form.

Figure 1.8

The toolbox will be docked at the side of the Visual Basic form when the Dockable command (right-mouse button) is selected.

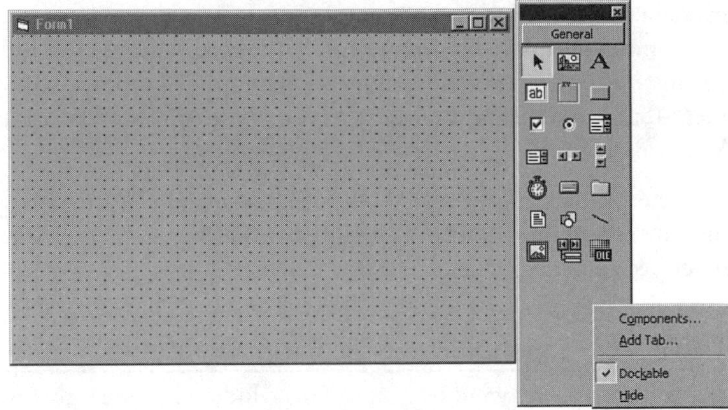

Given the capability to dock any window at any position in the Visual Basic MDI container, it is possible to customize the environment to your liking. Many developers prefer to dock the toolbox at the bottom of the container in order to make room for form design.

To view a window or toolbar that is not displayed, choose View from the Visual Basic menu to see a list of windows and design-time utilities.

The capability to dock windows within the MDI environment can be controlled through the Docking tab of the Options dialog box (choose Tools, Options). If the docking option for a window is not selected (see Figure 1.9), it will remain free floating.

Figure 1.9

The Docking tab of the Options dialog box presents many possibilities for docking the Visual Basic windows and design tools, such as the Color palette.

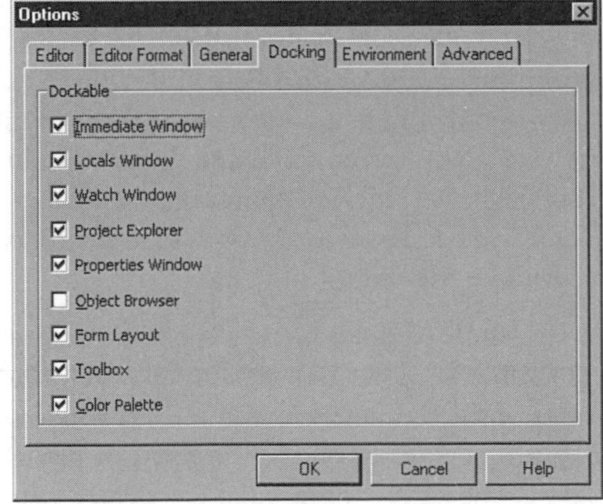

The SDI Design Environment: Using Free-Floating Windows

The single-document interface, the one preferred by the authors, essentially makes all windows free-floating outside of any container. This increases the amount of screen "real estate" available, which is important in the rich (and often times cluttered) design environment of Visual Basic. The SDI design environment is probably also more familiar to you from working with other Windows applications, including earlier versions of Visual Basic.

To move to an SDI environment, select the Advanced tab in the Options dialog box (by choosing Tools, Options), as shown in Figure 1.10. Once the SDI environment is selected, a dialog box will appear, informing you that your selection will take effect the next time Visual Basic is launched. To view the SDI environment, exit Visual Basic (by choosing File, Exit), and launch Visual Basic again. You will now be in an SDI environment similar to the one you see in Figure 1.11. To view individual windows and toolbars, you can make selections under the Visual Basic View menu.

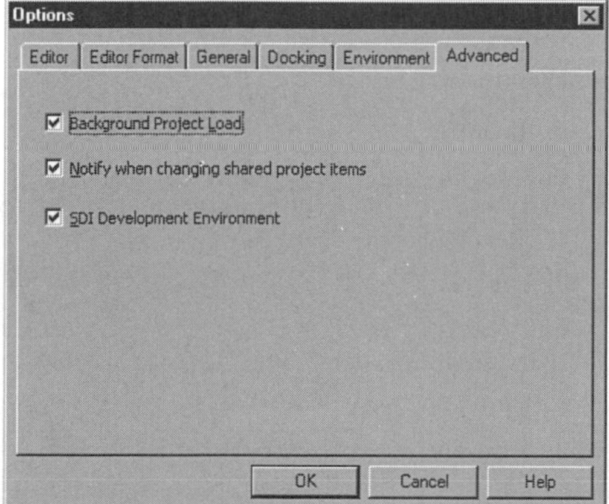

Figure 1.10

The SDI selection in the Options dialog box creates a design environment in which windows remain separate and are no longer docked.

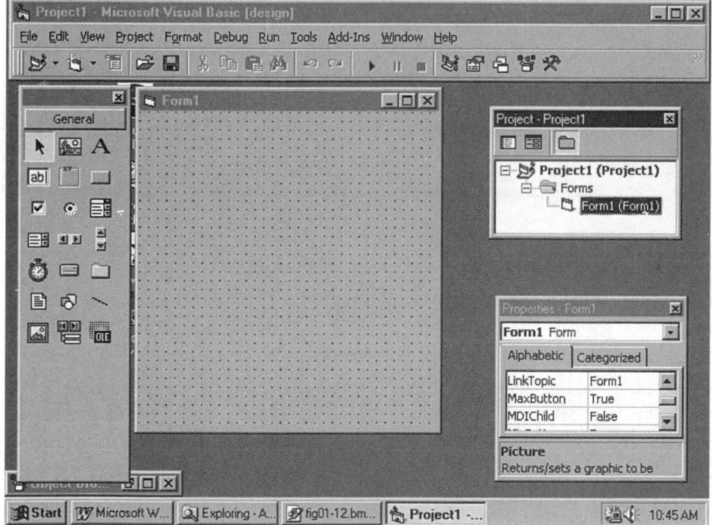

Figure 1.11

The Visual Basic single-document interface (SDI) features free-floating windows, and is good to use when you work on a small screen.

1.4.2 Running Visual Basic Applications

Now that you are a bit more familiar with the Visual Basic design environment, let's run several Visual Basic applications.

Exercise 1.3 Running the Screen Blanker Demo

Run the Screen Blanker Demo, which is named blanker.vbp and is located in the Graphics folder.

1 Start Visual Basic from Windows.

2 Use the mouse to select File, Open Project. If a dialog box appears asking you to save a file, click No.

3 Find your Visual Basic directory and the following subdirectory path:

   Visual Basic\Samples\Pguide\Blanker

   This program is also on the student disk.

4 Double-click blanker.vbp to open the application, or select the file and click the Open button.

5 Choose Run, Start from the Visual Basic menu, or press F5. This runs the Visual Basic application.

6 When the Screen Blanker Demo appears, click the Start Demo button. (This button is a control that begins an event. Can you tell what the event is?)

7 Click the Stop Demo button and experiment with the other choices found under the Options menu. Open Options and make a selection. Click Start Demo to begin another demonstration.

8 Choose Options, Exit to exit the application and return to the Visual Basic design environment.

What does this demo tell you about Visual Basic? With Visual Basic, you can write any number of graphics procedures. The demo should reinforce the fact that Visual Basic is a visual language.

Exercise 1.4 Running the Controls Demo

This application demonstrates some of Visual Basic's controls.

1 Use the mouse to choose File, Open Project. If a dialog box appears asking you to save a file, click No.

2 Find the following directory (assuming you have named your root Visual Basic directory "Visual Basic"):

   Visual Basic\Samples\Pguide\Controls

   You can also use the student disk.

3 Select the controls.vbp file, and click the Open button.

4 Select Run, Start from the Visual Basic menu, or press F5.

5 When the Controls application appears, click the various buttons of the Main application form to see key properties of different Visual Basic controls demonstrated.

6 Select File, Exit to exit the application and return to the Visual Basic design environment.

This demonstration illustrates some key features of Visual Basic's standard controls, including the use of a Command Button, Label, Text Box, Check Box, Option Button, and Image Box controls.

Exercise 1.5 Running the Calculator Demo

This final demonstration application illustrates how Visual Basic can be used to create a simplified calculator. It is the kind of application that you should be able to develop with this language within a relatively short period.

1 Open and run the Calculator.vbp file. The complete path is as follows:

   Visual Basic\Samples\PGuide\calc\calc.vbp

   You can also use the student disk.

2 Observe the design of this application. Numeric input keys and operator keys are placed on a form with a space for displaying output.

3 Run the application and experiment by making calculations. Notice that input is limited to mouse clicks rather than the pressing of numeric keys. Also notice that input is not permitted in the area that displays the results of calculation. You will soon learn the different kinds of controls that allow user input and those that are used primarily for display.

After experimenting with the calculator, you can close it by double-clicking the application's System menu icon (the box at the upper-left corner of the window) or by clicking the standard Close button in the upper-right corner.

Chapter Summary

Visual Basic is a high-level, event-oriented, high-visual, and compiled or interpreted language. As you will come to appreciate, Visual Basic uses a large number of English words, waits for an event to occur before taking any action, and features a GUI in support of its high-visual orientation.

Visual Basic has a close relationship with Windows. Visual Basic's main characteristic is the *encapsulation* of the complexity of Windows into *objects* that are easily manipulated with simple coded instructions.

Objects are manipulated by exposing properties, methods, and events. A property is an attribute of an object, while an event is an action recognized by an object. A method is code executed by an object that acts on data held by that object. The invocation of the `Form1.Hide` method, for instance, triggers code held by the form that hides it from view.

Properties, methods, and events serve as the object's public interface. The structure and means by which an object holds data or executes its methods is hidden from view (encapsulated), while its public interface makes it usable to other objects and to program instructions.

Visual Basic has both visible and invisible objects. Forms and most controls are examples of *visible* objects. Five system objects are examples of *invisible* objects: the `App` object, the `Clipboard` object, the `Screen` object, the `Debug` object, and the `Printer` object.

Objects are instances of classes. Classes define the default properties, methods, and events of an object. Objects in Visual Basic are created during design time when controls are added to forms and created dynamically by coded instructions. The Object Browser utility in Visual Basic permits you to examine the classes available for use by your application and the object properties, methods, and events defined by the class.

A statement is a reserved word that causes a computer program or the computer to do something. The `Beep` statement, for example, tells the computer to make a sound.

An instruction represents a line of code used in a Visual Basic application. Every instruction begins with a statement or a method.

Visual Basic's design environment can be designated as an MDI (multiple-document interface), featuring dockable windows inside a parent container, or an SDI (single-document interface), featuring free floating windows.

Skill-Building Exercises

1. Open and run the MDI (MDI\Mdinote) and the SDI (SDI\Sdinote) samples. Based on these samples and the material contained in this chapter, describe the difference between MDI and SDI.

2. Start Visual Basic in MDI or SDI mode, depending on the settings on your machine. Choose Tools, Options, and click the Advanced tab. Perform one of the following steps:

    - Click SDI Development Environment if it is not checked, and exit Visual Basic. Start Visual Basic to note the difference in the development environment.

    - Click SDI Development Environment if it is checked (to uncheck it), and exit Visual Basic. Start Visual Basic to note the difference in the development environment.

3. Open and run the Alarm Clock sample. When the display appears, click the form to set the time for the alarm. The path is as follows:

    Visual Basic\Samples\PGuide\alarm\alarm.vbp

    You can also use the student disk.

4. Open the Alarm Clock sample once again. Click the Object Browser button on the toolbar, and select the Project1 library (not the VB library). What are the members of the class `AlarmForm`, and what types of members are they: property, event, or method?

5. If your computer has audio speakers, turn them on. Open and run the ATM sample. (Go to the path or use the student disk.) When the display appears, enter 123456 as the PIN number. Withdraw $100.00 from the checking account. What message is stated to end the transaction?

Writing and Running Your First Visual Basic Program

In Chapter 1, "Running a Visual Basic Program," you learned how to run a program that had been made for you. You probably noticed that when working with Visual Basic, several types of windows could be inspected—even though you were not asked to examine any one window in detail. In this chapter, you learn about three types of windows:

- Project window
- Properties window
- Code window

You learn how to move from one window to the next in writing a Visual Basic program. Also, you are asked to write and run several simple Visual Basic programs.

In working through the exercises in this chapter, you master a number of new tasks, including how to

- open a new project;
- save a project;
- name a form and a project;
- remove a form from a project;
- add a form to a project;
- print a form, form text, and code;
- write a `click()` procedure;
- write a `KeyPress()` procedure;
- change font size, style, and type;
- change the window state.

Chapter Objectives

The tasks for this chapter are distilled into five objectives that teach you the fundamentals of writing and running a Visual Basic program, as follows:

1. Understanding the concept of a Visual Basic project
2. Learning how to manage Visual Basic projects
3. Opening the Properties window and changing form properties
4. Opening the Code window and writing Visual Basic instructions
5. Producing a compiled application from a Standard EXE project

Objective 2.1 Understanding the Concept of a Visual Basic Project

Every Visual Basic application is defined and saved as a *project.* Each project, in turn, consists of a collection of files. These files can include modules, insertable objects, and a single resource file. Projects themselves are tracked by a *project file,* which has a .VBP extension.

Form modules, as discussed in Chapter 1, are first seen as windows with a caption, but little else. Form modules contain coded instructions called *procedures* for the form itself and for the controls and insertable objects placed on the form. Other modules include *standard modules* and *class modules.* You will learn the differences among these types of modules as you progress through this book.

The tools of the toolbox represent controls and insertable objects. The toolbox can be made visible or invisible by choosing View, Toolbox. When visible, the toolbox contains a set of icons—each representing a standard or a custom control. A Visual Basic project always includes a set of standard controls and can also include custom controls. Each custom control has an .OCX extension. In this chapter, you are asked to write some beginning Visual Basic procedures for forms. In Chapter 3, "Adding Controls and Event Procedures to Form Modules," you will add procedures for standard controls.

Projects can also include a *resource file,* which can contain either *binary* data, such as images and sounds, or *string* data. Resource files need to be created by a Windows resource compiler (typically a C or C++ language compiler), or you can use the resource compiler bundled with Visual Basic.

2.1.1 The Files that Make Up a Standard EXE Project

In working with Visual Basic, you are asked to work with different types of files. Each file type is given a unique file extension, which enables you to identify it. Visual Basic contains over two dozen types of files, each of which has a different extension. Table 2.1 shows eight common file types contained within a project, their extensions, and their meanings.

Table 2.1 Common File Types

Type of File	*Extension*	*Description*
Standard module	.BAS	Each coded module is saved as a file.
Class module	.CLS	Each class module is saved as a file.
Form module	.FRM	Each form is saved as a file.
Binary	.FRX	Each Icon or Picture property value is saved by Visual Basic in .FRX form.
Log	.LOG	A file for logging load errors.
Resource	.RES	A project can have a single resource file for storing binary and string data.
Project	.VBP	Each project is saved as a file.
Workspace	.VBW	Each workspace for a project is saved as a file.

This file notation is used throughout this book. When you save a form, it is saved with a .FRM extension, and all event procedures written for that form are also saved in the .FRM file. Because procedures are not saved as separate files with their own unique extension, they become bound to forms. When you write and save a coded standard module, it is saved with a .BAS extension. Coded modules are thus different from event procedures written for forms. Coded modules are not bound to forms.

2.1.2 The Project Window (Project Explorer) in the Visual Basic Design Environment

Whenever you begin a Visual Basic program, you begin a new project. Each project can contain several forms, several standard and class coded modules, and custom controls. The Project window displays a list of all modules assigned to a project. Whenever you begin a new project, a default Project window is created (see Figure 2.1). If the window fails to appear or is hidden, choose View, Project Explorer; press Ctrl+R; or click the Project Explorer icon. In addition, your environment can be set to remove the docking of the Project Explorer. This setting is controlled by choosing Tools, Options, clicking the Docking tab, and turning off the Project Explorer and Object browser selections.

Figure 2.1 illustrates a Project window that contains three small icons on the left-hand side. The leftmost icon enables you to view code, while the middle one enables you to view the underlying object (which in this case is the form). Forms, the third icon, is important when projects become large and contain several forms or when projects are combined. With large projects, all forms are placed in one folder, and all modules are placed in a second. Toggling forms enables you to show or hide all forms for a project that are contained in the Forms folder.

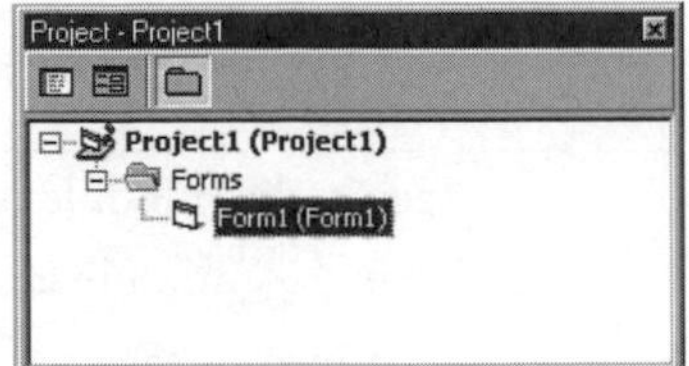

Figure 2.1

This Project window reveals that Project1 contains a forms folder that is limited to a single form.

Objective 2.2 Learning How to Manage Visual Basic Projects

In this section, you learn how to open and save a project, add and delete files from the Project window, give an object focus (further explained in section 2.2.5, "The Concept of Focus"), and use the other File commands provided with Visual Basic. You will also work through several exercises to test your knowledge of the concepts presented.

2.2.1 Opening, Saving, Adding, and Removing Projects

Whenever you open Visual Basic, you always start a new project. However, there are File commands that enable you to work with both old and new projects, and File and Project commands for working with old and new files. The File menu on the menu bar contains six commands designed for managing projects, as follows:

- **New Project**—Choose File, New Project to start a new project. If you are working on a project, this command prompts you to save the current project. It then closes the current project to enable you to start a new one.
- **Open Project**—Choose File, Open Project to open a saved project. Before opening a project, this command asks if you want to save a current project if one exists.
- **Add Project**—Choose File, Add Project to open a saved project and add it to another, provided one exists. A Visual Basic application can contain any number of projects.
- **Remove Project**—Choose File, Remove Project to remove the selected project from a project group.
- **Save Project As**—Choose File, Save Project As to save the current project to a disk. This option always prompts you for a project name. Use this command to save a project for the first time or save a copy of an existing project. When first saving a project, you are asked whether or not you want to save each form and module associated with a project.

  For each form or module, you can assign a name. You should name each form with a name other than the default name (such as `form1.frm` and `form2.frm`); otherwise, you will design over the form when you start a new project and find it impossible to retrieve the form you thought you had saved. You should also name each project with a unique name, or else Visual Basic will assign every project the same default name.

- **Save Project**—Choose File, Save Project to save the current project to a disk. This option saves all forms and modules as separate files. It creates a file with a .VBP (Visual Basic Project) extension. If the project is being saved for the first time, this command prompts you for the project name. If saved earlier, it updates the .VBP file.

The .VBP file is saved as text or in ANSI (American National Standards Institute) format. This allows you to inspect its contents with any text editor. Each project contains one—and only one—.VBP file, which provides a description of the project, showing the names of the stored files and other descriptive information.

2.2.2 Naming a Project

Each time you begin a new project, the default name, Project1, appears at top of the form. Unless you change this name, every project you work on will be named Project1. To change the name, choose Project, Project1 Properties (see Figure 2.2), and type a new project name. For example, you might enter Project2_1 for this initial project.

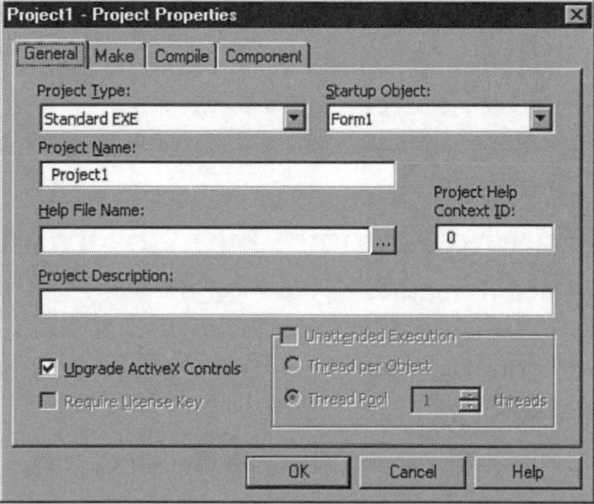

Figure 2.2

The Project Properties dialog box is used to change the name of the project from the default, Project1, to a name of your choosing.

Exercise 2.1 Using the Save Project As Command

This exercise asks you to do very little, but it does introduce you to the file naming conventions. Start Visual Basic, and perform the following steps:

1 Open a Standard EXE new project.

2 Choose Project, Project1 Properties and change the name of the project to **Project2_1**. Click OK. This new name should appear on the topmost title bar of Visual Basic.

continues

3 Choose File, Save Project As. A dialog box appears and asks if you want to save the file as Form1.frm. If you click Cancel, the form is not saved and you have to rebuild it if you need it. Imagine that you spent an hour perfecting a form. Unless you save it, your work will be lost.

4 Before you click Save, navigate to the location where you want to save the file, and assign a name to the form. Assign a unique name—something other than the default, Form1.frm. You can use the notation F2-1.frm to indicate that this form goes with Exercise 2.1 (not very clever, but it works). A more descriptive name would be better.

If you want to store the form on a diskette placed in the a: drive of the computer, navigate to the drive and type **F2-1.frm**.

5 After the form has been saved, a second dialog box appears asking you to assign a name to the project. Again, assign a unique name (not the default, Project1.VBP). You can call it **P2-1.VBP** to indicate that this project goes with Exercise 2.1. Here too, a descriptive project name would be better.

If you want to save the project on a diskette placed in the a: drive of the computer, navigate to the drive and type **P2-1.VBP**.

2.2.3 The Add File and Remove File Commands

Files can be added to, or removed from, the .VBP project file using the file commands contained on the Project menu. When adding or removing files, keep in mind that you are adding or removing files from a project. The following are two Project menu commands that you will use often in this book:

- **Add File**—Choose Project, Add File to open a dialog box (see Figure 2.4) that enables you to navigate the Windows file directory to identify a previously saved file and to add the file to a project, for example, an .FRM (form), .BAS (coded module), .CLS (class module), or .RES (resource file). This command is especially important when you use a file built for a different project. Simply identify the file name, and click OK.
- **Remove <Name of File>**—Choose Project, Remove <Name of File> to remove the file from the current project that is active (has the *focus*, a term that is described next).

Other Project menu file commands include Add Form, Add MDI Form, Add Module, and Add Class Module. These enable you to add new types of forms and modules to a project.

2.2.4 The Save File and Save File As Commands

The following are two File menu commands that are used to save a particular file rather than an entire project:

- **Save File**—Choose File, Save File to save the form or module that is active (has the focus). If the file has previously been saved, this command updates the file. If the file is new, this command prompts you for the name of the file and, once entered, adds the file to a project when the project file (.VBP file) is next saved.
- **Save File As**—Choose File, Save File As to save the current file to a disk. This option always prompts you for the file name. Like the File, Save Project As command, File, Save File As is used to save a file for the first time or to save a copy of an existing file.

You will find file commands most useful when you want to use the same forms and modules in different projects. There is nothing to prohibit you from using the same form in dozens of projects because each form is saved as a separate file.

2.2.5 The Concept of Focus

The concept of *focus* is important in Visual Basic. The object that is selected (such as a form, control, or menu bar selection) is said to have focus. In order for Visual Basic to know what file to save, it is necessary to click the object to give it focus, making it the active object. Likewise, to remove a file, the object must be given focus. To remove a file, open the Project window, click the file to remove, and choose File, Remove and the name of the file to remove will also appear. An alternative is to display the Project Explorer, click the file that you want to delete, and right-click to produce the shortcut menu shown in Figure 2.3. We discuss the concept of focus at length in this book. There are even unique events, such as `GotFocus` and `LostFocus`, that deal with this issue.

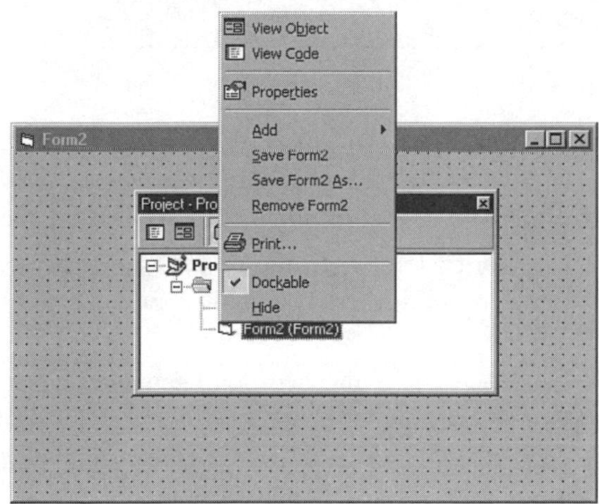

Figure 2.3

Display the Project Explorer and right-click to reveal several file options, including a command for removing a file from a project.

Exercise 2.2 Using the Remove and Add Commands

This exercise assumes that you completed Exercise 2.1 and saved the form as F2-1.frm.

1 Open a new Standard EXE project, and change the name of the project to **Project2_2** (click Project, Project1 Properties).

2 Choose Project, Remove Form1. `Form1` must have the focus because it is the only object assigned to the project.

3 Choose Project, Add File. Navigate through Windows to the location where you saved F2-1.frm, and add this form to your project (see Figure 2.4).

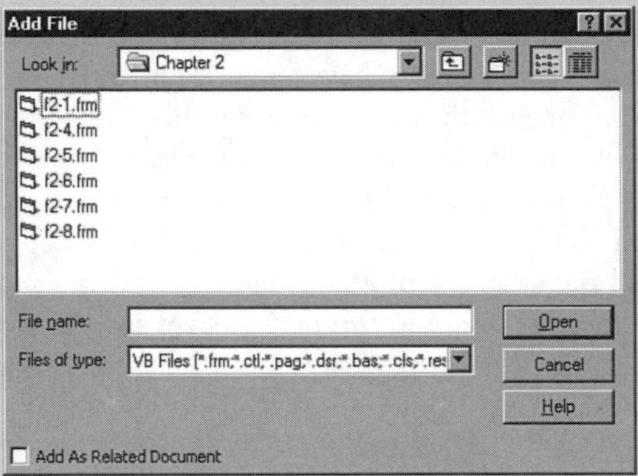

Figure 2.4

The Add File dialog box lists saved forms that can be added to a project.

4 Open the Project Explorer Window, and give form F2-1.frm the focus.

5 Choose File, Save File As (or click the right mouse button). When the dialog box opens, navigate to the location where you want to save the file, and name it **F2-2.frm**.

6 Open the Project Window again and observe what happens. The name of the form has been changed and assigned to the new project.

7 Choose File, Save Project As. In this instance, you are not asked to name the form because it has already been named and saved. Instead, you are asked only to name the project, which we named P2-2.vbp.

2.2.6 Other File Menu Commands

Besides the four Project menu commands and the two file commands found on the File menu, the latter also contains the Print and Print Setup commands and the Make Project1.Exe commands, as follows:

- **Print**—Choose File, Print to display several options for printing information about a project (see Figure 2.5). You can print the form image, code, or the form as text, and you can choose to print information for a module or for the entire project.
- **Print Setup**—Choose File, Print Setup to launch the standard Windows Print Setup dialog box.
- **Make Project1.Exe**—Choose File, Make Project1.Exe to create an executable application. Project modules are compiled into a single .EXE file. This file is either a P-Code interpreted executable or a native code executable file. (The two code formats are explained in section 2.5.1, "Interpreted P-Code," and section 2.5.2, "Compiled Native Code," later in this chapter.)

  Both types of executables require runtime files. The Visual Basic Setup Wizard can help you determine which runtime file your application requires. (See Chapter 23, "Creating an ActiveX DLL," for more information about the Visual Basic Setup Wizard.)

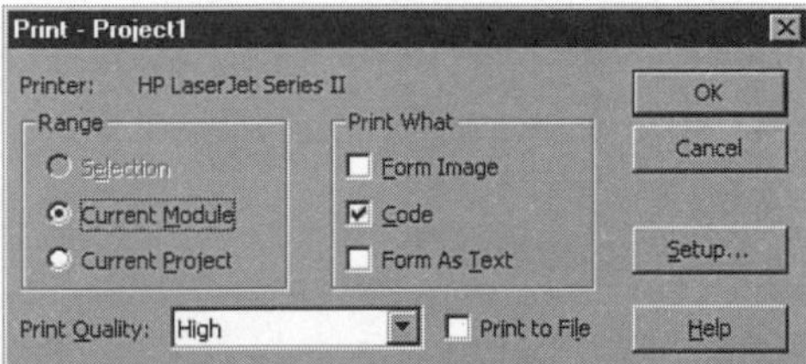

Figure 2.5

The Print-Project dialog box enables you to choose the parts of an application that you want to print.

Exercise 2.3 Using the Print Command

This exercise is designed to show you the type of information printed for a project.

1 Open a new Standard EXE project.
2 Choose File, Print.
3 When the dialog box appears, click Current Project, Form Image, Code, and Form as Text.
4 Click OK.

What is the difference between the Form Image and the Form as Text choices? Why is there no computer code?

Objective 2.3 Opening the Properties Window and Changing Form Properties

Besides a Project Explorer window, each form (or control) contains a Properties window. As stated in Chapter 1, "Running a Visual Basic Program," a property is a named attribute of an object. Properties are used to change the appearance or behavior of objects.

Figure 2.6 illustrates the default Properties dialog box for a form. If this box is not visible, you can bring it into view by pressing the F4 function key, or choosing View, Properties Window. With the form having the focus, press your right mouse button and drag down the Properties list.

Figure 2.6

The Properties dialog box of a form contains the built-in form attributes that you can change.

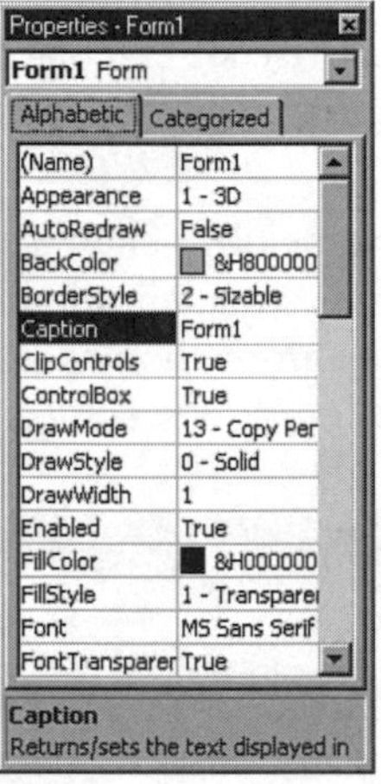

Over 50 properties can be set for a form. Although this number seems overwhelming at first, some properties are used so frequently that they represent an initial set to remember:

`Caption`	`FontName`
`Name`	`FontSize`
`BorderStyle`	`FontBold`
`MinButton`	`FontItalic`
`MaxButton`	`FontStrikethru`
`ControlBox`	`FontUnderline`
`WindowSize`	`ForeColor`
`BackColor`	`StartUpPosition`

To simplify this list somewhat, the items can be grouped by their uses:

- Seven properties deal with the way in which the form or text appears on the display screen (`Caption`, `BorderStyle`, `MinButton`, `MaxButton`, `ControlBox`, `WindowSize`, and `StartUpPosition`).
- The identification of an object is specified by the `Name` property.
- Font settings are controlled using the Fonts dialog box (`FontName`, `FontSize`, `FontBold`, `FontItalic`, and `FontStrikethru`).
- Color settings (`ForeColor` and `BackColor`) use a special palette.

There are three ways to change most property settings:

- Click the property to highlight it, and enter (type in text) the new property setting in the text box.
- Double-click the setting to view the alternative settings.
- Click the down arrow to open the drop-down menu for property settings.

Properties for fonts and colors launch a separate dialog box for you to enter settings.

2.3.1 Changing a Caption

The `Caption` property enables you to change the title that appears on the top of a form. Instead of `Form1`, use the `Caption` property to add a title. To change the caption, select Caption in the Properties dialog box and type the new title in the text box next to the property name.

2.3.2 Changing a Form Name

The `Name` property specifies the name of the form to be used in writing Visual Basic code. Remember the following rule: *The name, not the caption, is the identifier.* When writing coded instructions, the `Name` property appears as the object name. To change the name, select Name in the Properties dialog box and type the new name.

2.3.3 Changing the Border Style

The `BorderStyle` property enables you to change the appearance of the form's border. To change the border setting, double-click the current setting to reveal other choices, or click the down arrow to open the drop-down list of the six options.

If you are uncertain about the six border choices and their precise effects, click `BorderStyle` to give it the focus in the Properties dialog box, and press F1 (the Help function key). Visual Basic Help displays the six style choices and explains their effects, as shown in Table 2.2.

Table 2.2 The Settings for the *BorderStyle* Property

Setting	*Style*	*Description*
0	`None`	No border or border-related elements.
1	`Fixed Single`	Can include Control menu icon, title bar, Maximize button, and Minimize button. Resizable only using Maximize and Minimize buttons.
2	`Sizable`	The default. The size can be changed using any of the optional border elements listed for setting 1.
3	`Fixed Dialog`	Can include Control menu icon and title bar; can't include Maximize or Minimize buttons. Not resizable.
4	`FixedToolWindow`	Under 16-bit versions of Windows and Windows NT 3.51 and earlier, behaves like `Fixed Single`. Does not display Maximize or Minimize buttons. Not resizable. Under Windows 95, displays the Close button and displays the title bar text in a reduced font size. The form does not appear in the Windows 95 taskbar.
5	`SizableToolWindow`	Under 16-bit versions of Windows and Windows NT 3.51 and earlier, behaves like `Sizable`. Does not display Maximize or Minimize buttons. Resizable. Under Windows 95, displays the Close button and displays the title bar text in a reduced font size. The form does not appear in the Windows 95 taskbar.

The online Help is one of Visual Basic's strongest features. Remember, *the F1 key is your friend!* Use it often. The language is just too vast to remember all of it.

2.3.4 Disabling the Minimize Button

With the first two border styles, the Minimize button appears unless it is disabled by the `Minbutton` command. To disable the button (and remove it from the form), double-click the `MinButton` property to change the setting to `False`. (With this property, there are only two choices: `True` and `False`.)

2.3.5 Disabling the Maximize Button

With the first two borderstyles, the Maximize button appears, unless it is disabled. To disable the button, double-click the `MaxButton` property to change the setting to `False`. (Like `MinButton`, only two choices are available with `MaxButton`: `True` and `False`.)

2.3.6 Disabling the Control Box

With the first three border styles, the control box appears unless it is disabled. To disable the box, double-click the `ControlBox` property to change the setting to `False`. (The two settings for `ControlBox` are `True` and `False`.)

2.3.7 Changing the Window State

The `WindowState` property determines the appearance of the form at runtime. Table 2.3 shows its three settings.

Table 2.3 The Settings for the *WindowState* Property

Value	*Setting*	*Description*
`0`	`Normal`	The form appears in normal size (the default).
`1`	`Minimized`	The form is minimized and appears as an icon (either on the taskbar or above it) when the application is run.
`2`	`Maximized`	The form is maximized and expands to fit the entire screen when the application is run.

2.3.8 Changing the Background Color

Part of the enjoyment of working with Visual Basic is the ease with which colors can be added to a form. Color settings are expressed in *hexadecimal* notation (base 16). The designers of Visual Basic did not expect you to work with hexadecimal numbers and instead provided a color palette. To open the palette, double-click the `BackColor` setting and click Palette. When the Color palette appears, click a color to change the background color of the form. Following the click, the hexadecimal number assigned to the color appears in the text box.

2.3.9 Changing Font Properties

When working with text, Visual Basic enables you to change the font (or text style) you are working with. To determine which fonts are available on your system, open the Properties window, highlight Font, and click the ellipsis button to the right. A dialog box appears for setting numerous font properties, including Name, Bold, Italic, Size, StrikeThrough, and Underline.

2.3.10 Changing the Foreground Color

The ForeColor property determines the color of the text and graphics to be displayed. This differs from BackColor, which determines the background color of the object. To set this property, double-click ForeColor and click Palette. Make the color selection using the palette.

2.3.11 Changing the Startup Position

The StartUpPosition property specifies the position of the form when it first appears. Table 2.4 shows its four settings.

Table 2.4 The Settings of the *StartUpPosition* Property

Constant	*Value*	*Description*
vbStartUpManual	0	No initial setting specified.
vbStartUpOwner	1	Center on the item to which the UserForm belongs.
vbStartUpScreen	2	Center on the whole screen.
VbStartUpWindows	3	Position in upper-left corner of screen.

Objective 2.4 Opening the Code Window and Writing Visual Basic Instructions

The final type of window discussed in this chapter is the Code window. To open the Code window for a form, click the Project Explorer and click View Code, or choose View, Code. This opens the window shown in Figure 2.7. The object shown is the Form; the procedure is shown as Load(). The title bar reads Form1 because you have not changed the name of the form.

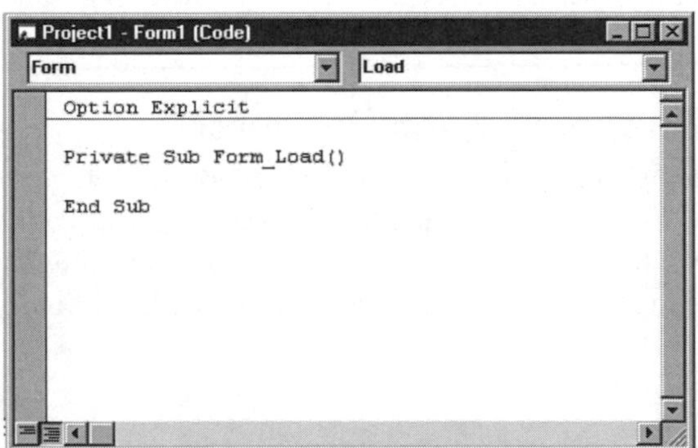

Figure 2.7

This Code window shows the start of a Form_Load() *procedure.*

The Code window contains the first two coded instructions, which are entered for you. The first combines the name of the object with the name of the event procedure. This coded instruction is shown in the following example, in which `Load()` is a special type of event:

```
Private Sub Form_Load()
```

It occurs when a form is first loaded. However, suppose you want to change the event from `Load()` to `Click()`. By selecting Click (click the procedure list box and scroll to Click, as shown in Figure 2.8), the first coded instruction changes to the following, in which `Click()` is another type of event:

```
Private Sub Form_Click()
```

It is triggered by a mouse click.

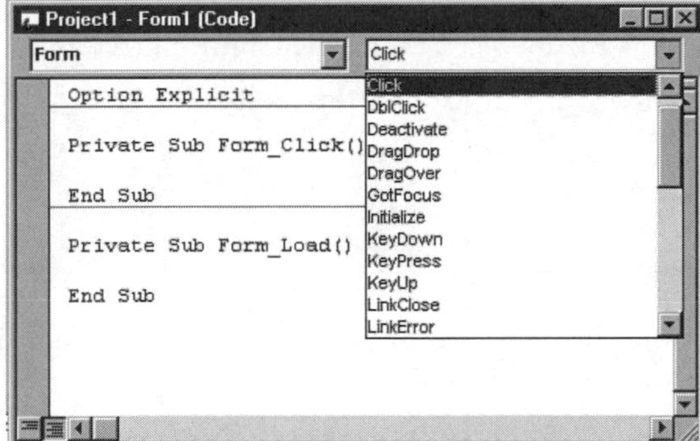

Figure 2.8

The Code window now shows the start of a `Form_Click()` procedure.

The event has now been changed from occurring at the time the form is loaded to the time when the form is clicked, with the user controlling the time of the event with the click of the mouse.

The ending instruction stays the same. It reads as follows, which means the end of the procedure (`Sub`):

```
End Sub
```

Exercise 2.4 Adding a *Click()* Procedure

In this exercise, you begin a new project.

1 Choose File, New Project and open a Standard EXE project. If the last project is still on-screen and has not been saved, you will be asked if you want to save changes to P2-3.vbp? Click No, unless you have already saved it. Either way will not lead to problems.

continues

2 When `Form1` appears, double-click to open the Code window.

3 Change the project name to **Project2_4**.

4 Change the procedure name from `Form()` to **`Click()`** in the procedure list box. Your code should read as follows:

```
Private Sub Form_Click()
End Sub
```

5 Type the following to create the procedure:

```
Private Sub Form_Click()
  ' This procedure will display a message on the form
  Print
  Print
  Print "     When you click, look at what happens!"
End Sub
```

6 Close the Code window by double-clicking the Control box icon.

7 Choose Run, Start; click the Start icon on the toolbar; or press F5 to run your procedure. When `Form1` appears, place the mouse pointer on the form and click. What happens?

8 Click the form again. What happens? It makes you want to click forever, doesn't it?

9 Choose Run, End from the menu bar or press the End button on the toolbar to return to the design of `Form1`.

10 Save this form and project. You can call the form **F2-4.frm** and the project **P2-4.vbp** if you want to use them in later exercises in this book; however, this is not necessary. Continue with the next exercise using the same form.

Exercise 2.5 Adding a *KeyPress()* Procedure

In this exercise, you add a second procedure to a form. Figure 2.9 shows the revised form. This exercise assumes you have completed Exercise 2.4 and are using that form.

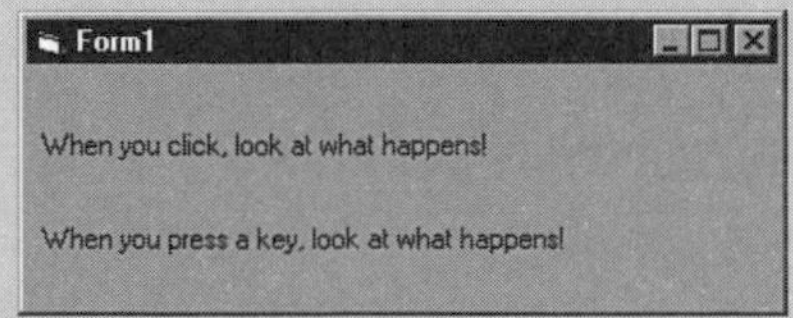

Figure 2.9

Adding a `KeyPress()` procedure to the form places more than one procedure on the same form.

1 Choose View, Code from the menu bar. The code you wrote for Exercise 2.4 should appear.

2 Choose KeyPress from the Procedure list box, and observe what happens to your code.

3 Type the following to complete this procedure:

```
Private Sub Form_KeyPress (KeyAscii As Integer)
'This procedure will display a different message on the screen
  Print
  Print
  Print "   When you press a key, look at what happens!"
End Sub
```

4 Close this procedure (using the Control menu icon) and run the revised program. There are four ways to run your program:

- Choose Run, Start.
- Choose Run, Start with Full Compile.
- Press F5.
- Press the Start button on the toolbar.

This time, press F5 to start your program; then, as a test, do the following:

1 Click the form.

2 Press a key.

3 Press a second key.

5 Click the End button on the toolbar to stop your program a different way, or choose Run, End to stop the program.

6 When you are returned to `Form1`, double-click the form to open the Code window. What code appears?

7 Scroll through the list of procedures in the Procedure menu. Coded procedures are displayed in bold.

8 Save this form and project at this time. You will be asked to work with the form and project in the next several exercises, but it is wise to save your work periodically. You can save your form as **F2-5.frm** and your project as **P2-5.vbp** to keep your examples in sequence. Name the project **Project2_5** before saving.

Exercise 2.6 Changing the Font Size Property

In this and the next two exercises, you change the properties of the form used in Exercise 2.5.

1. Choose File, Open Project, if necessary, and select View, Project Explorer, or click the Project Explorer icon to open the Project window. Click View Form.
2. Click the form to give it the focus, and press F4 to open the Properties dialog box.
3. Click font and change the font size to 18 points (or to another size if this size is not available on your system).
4. Run the program. Click to display the first sentence. Press a key to display the second sentence.
5. Exit to continue with the next exercise.

Exercise 2.7 Changing Font Style Properties

In this exercise, use the form from Exercise 2.5 once again.

1. Open the Properties dialog box and open the Font dialog box by double clicking the Font property. Change the Font to Bookman Old Style (or another font if your system does not have this font installed).
2. Change the style to bold and italic.
3. Run the revised program. Click to display the first sentence (it should be in bold and italic). Press a key to display the second sentence.
4. Exit and save your project at this time. The next exercise asks you to begin a new project.

Exercise 2.8 Adding a Message to a Form and Changing Property Settings

In this exercise, you write a new procedure and change properties.

1. Open a new project.
2. Add the following `Form_Click()` Visual Basic procedure to the form:

```
Private Sub Form_Click()
   'This procedure will display a Happy Birthday message
   Print "                              HAPPY"
   Print "                    BIRTHDAY"
End Sub
```

3 Change the following form properties:

Property	*Setting*
`Backcolor Palette`	`= &H00FF00FF& (light purple)`
`BorderStyle`	`= 1 'Fixed Single`
`Caption`	`= Today is the day`
`Font Style`	`= Italic`
`Font`	`= Arial`
`Font Size`	`= 24`
`ForeColor Palette`	`= &H00FF0000& (blue)`
`WindowState`	`= 2 'Maximized`

4 Run your program.

5 Exit and save your program. Give the form a unique name, such as **BDAY.FRM**. Give the project a similar name, such as **BDAY.VBP**.

Objective 2.5 Producing a Compiled Application from a Standard EXE Project

This final objective asks you create an executable file from a completed Visual Basic application. Two options are available: Under the Compile tab of the Project Properties dialog box, you can select Compile to P-Code or Compile to Native Code (see Figure 2.10). Compiling to native code is the default.

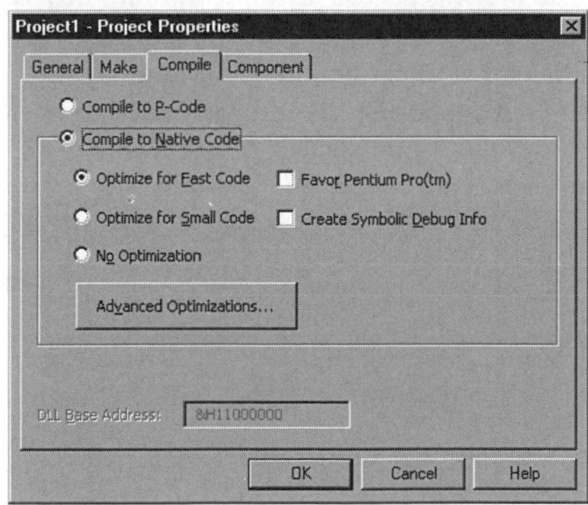

Figure 2.10

The Compile tab of the Project Properties dialog box enables you to control the way in which you compile your completed application.

2.5.1 Interpreted P-Code

By selecting P-Code, the modules of your Visual Basic application are compiled into what is called *pseudocode* or false code. An interpreter program that then produces machine-language instructions reads the coded instructions. All versions of Visual Basic prior to Visual Basic 5.0 were interpreted from P-Code.

Interpreted programs offer some advantage. Applications can be designed rapidly because new code can be tested immediately, without time-consuming compilation. They also offer some security advantages in networked applications. You can rely on the interpreter to serve as a buffer between an application and computer hardware. This is the security model built into the Java programming language, for example. To their disadvantage, interpreted programs tend to run slower than compiled programs.

2.5.2 Compiled Native Code

With Visual Basic 5.0, you can create native code compiled applications: The compiler takes the modules of your project and creates a machine language compiled application. Although the native code executable does not require interpretation, it still depends on other files to run. These files consist, mainly, of the object linking and embedding (OLE) libraries that make up the ActiveX architecture upon which Visual Basic is based.

One way to think about the difference between interpreted and compiled programs is to compare it to translating a letter from English to German (or another language). A compiled program translates the entire letter once, producing machine-language instructions. An interpreted program translates each letter every time the letter is read, producing P-Code, which is then translated into machine language.

2.5.3 The Make Tab of the Project Properties Dialog Box

When you decide to create an executable, the Make tab of the Project Properties dialog box comes into play. The Make tab presents general attribute options for your executable (see Figure 2.11), including the title of the application, the icon used to represent the application when it is minimized, and different version options. The Make tab also enables you to set command-line arguments for your application should you decide to use them.

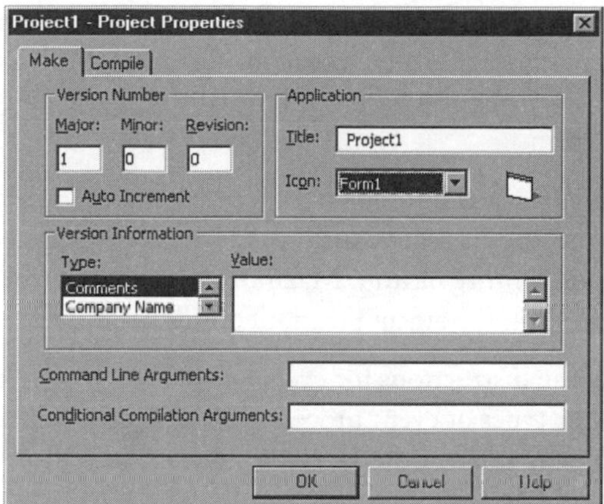

Figure 2.11

The Make tab enables you to set version numbers, version information, application attributes, command-line arguments, and conditional compilation arguments.

Exercise 2.9 Creating an Executable Application

This exercise asks you to create a make file from the Happy Birthday application created in Exercise 2.8.

1 Open Exercise 2.8 if it is not on the screen.
2 Choose File, Make Project2-8, if this reflects the name you gave your project.
3 When the Make Project dialog box opens, select a file name, such as **BDAY1.EXE**, and click OK. That's it. You have created a native code executable file. Remember that native code is the default.
4 Choose File, Make Project2-8 again, and when the dialog box opens, click Options.
5 Click the Compile tab.
6 Click Compile to P-Code, and click OK to return to the Make Project box.
7 Select a different file name, such as **BDAY2.EXE**.
8 Close Visual Basic, and start Windows Explorer. Locate both BDAY1.EXE and BDAY2.EXE. Click BDAY1.EXE to run the program. Close BDAY1.
9 Click BDAY2.EXE to run the program. Close BDAY2.

Chapter Summary

This chapter explains the Project Explorer window, the Properties window, and the Code window. In addition, it teaches you how to design, program, and run several Visual Basic programs and, in this way, how to manage projects.

By now, you should know that every Visual Basic application is defined and saved as a project. Each project, in turn, consists of a collection of forms, controls added to forms, and procedures. In this chapter, our attention has been limited to the use of a single form and writing procedures for that form.

The Properties window shows the attributes of an object (such as a form) that can be modified. In this chapter, you studied several form properties. You learned to change the caption, name, border style, Minimize button, Maximize button, control box, window state, background color, foreground color, start up position, and font properties of a form.

The Code window contains the instructions for event procedures attached to an object. In this chapter, you learned three types of event procedures: `Form_Load()`, `Form_Click()`, and `Form_KeyPress()`. These are only three of the 16 standard events that can be attached to a form.

Managing Visual Basic projects requires considerable care. Each project should be given a unique name, using the Project Properties dialog box. When saving a form, each and every form should be assigned a unique name. The same is true of a project. Forms are always stored as separate .FRM files, with controls bound to the form. When you decide to delete a form, all controls placed on that form are also deleted. Projects are stored as .VBP files. These are text files that can be edited with a text editor.

Although it was discussed only briefly in this chapter, the concept of focus must be remembered. The object that is active (such as a form, control, or menu bar selection) is said to have focus.

With Visual Basic, it is easy to create an executable file. Once an application is designed, choose File, Make Project1.EXE to create a file that can be executed outside the development environment.

Skill-Building Exercises

1. Change `WindowState` of Exercise 2.8 from `Maximized` to `Minimized`, and run the Happy Birthday application. What change occurs?

2. Design an application that contains two events: `Form_MouseDown()` and `Form_MouseUp()`. When the `MouseDown()` event occurs, display the following message on the form:

   ```
   When you press the mouse button, this message appears.
   ```

   When the `MouseUp()` event occurs, display this second message on the form:

   ```
   When you release the mouse button, this message appears.
   ```

3. Design an application that contains the events `Form_Activate()` and `Form_Click()`. For the `Form_Activate()` event, display the following message on the form:

   ```
   Please click the mouse.
   ```

   For the `Click()` event, display the following message:

   ```
   Thanks... That feels better!!!
   ```

4. Redesign the Happy Birthday application to display the following:

   ```
   Happy
   ___________________________________________
   Birthday
   ___________________________________________
   ```

   In this design, lines are placed across the form. In addition, use a `Fixed Dialog` border (setting `3`), and experiment with new colors.

5. Design an application that uses the `Form_Resize()` event. When the form is resized, display the following message:

   ```
   You resized me

   by making me bigger or smaller.
   ```

3

Adding Controls and Event Procedures to Form Modules

So far you have learned how to add controls to a form, but you have not yet done it. In this chapter, you add *controls* to form modules and attach *event procedures* (Visual Basic code) to a control. How can you do this before you understand the Visual Basic language? Whether you knew it or not, you wrote some Visual Basic code in Chapter 2, "Writing and Running Your First Visual Basic Program." The `Print` method, for example, is part of the Visual Basic language.

In this chapter, you will write several additional Visual Basic instructions. The instructions take you through the steps slowly to help you gain confidence in working with the language.

You will master several tasks in working through this chapter, including how to

- add a control to a form,
- change the properties of a control,
- add a procedure to a control,
- write an assignment statement,
- use the `Dim` and remark statements,
- use the `Print` and `Cls` methods,
- assign text to a text box,
- assign a caption to a label.

Chapter Objectives

The four objectives of this chapter introduce you to adding controls to a form and writing event procedures to display information:

1. Understanding standard toolbox control objects
2. Adding a control to a form and setting its properties
3. Coding an event procedure for a control
4. Changing object properties with coded statements

Objective 3.1 Understanding Standard Toolbox Control Objects

Remember from Chapter 1, "Running a Visual Basic Program," that a control is an object attached to a form module. The designers of Visual Basic decided to provide a number of standard controls (located on the toolbox) to represent the standard types of objects to be added to a form. You can also purchase custom controls, which have an .OCX extension when installed. To help you remember the differences among controls, tools are grouped into four initial categories:

- Text controls
- Containers for groups of controls
- Button controls
- Scroll bar controls

There are still other controls on the toolbox, but the preceding four categories represent a significant number. Remember that by letting your mouse cursor linger a moment over one of the controls in the toolbox, a ToolTip will appear to identify it.

3.1.1 Text Controls

The following four controls deal primarily with text. The left-hand side of Figure 3.1 shows how these controls appear on a form.

- **Text Box control**—Places text entered by the user on a form. Information that the user types appears in the text box. A text box can also contain default text to enable the user to bypass the control if the text does not need to be changed.
- **Label control**—Displays text entered by the program designer. The user cannot type or directly change a label.
- **List Box control**—Displays a list of items. The user can make one or more selections from the list. With a list box, the Visual Basic designer can add or remove items from a list.
- **Combo Box control**—Combines the features of a list box and a text box. This control gives the user the choice of selecting an item from a list or entering a value from the keyboard.

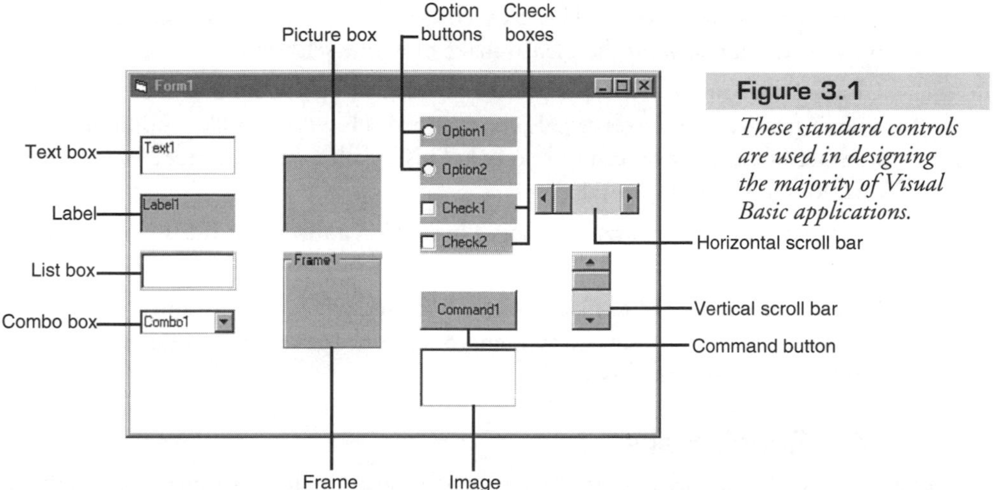

Figure 3.1

These standard controls are used in designing the majority of Visual Basic applications.

Two of the text control objects—the text box and the combo box—enable the user to enter data into a program. On the other hand, a user can make a selection from a list box, but can't enter data. Similarly, a label cannot be changed; however, because labels are read-only, they are used often to display the results of processing.

3.1.2 Containers

Containers house or contain (like a bucket) a set of controls. There are two types of container controls, as follows:

- **Picture Box control**—Used to store or contain a graphic image, such as a bitmap. It is typically used with a frame. It can also serve as a container for other controls, such as a group of images or a toolbar.
- **Frame control**—Used to provide a rectangular area on the form where other controls can be placed as a group. It generally serves as a container for a group of buttons, such as a set of option buttons.

The middle portion of Figure 3.1 shows how these controls appear on a form.

3.1.3 Buttons

Buttons are controls that users click to turn on or off. The button manipulation action (for example, a single mouse click) invokes an action called an *event.* The middle portion of Figure 3.1 shows option buttons, check boxes, a command button, and an Image control. The following describes the button controls:

- **Option Button control**—Enables the user to make one selection from a group of choices, similar to a multiple-choice question. Several option buttons are usually grouped inside a frame or picture box. When one option within a group is selected, all other options are canceled. This is called *exclusive choice.* You should use option buttons when you want the user to make only one selection.

- **Check Box control**—Enables the user to turn an option on or off, or set a value to true or false. Check boxes can also be grouped. However, unlike option buttons, any number of check boxes can be selected. This is called *non-exclusive choice:* Selecting one does not cancel the others.
- **Command Button control**—Places a button on a form that the user clicks to invoke an event.
- **Image control**—Used to place an image on the screen. A user can click the image like a command button to invoke an event. Several Image controls, arranged as a group, can be used to design a toolbar.

3.1.4 Scroll Bars

Scroll bars enable the user to examine a list when the entire list does not fit on a screen. Both horizontal and vertical scroll bars can be added to a form (see the right-hand side of Figure 3.1), as follows:

- **Horizontal Scroll Bar control**—Provides for easy navigation through a long list of information, or enables the user to set speed, quantity, or quality. Placement is across the screen.
- **Vertical Scroll Bar control**—Same as the horizontal scroll bar except that placement runs down the screen.

The 12 controls described in this section do not take into account all of the standard controls found in the toolbox. In this chapter, you will use only the Command Button, Text Box, and Label controls. In Chapter 4, "Learning to Think Visually," you will use check boxes, option buttons, a frame, a combo box, a list box, and scroll bars.

Objective 3.2 Adding a Control to a Form and Setting Its Properties

With the preceding brief overview of several different types of controls fresh in your mind, consider the procedures for placing a control on a form, and once positioned, changing the properties of a control. As a rule, the properties of a control should be set before event procedures (that is, coded procedures) are written. This practice enables you to prototype your design before you begin the code-writing process. A prototype in this instance is a partial, but not full, representation of your finished design.

3.2.1 Adding a Control to a Form

There is a long procedure and a short procedure for adding a control to a form, and each method has merit. The following is the long procedure:

1. Select the control you want from the toolbox.
2. Make the form active by clicking it to indicate where you want to place the control. This step changes the pointer to a crosshair.
3. Drag the handles on the control to size it the way you want.

The short procedure is simply the following two steps:

1. In the toolbox, double-click the control you want. This places the control on the active form.
2. Move and change the size of the control as needed.

Figure 3.2 shows how you move a control. Click the center of the control, and drag the control to a new position. To change the size of the control, click one of the handles (the small black boxes), and drag the control in a horizontal, vertical, or diagonal direction.

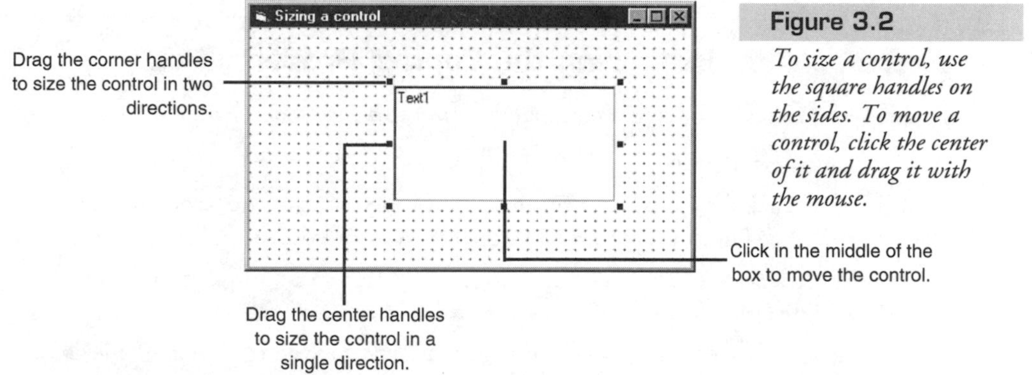
Sizing a control
Text1
Drag the corner handles to size the control in two directions.
Click in the middle of the box to move the control.
Drag the center handles to size the control in a single direction.

Figure 3.2

To size a control, use the square handles on the sides. To move a control, click the center of it and drag it with the mouse.

When placing option buttons within a frame on a form, the longer procedure binds the control to the frame container. (You are asked to use a frame in Chapter 4, "Learning to Think Visually.")

3.2.2 Setting Properties for a Control

After adding a control to a form, the properties of the control can be modified. Control properties work in the same way as form properties: They can be used to change the appearance or the behavior of the control. Figure 3.3 shows the Properties dialog box for a command button. Like a form, you can set a wide variety of options for a command.

Besides `Caption` and `Name`, command button properties include ways of changing the appearance of the text on the button, the size of the button, and even how you want the cursor to appear when it is placed over the button (the `MousePointer` property). You cannot change the color of the button, as the next exercise demonstrates.

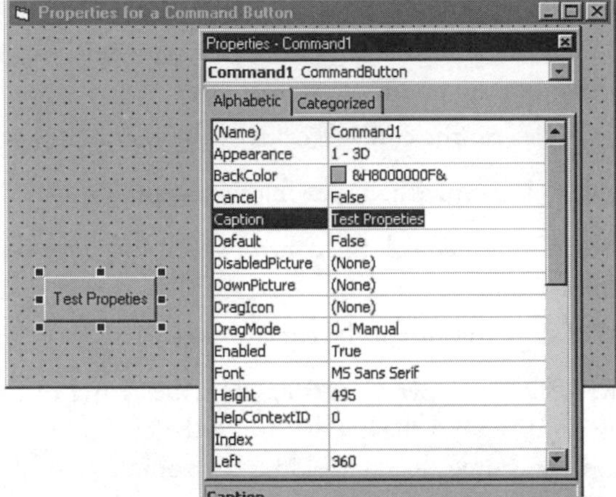

Figure 3.3
The properties box for a command button contains a `Caption` and a `Name` property and a host of other properties.

Exercise 3.1 Examining the Control Property Box

In this short exercise, you add a command button to a form and change its properties.

1 Open a new project.

2 Double-click the Command Button control. If you place the wrong one on the form, select Edit, Delete to remove the control from the form.

3 With the `Command1` button on the screen (see Figure 3.3), move it to the left-hand side of the form.

4 Change the `BackColor` property using the `BackColor` palette to change the color.

5 Run the program. What does the `BackColor` setting change?

6 Click the form and the `Command1` button. Can you determine why nothing happens? Because you have not indicated what action (that is, what event) is to take place when the button is clicked, the piece missing is a procedure.

7 Exit to return to the form.

Objective 3.3 Coding an Event Procedure for a Control

Adding a control to a form and setting the control properties set the scene for adding a procedure. Remember that a control will do nothing in an event-oriented environment unless an event procedure is attached to it. To add to this concept, more than one event can be associated with a single control, as with a form. In Exercise 3.2, you will examine the various event alternatives for a single control.

3.3.1 Controlling Naming Conventions

When adding an event procedure, a default name is always assigned to the form or control unless you decide to change it. For example, the first command button added to a form is named `Command1`, and the second is `Command2`. In this text, we ask you to follow a command syntax by adding a name prefix to a control (and many times to forms). For example, if you decide to add a command button to a form and change the command caption to `Start`, you would name the control `cmdStart`. The `cmd` is the command prefix. The prefix always tells you what specific control you are working with.

The exception to the above rule is when you are working with labels that have no effect on the input or output to a program. Because changing the name of each control takes considerable time, you should often leave label conventions (`Label1`, `Label2`, and so on) alone, unless the content of a label is determined by a procedure. This separates labels whose content is determined by a procedure from those whose content is not determined by a procedure.

Table 3.1 shows the controls, the default names, and the name prefixes that are used in this book for the 12 controls defined in Objective 3.1, "Understanding Standard Toolbox Control Objects."

Table 3.1 The 12 Controls

Control	*Default Name*	*Name Prefix*
Text Box	`Text1`	`txt`
Label	`Label1`	`lbl`
List Box	`List1`	`lst`
Combo Box	`Combo1`	`cbo`
Picture Box	`Picture1`	`pic`
Frame	`Frame1`	`fra`
Option Button	`Option1`	`opt`
Command Button	`Command1`	`cmd`
Check Box	`Check1`	`chk`
Image	`Image1`	`img`
Horizontal Scroll Bar	`HScroll1`	`hsb`
Vertical Scroll Bar	`VScroll1`	`vsb`

Exercise 3.2 Adding a Procedure to a Command Button

For this exercise, you write a procedure and attach it to a command button. Figure 3.4 shows how the form will look when you run the program.

Figure 3.4

In this display, the user is expected to click the Start button to display the message on the form.

1 Begin a new project or continue the last.

2 Add a command button to the form.

3 With the form active (not the command button), press F4 to reveal the properties box. Make the following change:

```
Caption =  A Printed Message
```

4 With the command button active, press F4 to reveal the properties box. Make the following changes:

```
Caption =  Start
Name = cmdStart
```

5 With the command button still active, double-click to open the Code window. Enter the following code, between the `Private Sub cmdStart_Click()` and the `End Sub` instructions:

```
Private Sub cmdStart_Click()
  ' The Start command button is used to display a message
  Dim message As String
  Print
  message$ = "Welcome to Visual Basic"
  Print   Spc(20); message$  ' prints 20 spaces followed by the message
End Sub
```

6 Save this project.

7 Close the Code window and run the program. Click the Start button.

8 Exit and open the Code window to examine the other types of events that can be assigned to the `cmdStart` control. For example, instead of writing a `cmdClick()` procedure, you might write a `MouseDown()` procedure, such that when the user presses the mouse button, the message is displayed. As another alternative, write a `MouseUp()` procedure.

3.3.2 Coded Statements and Methods

In the preceding procedure, you wrote several coded instructions, which introduced two different programming statements and a method: the `Dim` statement, an assignment statement, and the `Print` method. You also added a remark or comment to your code and used the `Private` keyword.

The *Dim* Statement

The `Dim` statement is used to declare variables and to allocate storage space. The statement begins with the keyword `Dim`, as in:

```
Dim message As String
```

It declares the variable message as a string of characters. As a string, it can contain characters, including blank spaces and punctuation, as well as numbers.

When typing statements, observe that the Visual Basic editor changes keywords to blue. If your statement is correct, the word `Dim` and the words `As String` appear in blue. This tells you that the syntax used in writing the statement is correct. If you make a mistake, as in the following code in which the word *String* is misspelled, the blue markings will not appear:

```
Dim message As Sting
```

The Assignment Statement

With an assignment statement, an equation is used. The value expression on the right side is assigned to the variable on the left. The syntax for the assignment statement is as follows:

```
[Let] variable = valueexpression
```

The syntax states that the keyword `Let` is optional. In this book, we ask you to follow a syntax that does not use the keyword `Let` in your coded instructions. The dollar sign ($) is also optional. You could have written the following:

```
message$ = "Welcome to Visual Basic"
```

Because the dollar sign ($) is optional, you could have written the following, for `message` has been declared as a variable of type `String`:

```
message = "Welcome to Visual Basic"
```

In general, use a dollar sign to help distinguish a string variable from another type.

A string expression must begin and end with quotation marks. If you write the following, omitting the quotation marks, you will be in error:

```
message = Welcome to Visual Basic
```

The Remark Statement

In Visual Basic, any statement that begins with a single quotation mark (') or the keyword `Rem` is a remark, as follows:

```
' The Start command is used to display a message
```

Remarks or comments are used to describe the purpose of a procedure as well as key features of a procedure. Use remark statements to document how your code works and to provide other types of important coding information. Also, be careful when writing the statement: Use the apostrophe (') which is the same as the single quotation mark and shares the quotation mark (") key, and not the grave accent (`), which can be found on the tilde (~) key. In writing the procedures in this book, we used the apostrophe rather than `Rem`. This enables us to place remarks to the right of an instruction, as in the following example:

```
Print     ' This instruction displays a blank line
```

When writing remarks, use the color code returned by the editor to determine whether your statements are correct or incorrect. A correct remark statement will, by default, be shown in green.

The *Private* Keyword

All control procedures are preceded by the `Private` keyword by default. This keyword makes the procedure accessible only in the form module in which it is declared. The meaning and uses of the `Private` keyword, as well as other keywords that can be used in procedure declarations, are discussed in detail in Chapter 10, "Defining the Scope of Variables and Procedures," when you consider the topic of variable and module scope. For the time being, all procedures you write will use the `Private` keyword.

The *Print* Method

The single method used in Exercise 3.2 was the `Print` method. A method differs from a statement, such as an assignment statement, by specifying the behavior of an object. The following statement features a period (.) to separate the object from the action to be taken:

```
Form1.Print "Message to be printed"
```

That, in turn, is followed by a space, so that the following instruction would be in error:

```
Form1.Print="Message to be printed"
```

The syntax used for the `Print` method itself is quite complex and deserves special attention. You consider that notation next. Once you understand this syntax, you will be able to understand the syntax designed for all methods.

Understanding Syntax Notation

The `Print` method enables you to display information directly on a form or in the Debug window. The method contains a complex syntax, which uses brackets, braces, pipes, and parentheses. Let's consider what this all means. The complete syntax is as follows:

```
[object.]Print [{Spc() | Tab() }] [expression list][{; | ,}]
```

Table 3.2 explains the components of the syntax of the `Print` method.

Table 3.2 The Syntax of the *Print* Method

Component	*Description*
`object`	Indicates the object (for example, the name of the form, not the caption) on which the expression list will be printed.
`object.Print`	Signifies a method rather than a statement, even though the object name is optional.
`[ ]`	Indicates an optional component. It does not have to be included in a `Print` statement. In the syntax above, everything is optional except the word `Print`.
`{ }`	Indicates choices. You can select one from the list. As an example, you can select a semicolon (;) or a comma (,), but not both.
`expression list`	The string to be printed. If no expression is indicated, a blank line will be printed.
`|`	Means *or*. For example, you can select `Spc()` or `Tab()`.
`Spc()`	Invokes the space function. For example, `Spc(20)` provides a blank space of 20 characters.
`Tab()`	Invokes the tab function. `Tab(2)` means to tab two columns to the right. When you use the tab function with `Print`, the print surface is divided into a set of fixed-width columns.
`;`	A semicolon places the cursor immediately after the last character displayed.
`,`	A comma places the cursor in the next print zone. Print zones begin every 14 columns on the screen.

In the preceding example, the following instruction produces a blank line:

```
Print
```

Therefore, the following instruction would print the message `Welcome to Visual Basic` 20 spaces to the right:

```
message = "Welcome to Visual Basic"
Print Spc(20); message$
```

Remember that `object` is optional. In Visual Basic, if an object is not specified, the default object is the form. Although the instruction `Print` can appear to be a statement, it is a method without the object specified (form in this instance).

The *Cls* Method

A second method that is often used when displaying text with the Print method is the clear or `Cls` method. This method clears printed text or other graphic output from the form. The method can also be used to clear graphical output in a picture box. The complete syntax for this method is as follows:

```
[object.]Cls
```

The name of the object is optional. Thus, to clear text from a form, you could have written the following instruction:

```
form1.Cls
```

Observe that if you specify the object, the name of the object is required, not the caption.

Exercise 3.3 Adding a Second Command Button and Procedure to a Form

For this exercise, you add a second command button and a control procedure to a form. Figure 3.5 shows the completed form.

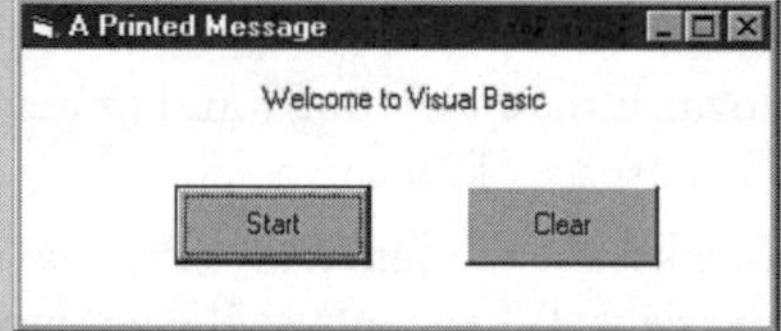

Figure 3.5

Adding a Clear button will enable the user to clear the text displayed on a form.

1. Begin a new project or continue with the last one.
2. Add the Start button and code required in Exercise 3.2, if necessary.
3. Activate the form, not the button, and press the F4 key if the form properties box is not visible. Change the caption to **A Printed Message**, if necessary.
4. Add a second command button and move it to the right of Start.
5. Activate the second command button, and press the F4 key to reveal the control properties box. Make the following change:

```
Caption    =  Clear
Name       =  cmdClear
```

6 With the command button still active, double-click to open the Code window. Enter the following code:

```
Sub cmdClear_Click()
  ' This single instruction clears the screen
  Cls
End Sub
```

7 Close the Code window and run the program. Click Start and click Clear. Click Start and click Clear again.

8 Save this project. You will start a new project in the next exercise.

Objective 3.4 Changing Object Properties with Coded Statements

Besides changing the value of properties using the properties box, the assignment of a property can be done using code. In this section, you examine the assigning of text to a text box and the assigning of a caption to a label using code.

3.4.1 The Text Box

Properties of objects can be set using the properties box or can be set within coded procedures using assignment statements. Visual Basic follows a specific syntax for each property setting. The following is the complete syntax for assigning the `Text` property to a combo box, list box, or text box:

```
[form.] {combobox | listbox | textbox}.Text [= stringexpression]
```

The syntax indicates that text can only be used with a combo box, list box, or text box, and that one of these choices is required.

With a text box named `txtDisplayIt`, you can use this syntax to write the following:

```
Form1.txtDisplayIt.Text = "Welcome to Visual Basic"
```

With this instruction, the string expression `Welcome to Visual Basic` is assigned to the `Text` property of the object `txtDisplayIt` (which is the name given to the text box).

The corresponding instruction can be used to assign an empty (or zero length) string expression to the `Text` property:

```
txtDisplayIt.text = ""
```

Exercise 3.4 Adding a Text Box to Display Output

In this exercise, you add a text box to a form to display output. (You will use labels in Exercise 3.5 to display output.) Figure 3.6 illustrates the completed form. Because this exercise is similar to the previous one, you can copy and rename the last exercise to save time.

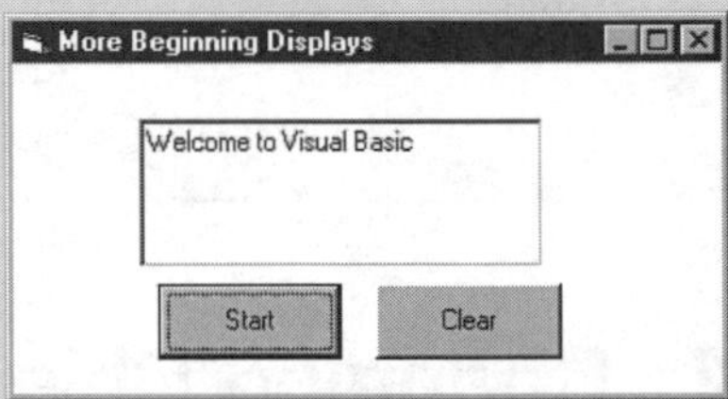

Figure 3.6

This display prints a message inside a text box rather than on the form.

1 Start a new project.
2 Change the caption of the new form to **More Beginning Displays**.
3 Add a text box to the screen, and move it and change its size to take the shape illustrated in Figure 3.6. The Text Box control has the letters `ab` on it.
4 Change the following properties of the text box, removing `Text1` from the `Text` property, leaving it blank:

```
Text  =
Name  =   txtDisplayIt
```

(The second statement changes the name of the object.)

5 Add a command button, and change the following properties:

```
Caption  =  Start
Name     =  cmdStart
```

Position it as shown in Figure 3.6.

6 Add a second command button, and change the following properties:

```
Caption   =  Clear
Name      =  cmdClear
```

7 Activate the `cmdStart` button, and open the Code window. Examine the objects that are now available for you to select from. Click `cmdStart` to open the pull-down menu. The objects `cmdStart`, `cmdClear`, and `txtDisplayIt` should now appear on the list.

8 Enter the following statements:

```
Private Sub cmdStart_Click()
   'The .Text property displays a message in the Text box
    txtDisplayIt.Text = "Welcome to Visual Basic"
End Sub
```

9 Close the Code window, activate the Clear button, and open the Code window.

10 Enter the following statements:

```
Private Sub cmdClear_Click()
  'This single instruction clears the Text box
   txtDisplayIt.Text  = ""
End Sub
```

11 Run and test your program. Click Start and click Clear. Perform this sequence several times.

3.4.2 The Text *MultiLine* Property

Open the text box and examine the 40 properties. Of these, the `MultiLine` property is important. Setting this property to `True` enables a text box to accept and display multiple lines of text at runtime (that is, when your program is running). Setting the `MultiLine` property to `False` restricts text to one line and ignores carriage returns. The syntax for the `MultiLine` property is as follows:

```
[form]textbox.MultiLine = {True | False}
```

3.4.3 The Text *Alignment* Property

The `MultiLine` property is important, in part, because of its effect on the `Alignment` property. Unless the `MultiLine` property is `True`, the `Alignment` property is ignored. The following is the syntax for the `Alignment` property:

```
[form.]control.Alignment [= numericexpression]
```

As an example, Table 3.3 shows the settings for a text box.

Table 3.3 Text Box Settings

Setting	*Description*
0	Text is aligned to the left (the default).
1	Text is aligned to the right.
2	Text is centered.

3.4.4 Assigning a Caption to a Label

A caption is assigned to a label, rather than text. The complex syntax for this assignment is as follows:

```
[form.][control.]Caption [= stringexpression]
```

Even though the syntax appears somewhat less complex than the syntax for the `Text` property of a text box, a caption can be used with a much greater number of objects. The Visual Basic Help contains an "Applies To" section for a caption, which indicates that a caption applies to a dozen standard objects, such as a form, command button, and label.

In Exercise 3.5, you use labels with borders rather than text boxes, for displaying output. This practice is desired when output is not to be changed by the user. Usually, you never have to be concerned about whether or not a caption will store a sufficient number of characters. When used with labels, the length of a caption is unlimited. When used with a form and all other controls, the length is limited to 255 characters.

3.4.5 The Label *AutoSize* Property

The properties associated with a label are also numerous, many of which fall outside the scope of this chapter. Like a text box, the text in a caption of a label can be aligned using the `Alignment` property. The three settings are the same as with a text box: `Left`, `Right`, and `Centered`.

A property unique to labels and picture boxes is `AutoSize`. Set this property to `True` when you want the size of the label to change automatically to fit the label caption. Set it to `False` to keep the size of the control constant. (The contents are truncated if the text exceeds the area of the control.) `False` is the default.

3.4.6 The Label *WordWrap* Property

Closely associated with the `AutoSize` property is the `WordWrap` property. When set to `True`, the text *wraps* and the label control expands or contracts vertically to fit the text and the size of the font. The horizontal size doesn't change. When set to `False` (the default), text doesn't wrap, and the label expands or contracts horizontally to fit the length of the text and vertically to fit the size of the font and the number of lines.

3.4.7 The Label *BorderStyle* Property

Another useful label property is `BorderStyle`. This property is used to add a border to a label to make it appear as a text box. The syntax is:

```
[form.][control.]BorderStyle = {0 | 1}
```

Set this property to `0` when you don't want a border (the default for the Image control and Label control), and set it to `1` for a fixed, single border (the default for the Text Box control and Picture Box control).

Exercise 3.5 Using Labels Instead of Text Boxes to Display Output

The purpose of this exercise is to introduce you to the use of labels. You use three labels to display a user's first name, middle name, and last name. In addition, you add three labels not affected by coded procedures, and the command buttons Start and Clear. Figure 3.7 shows the completed design.

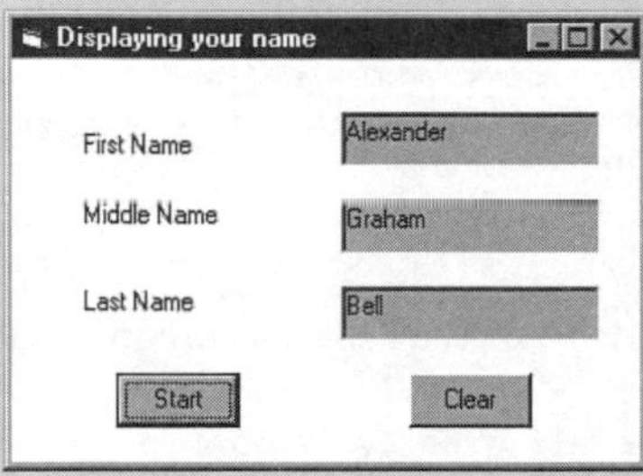

Figure 3.7

This display uses read-only labels rather than text boxes to display a person's name.

1 Begin a new project.

2 Change the following form properties:

```
Caption  =  Displaying your name
Name     =  frmName
```

Observe that you should use the prefix `frm` with a form.

3 Add the three labels, and change the captions to **First Name**, **Middle Name**, and **Last Name**.

4 Add a **First Name** display label, and change the following properties:

```
Caption      =
Name         =    lblFirst
BorderStyle  =    1 - FixedSingle
```

5 Add the **Middle Name** display label, and change the following properties:

```
Caption      =
Name         =    lblMiddle
BorderStyle  =    1 - FixedSingle
```

6 Add the **Last Name** display label, and change the following properties:

```
Caption      =
Name         =    lblLast
BorderStyle  =    1 - FixedSingle
```

continues

7 Add the first command button, and change the following properties:

```
Caption  =  Start
Name     =  cmdStart
```

8 Add the second command button, and change the following properties:

```
Caption   =   Clear
Name      =   cmdClear
```

9 Open the Code window for the `Start` command, and enter the following code:

```
Sub cmdStart_Click()
  'Program to display first, middle, and last name
   lblFirst.Caption = "Alexander"
   lblMiddle.Caption = "Graham"
   lblLast.Caption = "Bell"
End Sub
```

10 Open the Code window for the `Clear` command, and enter the following code:

```
Sub cmdClear_Click()
  'Program to clear all display labels
   lblFirst.Caption = ""
   lblMiddle.Caption = ""
   lblLast.Caption = ""
End Sub
```

11 Run your program. Test it by clicking the Start and Clear buttons several times.

12 Save your form and project.

3.4.8 The *With* Statement

At times, it might be desirable to change several properties of a control, such as a label, all at once. An abbreviated way to do this is by using the `With` statement. Consider the following code, which changes three properties of the label named `lblMine`:

```
Sub cmdChange_Click()
  'Program to clear all display labels
   With lblMine
       .Height = 200
       .Width = 300
       .Caption = "This is a new Caption"
   End With
End Sub
```

The syntax of this statement is quite straightforward. It reads as follows:

```
With object
  [.property = expression]
End With
```

Table 3.4 explains the components of the `With` statement.

Table 3.4 The *With* Statement Syntax

Component	*Description*
`object`	The name of an object or a user-defined type (required)
`property`	The property of the object to be executed (optional)

Exercise 3.6 Using the *With* Statement to Modify Several Properties of a Label

This exercise asks you to modify Exercise 3.5. (Steps 1 through 8 in this exercise are the same as those in Exercise 3.5.)

1 Begin a new project.

2 Change the following form properties:

```
Caption  =  Displaying your name
Name     =  frmName
```

Observe that you should use the prefix `frm` with a form.

3 Add the three labels, and change the captions to **First Name**, **Middle Name**, and **Last Name**.

4 Add a **First Name** display label, and change the following properties:

```
Caption      =
Name         =   lblFirst
BorderStyle  =   1 - FixedSingle
```

5 Add the **Middle Name** display label, and change the following properties:

```
Caption      =
Name         =   lblMiddle
BorderStyle  =   1 - FixedSingle
```

6 Add the **Last Name** display label, and change the following properties:

```
Caption      =
Name         =   lblLast
BorderStyle  =   1 - FixedSingle
```

7 Add the first command button, and change the following properties:

```
Caption  =  Start
Name     =  cmdStart
```

8 Add the second command button, and change the following properties:

```
Caption    =   Clear
Name       =   cmdClear
```

continues

9 Modify the `cmdStart_Click()` procedure as follows:

```
Private Sub cmdStart_Click()
  'Program to display first, middle, and last names
  With lblFirst
    .Caption = "Alexander"
    .BackColor = vbGreen
    .Alignment = 2
    .Height = 400
  End With
  With lblMiddle
    .Caption = "Graham"
    .BackColor = vbGreen
    .Alignment = 2
    .Height = 400
  End With
  With lblLast
    .Caption = "Bell"
    .BackColor = vbGreen
    .Alignment = 2
    .Height = 400
  End With
End Sub
```

10 Test your code. What are the advantages of the `With...End With` structure?

Chapter Summary

The objectives of this chapter introduced you to 12 controls found on the toolbox and instructed you in the use of three of them: a command button, a text box, and a label. The various types of controls are easier to remember if you group them, as follows:

- Four controls are used with text:
  - Text Box control
  - Label control
  - List Box control
  - Combo Box control
- Two controls are containers:
  - Picture Box control
  - Frame control
- Three controls are buttons:
  - Option Button control
  - Check Box control
  - Command Button control

- The Image control is used to display graphics and can receive mouse events, which allows it to be used as a button.
- Two controls are scroll bars:

  Horizontal Scroll Bar control

  Vertical Scroll Bar control

Adding a control to a form can be done by double-clicking the control, or by clicking the control and dragging it to change its size on a form. Control properties can be set once the control is placed on the form. Double-click the control to open the Code window and add an event procedure to a form.

Add name prefixes to forms and controls to help in the management of objects in a design. Adding the name `cmdStart` tells you that the object is a command button with a likely caption of Start.

The `Dim` and `Rem` statements, the `Print` and `Cls` methods, and assignment statements were used in writing the exercises for this chapter. The `Dim` statement is used to declare variables. Use a single quotation mark (') to add comments (remarks) to code. The `Print` method displays characters on an object, such as a form. The `Cls` method clears text or graphics from a form or picture box.

Assignment statements are used frequently to assign property values to an object within a coded procedure. When writing an assignment statement, the syntax associated with the object becomes important. The complex syntax for the `Text` property of a text box is as follows:

```
[form.] {combobox | listbox | textbox}.Text [= stringexpression]
```

The following is the complete syntax for the `Caption` property:

```
[form.][control.]Caption [= stringexpression]
```

Text boxes and labels are used for different purposes. A text box is used when you require the user to enter values to be used by a procedure. A label is used to display output when the user is not allowed to change the display.

Skill-Building Exercises

1. Modify Exercise 3.2 using the `MouseDown()` and `MouseUp()` events. When you click the Start button, change the `Click()` event to a `MouseDown()` event to display the following:

   ```
   Welcome to
   ```

   Write a `MouseUp()` event to display the following:

   ```
   Visual Basic
   ```

   When you run this program, the `MouseDown()` action displays the first part of the message. Releasing the mouse button completes the message. Display this as the caption of the form: **A Printed Message**.

2. Place two labels on the screen, changing the caption of the first to **Click here**. For both labels, make the following property changes:

   ```
   Alignment = Center
   BorderStyle = 1 - Fixed Single
   ```

   Write a `Click()` event for the first label. When you click the `Click here` label, display the message `Welcome to Visual Basic` in the space set aside for the second label. Change the caption of the form to **Working with Labels**.

3. Modify Exercise 3.5 using Figure 3.7 as a guide. Instead of using a Start button to display the name, modify the design so the user must click the label `First Name` to display the first name of the person, the label `Middle Name` to display the middle name, and the label `Last Name` to display the last name. Remove the Start button from the design, but keep the Clear button.

4. Design a form with two multiple-line text boxes and two command buttons, one named **Start** and one named **Clear**. In the first text box, display a person's name, such as Alexander Graham Bell. Place part of the name on one line of the text box and part on a second line. In the second text box, display the person's address, such as 7677-5678 Telephone Drive. Once again, display part of the address on one line and the remainder on the second line. Use a larger font size, such as 12. Change the caption of the form to **Name and Address**.

5. Design another form with two multiple-line boxes, but no command buttons. Above the first text box, add the label **My name is**. Above the second text box, add the label **My address is**. Write a `Click()` event such that when you click the name text box, the person's name appears. Then, write a double-click event to clear the text box. Write a second `Click()` event to add the address in the second text box; write a double-click event to clear the address in the second text box. Use a larger font size, such as size 12. Add the form caption **Name and Address II**.

4

Learning to Think Visually

Visual Basic requires you to give special attention to the design of forms before beginning to write computer code. This might be a somewhat different task for those who have programmed before. The Visual Basic designer must be much more concerned with the user interface, compared to earlier programming languages. Some might say that the designer must be *artful* as well as *logical* in the design of a computer application.

In this chapter, you continue to learn about controls in the design of Visual Basic applications and to practice working with the controls contained in the toolbox. The tasks include

- adding a combo box to a design,
- adding a frame to a design,
- adding check boxes to a design,
- adding a list box to a design,
- setting the tab order in a design,
- disabling a control,
- adding a menu bar to a design,
- writing menu procedures,
- adding scroll bars to a design.

Chapter Objectives

The objectives for this chapter involve working with additional controls in the toolbox. However, the discussion is framed in terms of a short (and rather silly) problem. In solving the problem, we ask you to think visually about how a form should look before you begin to add controls to that design. The objectives are as follows:

1. Thinking about a problem
2. Using design-time and runtime properties to manipulate list boxes and combo boxes

3. Using frame controls to improve user-interface design
4. Controlling the tab order in a design
5. Learning to enable and disable a control
6. Creating a menu bar and writing code for menu procedures
7. Adding a scroll bar to a design

The following are the specific controls introduced in this chapter:

Combo boxes
List boxes
Option buttons
Check boxes
Scroll bars

This set completes the standard Windows controls.

Objective 4.1 Thinking About a Problem

Visual interfaces are easy to create if you are handed the problem and the interface. They are most difficult when you are facing an actual ill-defined problem and when you have no knowledge of the type of interface that is required. As a way of simulating the design process, you can step through a problem, taking the role of both the program designer and the user.

Better yet, ask a colleague to assume the role of the user while you act as the designer, and employ role playing to arrive at a solution using the hypothetical scenario described in the following sections.

4.1.1 Role 1: The User

"I need a way of entering personal information about our students," mutters the admissions officer. "People in data processing are telling me about their new-found ability to solve computing problems faster than in the past, but I'm still suspicious. I just don't know enough about the computer to tell smoke from fire."

"What I want is a simple system," continues the admissions officer. "The system should permit me to enter into the computer a person's name and sex, the student's year in school, some special category if applicable (we have several), and what the student thinks of himself or herself. Because I hate typing, I want a system that is as simple to use as possible. I also do not want to move too fast."

"I would like to review the design to be used in entering one student into the system. I want to display a student's name and sex; determine whether the student is a freshman, sophomore, junior, or senior; and conduct a self-image survey, which asks the student whether or not he or she is wise, kind, cute, or sturdy. The student should check all that apply."

4.1.2 Role 2: The Program Designer

"I wish I had some idea of what the admissions officer wants," the designer contemplates. "Perhaps after my meeting today, I will have something to go on. I guess I'd better write down some questions to ask."

4.1.3 The Meeting

Read this section after role playing in the preceding scenario or thinking about the topics of conversation that might take place at a meeting.

In speaking with the admissions officer, the program designer should ask questions to determine the nature of the problem, and determine the types of input to be entered into processing and the types of output to expect. Let's assume that a tape recording was made of the meeting between the admissions officer and the program designer, and the following is a transcription of that tape.

Adm. officer:	Good morning. Glad you could make it.
Designer:	My pleasure. Are you ready to begin?
Adm. officer:	Yup. Go ahead.
Designer:	Perhaps you could tell me what type of application you are interested in.
Adm. officer:	I would like a small student admissions system limited to four or five types of information. I need to record the name and sex of the student, his or her year in school, and the student's self-image assessment.
Designer:	Do you want the student's full name?
Adm. officer:	Yes.
Designer:	Will you have to type in the name, or can you pull this information from another computer file?
Adm. officer:	We'll have to type in the name, I'm afraid. This information is to be used for gaining insight about the self-image of our students. If possible, we would like to enter all the information when speaking with each student.
Designer:	Okay. I just wanted to know how you intended to collect the data. Let me ask another question. Can we limit the student's choices of year in school to freshman, sophomore, junior, and senior?
Adm. officer:	For 90 percent of our students, the answer is yes; however, we have other categories, such as *adult special.*
Designer:	Does that mean you want to be able to enter the name of the category or add the word *other* as a choice?

Adm. officer:	Let's use *other.*
Designer:	How do you plan to collect the self-image information of a student?
Adm. officer:	For now, we want to ask each student to choose *wise, kind, cute,* or *sturdy* in a survey question.
Designer:	Only these four choices?
Adm. officer:	We might add others later, but for now we only want to check off these four. One more thing: We don't want to get in too deep to begin with. Would it be possible for you to show me a design that handles only one student? This would help us determine how we want to move ahead and what we want from a larger system.
Designer:	Seems simple enough. Let me get back to you later today.

4.1.4 Arriving at a Solution

After reading the preceding conversation, can you picture in your mind what type of application the admissions officer needs? Can you picture a possible form and the types of controls needed on this form? If you can, you are beginning to think visually.

More difficult to determine is the exact nature of the problem the admissions officer is attempting to solve. If you get the impression that the admission's staff is not very sure of their requirements, you are correct. The staff wants some type of system to support the study of student profiles, but the specifics of such a system are still quite vague.

Objective 4.2 Using Design-Time and Runtime Properties to Manipulate List Boxes and Combo Boxes

Before you continue the case study, you need to learn about two of the controls that you might choose for your admissions design. These are *list boxes* and *combo boxes.* Both controls are designed to provide a list of choices, but in different ways. A list box is limited to a list of choices. From this list, a user can make one or several choices. A combo box restricts the user to one choice or entry. It is called a combo box because it is a combination of a list box (used for selecting items from a list) and a text box (used for entering text).

4.2.1 List Box Properties

Add a list box to a blank Visual Basic form to study the properties of this control. Besides the normal font and size properties, two properties of particular interest at this time are the `MultiSelect` property and the `Columns` property.

The *MultiSelect* List Box Property

The `MultiSelect` property enables the user to make more than one selection from a list. The syntax for this property is as follows:

```
[form.] {filelistbox | listbox}.MultiSelect = {0 | 1 | 2}
```

Table 4.1 shows the settings of the `MultiSelect` property.

Table 4.1 *MultiSelect* Settings

Setting	*Description*
`0`	Multiple selections are not allowed (the default).
`1`	Simple multiple selection. Clicking the mouse or pressing the spacebar selects or deselects an item in a list.
`2`	Extended multiple selection. Shift+click or Shift+arrow key extends the selection to allow a block of items to be selected. Ctrl+click selects or deselects an item from the list.

Returning to the admissions project, can you determine what setting is needed for that design?

The *Columns* List Box Property

The `Columns` property determines whether the list box is to contain more than one column, and how scroll bars are shown if the list box needs to be extended. The syntax for this property is as follows:

```
[form.] listbox.Columns = positivenumber
```

Table 4.2 explains the settings for the `Columns` property.

Table 4.2 *Columns* Property Settings

Setting	*Description*
`0`	Items are arranged in a single column (the default). The list box scrolls vertically.
`1` to *n*	Items arranged in a series of columns, filling each column from left to right. The list box scrolls horizontally. Use this setting when you want to display a table of values.

4.2.2 The Combo Box *Style* Property

Combo box properties are similar, but certainly not identical, to list box properties. An important combo box property is the `Style` property. Also, you need to use the `AddItem` method and the `Form_Load()` procedure to describe how a combo box or list box is filled with data and displayed.

There are three styles of combo boxes (see Figure 4.1). The following list contains a description of each:

- **Style 0**—A drop-down combo box. The user clicks the detached arrow to open (and drop down) the list of choices. Selecting one of the choices places it in the text portion of the combo box. Use Style 0 when you want the user to select from a list and you want to conserve screen space. This style of combo box enables a user to type in a choice that doesn't appear on the list.
- **Style 1**—A simple combo box. Observe that the detached arrow is missing from this combo box. This leads to the display of all items, provided there is space. If additional space is required, a vertical scroll bar is automatically inserted. Similar to Style 0, this style of combo box enables a user to type in a choice that doesn't appear on the list.
- **Style 2**—A drop-down list box. Like the drop-down combo box, a drop-down list box requires the user to click an arrow (attached, not detached) to open the list. Selecting one of the choices places it in the text portion of the combo box. This style saves space, but *does not* enable the user to enter text. The user can make list selections only.

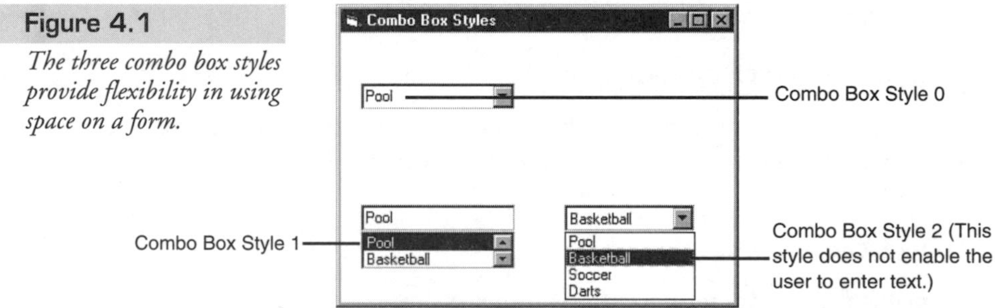

Figure 4.1

The three combo box styles provide flexibility in using space on a form.

4.2.3 The *AddItem* Method

Adding data to a combo box or list box is not as obvious as you might expect. To add items to a combo box or list box, use the `AddItem` method. The syntax for this method is as follows:

```
{ComboBox | ListBox}.AddItem item [, index]
```

Table 4.3 explains the components of the `AddItem` syntax.

Table 4.3 The Syntax of the *AddItem* Method

Component	*Description*
`AddItem`	A built-in Visual Basic method.
`item`	The string expression to add to the list. If the expression is a literal string, it must be enclosed in quotation marks.
`index`	Indicates the placement of the item in the list (and is optional). An index of `0` represents the first position.

As an example, if you want to add the word *Freshman* to `cboMyCombo` and to the first position in the list, you would type the following:

```
cboMyCombo.AddItem "Freshman", 0
```

If an index is not added, the `AddItem` method adds items in the order indicated by the code.

4.2.4 The *Form_Load()* Event

A `Form_Load()` event occurs when a form is loaded into memory. Among other functions, this event is used to fill combo boxes and list boxes with data. When the form is loaded at runtime, the code in the `Form_Load()` event populates list boxes and combo boxes with list items. As an example, examine the following procedure:

```
Private Sub Form_Load()
   lstSport.AddItem "Pool"
   lstSport.AddItem "Basketball"
   lstSport.AddItem "Soccer"
   lstSport.AddItem "Darts"
End Sub
```

This procedure fills the list box `lstSport` with the words *Pool, Basketball, Soccer,* and *Darts* when the form is loaded at runtime. Because the index is not indicated, the list will be loaded as entered.

4.2.5 Manipulating Items in a List Box or Combo Box

After items have been added to a list box or combo box, they can be sorted by setting the `Sorted` property, or removed one at a time using the `RemoveItem` method. There is also a `Clear` method to clear a list box or combo box, removing all items with a single instruction.

Sorting a List Box or Combo Box

If you want items sorted in a list box or combo box, set the `Sorted` property to `True`, omitting the index. This sorts the list in alphabetical order. Because the order is not case sensitive, the words "Darts" and "darts" will be treated the same. The syntax for the `Sorted` method is:

```
[form.] {ComboBox | ListBox }.Sorted
```

The `Sorted` properties are `True` (list items only alphabetically) or `False` (do not list the sorted items alphabetically). Sorting is always in ascending order (*A* through *Z*). In addition, the `Sorted` property has as the default the value `False`. This can be changed by assigning `True` to the `Sorted` property.

Removing Items from a List Box or Combo Box

Removing an item from a combo box or list box uses the built-in `RemoveItem` method. The syntax for this method is as follows:

```
{ListBox | ComboBox}.RemoveItem index
```

For example, the following instruction would remove `"Pool"` from the list box shown earlier in section 4.2.4, "The `Form_Load()` Event."

```
lstSport.RemoveItem 0
```

Clearing Items from a List Box or Combo Box

Clearing all items from a combo box or list box requires the `Clear` method. The syntax in this instance is as follows:

```
{ListBox | ComboBox}.Clear
```

For example, the following instruction clears all the items from `lstSport` (see section 4.2.4, "The `Form_Load()` Event").

```
lstSport.Clear
```

4.2.6 Using Runtime Properties to Manipulate List Boxes or Combo Boxes

Not all the properties of List or Combo Box controls are available during design time. Like other controls, some are only available at runtime and are known as *runtime properties.* Some of these runtime properties are read-only. That means that only the property's value can be retrieved during runtime. You can't assign it a value with code. Runtime properties that can be assigned values are known as *read/write properties.* That means that their values can be both retrieved and assigned with code.

Both list boxes and combo boxes have a runtime property called `ListCount`, which is read-only. It returns the number of items in a list box or combo box. You will not see the property in the Properties window during design time—nor can you assign it a value during runtime. Its value changes as items are added or removed from a list box or combo box. Even though it is read-only, the `ListCount` value can be retrieved and displayed. For instance, to learn the number of items in the `lstSport` example (see section 4.2.4, "The `Form_Load()` Event"), you might write the following code fragment:

```
Print lstSport.ListCount
```

Assuming nothing had been added or deleted after the `Form_Load()` procedure, the number `4` would be printed on the form. Because lists always start with `0`, the index value of the last item in a list is not equal to the `ListCount`. Instead, it is equal to `ListCount -1`.

List boxes and combo boxes have two other runtime properties named `List` and `ListIndex`. The `List` property, which is read/write, retrieves or assigns the expression in the list box or combo box. (This differs from the `AddItem` method, which populates the list box or combo box.) Its syntax is as follows:

```
[form.]control.List(index)[ = stringexpression ]
```

If you wanted to print the fourth item in a combo box, you might enter the following code fragment:

```
Print cboProducts.List(3)
```

If the fourth item was `"men's gloves"` and you wanted to change it to `"women's gloves"`, you would enter the following code fragment:

```
cboProducts.List(3) = "women's gloves"
```

The `ListIndex` property, a read/write property, indicates the index value of the currently selected item in a list box or combo box. It is frequently used to ensure that the first item in a list box is displayed. The following code fragment indicates that the first item is displayed in a combo box rather than an empty line:

```
cboProducts.ListIndex = 0
```

If you wanted to have the tenth item in a 20-item combo box displayed, enter the following code:

```
cboProducts.ListIndex = 9
```

To go to the last item in a list or combo box, enter the following code:

```
cboProducts.ListIndex = cboProducts.ListCount -1
```

Exercise 4.1 Adding a List Box to a Form

With your beginning knowledge of how combo and list boxes work, you'll add a list box to a form and fill it with data in this exercise. To continue with the admissions project in Objective 4.1, "Thinking About a Problem," you will provide an initial design for the admissions officer. After you find out more about what the admissions staff wants, you will add features to your design. Figure 4.2 illustrates the design this exercise asks you to complete.

continues

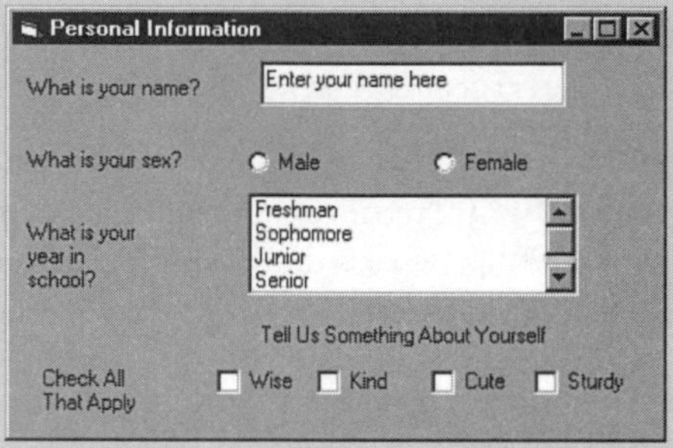

Figure 4.2

The initial student profile display features three new controls: option buttons, a list box, and check boxes.

Is this the design you visualized? Does it contain a text box, option buttons, a list box (or combo box), check boxes, and labels?

1 Begin a new project.
2 Change the caption of the form to **Personal Information.**
3 Add a **`txtName`** text box, as shown in Figure 4.2.
4 Add two option buttons, as shown in Figure 4.2. Type the captions **Male** and **Female**. Use the names **`optMale`** and **`optFemale`**.
5 Add a list box, as shown in Figure 4.2. Name the list **`lstSchoolYr`**.
6 Add four check boxes: **`chkWise`**, **`chkKind`**, **`chkCute`**, and **`chkSturdy`**. Type the captions **Wise**, **Kind**, **Cute**, and **Sturdy**.
7 Add the following five labels:

   What is your name?

   What is your sex?

   What is your year in school?

   Tell us something about yourself.

   Check all that apply.

8 Fill the list with item information. Place the item information in the `Sub Form_Load()` event procedure, as follows:

```
Private Sub Form_Load()
    'Warning:  A list box begins with the lowercase letter L,
    'not the number 1
    lstSchoolYr.AddItem "Freshman"
    lstSchoolYr.AddItem "Sophomore"
    lstSchoolYr.AddItem "Junior"
    lstSchoolYr.AddItem "Senior"
    lstSchoolYr.AddItem "Other"
End Sub
```

9 Close the code window, and run the program. The list box should show the five items as entered. Use the Tab key to move from one control to another. If the tab order is not correct, don't worry: You will learn how to change it later in Exercise 4.3.

10 Save this project for use with the next exercise.

Objective 4.3 Using Frame Controls to Improve User-Interface Design

Exercise 4.1 might have helped you to discover that frames were needed for the option buttons, if you did not add them to your design in sequence. As a test, place two option buttons on a form, followed by a check box and another set of option buttons. You will discover that only one option can be selected, when you wanted to enable the user to make two selections.

Frames are used to define a group of options. Stated in a broader context, the `Frame` control provides a visual and functional container for a set of other controls, such as option buttons and check boxes.

Adding a frame to a form is somewhat difficult when you want to place other controls inside the frame. The important lesson to remember when adding a frame to a form is to add the frame first, then drag (not double-click) or cut and paste the controls to be contained onto the frame. In brief:

1. Create the frame.
2. Select the control to be added in the toolbox and then "draw" the control in the frame or cut existing controls and, with the frame having the focus, paste the controls into the frame.

Exercise 4.2 Adding Frames to a Form

For this exercise, you add two frames to the admissions form designed in Exercise 4.1 and replace the list box with a Style 2 combo box. Figure 4.3 shows the revised design.

1 Open the project and add a frame to your design. Attempt to drag the two option buttons into the frame. What happens?

2 Try again: Cut and, with the frame having the focus, paste the two option buttons into the frame. After pasting the buttons, adjust the frame as needed.

continues

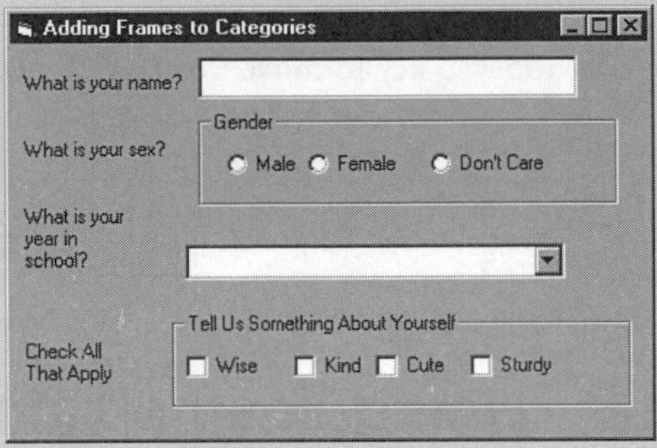

Figure 4.3

This design adds two frames and replaces the list box with a combo box.

3 Draw a third option button inside the frame.

4 Add the caption **Gender** to `Frame1`. Name the frame **`fraGender`**.

5 Add the captions **Male**, **Female**, and **Don't Care** to the three option buttons. (Sorry for the last category. We're just trying to keep you awake.)

6 Cut and paste into a second frame the four check boxes from your design, and add the label **Tell Us Something About Yourself.** Name the frame **`fraTellUs`**.

7 Replace the list box with a Style 2 combo box. Name the box **`cboSchoolYr`**. Change the `Form_Load()` procedure to load the combo box using this name.

8 Test your design and write down what you feel is wrong with your design. What purpose does the frame placed around `Gender` serve? What purpose is served by the frame placed around `Tell Us Something About Yourself`?

9 Save this project for use with the next exercise.

Objective 4.4 Controlling the Tab Order in a Design

When designing a form, such as the one for the admissions officer, you can view such a form as a *record* for one student. The record, in turn, contains *fields,* such as student name, sex, year in school, and self-evaluation scores. When placing information on this form, users want to begin in the upper left-hand corner of the form (the first field), enter information at that point, then press Tab on the keyboard to jump to the next field—either the field to the immediate right or the field on the next line. If you try this with your design, one of the problems you might discover is that the tab order is incorrect. When you press Tab after making an entry, you do not jump to the next field (to the right or down), as you would like.

The initial tab order is based on the sequence you followed in adding controls to a form. In the admissions example, if you added the text box first followed by an option button, the text box should have the focus first. The Tab key then moves the focus from the text box to the first option button.

4.4.1 The *TabIndex* Property

If some controls are deleted on a form and others added, the tab order you desire might be lost. To regain this order, set the `TabIndex` property for a control. The syntax for this change is as follows:

```
control.TabIndex = value
```

A `value` of `0` indicates the starting point of the order. This is followed by the values `1`, `2`, and so on.

When the tab index for one control is changed, all others are adjusted automatically, with following controls adjusted upward. Suppose you have the following tab order in your design:

```
chk1  TabIndex = 0
chk2  TabIndex = 1
chk3  TabIndex = 2
```

Changing the tab index for `chk3` from `2` to `0` leads to the following order:

```
chk1  TabIndex = 1
chk2  TabIndex = 2
chk3  TabIndex = 0
```

4.4.2 The *TabStop* Property

Besides changing the tab order, it is also possible to remove a control from the tab order. To do this, set the `TabStop` property to `False`, as in the following example:

```
control.TabStop = False
```

Once set to `False`, the tab order is maintained; however, the control is skipped when the user cycles through the order.

The `TabStop` property explains how a set of option buttons is able to work when contained by a frame. When the user makes a selection, the `TabStop` property is set to `True`, as in the following example:

```
Option1.TabStop = True
```

However, once set to `True`, all other option buttons are set to `False`:

```
Option2.TabStop = False
Option3.TabStop = False
```

This leads to skipping the remaining options within the frame. To move among option buttons in a frame, use the arrow keys.

Exercise 4.3 Controlling the Tab Order

In this exercise, you will control the tab order for the admissions office application.

1 Open the project, and check the tab order in your design.
2 Change the tab index property of the last check box. Set the `TabIndex` to `0` for `chkSturdy`, and run the program. What happens?
3 Return to design mode and set the `TabIndex` to `0` for `chkWise`.
4 Continue to move up the design, always setting the next control in the tab order to `0`. In so doing, you will move up the design, moving from right to left and bottom to top.
5 Test your design. What happens when you make one sex choice and press Tab?
6 Save your project at this time for use with the next exercise.

Objective 4.5 Learning to Enable and Disable a Control

Controls in Visual Basic can be enabled or disabled by setting the `Enabled` property to `True` or `False`, respectively. The syntax for this property is as follows:

```
[form.][control.]Enabled [= Boolean]
```

The `Boolean` value is either `True` or `False`. The following expression enables the object to respond to events:

```
option3.Enabled = True
```

The following expression prevents the object from responding to events:

```
option3.Enabled = False
```

Disabling an option button dims (grays) the color of the button to indicate that the button cannot be used.

Exercise 4.4 Disabling an Option Button

In this exercise, you will disable the option button for the sex choice shown as Don't Care (see Figure 4.4).

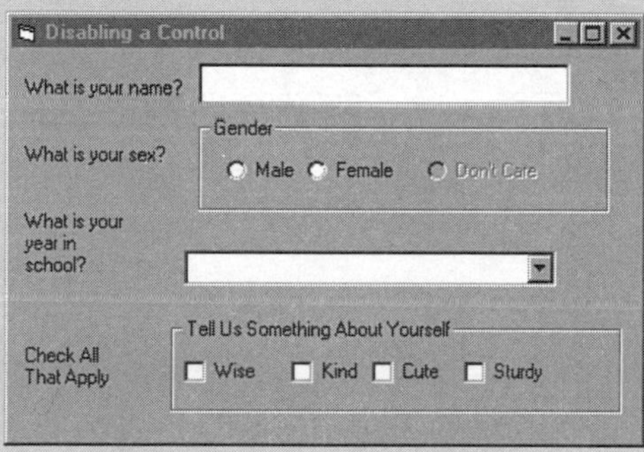

Figure 4.4
This design disables the control entitled Don't Care.

1 Open the project and move to the Don't Care button.
2 Press F4 to open the Properties window.
3 Change the `Enabled` property to `False`.
4 Run and test your design. What happens when you try to click the Don't Care option?

Objective 4.6 Creating a Menu Bar and Writing Code for Menu Procedures

Visual Basic provides a menu bar builder, which enables the Visual Basic programmer to create a menu for an application at design time. Menus are a different type of control. They are used to provide an application design with a set of commands that is easy to find and to use. They have the advantage of adding functionality to a program without sacrificing much screen "real estate." The ease with which a developer can add a menu bar to a Visual Basic application is one of the language's strongest features.

4.6.1 The Menu Design Command

To add a menu bar to a design, give the form focus and choose Tools, Menu Editor, or click the Menu Design button on the toolbar. The dialog box shown in Figure 4.5 opens.

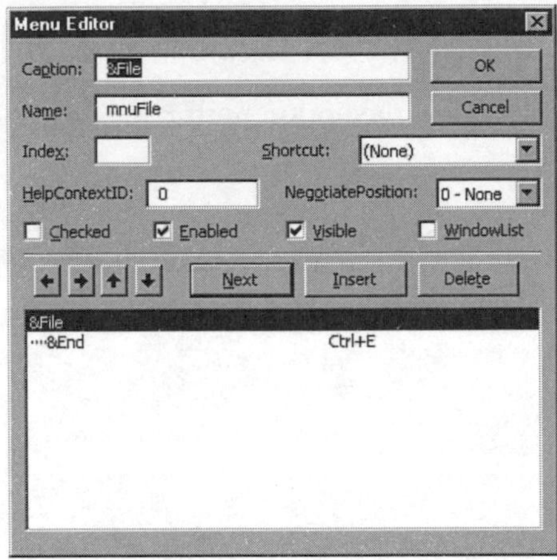

Figure 4.5

The Menu Editor serves as a tool for adding menu selections to the top of a form.

The two most important properties for the Menu control are the first and second text boxes. Use these two boxes as follows:

1. Add the caption for the control in the Caption text box. The caption is the text that will appear on the menu bar. For example, type **`&File`**. (The purpose of the ampersand is described in the next set of steps.) Observe that the word *File* appears both in the caption text box and as the first entry in the Menu control list box. When you run the program, the word *File* will appear as the first name on the menu bar.
2. Press Tab and enter the name of the Menu control. This is the name you use to refer to the control from code. For example, type the name **`mnuFile`**.

After the menu caption and name have been entered, it is possible to add the subordinate command to fall under the command category. To add a subordinate command, do the following:

1. Click Next to advance the list box to the next entry, clear the caption and name boxes, and advance to the next line.
2. Enter the caption for the control in the Caption text box. For example, enter the caption **`&End`**.
3. Following the entry of this name, click the right arrow. The list box should look as follows:

   ```
   &File
   ....&End
   ```

   This indicates that the command `End` is subordinate to the `File` command. The ampersand in front of both menu commands enables standard Windows keyboard navigation to menu choices. Placing the ampersand in front of the letter *F* enables the user to press Alt+F to access File menu choices.
4. Enter the name of the command. For example, enter the name **`mnuFileEnd`** (which means the End choice under the File menu).

5. Tab to Shortcut to set the shortcut key. Open the combo box to select Ctrl+E as the shortcut key combination. By pressing the Ctrl+E on the keyboard, the End command is triggered. Alternatively, the user could use Alt+F and then press the E key to execute the `mnuFileEnd` instruction.
6. Click OK if you have completed building the menu bar.

You will return to the menu bar in other applications. For now, you should become familiar with adding just the File, End command sequence to an application.

4.6.2 Writing Code for a Menu Control

Adding the word *File* to the menu bar and the word *End* to the list of File commands does little more than add the captions to the menu bar. The next step is to write a `Click()` event procedure for a command. When the user chooses a Menu control, a `Click()` event triggers the menu event procedure. This event procedure is one that the Visual Basic programmer must write.

Writing the code for a Menu control is done in the same way as writing the code for other controls.

1. Click the form to open the code window.
2. Find the name of the menu command in the Object list.
3. Find the Click procedure in the procedure list.
4. Add the instructions for the `Click()` event. For example, to end a program, type the word `End`. The completed procedure is as follows:

```
Private Sub mnuFileEnd_Click()
    End
End Sub
```

Exercise 4.5 Adding Menu Commands to Your Application

In this exercise, you add a File menu and End command to the admissions office application. Figure 4.6 shows the modified display.

1 Open the project and choose Tools, Menu Editor Design (or click the Menu Design icon on the toolbar).

2 Type **`&File`** for the command Caption and **`mnuFile`** for the Name.

3 Click Next.

4 Type **`&End`** for the command Caption and **`mnuFileEnd`** for the Name. Add a shortcut key of Ctrl+E. Click the right arrow key to make this command subordinate to the File command.

5 Click OK to close the dialog box.

continues

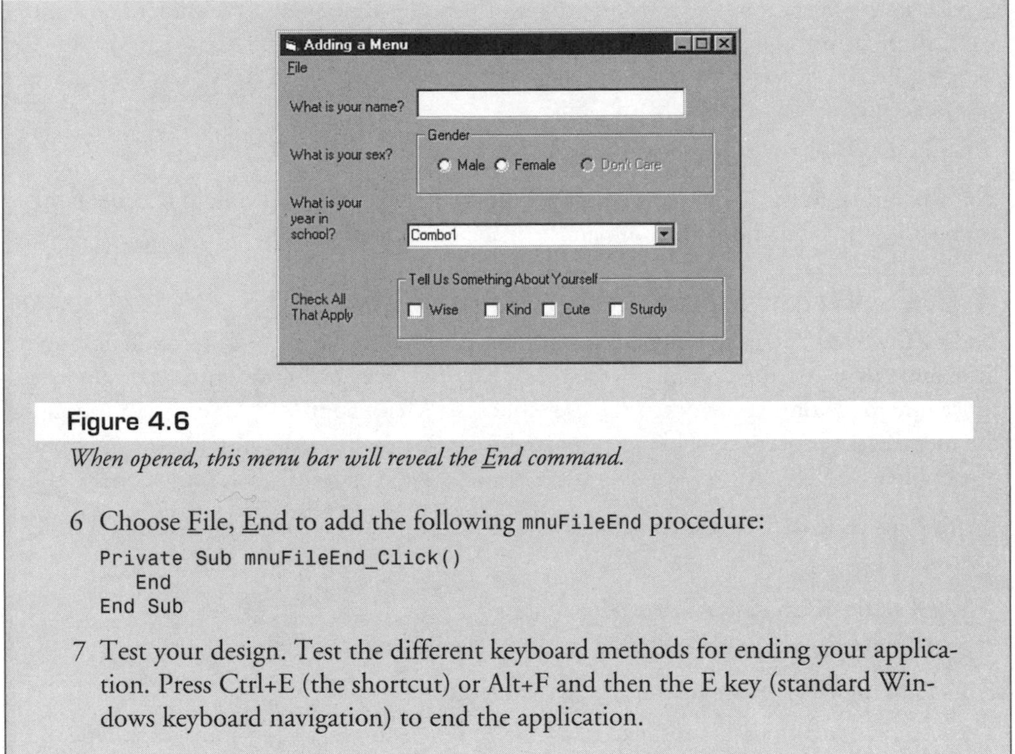

Figure 4.6

When opened, this menu bar will reveal the End command.

6 Choose File, End to add the following `mnuFileEnd` procedure:

```
Private Sub mnuFileEnd_Click()
   End
End Sub
```

7 Test your design. Test the different keyboard methods for ending your application. Press Ctrl+E (the shortcut) or Alt+F and then the E key (standard Windows keyboard navigation) to end the application.

Objective 4.7 Adding a Scroll Bar to a Design

Before concluding this chapter, you should visualize a second design situation. It's time to move away from the admissions project and begin working on a new problem. Consider the following scenario.

4.7.1 Designing a Demonstration Scroll Bar

"I want a simple application to explain to my students how a scroll box works. Named Measuring Speed, the application should have a minimum range of 0 miles per hour (mph) and a maximum range of 120 mph. As a student uses the scroll bar, it should be possible to click the ends of the bar or the middle to demonstrate a slow and a fast scroll. Some kind of Help system should be provided to explain the technique to follow in a fast and a slow scroll. The user should also be able to click to stop the application."

Given only this information, make a sketch of how your design will look.

4.7.2 Scroll Bar Basics

Whether you know it or not, you are quite familiar with scroll bars, having worked in a Windows environment. In the Visual Basic toolbox, you'll find a Vertical Scroll Bar control and a Horizontal Scroll Bar control. As shown by Figure 4.7, each scroll bar contains a scroll arrow at the ends and a scroll box that moves from end to end when you drag it using the mouse. The scroll box is used to indicate the current value specified by the bar.

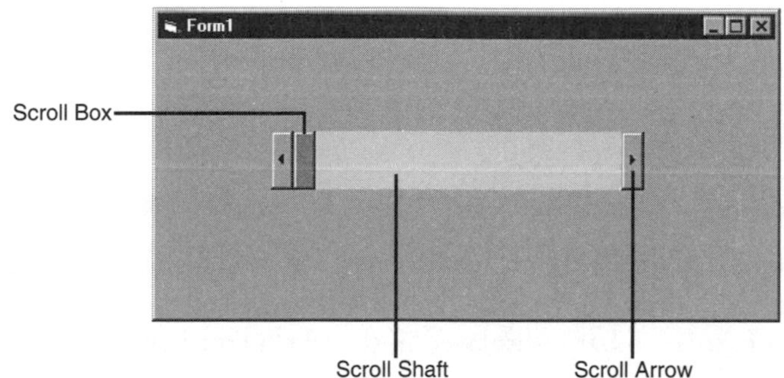

Figure 4.7

A scroll bar requires two procedures: one for the scroll box and another for the scroll shaft and scroll arrows.

Different mouse actions determine how rapidly values change on a scroll bar. Click the arrow to move an incremental unit toward the end. Use the scroll box to drag through a range of values. Click the scroll shaft between the arrow and the scroll box to move a larger increment between the scroll box and the end.

The five most important properties of the scroll bar are those that set the upper and lower range values (`Max` and `Min`), the large and the small change values (`LargeChange` and `SmallChange`), and the value (`Value`). Table 4.4 explains these properties.

Table 4.4 Scroll Bar Properties

Property	*Description*
`Max`	Sets the maximum value for a scroll bar. For example, the maximum for a temperature scroll bar might be 212.
`Min`	Sets the minimum value for a scroll bar. For example, the minimum for a temperature scroll bar might be –32.
`LargeChange`	Determines the amount of change to the value property when the user clicks the space between the scroll box and the scroll arrow.
`SmallChange`	Determines the amount of change to the value property when the user clicks a scroll arrow.
`Value`	Specifies the return value represented by the current position of the scroll bar.

By default, `LargeChange` and `SmallChange` are set to `1`; however, it is possible to change this setting to a number from `1` to `32767` (inclusive). The default settings for `Max` and `Min` are `32767` and `0`, respectively. The possible range for these settings is from `-32768` to `32767`. `Max` and `Min` determine maximum and minimum values, respectively; the actual value depends upon the position of the scroll box.

4.7.3 The *Scroll()* and *Change()* Event Procedures

Two event procedures are associated with the design of a scroll bar: the `Change()` event and the `Scroll()` event. The `Change()` event occurs whenever the `Value` property changes. The code associated with the `Change()` event might read as follows:

```
Private Sub HScroll1_Change()
   lblValue.Caption = HScroll1.Value
End Sub
```

The purpose of this procedure is to report the change whenever the value changes. That is, whenever the user clicks the space between the scroll box and the scroll arrow or selects the scroll arrow, the new value of the scroll bar is displayed in the label named `lblValue`.

The `Scroll()` event is unique to scroll bars. Even so, the code associated with the `Scroll()` event is similar to the code for the `Change()` event:

```
Private Sub HScroll1_Scroll()
    lblValue.Caption  = Hscroll1.Value
End Sub
```

The `Scroll()` event occurs whenever the user drags the scroll box. Like the `Change()` event, this procedure immediately displays the changed value of `HScroll1` as the user drags the scroll box.

You might consider the difference between the two events as a jump in values, say from 0 to 30, with the `Change()` event, and a continuous updating of the value while the scroll bar is being dragged with the `Scroll()` event.

Exercise 4.6 Using Scroll Bars in an Application

In this exercise, you design the speed-measuring application you were asked to visualize in section 4.7.1, "Designing a Demonstration Scroll Bar."

1 Open a new project and change the form caption to **Measuring Speed**.

2 Add a text box and remove the text caption. Change the name to **`txtSpeed`**.

3 Add the Horizontal Scroll Bar control. Name it **`hsbSpeed`**.

4 Set the following properties for the scroll bar:

```
Max = 120
Min = 0
LargeChange = 10
SmallChange = 1
```

5 Add a command button, change the caption to **Stop**, change the name to **cmdStop**, and write an `End()` event procedure.

6 Add a second command button, change the caption to **Help**, and change the name to **cmdHelp**.

7 Write the following code for the `cmdHelp` button:

```
Private Sub cmdHelp_Click()
   Dim Message as String
   Dim Title as String
   Message = "Click the ends to make a small change or the middle to make a
   ➥large change."
   Title = "Help with the scroll bar"
   MsgBox Message, , Title
End Sub
```

8 Double-click the scroll bar to open the code window and enter the following code:

```
Private Sub hsbSpeed_Change()
    txtSpeed.Text = hsbSpeed.Value & " mph"
End Sub
```

9 Write the `Scroll()` procedure for the scroll bar. This code is as follows:

```
Private Sub hsbSpeed_Scroll()
    txtSpeed.Text = hsbSpeed.Value & " mph"
End Sub
```

10 Test your design. Experiment by changing the `Max` and `Min` values and the `LargeChange` and `SmallChange` values.

This exercise included some code statements with which you are unfamiliar. Don't worry. It will become clearer when we review how to assign values to variables, and the syntax and usage of the `MsgBox` statement. For now, think of how far you have come!

Chapter Summary

The objectives of this chapter were developed to help you think visually in the design of a Visual Basic program, and to instill in your mind the process followed in design. You should think about the problem and visualize in your mind the design requested by the user. In doing this, you might recognize that different types of text boxes and buttons become important, as do frames and the tab order. Placing commands on a menu bar is another possibility. Scroll bars add still another form of functionality to a design.

List boxes and combo boxes present to the user a list of items. With a list box, the `MultiSelect` property permits more than one selection. A list box can also have multiple columns. A combo box is a combination of a text box and list box. The `Style` property determines the type of combo box. Currently, there are three styles.

Methods are required to add, delete, or clear items from a list or combo box. The `AddItem` method serves to add items; the `RemoveItem` method is used to delete an item. The `Clear` method removes all items from a list. The `Sorted` list box and combo box property lists items alphabetically.

Items can be loaded into a list box or combo box using the `Form_Load()` procedure.

Frames are used to store a set of option buttons and other types of controls. When enclosed by a frame, only one option button can be selected within the frame. Option buttons placed outside the frame are not affected.

The tab order of a design is controlled by the `TabIndex` setting. An index value of `0` determines the first tab in a tab-order sequence. The `TabStop` property is used to skip a control in a tab order sequence.

The Menu Editor is used to build a menu for a design. Each item placed on a menu must have a caption and a menu name. A menu hierarchy is determined by placing subordinate menu items to the right of a menu name. After the menu bar is built, procedures must be written for each menu command.

In the design of an interface, scroll bars provide an interesting visual picture. The `Value` property determines the return value represented by the current position of the scroll bar. The range of returned values is set by the `Min` and `Max` properties. The speed with which values change is determined by the `SmallChange` and `LargeChange` properties.

`Change()` and `Scroll()` event procedures are both required with scroll bars. The `Change()` event occurs whenever the user clicks a scroll arrow or the space between the scroll box and the scroll arrow. The `Scroll()` event occurs whenever the user drags the scroll box.

Skill-Building Exercises

1. Test the differences between the three styles of combo boxes. Place all three combo boxes on the screen (see Figure 4.1). Place the list items *Pool, Basketball, Soccer,* and *Darts* in each combo box. Use styles 0, 1, and 2. Remove the text from the styles. Add the labels **`Combo Box - Style 0`**, **`Combo Box - Style 1`**, and **`Combo Box - Style 2`**. Test your design. What conclusions can you draw?

2. Create horizontal and vertical scroll bars. Set the scroll bar range for both bars for values between 1 and 45. Set the `SmallChange` value to `1` and the `LargeChange` value to `5`. Use a label to display the scroll bar value. Change the background color (`Backcolor`) for the label and the form. Change the caption on the form. Write both `Change()` and `Scroll()` procedures. In writing these procedures, add the following instructions, changing the name of the `Label1` to **`Hscroll1`** and **`Vscroll1`** as necessary:

   Label1.Caption = {H | V}scroll1.Value

3. Place a Style 2 combo box on a form. Add a text box, a label, and a command button. Write a `Form_Load()` procedure that adds the following cities to the combo box: **New York**, **Los Angeles**, **Chicago**, **San Francisco**, **Atlanta**, and **Houston**. Place the label over the text box, and change its caption to read **Enter Additional City**. In the command button with the caption Add City, add a procedure that appends whatever city the user enters into the text box to the top position of the combo box.

4. Extend Exercise 4.4 by adding two more command buttons. Make **Clear** the caption for one command button, and write a procedure in that button's `Click()` event to clear all items in the combo box. Make **Delete** the caption of the other command button, and write a procedure that will delete the currently selected item.

5

Introduction to the Graphic Controls and Graphic Methods of Visual Basic

Windows owes its tremendous popularity, in part, to the appeal of its interface. As with other *graphical user interfaces* (GUIs), users quickly come to appreciate the visual cues that Windows provides to guide them through a program. They especially like interface colors, shapes, and images. If given a choice, most users prefer to click a mouse button to enter instructions rather than type commands. Standard GUI features, such as buttons, pull-down list boxes, sizable forms, and drag-and-drop functionality, have greatly eased the introduction of computing to millions of new users.

This chapter introduces several *controls* that can assist you in adding visual interest and definition to your program. These graphic controls—the Line control, Shape control, and Image control—provide flexible ways to enhance the visual appeal of an application. By exploiting the graphic richness of Windows, these controls enable you to easily add geometric shapes, colors, and images to the user interface. The Label control, which you have already used, is also a graphic control.

This chapter introduces several of Visual Basic's graphic *methods,* which are coded instructions that directly "draw" upon a Visual Basic form, picture box, or `Printer` object. Graphic methods differ from graphic controls by performing operations at a *lower* level. A good illustration of the difference between the two is to compare using the `Print` method (a graphic method) on a form and assigning a new caption to a label that has been placed on a form. Both actions produce text output, but the `Print` method operates directly on the form. Assigning a new caption to a Label control triggers code assigned to the Label control, which then "draws" the text. The differences will become clearer as you work through the exercises in this chapter, which show you how to

- manipulate the Line control in design time and runtime;
- add a Shape control to a project and manipulate its properties in code;
- learn to adjust layering for graphic controls, including the Label control, in design time and runtime;
- add images to the Image control in both design time and runtime;
- work with the `Line`, `PSet`, and `Print` graphic methods.

Chapter Objectives

This chapter addresses all of the Visual Basic graphic controls (other than the Label control): the Line control, Shape control, and Image control. The objectives are as follows:

1. Learning the basics of the Visual Basic coordinate system
2. Manipulating the Line and Shape controls in design time and runtime
3. Layering controls in design time and runtime
4. Using the design-time and runtime properties of the Image control
5. Introducing the Visual Basic graphic methods

Objective 5.1 Learning the Basics of the Visual Basic Coordinate System

Before you use graphic controls, it is important to understand how these controls differ from other Visual Basic controls and how the coordinate system works. The coordinate system is used in the precise placement of controls on a form or within a container, such as a frame or a picture box.

5.1.1 How Graphic Controls Differ from Other Visual Basic Controls

The graphic controls are added to a form like other controls: You can double-click the toolbox or select the control in the toolbox and then "draw" it on the form. Two of these controls, the Line control and Shape control, have no events associated with them. Their properties can be adjusted with code, but they cannot, for instance, recognize a mouse click. In the code window under the Object combo box, the line or shape objects are not listed because they have no events to code. Internally, Windows treats the Line and Shape controls as part of the form.

The Image control—which can display a bitmap image—can recognize mouse events, such as clicking, double-clicking, or being dragged over with the mouse. However, the Image control, like the Line and Shape controls, cannot receive focus. Although an event can be triggered by clicking an Image control, the Image control does not then retain the focus the way a command or option button does. The Label control has similar behavior: It can receive mouse events, but it cannot retain focus. (You will work with the Image control's mouse events in Chapter 14, "Programming User Events.")

5.1.2 The Relationship of Graphic Controls to a Form

Every form has a coordinate system. No matter where a control is on a form, its location can be described by *x* and *y* coordinates of the form. These coordinates are expressed in the `Left` (x) and `Top` (y) properties of controls. The `Width` and `Height` properties of controls measure, appropriately enough, the value of the controls' width and height relative to the coordinate system of the form. The x coordinate of the right side of a control, therefore, would be approximately the value of the control's `Left` property plus the value of the control's `Width` Property. This will become clearer as you work with the controls' locations.

Like other x,y coordinate systems, Visual Basic's coordinate system is a two-dimensional grid. In Figure 5.1, you see highlighting where the x- and y-axes values begin for a form. By default, the 0 values of the x-axis (horizontal) and the y-axis (vertical) are in the extreme upper-left corner.

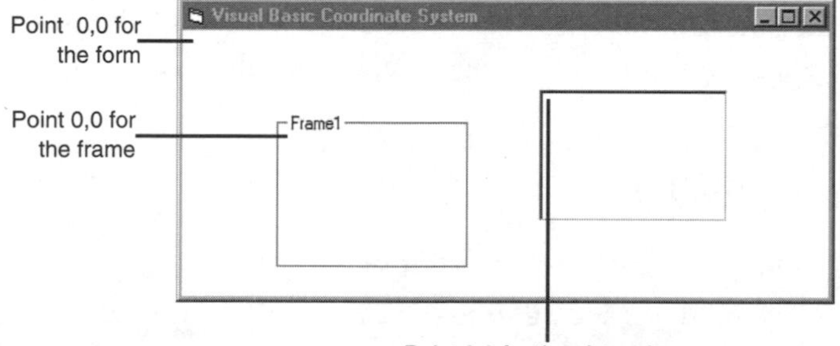

Figure 5.1
The Visual Basic coordinate system is applied to a form and containers placed on a form.

Unlike the coordinate system typically used in graphs, the value for the y-axis increases as you approach the bottom of a form. As Figure 5.1 shows, a separate coordinate system is maintained for Visual Basic's two container controls: the Frame control and the Picture Box control. You have used the Frame control to group option buttons, and you will briefly read about the Picture Box control in this chapter. Remember, a control's `Left` and `Top` properties are determined by the coordinate system for the container in which the control is placed. Therefore, the `Left` (x-axis value) and `Top` (y-axis value) properties of an option button within a frame are determined by the coordinate system of the frame, not the form.

The default scale mode of forms and controls is *twips*. How do twips relate to a measurement system you can understand? A twip is 1/20 of a printer's *point*. There are 72 points per inch and, therefore, 1440 twips, which makes a twips-to-inches association somewhat difficult. You will eventually become used to twips notation. For example, examine the `Top`, `Left`, `Width`, and `Height` properties of a blank form. Their values are expressed in twips.

Objective 5.2 Manipulating the Line and Shape Controls in Design Time and Runtime

This section examines the Line and the Shape controls and how their properties can be set during design time or runtime. Both controls are found on the toolbox; as such, they contain sets of easy-to-find properties. Before starting this section, find both controls on the toolbox.

5.2.1 The Line Control Properties

The Line control is actually one of the few controls that does not have `Left`, `Top`, `Width`, or `Height` properties. Instead, it has four properties to locate it on a form or inside a container control: `X1`, `X2`, `Y1`, and `Y2`. The coordinates `X1` and `Y1` define one endpoint of the line, and the coordinates `X2` and `Y2` define the other endpoint. The Line control also has three border properties: `BorderColor`, `BorderStyle`, and `BorderWidth`.

Exercise 5.1 Manipulating the Line Control in Design Time and Runtime

In this exercise, you use the Line control to set properties in design time and during runtime. You begin with design-time property changes and observe the change after each new property setting.

Part 1: Design-Time Changes

1. Open a new project and double-click the Line control.
2. In the Property window for `Line1`, add **1000** to the current value of the `X2` property of `Line1`.
3. Add **1000** to the value of the `Y2` property of `Line1`.
4. Change the `Line1.BorderWidth` property to **20**.
5. In the `BorderColor` property, select a new color from the color options.

Part 2: Experiment with Runtime Changes

1. Open a new project, and add two lines approximately one-half inch long near the upper-left corner of the form. Keep the lines separate and drag them into position. Position `Line1` horizontally and `Line2` vertically with their ends touching.
2. Make sure that the `X1` point of `Line1` (the horizontal line) and the `Y1` point of `Line2` (the vertical line) are at the intersection of the lines.

3 In the `Form_Click()` event, add the following code:

```
Private Sub Form_Click()
   Line1.BorderWidth = 4
   Line2.BorderWidth = 4
   Line1.X2 = Line1.X2 + 500
   Line2.Y2 = Line2.Y2 + 500
End Sub
```

4 Run and test the program. Click the form more than once. With each mouse click on the form, the horizontal and vertical lines should increase their sizes by 500 twips as you add to the `Line1.X2` and `Line2.Y2` property values.

5.2.2 The Shape Control Properties

The Shape control is even more flexible than the Line control. A single control can be made into one of six shapes: a rectangle, a square, an oval, a circle, a rounded rectangle, or a rounded square (see Figure 5.2).

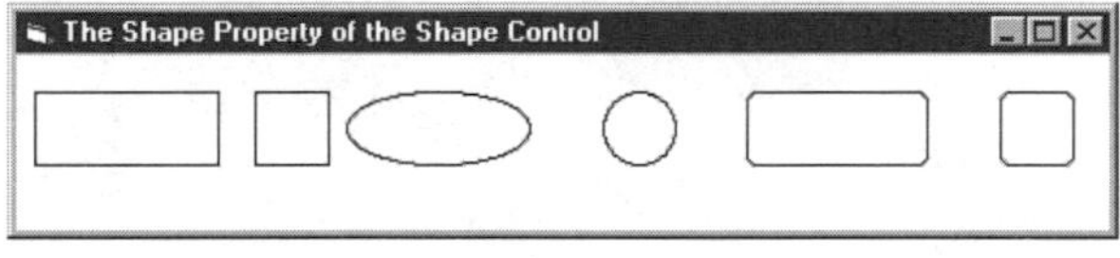

Figure 5.2

The size of each of the shape selections can be changed to meet the needs of your application.

These shapes are controlled by the Shape control's `Shape` property and can be set in either design time or runtime. The Shape control is positioned and sized on a form or inside a container control with the familiar `Left`, `Top`, `Width`, and `Height` properties. Its other properties are similar to that of the Line control, except it also has `FillStyle` and `FillColor` properties. By clicking `FillColor` in the Property window, a color palette appears for your selection. The appearance of the interior of a shape depends on the `FillStyle` setting. The eight `FillStyle` settings are `solid`, `transparent`, `horizontal`, `vertical`, `upward`, `downward`, `cross`, and `diagonal` (see Figure 5.3).

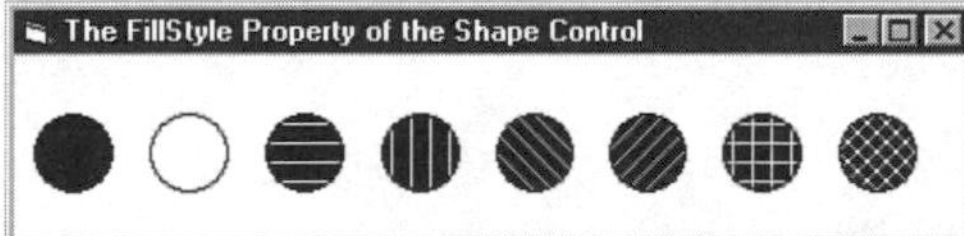

Figure 5.3

The `FillStyle` property is used to distinguish one shape from another.

The interior appearance is also affected by the Shape control's `BackStyle` and `BackColor` properties. If the `BackStyle` is set to `Transparent(0)`, the color of the object behind the shape, such as a form, will be the shape's `BackColor`. If `BackStyle` is set to `Opaque`, the

`BackColor` will be that set by the `BackColor` property. By mixing `BackColor`, `FillColor`, and `FillStyle`, a wide variety of pleasing visual effects can be achieved.

The Shape control can be employed to perform numerous functions. It can be used as a marquee border, in simple animation, or to give the user visual feedback of the processing status. One of the best ways to get to know the Shape control is to simply experiment with its properties during design time. All entered property changes are immediately reflected on the Shape control on the form.

Exercise 5.2 Changing the Shape Control's Properties During Runtime

This exercise is similar to the last one. It changes the Shape control during runtime to demonstrate the flexibility available to you by simply changing object properties. Figure 5.4 shows the design that you are to create.

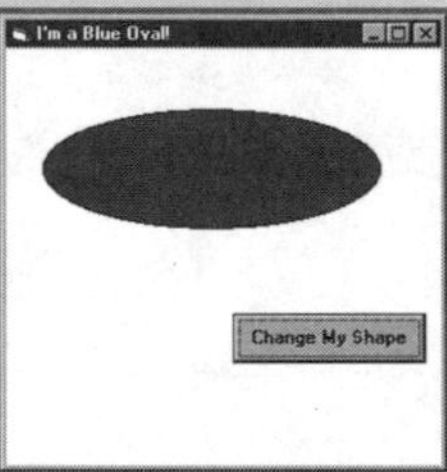

Figure 5.4

The Change My Shape button will change the shape from an oval to a rectangle.

1 Open a new project, and add a Shape control and a command button to the form.
2 Change the Form's caption to **I'm a blue oval**.
3 Name the command button **`cmdChangeMe`**, and change its caption to **Change My Shape**.
4 Change the Shape control's properties so that it appears as a royal blue oval with the following properties:

```
FillStyle  = Solid
Shape      = Oval
FillColor  = royal blue
```

5 In the cmdChangeMe_Click() event, add the following code:

```
Private Sub cmdChangeMe_Click()
   Shape1.Shape = 0 ' rectangle
   Shape1.FillColor = vbRed   ' bright red
   Shape1.Width = 1500
   Shape1.Height = 3500
   Form1.Caption = "Now, I'm a Red Rectangle!"
End Sub
```

In this code, vbRed is a constant built-in to Visual Basic. (Constants, both built-in or defined by you, are discussed in Chapter 6, "Variables and Constants.")

6 Run and test the program. The Click() event should generate a red vertical rectangle.

Objective 5.3 Layering Controls in Design Time and Runtime

You may have noticed that Visual Basic has different graphic layers. A command button, for instance, always appears in front of a graphic control, as does an option button or a combo box. The output of the Print command on a form appears behind a graphic control.

5.3.1 The Design-Time Layering of Controls

If two controls of the same graphic level overlap, it is necessary to designate which control appears in front. This can be set in either design time or runtime. Labels, the Line control, and the Shape control are all in the *middle* graphic level. If you want a label to overlap a Shape control, you must place it on top of the Shape control. In design time, you can accomplish this in one of two ways:

- You can place the controls on the form in the order that you want them to appear, placing the bottom control first. Any subsequent control goes *on top* of the one that preceded it within its graphic level.
- You can determine the placement of the controls by choosing Format, Order (also available by right-clicking), and selecting either Bring to Front or Send to Back. By selecting controls and using these Format, Order menu commands, you can arrange the layering of controls within a graphic level as you want.

Exercise 5.3 Welcoming the Shape Control

In this exercise, you design an interface to explore the three graphic layers. Figure 5.5 shows the interface during design time; Figure 5.6 shows the interface during runtime.

Figure 5.5

Use a rectangle with a small label placed on top of it.

Figure 5.6

The label control in this design represents the topmost graphical layer, while the shape control represents the middle graphical layer.

1. Create the interface you see in Figure 5.5. Draw the rectangle first and then place the small label you see on top of it. Name the command button **`cmdWelcome`** and the label **`lblWelcome`**.
2. Set the following `lblWelcome` properties:

```
Alignment = Center
BackStyle = Transparent
ForeColor = White
AutoSize = True
WordWrap = True
```

3 Select the label, choose Format, Order, and select Send to Back. (Alternatively, select the label and right-click your mouse. The Bring to Front and Send to Back menu choices should be available to you.) Now bring the label to the front again by choosing Format, Order, and selecting Bring to Front.

4 In the `cmdWelcome_Click()` event, enter the code:

```
Shape1.Shape = 4 ' rounded rectangle
Shape1.Top = 200
Shape1.Left = 200
Shape1.Width = 4000
Shape1.Height = 3000
Shape1.BorderWidth = 6
lblWelcome.Caption = "Welcome to the Shape Control!!"
lblWelcome.FontSize = 24
lblWelcome.BorderStyle = 0 ' set to no border
lblWelcome.Top = Shape1.Top + Shape1.Height / 8
lblWelcome.Left = Shape1.Width / 4
```

5 Run and test the program. It should produce an image similar to that in Figure 5.6 (note the alignment).

This code is a little more complicated than that in previous exercises. The last two lines of the procedure might be difficult to understand at first. You set the top and left position of `lblWelcome` by making reference to `Shape1` properties. The next-to-last line of code positions the top of `lblWelcome` 1/8 of the way from the top of `Shape1`.

Do you understand how this is done? First, take the top of `Shape1` and then add to it 1/8 of the value of `Shape1.Height` (remembering that the y-axis increases as you approach the bottom of a form). Similarly, the left value of `lblWelcome` is set 1/4 of the way in from the left side of `Shape1`.

5.3.2 The *ZOrder* Method

Runtime control of the graphic layers is more complex than control during design time. In runtime, you can designate the order of controls and forms by using the `ZOrder` method. It is called `ZOrder` because it controls placement of controls in the *Z* dimension. The syntax of this method is as follows:

```
[object.]ZOrder [position]
```

The `object` component can be any form or control except the timer or menu, and `position` can be either `0` or `1`. If no position argument is specified, `0` is assumed, and the control is brought to the topmost position. If the position argument is `1`, then the control is sent to the rear of the `ZOrder`.

Exercise 5.4 Toggling *ZOrder*

In this exercise, you use the `ZOrder` method to manipulate the *Z* dimension position of two command buttons and two Shape controls. Figure 5.7 shows the interface during design time.

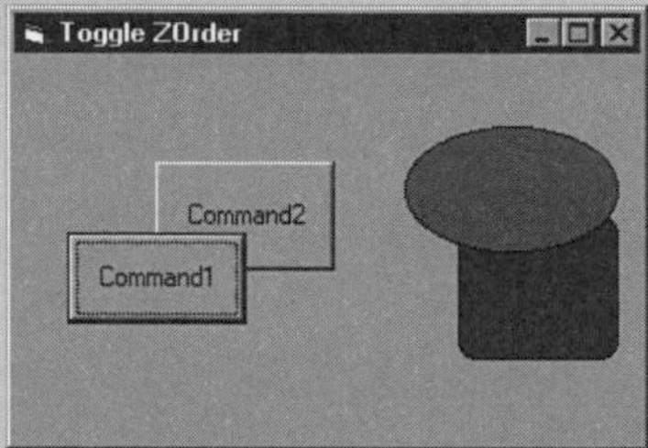

Figure 5.7

The ZOrder method is used to arrange the control and the order of shapes within the same graphical layer.

1 Create the interface you see in Figure 5.7. Add two command buttons, **`Command1`** and **`Command2`**, to a form. Choose Format, Order, and make sure that `Command1` has the topmost position.

2 Add two Shape controls, **`Shape1`** and **`Shape2`**, to a form. Make sure `Shape1`, a red oval, has the topmost position and that `Shape2`, a blue, rounded rectangle, has the bottom-most position.

3 In the `Command1_Click()` event, add the following code:

```
Command1.ZOrder 1
Shape1.ZOrder 1
```

4 In the `Command2_Click()` event, add the following code:

```
Command2.ZOrder 1
Shape2.ZOrder 1
```

5 Run and test the program. Observe the toggling of the `ZOrder` as `Command1` and `Command2` are clicked.

Objective 5.4 Using the Design-Time and Runtime Properties of the Image Control

The Image control displays bitmap images in a flexible, resource-efficient way. Unlike the Picture Box control, which is a container control for storing controls and images and receiving graphics method commands, the purpose of the Image control is to display images and be available to be used, if needed, as a button. This makes it much less taxing on system resources than the Picture Box control. If you need to display an image and do not need the added functionality of the Picture Box control, use the Image control.

5.4.1 The Image Control's Special Properties

The Image control is simple and powerful. Having most of the properties of the other graphic controls, it also has the `Picture` and the `Stretch` properties.

The *Picture* Property

One of the most powerful properties is the `Picture` property. If you simply designate the path of the bitmap image to be displayed, the Image control will, by default, display the image in its original size. You can either enter the path, or, by double-clicking the `Picture` property in the property box, you can browse and select an appropriate image file in the dialog box that appears. The image file must have a .BMP, .JPG, .GIF, .DIB, .WMF (Windows Metafile Format), or .ICO (icon format) extension.

The *Stretch* Property

The Image control also has a `Stretch` property. Its default value is `False`, which means that the Image control will take on the size of the image that is loaded into it. If you change the size of the image after loading it, the image will be cropped. By setting the `Stretch` property to `True`, you make the size of the image conform to whatever size you set the Image control.

Exercise 5.5 Experimenting with the Image Control

In this exercise, you experiment with the Image control during design time and during runtime. In part one, you experiment with the `Stretch` property; in part two you add a `Click()` event to an image box.

Part 1: Changing Image Control Properties

1. Open a new project and add an Image control to your form. Name it **`imgTest`**.
2. Open the Properties window and double-click the `Picture` property. The Load Picture dialog box should appear. Select an image on your hard disk. If you have no files with the .BMP or .WMF extensions, you can find small icon images with the .ICO extension in the \icons subdirectory under your Visual Basic directory.
3. Leaving the `Stretch` property of `imgTest` to its default value of `False`, select the Image control and then change its size with the newly loaded image. If you shrink the image, you will see the image cropped. If you enlarge the image, transparent space is added.

continues

4 Change the `Stretch` property to `True`, and experiment with changing the size of the image.

5 Change the `imgTest Stretch` property back to `False`, and reload your image by double-clicking in the `imgTest.Picture` property and selecting your image again.

Part 2: Coding an Image *Click ()* Event

1 Add two command buttons to your form (but no code).

2 Add the following code to the `imgTest_Click()` event.

```
MsgBox "You clicked the image"
```

3 Run and test the program. Notice that the `Click()` event triggers a `msgBox`, but the program's focus remains with whatever command button had focus before the `Click()` event.

5.4.2 Using the *LoadPicture* Function During Runtime

So far, you have manipulated the design-time property settings. With the `LoadPicture` function, you can load a new image into the Image control during runtime, or clear it. The syntax for the `LoadPicture` function is straightforward:

```
LoadPicture([stringexpression])
```

The `stringexpression` is the path to the image file. The following is an example:

```
Image1.Picture = LoadPicture("C:\images\City.BMP")
```

If the image was in the root directory of the application, only the image file name would need to be used as an argument.

```
Image1.Picture = LoadPicture("City.BMP")
```

To clear an image from an Image control, simply assign a null string to the `LoadPicture` function argument:

```
Image1.Picture = LoadPicture("")
```

The `LoadPicture` function can be used to assign values to `Picture` properties dynamically. Besides images, forms have `Picture` properties that will load a bitmap image into the form's background. Picture boxes also have a `Picture` property.

Exercise 5.6 Using the *LoadPicture* Function

In this short exercise, you design three `Click()` events: two to load different icon files (.ICO) in an image box and one to clear the image box.

1 Open a new project, and draw the interface you see in Figure 5.8.

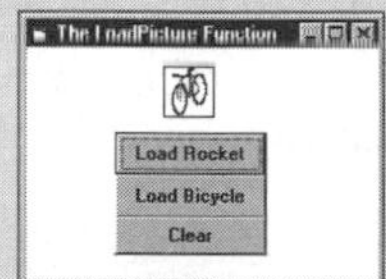

Figure 5.8

This display uses the LoadPicture function to place an image in the image box.

2 Name the Image control **`imgLoad`**. Name the command buttons **`cmdRocket`**, **`cmdBicycle`**, and **`cmdClear`**. The bicycle image is bicycle.ico, found under the vb\icons\industry subdirectory (and on the student disk).

3 In the `cmdRocket_Click()` event, enter the following code:

```
Dim Path As String
Path = "C:\vb\icons\industry\"   ' change path as necessary
imgLoad.Picture = LoadPicture(Path & "Rocket.ico")
```

If the path to the Visual Basic subdirectories is different, make the necessary changes when assigning an expression to the `Path` variable.

4 In the `cmdBicycle_Click()` event, enter the following:

```
Dim Path As String
Path = "C:\vb\icons\industry\"
imgLoad.Picture = LoadPicture(Path & "Bicycle.ico")
```

5 In the `cmdClear_Click()` event, enter the following:

```
imgLoad.Picture = LoadPicture("")
```

6 Run and test the program.

As a rule, it is unwise to code a specific `Path` statement; instead, you should try to determine at runtime the location of your application. Visual Basic provides the `App` system object, which has a `Path` property that can be read. (The `App` object and its `Path` property are addressed in Chapter 20, "File-Processing Controls and Sequential File Processing.")

Objective 5.5 Introducing the Visual Basic Graphic Methods

Graphic methods, such as the `Print`, `Line`, and `Circle` methods, can be said to "draw" on the bottom layer of forms and picture boxes. Graphic methods give you great flexibility and are very efficient in keeping file sizes small. Their flexibility comes from their being much more closely linked than graphic controls to the rich graphic capabilities inherent in the Windows operating system. To their disadvantage, graphic methods typically take longer to code than graphic controls: They require a more complex syntax than the code required to manipulate graphic control properties.

As with most things, the best way to learn the Visual Basic graphic methods is to experiment with them. The more you practice with graphic methods, the more fun you will have with them.

5.5.1 *CurrentX* and *CurrentY* Runtime Properties

When using graphic methods, such as the `Print` method, you can designate where the drawing operation begins by designating the `CurrentX` and `CurrentY` properties. These properties set the x and y values on a form, picture box, or Printer object. By setting these properties immediately before invoking the `Print` or some other graphic method, you can specify where on a form or picture box you want drawn output to appear.

`CurrentX` and `CurrentY` are *runtime* properties. They do not appear in the properties window during design time and can be specified only with coded instructions.

Exercise 5.7 Using *CurrentX* and *CurrentY* for Printing

In this exercise, you use the Picture Box control and `CurrentX` and `CurrentY` to display a message on a form and inside a picture box at a specified location. Figure 5.9 shows the design you are asked to create.

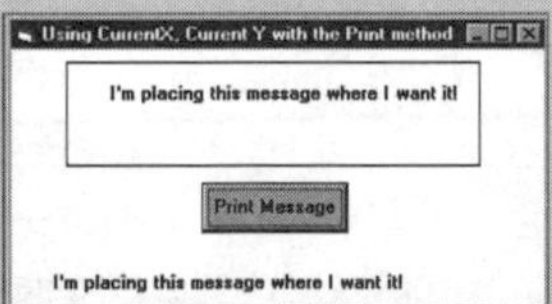

Figure 5.9

`CurrentX` *and* `CurrentY` *properties specify where graphics methods, such as* `Print`, *begin.*

1. Design the interface. Name the picture box `pic1` and the command button `cmdPrintIt`, and set the form caption to **Using CurrentX, CurrentY with the Print method**. Add the caption **Print Message** to the command button. Make sure the form `ScaleMode` for the form is set to twips. Set the `Form.Height` property to **3000** and the `Form.Width` property to **5000**.
2. In the `cmdPrintIt_Click()` event, enter the following code:

```
Private Sub cmdPrintIt_Click()
   Dim message As String
   message="I'm placing this message where I wanted!"
   Form1.Font.Bold = True
   Form1.CurrentX = 400
   Form1.CurrentY = 2200
   Form1.Print message  'This prints the message at the bottom of the form
   Pic1.Font.Bold = True
   Pic1.CurrentX = 400
   Pic1.CurrentY = 200
   Pic1.Print message  'This prints the message inside the picture box
End Sub
```

3. Run and test the program. Notice that the 0,0 starting point is unique for each container—whether it's a form or a picture box.

5.5.2 The *Line* Method

The `Line` method is one of the more useful, if somewhat complicated, graphic methods to use. It can be used to draw either lines or rectangles. It can control the color of a line as well as the fill, if any, of a rectangle. The complete syntax of this method is as follows:

```
[object.] Line[[Step](x1,y1)]-[Step](x2,y2),[color][,B[F]]]
```

Table 5.1 explains the components of the `Line` method syntax.

Table 5.1 Syntax of the *Line* Method

Component	*Description*
`object`	Object on which the line or rectangle is drawn.
`Step`	Keyword that specifies that the starting points are relative to the current graphics position, given by `CurrentX` and `CurrentY`.
`x1,y1`	Values (single-precision) indicating the coordinates of the starting point for the line or rectangle. The unit of measurement is determined by the `ScaleMode` of the object. If omitted, the line begins at the position indicated by `CurrentX` and `CurrentY`.
`Step`	Keyword that specifies that the endpoint coordinates are relative to the line starting point.

continues

Table 5.1 continued

Component	*Description*
x2,y2	Values (single-precision) indicating the coordinates of the endpoint of the line to draw.
color	Long integer value indicating the RGB color used to draw the line. If omitted, the ForeColor property is used. You can use the RGB function or QBColor function to specify the color.
B	Option that causes a box (rectangle) to be drawn using the xl,yl and x2,y2 coordinates to specify opposite corners of the box.
F	If the B option is used, the F option specifies that the box is filled with the same color used to draw the box. You cannot use F without B. If B is used without F, the box is filled with the current FillColor and FillStyle; the default value for FillStyle is transparent.

This is a complicated syntax, but it should become clearer when you move to the examples below. Remember the following additional key points:

- The width of the line is controlled by the DrawWidth property of the object (Form, PictureBox, or Printer) upon which the drawing occurs. The setting of the DrawMode and DrawStyle affects the way the line is drawn on the background.
- When the Line method executes, CurrentX and CurrentY are set to the endpoint specified by the x2,y2 arguments. When drawing lines that are connected, subsequent lines should have as their starting point the endpoint of the previous line.
- The DrawMode determines the "pen" of the drawing. There are 16 different settings, some of which can be quite complex. (Refer to Visual Basic Help for more information about DrawMode properties.)

Exercise 5.8 Drawing a Triangle with the *Line* Method

Look again at the syntax of the Line method. If you delete the X1,Y1 coordinates from the method, the line will be drawn from the CurrentX and CurrentY position to the endpoints. The Step keyword enables you to describe the length of the line drawing. This exercise asks you to draw a simple triangle using this method.

1 Add a button named **cmdDraw** to a form.

2 Set the ScaleMode of the form to pixels. This helps insure that most line drawings with angles, such as triangles, do not have small gaps at line intersections.

3 In the `cmdDraw_Click()` event, enter the following code:

```
Private Sub cmdDraw_Click()
   DrawWidth = 2
   CurrentX = 150
   CurrentY = 175
   Line -Step(300, 0)      ' draw the base line
   Line -Step(-150, -150) ' draw the right side of the triangle
   Line -Step(-150, 150)  ' draw the left side
End Sub
```

4 Run and test the program.

`CurrentX` and `CurrentY` change when each `Line` method is executed. In this instance, you drew a triangle in a counter-clockwise direction. If you were drawing the triangle in a picture box, the object would have to be referenced, as in the instruction `Pic1.Line-Step (300,0)`.

Exercise 5.9 Using the *Line* Method with the *B* and *F* Arguments

The `B` and `F` options of the `Line` method enable the developer to construct rectangles with different `ForeColor`, `FillColor`, and `FillStyle` attributes (see Figure 5.10). An endless combination of rectangles can be created. This exercise provides a simple demonstration of these methods. Before you begin, review the syntax of the `Line` method to remind yourself how the `B` and `F` arguments work together and in conjunction with the `FillColor`, `ForeColor`, and `FillStyle` properties. This exercise demonstrates the interaction of these properties.

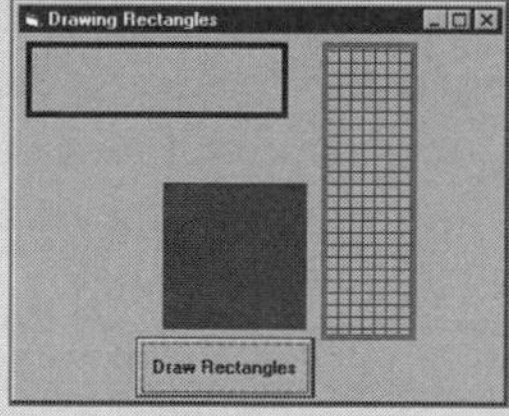

Figure 5.10

The `Line` method is also used to draw rectangular shapes.

1 Draw the interface you see in Figure 5.10. Set the `Width` property of the form to **5000** twips and the `Height` property to **4000** twips. Add a command button, **`cmdDraw`**, and place it at the bottom of the form.

continues

2 In the `cmdDraw_Click()` event, enter the following code:

```
Private Sub cmdDraw_Click()
   Form1.DrawWidth = 4
   Form1.FillStyle = vbFSTransparent
   Form1.ForeColor = vbBlack
   Form1.Line (100, 100)-(2700, 800), , B
   Form1.FillStyle = vbFSSolid
   Form1.ForeColor = vbBlue
   Form1.FillColor = vbYellow
   Form1.Line (1500, 1500)-(2900, 2900), , BF
   Form1.ForeColor = vbRed
   Form1.FillColor = vbBlue
   Form1.FillStyle = vbCross
   Form1.Line (3100, 100)-(4000, 3000), , B
End Sub
```

3 Run and test the program. You should see output similar to that in Figure 5.10.

The `FillStyle`, `ForeColor`, and `FillColor` properties and the `B` and `F` arguments determine the appearance of each of the rectangles. These properties are set *before* the `Line` method is invoked.

The first rectangle has a transparent `FillStyle` and a black `ForeColor`. The second has its `FillColor` set to yellow, but observe that the `F` argument overrides this assignment. The rectangle is blue, taking on the color of the specified `ForeColor` property. The last rectangle has different `ForeColor` and `FillColor` property settings. A blue cross pattern is drawn inside a rectangle with a red border color.

Experiment with this exercise. You will be surprised by the wide variety of effects that can be created!

5.5.3 The *PSet* Method

With the `PSet` method, you can paint individual pixels on your screen with a color you specify. It is very flexible, and is particularly useful for coloring irregularly shaped objects or groups of pixel points. The size of the points you draw with the `PSet` method depends on the `DrawWidth` property of the object upon which you are drawing. As with other graphic methods, `CurrentX` and `CurrentY` are set when `PSet` is executed. The syntax is as follows:

```
[object.]PSet[Step](x,y)[,color]
```

Table 5.2 explains the components of the `Pset` method syntax.

Table 5.2 The Syntax of the *Pset* Method

Component	*Description*
`object`	Object on which the point is to be drawn.
`Step`	Keyword specifying that the coordinates are relative to the current graphics position given by `CurrentX` and `CurrentY`.
`x,y`	Values (single-precision) indicating coordinates, which depend on the unit of measurement of the `Scale` properties of the object.
`color`	Color specified for a point. If omitted, current `ForeColor` is used. `RGB` or `QBColor` functions can be used.

Exercise 5.10 Making Confetti

This exercise draws 1,000 pieces of "confetti" on a form using the `PSet` method (see Figure 5.11). The program draws colors in four-pixel squares, and the procedure stops when 1,000 pieces of "confetti" have been drawn. The code uses several more advanced features of the Visual Basic language:

- It uses a built-in Visual Basic function, the `Rnd` function, to randomly generate numbers. These numbers are used to specify `CurrentX` and `CurrentY` locations and to specify colors for drawing the "confetti."
- Colors are specified by the `QBColor` function, which can generate 16 standard Windows colors.
- A loop is used, which causes code to be repeatedly executed until a certain condition is met. Loops are introduced in Chapter 11, "Working with Loops." Nevertheless, this code illustrates the importance of a loop structure, and it illustrates how the `PSet` method can be used.

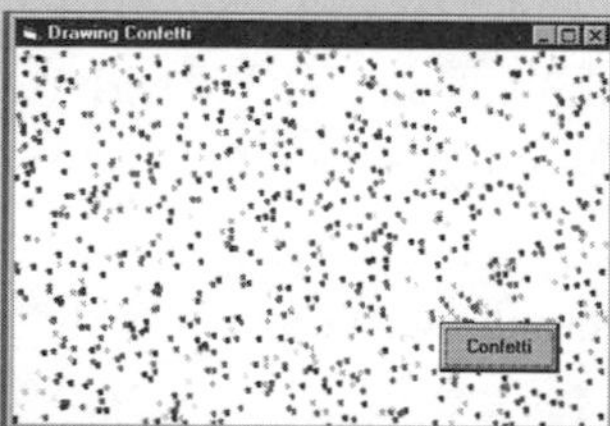

Figure 5.11

The `PSet` method can be used to draw one pixel at a time on the form.

continues

1 Draw the interface you see in Figure 5.11. The interface consists of a single form (**Form1**) and a command button (**cmdConfetti**).

2 In the cmdConfetti_Click() event, enter the following code:

```
Private Sub cmdConfetti_Click()
   Dim XPos As Single
   Dim YPos As Single
   Dim counter As Integer
   Dim color As Integer
   Form1.DrawWidth = 4
   Do    'Beginning of the loop
      XPos = Rnd * Form1.ScaleWidth
      YPos = Rnd * Form1.ScaleHeight
      color = Rnd * 15     'Random number function
      PSet (XPos, YPos), QBColor(color)
      counter = counter + 1
  Loop Until counter >= 1000  'end of the loop
End Sub
```

3 Run and test the program.

Observe that three numeric values are determined by using the Rnd function: the *x* position and *y* position for the PSet method, and the color. These three values are then used by the PSet method to draw a four-pixel point. All of this code is placed inside a loop that executes 1,000 times, thereby generating 1,000 pieces of "confetti."

5.5.4 The *RGB* Color Function

In the previous exercise, you used the QBColor function. This function is a holdover from earlier versions of Basic, and was used extensively when most color monitors handled just 16 colors. Until a few years ago, 16-color VGA was the default resolution and color mode for Windows. More powerful hardware, especially improved video systems, has made 65,000 colors or even 16 million colors standard on new Windows machines, although 256 colors is still considered to be somewhat "standard."

Visual Basic takes advantage of this video capability with the RGB function, which enables the programmer to specify each of the color combinations—red, green, and blue—to designate a color.

The function has the following simple syntax:

```
RGB(red, green, blue)
```

You assign values between 0 and 255 for each of the colors with 255 designating the most intensity. If the function reads RGB (0,0,0), it will return black. The expression RGB(255,255,255) would return white. Red, green, and blue colors have 255 in one argument with the other two values set to 0, as in RGB(255,0, 0). The expression RGB(255,255,0) returns yellow. Assuming your computer has the graphics capability, it is possible to designate 16,777,216 colors using the RGB Color Function (256*256*256), as shown in Figure 5.12.

Figure 5.12

This design enables you to test the RGB method to examine over 16 million colors.

Exercise 5.11 Surveying RGB Mixtures

This application replicates some of the functionality in a standard Windows Color dialog box.

1 Draw the interface you see in Figure 5.12. The interface includes three scroll bars (**RScroll**, **GScroll**, **BScroll**), a picture box (**picColor**), and three labels (**lblRed**, **lblGreen**, and **lblBlue**).

2 Set the `Min` value for the scroll bars to **0** and the `Max` value at **255**.

3 Declare `Red as Integer`, `Blue as Integer`, `Green as Integer` in the Code window under `(General)`, as follows:

```
Option Explicit

Dim Red As Integer

Dim Green As Integer

Dim Blue As Integer
```

You are not declaring the variables in a procedure; instead, you are declaring them at the *module level.* This kind of declaration is addressed in Chapter 10, "Defining Variable and Procedure Scope."

4 In the `Rscroll_Change()` event, enter the following code:

```
Private Sub RScroll_Change()
  Red = Rscroll.Value
  Green = Gscroll.Value
  Blue = BScroll.Value
  picColor.BackColor = RGB(Red, Green, Blue)
  lblRed.Caption = "Red " & Rscroll.Value
End Sub
```

5 Copy this code, and paste it into the `Change` and `Scroll` events of all the scroll bars. Change the assignments to the caption properties of the appropriate label.

6 Run the program and experiment with the color mixtures. Depending upon your video driver settings, the patterning of non-solid colors in the picture box will vary.

5.5.5 Topics for Further Study

The preceding exercises have illustrated the differences between graphic controls and graphic methods. Consider Exercise 5.8, for which you drew a triangle with the `Line` method.

Similar output could also have been achieved by using three Line controls. By using graphic methods, however, your program had less impact on system resources.

Taking full advantage of the Windows rich graphic environment with Visual Basic requires becoming familiar with a broad and deep body of knowledge. Your working through the exercises of this chapter made a good "down payment" on gaining that knowledge. However, space limitations prevent exploring the topic in greater depth. As a guide to your future study, the following are some important topics to review on your own:

- **`ScaleMode` property**—You have worked mainly in twips, but Visual Basic allows you to set other scale modes for the *interior* dimensions of a form. You can also customize the scale mode for a picture box. A form's `Top`, `Left`, `Width`, and `Height` properties are always expressed in twips, but the interior portion of the form—that is, the space inside the form border—takes on the scale specified in the form's `ScaleMode` property. The scale might be in pixels, centimeters, or inches, or can even be user-defined.

  There are four additional scale properties—`ScaleTop`, `ScaleLeft`, `ScaleWidth`, and `ScaleHeight`—which express the values of the form's interior dimensions. Controls placed on a form take on the `ScaleMode` of the form. The same is true for controls placed inside a picture box. For instance, if the `ScaleMode` is set to pixels, a command button's `Width` property would be expressed in pixels rather than twips.
- **`Circle` method**—To draw circles, ovals and arcs, you can use the `Circle` method on either a form or picture box. The syntax of the `Circle` method is somewhat complex, but the method is very flexible. See the Visual Basic online Help for details.
- **Persistent graphics**—When using graphics methods, there are two primary ways of creating persistent graphic output:
  - The code for the graphic output can be placed in a form's `Paint` event. This means the code will execute every time the form refreshes itself. A form refreshes, or repaints, itself every time a window in front of it is moved away, a window is moved or its size changed, or the `Refresh` method of the form is invoked in code.
  - The Form's `AutoRedraw` property can be set to `True`. A persistent bitmap of the form is created in memory, and graphic methods generate output on this persistent bitmap. When the form refreshes itself, it uses this persistent bitmap. Consult the Visual Basic online Help for more information.

Chapter Summary

Visual Basic's graphic controls include the Line control, Shape control, Image control, and Label control. Graphic controls differ from other Visual Basic controls in that they cannot receive focus. The Image control and the Label control, however, can receive events, such as `Click()` events.

Graphic controls, like all Visual Basic controls, have positioning properties, such as `Top`, `Left`, `Width`, and `Height`, and, in the case of the Line control, `X1`, `Y1`, `X2`, and `Y2`. The positioning properties are determined by the coordinate system of the container in which the graphic control is placed. Containers can be a form, picture box, or frame. The coordinate system for all containers starts at the x,y coordinate 0,0 in the upper-left corner of the container. To specify the point for a `Print` method to begin, you can designate the exact `CurrentX` and `CurrentY` position for a form or picture box.

Graphic controls exist in a middle graphic layer between the background of the container in which they are placed and other non-graphic controls, such as command buttons. The *layering* of graphic controls within their layer can be adjusted in design time by the Bring to Front and Send to Back commands. The `ZOrder` method can be used to reorder the layering of controls during runtime.

The Image control enables bitmaps (.BMP, .GIF, .DIB, .ICO), JPEG compressed image files (.JPG) and Windows Metafile Format files (.WMF) to be displayed in your application without the high resource requirements of a picture box or an additional form. Images can be loaded into the Image control in design time by double-clicking the `Picture` property of the Image control and assigning the path of the bitmap to the control. During runtime, bitmaps can be loaded into an Image control using the `LoadPicture` function. The size of the control can be changed to fit the image, or the image resized to fit the control.

Besides `Print`, graphic methods include the `Line`, `Pset`, and `Circle` methods. The `Line` method is used for drawing both lines and rectangles. The `Circle` method is used for drawing circles, ovals, and arcs. The `PSet` method draws individual pixels. All three of these graphic methods include arguments for determining the color of graphic output.

Color arguments in graphic methods can take the result of two color functions: `QColor` or `RGB`. The `QColor` function returns one of 16 colors, while the `RGB` function, depending upon the computer's video system, can return up to 16 million colors.

Skill-Building Exercises

1. Place a rectangle on a form. Draw two diagonal lines. One line should appear in the top-left corner and extend to the bottom-right corner; the second should appear in the top-right corner and extend to the bottom-left corner. Add the title **Shapes and Lines** to this form.

2. This exercise is an extension of the first exercise. Once again, place a rectangle on a form, two lines, and a command button with the caption **Resize me**. When the user clicks the button, have the rectangle move to a different position on a form, along with the two diagonal lines, running from the top-left to bottom-right corners of the form and from the top-right to the bottom-left corners of the form. Name this form **Resizing Lines and a Shape**. To solve this exercise, assign edges of the shape and the x,y coordinates of the lines during runtime.

3. Place a circle on a form and a label on top of the circle. Add a label `Click()` event. When the user clicks the label, have it display the following:

   ```
   I am the sun, and I am bright.
   ```

   Name this form **Make the Sun Shine**.

4. Add three ovals to a form and place labels over each oval. Name the labels **`lblGreen`**, **`lblRed`**, and **`lblYellow`**. Add a command button named **Stop Lights**. When the user clicks the button, display GO on the green label, BRAKE on the yellow label, and STOP on the red label. Add the appropriate colors to each label (red, yellow, and green). Name the form **What are the colors?**

5. Place two picture boxes on a form. In the top picture box, place an image box. Display a picture of a bell in this image box. Set the `Stretch` property to `True` to make a good-looking bell. In the bottom picture box, add a label. Finally, add an image `Click()` event. When the user clicks the bell, use the label to display the following:

   ```
   Ding!   Dong!   Ding!
   ```

6. Using the `Line` method, write a program that creates three empty squares that are diagonally arranged from the top left of a form to the bottom right.

7. Change the preceding exercise so that the border colors of the three squares are different, and the middle square has a blue interior color.

Part II Programming

6

Variables and Constants

When you first begin to write programs in Visual Basic, it becomes important that you learn the difference between a *variable* and a *constant* and how each is used in an application. Computers store computation results and data in variables or constants. Variables require locations in memory that can assume any value. The information stored in a variable can be recalled in another part of an event procedure by using the variable name associated with it. The information stored in a constant is data that remains the same throughout the program. It cannot be altered during the running of a program.

In this chapter, you learn about the type of data assigned to either a variable or a constant, and how to

- declare a constant;
- name a variable;
- make variable declarations explicit;
- use fixed- and variable-length strings;
- work with integer variables;
- work with Boolean variables;
- work with floating-point variables;
- work with variables of type `Currency`;
- work with variables of type `Date`;
- work with variables of type `Variant` and the `VarType` and `TypeName` functions;
- work with `Time`, `Date`, `Now`, and `DateDiff` functions.

Chapter Objectives

The objectives for this chapter are the following:

1. Learning the difference between a variable and a constant
2. Declaring a variable using a data type

3. Declaring and using the five types of numeric variables
4. Working with fixed- and variable-length string variables
5. Using functions for testing variables, including type `Variant`
6. Declaring and using variables of type `Date`

Objective 6.1 Learning the Difference Between a Variable and a Constant

When first starting to learn to program, the meaning of a *variable* as opposed to a *constant* can be troublesome. To keep matters simple, think of a value that does not change over time, such as the date of your birth. Because this date will not change, it is called a constant. When programming, use constants to store values that do not change over the course of the program. Table 6.1 shows the types of constants.

Table 6.1 Constant Types

Type	*Example*
Literal number (or expression)	`1 + 2 * 5`
String	`"This is a happy day"`
Symbolic constant (used with `Const`)	`Const PI = 3.141592654`

6.1.1 Declaring a Constant

When assigned to a symbolic name, a constant must be declared using the keyword `Const`. Suppose the following is placed at the top of a procedure:

```
Const HEADING = "Paper 6201"
```

Later in the procedure, the instruction

```
Print HEADING
```

would produce the message:

```
Paper 6201
```

The new version of Visual Basic has numerous additional constants to make programming tasks easier. Most of these constants contain the `vb` prefix. Table 6.2 shows some of the color constants in Visual Basic.

Table 6.2 Color Constants

Constant	*Value*	*Description*
`vbBlack`	`0x0`	Black
`vbRed`	`0xFF`	Red
`vbGreen`	`0xFF00`	Green
`vbYellow`	`0xFFFF`	Yellow
`vbBlue`	`0xFF0000`	Blue
`vbMagenta`	`0xFF00FF`	Magenta
`vbCyan`	`0xFFFF00`	Cyan
`vbWhite`	`0xFFFFFF`	White

In addition, there are numerous system colors specified by user choices in the Windows Control Panel. These also have constant designations, such as `vbDeskTop` that represents the user's choice for desktop color. These constants are easily accessible in Visual Basic's online Help file.

Other built-in constants are too numerous to include in this text, but you can examine them in the online Help file. For example, one can use integers to specify a form's `WindowState`, but using the built-in constants is easier and makes your code much more readable. Table 6.3 shows the `WindowState` constants and their meanings.

Table 6.3 *WindowState* Constants

Constant	*Value*	*Description*
`vbNormal`	`0`	Normal
`vbMinimized`	`1`	Minimized
`vbMaximized`	`2`	Maximized

Among the most frequently used constants are `True` and `False` for assigning values to Boolean data types. For example, the value for the `MaxButton` on a form is either `True` or `False`. If set to `True`, the Maximum button will appear on the form. If set to `False`, the Maximum button will not appear.

6.1.2 A Variable

A *variable* contains a value that is expected to change, hence the name. An example is your age. Whether you like it or not, your age is variable. With every day, the number of days since your birth changes. In programming, use variables to store values that are expected to change over the course of the program.

In the code thus far, you have assigned values to object properties, such as `label.caption` and `textbox.text`. In doing this, you have assigned a value to a variable. In Visual Basic, object properties are variables because they can take on more than one value.

Objective 6.2 Declaring a Variable Using a Data Type

A data type specifies the type of data assigned to a variable or a constant. Data type declarations specify the data required by the program. Before you use a variable in Visual Basic, you must declare it—provided `Option Explicit` is turned on. (See section 6.2.2, "`Option Explicit` Declaration," later in this chapter for more information.) It is also good programming practice to specify the type of data to be stored for a variable. The data types supported by Visual Basic, along with the associated type-declaration suffix, storage size, and range, are shown in Table 6.4.

Table 6.4 Visual Basic Data Types

Data Type	*Suffix*	*Storage Size*	*Range*
`Byte`	none	1 byte	0 to 255
`Boolean`	none	2 bytes	`True` or `False`
`Integer`	`%`	2 bytes	–32,768 to 32,767
`Long` (long integer)	`&`	4 bytes	–2,147,483,648 to 2,147,483,647
`Single` (single-precision, floating-point)	`!`	4 bytes	–3.402823E38 to –1.401298E–45 for negative values; 1.401298E–45 to 3.402823E38 for positive values
`Double` (double-precision, floating-point)	`#`	8 bytes	–1.79769313486232E308 to –4.94065645841247E–324 for negative values; 4.94065645841247E–324 to 1.79769313486232E308 for positive values
`Currency` (scaled integer)	`@`	8 bytes	–922,337,203,685,477.5807 to 922,337,203,085,477
`Decimal`	none	14 bytes	+/–7.9228162514264 to 337,593,543,950,335
`Date`	none	8 bytes	January 1, 100 to December 31, 9999

Data Type	*Suffix*	*Storage Size*	*Range*
`Object`	none	4 bytes	Any object reference
`String` (variable-length)	$	10 bytes + string length	0 to approximately 2 billion (65,400 for Microsoft Windows version 3.1 and carlicr)
`String` (fixed-length)	$	Length of string	1 to approximately 65,400
`Variant` (with numbers)	none	16 bytes	Any numeric value up to the range of a `Double`
`Variant` (with characters)	none	22 bytes + string length	Same range as for variable-length `String`
User-defined (using `Type` structure)	none	Number required by elements	The range of each element is the same as the range of its data type

As shown in Table 6.4, eight of these 15 data types are used with numbers, including the data types `Byte` and `Boolean`; two are used with strings; one is of type `Date`; and one is of type `Object`. The type `Variant` is a combination data type. It can contain numeric, string, date, and object types. However, it cannot contain a user-defined type, which is typically a mixture of different standard data types. The `Decimal` data type can only be used within type `Variant`.

For now, study these data types, but do not worry if you have questions because you will return to most of them. Some, such as the `Object` reference data type, will not be discussed until you deal with `Object` variables. Do observe, however, the number of bytes required by each type of variable. A long integer requires twice as much memory as an integer. Double-precision and currency require four times as much memory as an integer. The type declaration characters placed alongside some of the standard types can be substituted for the data type name shown when declaring a variable. They are added as suffixes. For example, the following declarations mean the same thing:

```
Dim number As Integer
Dim number%
```

6.2.1 Naming a Variable

When naming a variable (or a symbolic constant), choose a meaningful name that reflects the type of data to be stored. This improves the readability of the program and makes changes to the program easier. With Visual Basic,

- each variable must have a unique variable name,
- variable names must be limited to 255 characters,

- variable names must begin with an alpha character,
- variable names cannot contain an embedded period,
- variables must be unique within the same scope.

The following are valid variable names:

```
NameScreen
HighValue
```

However, the following examples are invalid variable names:

Name	*Reason the Name Is Invalid*
`5High`	The name begins with a digit.
`Form`	This name is not unique.
`Happy day`	Blank spaces are not allowed.
`Good.day`	An embedded period is not allowed.

6.2.2 The *Option Explicit* Declaration

If the Require Variable Declaration property is set to Yes (by selecting the check box), all variables in Visual Basic must then be declared. To set this property, choose Tools, Options, and select the Editor tab (see Figure 6.1). By default, all variables in Visual Basic are of type `Variant` unless explicitly typecast in the variable declaration. In this book—other than in a few circumstances—variables are explicitly declared as specific data types. This is called declaring a variable. Variable declaration makes your code execute faster and easier to read.

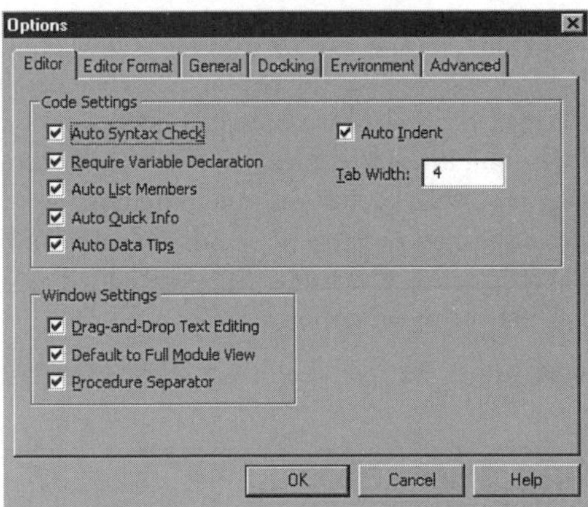

Figure 6.1

Selecting the Require Variable Declaration check box means you must explicitly declare each variable in your program.

Once checked, the following statement appears in the Declarations section of a module:

```
Option Explicit
```

When you use `Option Explicit`, all variables in a module must be explicitly declared before they can be used.

The following statement declares the variable `message` as a variable of type `String`:

```
Dim message As String
```

The following statement declares the variable `count` as a variable of type `Integer`:

```
Dim count As Integer
```

In the programs that follow, you will declare all variables before they are used. This will help you to recognize the different types of variables. It will also help you to avoid mixing types, such as trying to assign a string to a numeric variable. If a variable is not declared, a syntax error will occur.

Objective 6.3 Declaring and Using the Five Types of Numeric Variables

There are five types of numeric variables:

- Type `Integer` and `Long` (long integer) are used in declaring whole numbers (those without a decimal point).
- Type `Single` (single precision) and type `Double` (double precision) are used in declaring floating point numbers (those with a decimal point).
- Type `Currency` is used in declaring monetary numbers.

In this section, you briefly examine each type.

6.3.1 Integer Variables

Integer variables are stored as 16-bit numbers ranging in value from –32,768 to 32,767. They are used when integer numbers (numbers with no decimal places) are required in a Visual Basic program. When working with numbers with no decimal point, you must determine whether the number will be less than or greater than the appropriate range of values for type `Integer` to be used. If the number is outside that range, type `Long` is necessary.

Exercise 6.1 Working with Integers

In this exercise, you enter a person's age, height in inches, and weight in pounds. Figure 6.2 shows the completed solution to this exercise.

continues

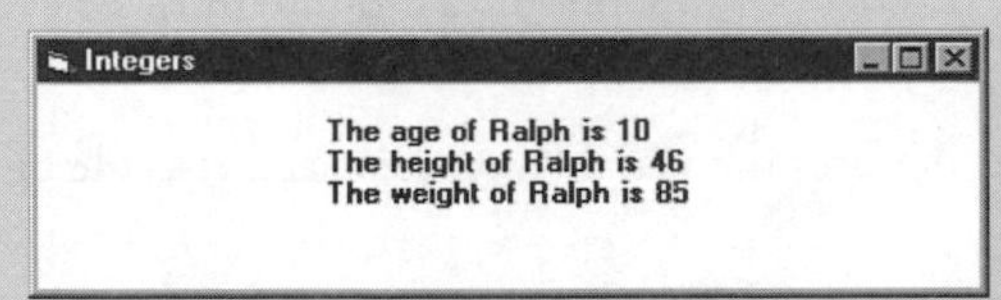

Figure 6.2

This display enables you to show three strings followed by three integer values.

1 Start a new project.

2 Change the `Caption` property of the form to **Integers** and the `Name` property to **`frmInt`**.

3 Write the following procedure:

```
Private Sub Form_Click()
  'Program with three integer variables
  'Declare all variables
  Dim person As String
  Dim age As Integer
  Dim height As Integer
  Dim weight As Integer

  'Assign values to all variables
  person$  = "Ralph"
  age = 10
  height = 46
  weight = 85

  'Display the results
  Print
  Print Spc(20); "The age of "; person$ " is"; age
  Print Spc(20); "The height of "; person$ " is"; height
  Print Spc(20); "The weight of "; person$ " is"; weight
End Sub
```

4 Run the program. Test to determine whether the suffix notation works. Change the declarations as follows, and run your program again.

```
Dim person$
Dim age%
Dim height%
Dim weight%
```

6.3.2 Long Integer Variables

Long integer variables are stored as signed 32-bit numbers ranging in value from –2,147,483,648 to 2,147,483,647, or about plus or minus two billion. This type must be used whenever a number exceeds the range of type `Integer`. Consider a ZIP code, such as 97405. Because this is larger than 32,767, type `Long` is required. A string could also be used, and would be required for postal codes that contain letters of the alphabet (such as Canadian postal codes).

Exercise 6.2 Adding a Long Integer

This exercise asks you to display a person's name and address. It requires the use of a long integer.

1 Begin a new project.

2 Change the `Caption` property to **Using a Long Integer** and the `Name` property to **`frmLongInt`**.

3 Write the following procedure:

```
Private Sub Form_Click()
  'Program to display a name and address
  'Declare all variables
  Dim person As String
  Dim street As String
  Dim city As String
  Dim state As String
  Dim zip As Long

  'Assign values to all variables
  person$ = "A.B. Waxworth"
  street$ = "1234 Mittlebox Canyon"
  city$ = "Lost Canyon"
  state$ = "OR"
  zip = 97405         'This number is greater than 32767

  'Display the results
  Print Tab(5); person$
  Print Tab(5); street$
  Print Tab(5); city$
  'Place state and zip on the same line
  Print Tab(5); state$; zip
End Sub
```

4 Test this program. Use type `Integer` rather than type `Long` for the ZIP code to observe the error.

6.3.3 Single-Precision Variables

Single-precision variables are used to store values containing decimal places. Type `Single` (which stands for *single precision*) represents a rational number that has both a whole and a fractional part. Due to this fractional part, this type is used to store a floating point number.

The following are examples of floating point numbers:

```
50.33435     5.5     -503.88865
```

A floating point number provides greater precision than an integer. This precision is indicated by the number of significant digits that can be represented in storage. A calculation using type `Single` might return a value of `23.25`. The same calculation using a variable of type `Integer` would *truncate* (cut off) the fractional portion and return a value of `23`. As another example, type `Single` might return the value `23.5`, whereas type `Integer` would round the value (.5 and above), and return `24`. The next exercise demonstrates this concept.

Exercise 6.3 Using a Single-Precision Variable

This short exercise asks you to study how rounding a number works.

1 Start a new project.

2 Change the `Caption` property to **Single Precision** and the `Name` property to **`frmSPrec`**.

3 Write the following procedure:

```
Private Sub Form_Click()
  'Program to compare single precision to an integer

  'Declare all variables
  Dim precise As Single
  Dim truncated As Integer
  Dim rounded As Integer

  'Make the first set of assignments
  precise = 50.33435
  truncated = 50.33435

 'Display them
  Print Tab(5); precise
  Print Tab(5); truncated

  'Make the second set of assignments
  precise = 5.5
  rounded = 5.5
  'Display them
  Print Tab(5); precise
  Print Tab(5); rounded

  'Make the third set of assignments
  precise = -503.88865
  rounded = -503.88865
  'Display them
  Print Tab(5); precise
  Print Tab(5); rounded
End Sub
```

In this program, a variable (such as `precise`) can be assigned different values. The first assignment sets the value of `precise` to 50.33435. After this value is printed, a second value can be assigned to `precise` and printed. Could this second value be assigned before the first value is printed? It could, but the first assigned value would be lost and could not be printed.

6.3.4 Double-Precision Variables

Double-precision variables store larger floating point numbers, but require twice as many bytes of storage compared to type `Single`. Type `Double` is used for very large or very small numbers (a number with 308 places to the right using scientific notation).

In choosing between type `Single` and type `Double`, use type `Single` unless the calculation requires greater precision. Besides needing greater storage, type `Double` requires more time to run than type `Single`.

Exercise 6.4 Comparing Single- and Double-Precision Variables

This brief exercise asks youto compare a single-precision variable to a double-precision variable.

1 Place a command button on a form, named **`cmdDivide`**, and add the `Caption` property **Print 1/3**. Add the form caption **Test of Precision**.

2 Write the following `cmdDivide_Click()` procedure:

```
Private Sub cmdDivide_Click()
  Dim test1 As Single
  Dim test2 As Double
  'Enter 15 number 3s

  test1 = .333333333333333
  test2 = .333333333333333
  Print "Single precision is = "; test1
  Print "Double precision is = "; test2
End Sub
```

3 Run and test the program. You will see the different degrees of precision in the displayed results.

6.3.5 Currency Variables

Currency variables are used for calculations involving money and for fixed-point calculations in which accuracy is very important. Type `Currency` variables are stored as 64-bit numbers (eight bytes) and are scaled by 10,000 to provide a fixed-point number with 15 digits to the left of the decimal and four digits tc the right.

Exercise 6.5 Using Currency Variables

This short exercise helps to clarify why currency variables are needed.

1 Place a command button on a form, named **cmdLoan**, and add the Caption property **Show Loan**. Add the form caption **Test of Currency**.

2 Write the following cmdLoan_Click() procedure:

```
Private Sub cmdLoan_Click()
  Dim Amount As Currency
  'Try this one next.  Comment out the above statement
  'Dim Amount As Single
  Dim IntRate As Single
  Dim SimpleInt As Currency

  Amount = 100000000
  IntRate = .08
  SimpleInt = 8000000

  Print "The amount of the loan is $"; Amount
  Print "The interest rate is "; IntRate; "%"
  Print "The simple interest for 1 year is $"; SimpleInt
End Sub
```

3 Run and test this program, using 100,000,000 (one-hundred million) as the loan amount. In testing, change the data type for the variable Amount from type Currency to type Single, and run the program again. Observe the result.

6.3.6 Boolean Variables

Boolean data types can be used in circumstances for which you simply want to determine whether a variable or condition is True or False. It is frequently used in logical tests and with If...Then statements (which are covered in Chapter 8, "If...Then...Else Logic and the Select Case Statement"). For now, you should know Boolean variables can act like a switch, becoming either True or False.

Exercise 6.6 Using the *Boolean* Data Type

This short exercise asks you to set to True and to False a variable of type Boolean. Figure 6.3 shows the interface you are asked to create.

1 Create the interface as shown. Name the label on the left **lblWrap**, and set its WordWrap property to True.

2 Name the label on the right **lblNoWrap**, and make sure its WordWrap property is set to False.

3 Enter the following code in the `cmdTest_Click()` event:

```
Private Sub Command1_Click()
  Dim wraptest As Boolean
  wraptest = lblWrap.WordWrap
  lblWrap.Caption = wraptest
  wraptest = lblNoWrap.WordWrap
  lblNoWrap.Caption = wraptest
End Sub
```

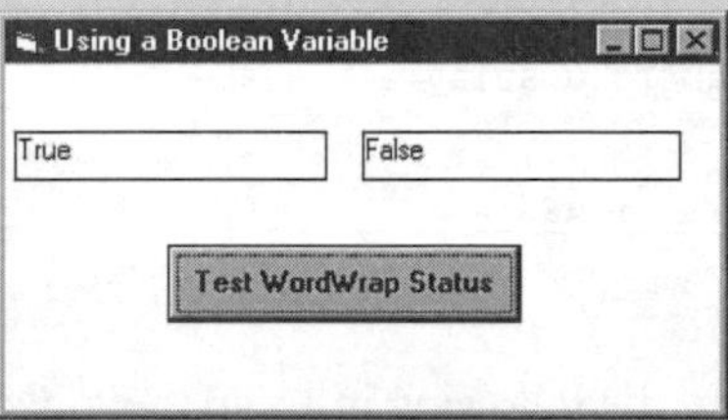

Figure 6.3

This interface requires the use of Boolean variables.

In this last procedure, the variable `wraptest` takes on two values within a single procedure. Initially, it is assigned the Boolean value for the `WordWrap` property in `lblWrap`, which is the value `True`. After its `True` value is displayed, `wraptest` is assigned the value of the `lblNoWrap.WordWrap` property, which has the value `False`. This value is displayed in the `lblNoWrap` label.

Objective 6.4 Working with Fixed- and Variable-Length String Variables

A string variable can hold any character, word, or phrase that the keyboard can produce. In Visual Basic, there are two types of string variables:

- **Variable-length**—Strings in which the number of characters in the string is not specified in advance. The maximum length of a string in 32-bit Windows operating systems is nearly two billion characters. In 16-bit Windows, it is around 65,500 characters. In a variable-length string, each character requires one byte of storage, and there is an additional 10 bytes of string information in 32-bit Windows.
- **Fixed-length**—Strings in which the number of characters in the string is specified in advance.

A string variable is defined as type `String`. Consider the following declaration:

```
Dim YourName As String
```

This instruction declares YourName as a variable of type String. The String suffix $ cannot be used in this declaration. For example, the following returns an error:

```
Dim YourName$ As String
```

However, the following does not return an error:

```
Dim YourName$
```

Consider the following code:

```
Private Sub cmdStart_Click()
  'The Start command is used to display a message
  Dim message As String
  Print
  message$ = "Welcome to Visual Basic"
  Print Spc(30); message$
End Sub
```

Why use a string variable when the following instruction would display the same thing?

```
Print "Welcome to Visual Basic"
```

String variables are advised when a string is to be used several times in a program, or when the contents of the string need to be manipulated. Visual Basic provides a large number of string functions and statements designed to join strings, compare strings, use strings with variants, or convert strings to numbers, for example. See Chapter 19 “String Functions” for more information.

6.4.1 String Variables

String variables can be assigned a series of characters or a null string. In addition, the value of a string can be assigned to another string variable.

A null string is a string without characters. It is written as follows:

```
Person$ = ""
```

This specifies a string of zero length. A text box or label is cleared with a null string. The following instruction clears the text from the text box:

```
txtName.Text = ""
```

The following instruction clears the text from the label caption:

```
lblName.Caption = ""
```

6.4.2 Assigning One String to Another

Assigning one string to another is a common practice. The following instructions illustrate this type of assignment:

```
MyName$ = "Marvelous"
Person$ = MyName$
```

In this example, the string MyName and the string Person both store the name Marvelous.

6.4.3 String Concatenation

The ampersand (&) can be used with strings to combine one string with another. This is called *string concatenation*, and the ampersand is called the *string concatenation operator*. Consider the following code:

```
First$ = "Robert"
Last$ = "Rogers"
MyName$ = First$  &  " " &  Last$
Print MyName$
```

This leads to the following display:

```
Robert Rogers
```

Observe the middle portion of the MyName$ instruction. The " " is a string that contains a single blank space. Lacking the space, Print MyName$ would display the following:

```
RobertRogers
```

Likewise, the code

```
First$ = "Robert"
Last$ = "Rogers"
MyName$ = Last$  &  ", "  &  First$
Print MyName$
```

leads to the display

```
Rogers, Robert
```

In older versions of Visual Basic, string concatenation could be performed with either the + operator or the & operator. Visual Basic 5, however, refers to string concatenation with the & operator. This makes code easier to read and doesn't confuse string concatenation with arithmetic operations.

In displaying string values, a string variable can be concatenated with itself, as the following example illustrates:

```
MyName$ = "Robert"
MyName$ = MyName$ & "Rogers"
Print MyName$
```

This works *if* you add spacing, as the next exercise demonstrates.

Exercise 6.7 Entering Variable String Data on a Form

This exercise asks you to design a form to record your first name, middle initial, and last name, and to display your name after all values have been entered. Use text boxes for all data entry. Display the name on the form—not in a text box (see Figure 6.4).

continues

Figure 6.4

This application features three strings to display a person's name.

1 Add three text boxes and two command buttons. Name the text boxes **`txtFirst`**, **`txtMiddle`**, and **`txtLast`**. Name the command buttons **`cmdDisplay`** and **`cmdQuit`**. Add three labels: **First Name**, **Middle Initial**, and **Last Name**.

2 Double-click the Line tool to place a line on the form. Click the edge of the line to position the endpoint on the screen.

3 Write a `cmdDisplay_Click()` procedure as follows, leaving `Form1` as the name of the form:

```
Private Sub cmdDisplay_Click()
  Dim MyName As String

  MyName = txtFirst.Text
  MyName = MyName & " " & txtMiddle.Text
  MyName = MyName & " " & txtLast.Text
  Form1.CurrentX = Line1.X1
  Form1.CurrentY = Line1.Y1 + 150
  Form1.Print "       Your name is: "; MyName
End Sub
```

4 Write a `cmdQuit_Click()` procedure, and name the form **Entering a Name**.

5 Run and test your program.

Exercise 6.8 Displaying a Name in a Different Order

Modify Exercise 6.7 to display your last name, followed by your first and middle initial. To do this, change only the `Print` statements used to control the way in which your name is printed. Change the form caption to **Entering a Name II**.

6.4.4 Fixed-Length String Variables

Fixed-length string variables hold a string to a specified size. A fixed-length string is declared using the following syntax:

```
String * size
```

For example:

```
Dim YourName As String * 50
```

Exercise 6.9 Using Fixed-Length Strings

Modify the first exercise to use fixed-length strings: 10 for the first name, 1 for the middle initial, and 15 for the last name. Using Figure 6.5 as a guide, do the following:

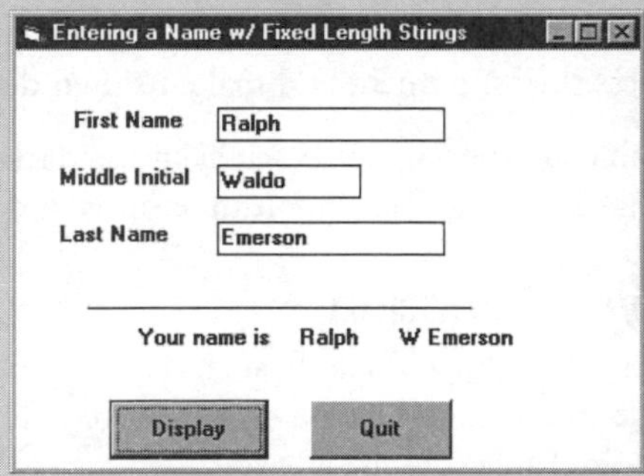

Figure 6.5

This application features fixed-length strings rather than variable-length strings.

1 Change the three declaration statements to show all three strings as fixed-length, and name the strings **`first`**, **`middle`**, and **`last`**, as follows:

```
Dim first As String * 10
Dim middle As String * 1
Dim last As String * 15
```

2 Modify the code so that the `Print` statement displays the `first`, `middle`, and `last` strings:

```
first = txtFirst.text
middle = txtMiddle.text
last = txtLast.text
Form1.CurrentX = Line1.X1
Form1.CurrentY = Line1.Y1 + 150
Form1.Print "       Your name is ";
Form1.Print "   " & first & " " & middle & " " & last
```

continues

In this code, the $ is not used for the three strings. Remember: The $ is optional for all strings, both fixed- and variable-length strings.

3 Run and test your design. In testing, add a name larger than the fixed-length size. Also, examine the output. How do fixed-length strings affect the way in which output is displayed?

Objective 6.5 Using Functions for Testing Variables, Including Type *Variant*

Type `Variant` is the default in Visual Basic (the type that all variables become if they are not explicitly declared). You can, for instance, simply declare a variable without a specific data type, such as in the declaration `Dim x`. The variable `x` could then be assigned any data type, and Visual Basic would assess the assignment and make its own determination of data type.

You can also declare a variable as type `Variant`, such as in the declaration `Dim x As Variant`. As such, it can contain numeric, string, date, Boolean, empty, and null values.

6.5.1 The *VarType* Function

The `VarType` function serves to test how Visual Basic has determined a variable assignment. The function returns a value to indicate the type of data stored. The return values of the `VarType` function are shown in Table 6.5.

Table 6.5 The Return Values of the *VarType* Function

Constant	*Value*	*Description*
vbEmpty	0	Empty (uninitialized)
vbNull	1	Null (no valid data)
vbInteger	2	Integer
vbLong	3	Long integer
vbSingle	4	Single-precision, floating-point number
vbDouble	5	Double-precision, floating-point number
vbCurrency	6	Currency
vbDate	7	Date
vbString	8	String
vbObject	9	Object
vbError	10	Error
vbBoolean	11	Boolean

Constant	*Value*	*Description*
`vbVariant`	12	Variant (used only with arrays of Variants)
`vbDataObject`	13	Data-access object
`vbDecimal`	14	Decimal
`vbByte`	17	Byte
`vbArray`	8192	Array

As indicated, `Empty` and `Null` do not mean the same thing. `Empty` means the variable has not been initialized (and hence the value of the variable is not known). This would be the case if you had declared the variable without an explicit type declaration, as in the simple declaration `Dim x`. `Null` means the variable has been initialized, but contains no valid data.

Although the `Variant` data type is flexible and can accommodate many programming needs, it is the largest data type you can use, so it can impose a marked performance penalty. Code can be more difficult to debug if type mismatches are encountered. By defining variables explicitly, code is easier to follow, program speed improves, and less memory is required.

Type `Variant`, as a rule, should be avoided. On a few occasions, functions will return values that you are uncertain of; you can use a `Variant` data type and then test for the return value using the `VarType` function. The `Variant` data type must also be used in applications that use *object linking and embedding,* referred to as *OLE.* (For more information about OLE, see Chapter 26, "Working with Objects from Microsoft Office.") If you are called upon to maintain code that has used `Variant` data types, you will also want to be able to employ the `VarType` function.

Exercise 6.10 Using the *VarType* Function

This short exercise asks you to determine the return values of variables when they are declared to be of type `Variant`. The design consists of a list box (`lstDisplay`) and a single command button (`cmdTest`).

1 On a form, add a list box named **`lstDisplay`** and a command button named **`cmdTest`**.

2 In the `cmdTest_Click()` event, add the following code:

```
Dim sVar As Variant 'Will be used to store a string
Dim iVar As Variant 'Will be used to store an integer
Dim lVar As Variant 'Will be used to store a long integer
Dim DVar As Variant 'Will be used to store type double
Dim ret As Variant
```

continues

```
sVar = "Using the VarType Function"
iVar = 980
lVar = 1000000
DVar = 0.09878654321

ret = VarType(sVar)
lstDisplay.AddItem ret
ret = VarType(iVar)
lstDisplay.AddItem ret
ret = VarType(lVar)
lstDisplay.AddItem ret
ret = VarType(DVar)
lstDisplay.AddItem ret
```

3 Run and test the program. In the list box, you should see a listing of the integer values consistent with the return type values shown in Table 6.5.

6.5.2 The *TypeName* Function

Visual Basic also includes a function named `TypeName`, which returns a string describing the variable used as an argument in the function. Like the `VarType` function, `TypeName` has the simple syntax of `TypeName(variablename)` in which `variablename` is of any type except a user-defined type.

Exercise 6.11 Using the *TypeName* Function with the *VarType* Function

This exercise asks you to extend the previous one. The design is limited to a list box and a single command button.

1 Extend the previous exercise by changing the relevant section of code in the `cmdTest_Click()` event to read as follows:

```
ret = VarType(sVar) & "  " & TypeName(sVar)
lstDisplay.AddItem ret
ret = VarType(iVar) & "  " & TypeName(iVar)
lstDisplay.AddItem ret
ret = VarType(lVar) & "  " & TypeName(lVar)
lstDisplay.AddItem ret
ret = VarType(DVar) & "  " & TypeName(DVar)
lstDisplay.AddItem ret
```

2 Run and test the program. The string expression of the variable type is returned by the `TypeName` function and displayed in the list box.

Objective 6.6 Declaring and Using Variables of Type *Date*

The `Date` data type is used for both dates and times, and is employed with Visual Basic's versatile and extensive `Time` and `Date` functions.

6.6.1 The *Time* Function

A function commonly used with type `Date` is the `Time` function. `Time` returns the time as found on your computer's clock. The syntax is simply the following:

```
Time
```

This function returns a `Variant` of `Type 7` (see Table 6.5). Time returns an eight-character string, formatted as hh:mm:ss, where *hh* is the hour, *mm* is the minute, and *ss* is the second.

6.6.2 The *Date* Function

A second function commonly used with type `Variant` is the `Date` function. `Date` returns the system clock date. The syntax is simply as follows:

```
Date
```

This function also returns a `Variant` of `Type 7`.

6.6.3 The *Now* Function

The `Now` function combines the time and date, and uses the `Date` variant. `Now` returns the date and time as found on the computer's clock. The syntax is simply the following:

```
Now
```

This function returns a `Variant` of `Type 7`. It is a double-precision number used to represent a date between January 1, 100 through December 31, 9999.

Exercise 6.12 Displaying the Time and Date of the System Clock

This short exercise asks you to use type `Date` to display the time and date found on the system clock (the clock maintained by your computer). It uses the `Time`, `Date`, and `Now` functions. If the time and date are not correct, you might want to change them, using the time and date DOS commands or changing the time and date in the Windows Control Panel. Figure 6.6 shows the interface that you are to create.

continues

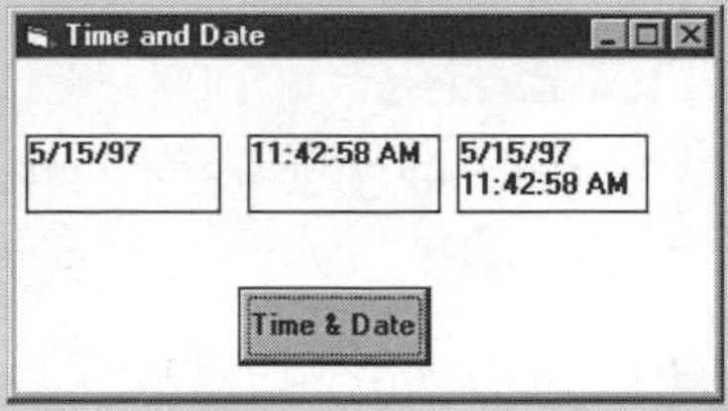

Figure 6.6

This application illustrates the use of the `Date` function, the `Time` function, and the `Now` function.

1 Add three labels, **`lblDate`**, **`lblTime`**, and **`lblNow`**, and a single command button named **`cmdShowTime`**. Place a `Caption` property of **Time & Date** on `cmdShowTime`.

2 Write the following `cmdShowTime_Click()` procedure:

```
Private Sub cmdShowTime_Click()
  Dim Today As Date
  Today = Date
  lblDate.Caption = Today
  Today = Time
  lblTime.Caption = Today
  Today = Now
  lblNow = Today
End Sub
```

In this procedure, you could have simplified the code to assign the returns of the different functions directly to the label. However, this procedure illustrates how `Time` and `Date` function returns can be assigned to a variable.

3 Run and test your program. Declare `Time` and `Today` as variables other than type `Date` or type `Variant`, and observe what happens.

6.6.4 The *DateDiff* Function

If you explore Visual Basic to any extent, you will discover that it contains a large number of functions that simplify the handling of dates. One of these functions is `DateDiff`. (Additional functions are discussed in Chapter 18, "Numeric Functions.") This function returns a numeric data type containing the number of intervals between two dates.

The syntax is as follows:

```
DateDiff(interval, date1, date2[, firstdayofweek[, firstweekofyear]])
```

The following list explains the elements of the syntax:

> `interval` is a string expression representing the interval between the dates. The settings are year (`"yyyy"`), quarter (`"q"`), month (`"m"`), day of year (`"y"`), day (`"d"`), weekday (`"w"`), week (`"ww"`), hour (`"h"`), minute (`"n"`), and second (`"s"`).

`date1, date2` are the two dates you want to compare.

`firstdayofweek` is a constant that specifies the first day of the week.

`firstweekofyear` is the first week of the year. If not specified, Sunday is assumed to be the first day of the week, and the week in which January 1 occurs is assumed to be the first week.

The `firstdayofweek` argument has the settings shown in Table 6.6.

Table 6.6 Settings for the *firstdayofweek* Constant

Constant	*Value*	*Description*
`vbUseSystem`	0	Usc application sctting if onc cxists; othcrwisc, usc NLS API setting
`vbSunday`	1	Sunday (default)
`vbMonday`	2	Monday
`vbTuesday`	3	Tuesday
`vbWednesday`	4	Wednesday
`vbThursday`	5	Thursday
`vbFriday`	6	Friday
`vbSaturday`	7	Saturday

The `firstweekofyear` argument has the settings shown in Table 6.7.

Table 6.7 Settings for the *firstweekofyear* Constant

Constant	*Value*	*Description*
`vbUseSystem`	0	Uses the application setting if one exists; otherwise, uses NLS API setting
`vbFirstJan1`	1	Starts with the week in which January 1 occurs (default)
`vbFirstFourDays`	2	Starts with the first week that has at least four days in the new year
`vbFirstFullWeek`	3	Starts with the first full week of the year

For example, if you want to find the number of days between today and a date entered into processing, you could write the following (in which `Tomorrow` is a variable of type `Date`):

```
DateDiff("d", Date, Tomorrow)
```

Exercise 6.13 Finding the Number of Shopping Days Until Christmas

In this short exercise, you determine the number of shopping days between today and December 25. Figure 6.7 illustrates the interface.

Figure 6.7

This application uses the `DateDiff` *function to determine the number of shopping days between today's date and 12/25/97.*

1 Add two command buttons, **`cmdTellMe`** and **`cmdEnd`**, and two labels (not text boxes), **`lblToday`** and **`lblDays`**. Add the form caption and the two descriptive labels, as shown in Figure 6.7. Write the `cmdEnd` procedure.

2 Write the following `cmdTellMe` procedure, changing `Date2` as necessary:

```
Sub cmdTellMe_Click()
  Dim Date2 As Date
  Date2 = "12/25/97" 'Change to fit the year
  lblToday.Caption = Date
  lblDays.Caption = DateDiff("d", Date, Date2)
End Sub
```

3 Run and test your program. Determine what happens if you add a string constant for `Date2`.

Chapter Summary

The purpose of this chapter is to introduce you to the different data types permitted by Visual Basic. By now, you should have a good understanding of the data types: type `Integer`, `Long`, `Single`, `Double`, `Currency`, `String` (fixed- and variable-length strings), `Date`, and `Variant`.

You should also have a good understanding of the difference between a constant and a variable. A *constant* stores values that are not expected to change over the course of a program. Contrary to that, a *variable* stores values that are expected to change. Object properties represent variables because they can be changed.

When naming variables, use a unique name of 255 characters or less that begins with an alpha character and contains no embedded periods. Variable names must be unique within the same scope.

To force the declaration of variables, select Require Variable Declaration. This will add `Option Explicit` in the environment options settings.

Remember that type `Integer` and type `Long` are used for declaring whole numbers, and type `Single` and type `Double` are used for floating point numbers. Use type `Currency` when dealing with dollars and cents. This avoids scientific notation.

`String` variables can be of variable length (the default) or of fixed length. To declare a fixed-length string, add the size of the string, as shown in the following example:

```
Dim YourName As String * 50
```

Type `Variant` is the default in Visual Basic when a variable is not declared. This type can contain numeric, string, date, empty, and null values. When writing procedures, type `Date` variables should be used to store date and time values.

There are a large number of date and time functions used with date variables. The functions described in this chapter are `Date`, `Time`, `Now`, `DateDiff`, and two special functions, `VarType` and `TypeName`. `VarType` returns the type of variant represented by a variable, and `TypeName` returns a string that names the variable.

Skill-Building Exercises

1. Create a display such that when the user clicks a button named Now Is, a time and date stamp appears in one label, and the following message appears in a second label (using two lines):

```
The Date is <insert the current date here>
The Time is <insert the current time here>
```

2. Use option buttons to confine user input to clicking either The sun is how many miles from the earth or The moon is how many miles from the earth. When the user makes a choice, use one of two labels to display the result. For the sun, the label should display the following:

```
The sun is 9.3E+07 miles from the earth.
```

For the moon, the label should read as follows:

```
The moon is 238000 miles from the earth.
```

3. Modify the fixed-string exercise. Create a display for the entry of a first name, middle name, and last name; however, display only the last name followed by the initials of the first and middle names. Place a comma after the last name, and add one period each after the first and middle initials, as in the following example:

```
Fox, P. E.
```

4. Place a frame on the screen, and add eight option buttons. Use the following table to add a caption to each button and to determine the distance of each planet from the earth:

Planet	*Distance from Earth*
Mercury	53,000,000
Venus	25,000,000
Mars	35,000,000
Jupiter	390,000,000
Saturn	793,000,000
Uranus	1,700,000,000
Neptune	2,678,000,000
Pluto	2,700,000,000

Use a label to display the distance from the earth when the user clicks a planet. Use constants for the first three planets; use variables for the remaining planets. Do not use scientific notation, but rather show the entire number without commas.

5. This somewhat morbid project asks you to determine the number of days you have left on earth. The only date to be entered into processing is your birthday (see `txtBirthdate`). In writing the `cmdBadNews` event, declare a constant as follows:

```
Const ITSOVER = 27375          '75 years times 365 days
```

This assumes that you will live until you are 75 years old. Determine the number of days you have lived so far, and display the results in the label named **`lbluhoh`**. After this, determine the days you have left to live to display the following:

```
Print "If you live to age 75 you have "; OhNo; " days left."
```

Calculating one's own demise is sobering, isn't it?

7

Math Operators and Formulas

In Visual Basic, you can use mathematics and formulas when writing instructions. If you are wondering how all this works, consider how you perform mathematics using a calculator: You enter a number, tell the calculator what to do with the number (add, subtract, multiply, or divide), and enter a second number. After the second number is entered, the operation (addition, subtraction, multiplication, or division) is performed.

This chapter explains how mathematical expressions are written using Visual Basic. Much like using a calculator, you enter a number, a mathematical operator, and a second number. In Visual Basic, there are three types of mathematical operators: *arithmetic, unary,* and *assignment.* When you complete this chapter, you will know how to

- use the seven arithmetic operators,
- write arithmetic expressions,
- arrange an arithmetic expression using operator precedence,
- use single and nested sets of parentheses,
- use a unary operator,
- use the assignment operator,
- work with the `Val()` and `Str()` functions.

Chapter Objectives

The objectives of this chapter are derived from the skills outlined in the preceding list, as follows:

1. Understanding the seven arithmetic operators
2. Writing arithmetic expressions using arithmetic operators
3. Understanding the unary and assignment operators
4. Converting strings to numbers and numbers to strings

Objective 7.1 Understanding the Seven Arithmetic Operators

Many types of programming problems require the manipulation of numbers. For example, in computing the interest on a loan, you multiply the daily interest rate by the principal amount and then by the number of days. Still, other uses of arithmetic include taking the power of a number, such as 4 raised to the fourth power, or finding the remainder of 10 divided by 3.

7.1.1 Operators and Operands

Operators and *Operands* are troublesome terms, but you need to understand them. An *operator* is a symbol used to designate an action performed on two numbers. Arithmetic operators specify a mathematical action, such as addition, subtraction, multiplication, and division. Arithmetic operators manipulate two operands placed one on each side of the operator. An *operand* is a constant, such as the number 5; a variable, such as `var1`; or an expression, such as `(5x + y)`, for which the expression is enclosed by parentheses. Remember this relationship:

```
<Operand1> operator <Operand2>
```

An *operator* is used to join two *operands.*

7.1.2 The Seven Types of Operators

In Visual Basic, there are seven types of arithmetic operators. Table 7.1 lists each operator, its symbol, and an example of its use.

Table 7.1 Binary Mathematical Operators

Operator	*Symbol*	*Example*	*Result*
Addition	+	4 + 5	9
Subtraction	-	9 - 6	3
Multiplication	*	3 * 4	12
Division	/	8 / 2	4
Exponentiation	^	2 ^ 3	8
Integer division	\	5 \ 2	2
Modulus	Mod	5 Mod 2	1

Besides the familiar addition and subtraction operator symbols (+ and –, respectively), the symbol for multiplication (*) differs from the familiar symbol (×), and division (/) is not shown by the more familiar sign (÷). The following are examples of these four operators (see Table 7.1):

```
4 + 5 =  9
9 - 6 =  3
3 * 4 = 12
8 / 2 =  4
```

The exponentiation operator (^) is used to raise a number to a power. The following example shows 2 raised to the third power:

```
2 ^ 3 = 8
```

Integer division is generally performed along with the modulus operator. With integer division, no remainder is provided. Instead, the difference is *truncated* (that is, the remainder is simply chopped off), as follows:

```
5 \ 2 = 2
```

The modulus operator then is used to compute the remainder. By using the preceding example but substituting the modulus operator for integer division, the remainder of the difference is computed:

```
5 Mod 2 = 1
```

If the modulus operator finds no remainder, as would be the case with `5 Mod 5`, the result of the operation is `0`.

With Visual Basic, addition, subtraction, multiplication, division, and exponentiation can all be used with `Integer`, `Long` (long integer), `Single` (single precision), `Double` (double precision), and `Currency` data types. Addition and subtraction can be used with the `Date` data type. Integer division and modulus can only be used with `Integer` and `Long`.

Exercise 7.1 Computing the Interest on a Loan

This short exercise asks you to compute the balance and the interest on a loan, and indicate the interest a person must pay. Use a daily interest rate of 0.002328, a principal of $5,000, and 50 days as the number of days since taking out the loan.

1 Write the following `Form_Click()` procedure:

```
Private Sub Form_Click()
  'Program to compute the new balance on a loan
  Dim days As Integer
  Dim principal As Integer
  Dim dailyInterest As Single
  Dim interest As Single
  Dim balance As Single

  days = 50
  principal = 5000
  dailyInterest = .002328
```

continues

```
    interest = principal * dailyInterest * days
    balance = principal + interest
    Print "The balance you owe is "; balance
    Print "The interest you must pay is "; interest
End Sub
```

2 Run and test your program. Find out what happens when you use (×) to multiply instead of (*).

Exercise 7.2 Computing Exponentiation

This initial exercise asks you to write a program to compute the number 2 raised to the 32nd power.

1 Start a new project, and write the following `Form_Click()` procedure:

```
Private Sub Form_Click()
    'Program to raise the power of a number
    Print "2 to the 32 power is"
    Print (2 ^ 32)
End Sub
```

2 Test your program. Try other relationships, such as `32 ^ 2`.

Exercise 7.3 Using Integer Division and the Modulus Operator

This short program asks you to divide 5,000 by 12 and to display the integer result and the remainder.

1 Write a `Form_Click()` procedure as follows:

```
Private Sub Form_Click()
    'Program to demonstrate integer division and modulus
    Dim sum As Integer
    Dim result As Integer
    Dim remainder As Integer

    sum = 5000
    result = sum \ 12
    remainder = sum Mod 12
    Print "The result is "; result;
    Print " and the remainder is "; remainder
End Sub
```

2 Run and test your program. In testing, divide 5,000 by 5,000 to observe what happens when there is no remainder.

Objective 7.2 Writing Arithmetic Expressions Using Arithmetic Operators

An arithmetic expression is two or more operands connected by arithmetic operators. An arithmetic expression is written to return a value. The following is an example of an arithmetic expression:

```
x + y * 2 ^ 3
```

With an expression such as the preceding, each operator links two operands to form a *binary linking* (an operator is placed between two operands). In processing this expression, Visual Basic examines the type of action to be performed and processes the one with the highest order or the highest precedence first. In Visual Basic, the precedence for the seven binary operators and the *negation* operator (discussed in the next section) is shown in Table 7.2

Table 7.2 Order of Precedence of the Mathematical Operators

Order	*Operator*	*Symbol*
1	Exponentiation	^
2	Negation	-
3	Multiplication and division	*, /
4	Integer division	\
5	Modulus	Mod
6	Addition and subtraction	+, -

7.2.1 Operator Precedence

What is meant by *precedence* and how does it work? Precedence provides a priority (or order of importance) to the different types of arithmetic actions. If an expression contains operators of different types, the operator with the highest precedence (or rank order) is performed first. In the following expression, exponentiation is performed first:

```
x + y * 2 ^ 3
```

Why? Exponentiation has a higher precedence than either addition or multiplication. After this action, the expression would be reduced to read as follows:

```
x + y * 8
```

Suppose you know that y = 5. If processing continues, multiplication would be performed next because that operator has higher precedence than addition. The expression would be reduced to read as follows:

```
x + 40
```

Finally, if you know that x = 3, then the value of the completed expression can be resolved. That is, 3 + 40 = 43.

If an expression contains operators with the same order of precedence, then left-to-right processing occurs. In the following expression, multiplication and division have the same precedence:

```
5 + 4 * 15 / 3
```

Thus, the expression would be reduced to the following:

```
5 + 60 / 3
```

Next, division is performed because division has a higher precedence than addition. The reduced expression would read

```
5 + 20
```

Finally, the value of 25 could be determined by addition.

7.2.2 Use of Parentheses

The order of precedence can be overruled by using parentheses, as follows:

```
(5 + 4) * (15 / 3)
```

Because the expressions inside the parentheses are resolved before the symbol to multiply is reached, this example is reduced to

```
9 * 5
```

The final operation produces 45. What this suggests is that you make good use of parentheses to leave nothing to chance. It also makes debugging of your program much easier.

Exercise 7.4 Computing an Average of Three Numbers

This simple exercise asks you to compute the average of the following numbers: 150.10, 175.20, and 200.30. All values are of type `Currency`.

1 Write a `Form_Click()` procedure as follows:

```
Private Sub Form_Click()
   'Program to demonstrate the need for parentheses
   'Computing an average
   Dim amount1 As Currency
   Dim amount2 As Currency
   Dim amount3 As Currency
   amount1 = 150.1
   amount2 = 175.2
   amount3 = 200.3
   Print "The average is "; (amount1 + amount2 + amount3) / 3
End Sub
```

2 Run and test your program. In your testing, remove the parentheses and inspect the results. Are you surprised? Also observe that the three values of type `Currency` are divided by an integer; however, the result is of type `Currency`. Visual Basic converts the result into the larger type. Because type `Currency` is larger than type `Integer`, the result is type `Currency`.

7.2.3 Nested Parentheses

Parentheses can be *nested,* that is, one set of parentheses placed inside another:

```
(5  + (4  *  15)) / 3
```

The innermost set of parentheses is evaluated first, in this instance leading to the following reduced expression:

```
(5  +  60) / 3
```

After this, the values within the remaining set of parentheses are evaluated, leading to

```
65 / 3
```

7.2.4 Adding an Arithmetic Expression to a Print Statement

An arithmetic expression need not always be assigned to a variable. Instead, if all values are known, they can be used directly in an instruction, such as a `Print` instruction. The instruction

```
Print (5 + (4  *  15)) / 3
```

leads to the display

```
21.66667
```

Objective 7.3 Understanding the Unary and Assignment Operators

Numeric operators (besides arithmetic operators) include the *unary* operator and the *assignment* operator. These operators are easy to understand because they require only a single operand.

7.3.1 The Unary Operator

Arithmetic operators are binary and require two operands. The Visual Basic language also features a *unary* operator, which requires a single operand. As shown in Table 7.2, this operator is called *negation.* It acts to change a value from positive to negative or negative to positive.

For example, if `TestNum` is equal to 5, the following expression changes the value of 5 from +5 to –5:

```
- TestNum
```

Likewise, if `TestNum` is equal to –5, the following expression changes the value of –5 to +5:

```
- TestNum
```

7.3.2 The Assignment Operator

The *assignment* operator is a binary operator that assigns or copies the value of the right operand to the left operand. It moves the value of an expression into a memory location. You employed this operator in Chapter 3, "Adding Controls and Event Procedures to Form Modules," using the following syntax:

```
variable = expression
```

Visual Basic uses the equal sign operator in making these right-to-left assignments. The following expression assigns the value of 11 to the variable `y` and stores this value in memory:

```
y = 5 + (3 * 2)
```

The assignment operator is used to copy the value of a constant, a variable, an arithmetic expression, or a function, as shown in the following table:

Expression	*Type*
`high_mark = 98`	Constant
`weight = measure1`	Variable
`cost = price * quantity`	Arithmetic expression
`miles = travel(time)`	Function

7.3.3 A Counter Variable

An assignment instruction becomes more meaningful with an instruction like the following:

```
count = count + 1
```

From mathematics, you would say that this equation is invalid because the two instances of `count` cancel each other out. However, in programming, an assignment is made rather than the determination of equality. In the preceding example, the value of `count + 1` (the expression) is assigned to the variable `count`. If the old `count` is 5, the expression `count + 1` leads to a new `count` of 6. A count increment occurs when a variable is increased by a constant, such as 1. A count decrement occurs when a variable is reduced by a constant amount, as in the following instruction:

```
count = count - 1
```

7.3.4 Adding the Object Name to the Variable Name

The left-hand side of an assignment instruction can be altered using the following syntax:

```
object.property = expression
```

For example, the following instruction assigns the constant value `vbBlue (&HFF0000)` to the `BackColor` property of `Form1`.

```
Form1.BackColor = vbBlue
```

Exercise 7.5 Fahrenheit to Celsius Conversion

In this exercise you are asked to write a program to convert degrees Fahrenheit into degrees Celsius and Celsius into Fahrenheit, for which Fahrenheit is equal to 32 degrees and Celsius is equal to 0 degrees. The formulas to remember are as follows:

- Subtract 32 from degrees Fahrenheit, and multiply by 5/9 to determine degrees Celsius.
- Multiply degrees Celsius by 9/5, and add 32 to determine degrees Fahrenheit.

1 Open a new project and write the following `FormClick()` procedure:

```
Private Sub Form_Click()
    'Program to convert Fahrenheit to Celsius
    'and Celsius to Fahrenheit

    Dim Celsius As Single
    Dim Fahrenheit As Single
    Fahrenheit = 32
    Celsius = (Fahrenheit - 32) * (5 / 9)
    Print Fahrenheit; "degrees Fahrenheit equals "; Celsius; "degrees
    ➥Celsius"
    Print

    Celsius = 0
    Fahrenheit = (9/5 * Celsius) + 32
    Print Celsius; "degrees Celsius equals "; Fahrenheit; "degrees
    ➥Fahrenheit"
End Sub
```

2 Run and test your program. As one test, try running this program without parentheses.

Exercise 7.6 The Money Exchange

You are asked to create an application titled Money Exchange. The application must be able to convert dollars into Deutschemarks. The relative value of currency continuously fluctuates, so for this exercise, one dollar is equal to 1.667 Deutschemarks, and one Deutschemark is equal to 0.6 dollars.

Figure 7.1 shows the display you are asked to create. During runtime, however, the display does not work as shown; rather, when the user enters a dollar amount, the `Calculate Dollars` command button disappears, leaving only the `Calculate Deutschemarks` button visible. Likewise, when the user enters a Deutschemark amount, the `Calculate Deutschemarks` command button disappears, leaving only the `Calculate Dollars` button visible.

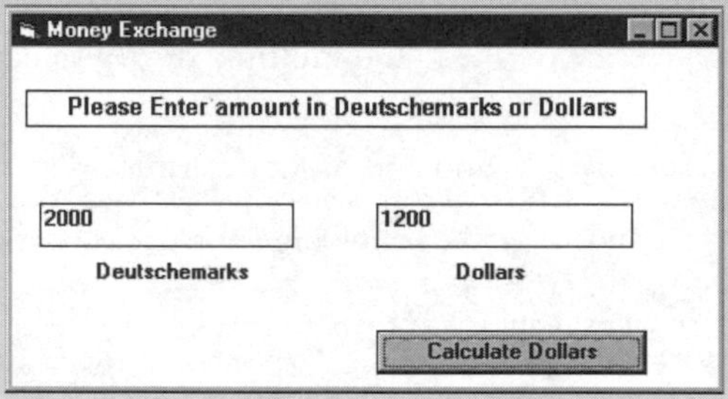

Figure 7.1

The Money Exchange display can be used to convert currency.

1 Design the form as shown in Figure 7.1. Use two text boxes and two command buttons. Add the three labels: **Please Enter Amount in Deutschemarks or Dollars**, **Deutschemarks**, and **Dollars**. Add the form caption **Money Exchange**. Add the captions **Calculate Dollars** and **Calculate Deutschemarks** to the two command buttons (one is invisible and not shown).

2 Name the `Deutschemarks` text box **`txtMarks`**; name the `Dollars` text box **`txtDollars`**.

3 Name the `Deutschemarks` command button **`cmdMarks`**; name the `Dollars` command button **`cmdDollar`**.

4 Set both buttons to invisible and disabled.

5 Write the `cmdDollar_Click()` procedure as follows:

```
Private Sub cmdDollar_Click()
   Dim marks As Currency
   Dim dollars As Currency
   marks = txtMarks.Text
   dollars = marks * .6
   txtDollars.Text = dollars
End Sub
```

6 Write the `cmdMarks_Click()` procedure as follows:

```
Private Sub cmdMarks_Click()
   Dim marks As Currency
   Dim dollars As Currency
   dollars = txtDollars.Text
   marks = dollars * 1.666667
   txtMarks.Text = marks
End Sub
```

7 Write a `txtDollars_KeyDown()` procedure. When text is entered in the `Dollars` text box, remove all text from the `Deutschemarks` text box. In addition, set `Calculate Deutschemarks` visible and enabled; set `Calculate Dollars` invisible.

```
Private Sub txtDollars_KeyDown (KeyCode As Integer, Shift As Integer)
   txtMarks.Text = ""
   cmdMarks.Visible = True
   cmdMarks.Enabled = True
   cmdDollar.Visible = False
   cmdDollar.Enabled = False
End Sub
```

Observe what the assignments in this procedure do. The first assignment clears the `txtMarks` text box. The next four assignments make the `cmdMarks` button visible and enabled and the `cmdDollar` button invisible and disabled.

8 Write a `txtMarks_KeyDown()` procedure. When text is entered into the `Deutschemarks` text box, make the `Calculate Dollars` command button visible and enabled.

```
Private Sub txtMarks_KeyDown (KeyCode As Integer, Shift As Integer)
   txtDollars.Text = ""
   cmdDollar.Visible = True
   cmdDollar.Enabled = True
   cmdMarks.Visible = False
   cmdMarks.Enabled = False
End Sub
```

This procedure is the reverse of the previous one. The assignments clear text from `txtDollars`, make the dollar button visible and enabled, and make the marks button invisible and disabled.

9 Run and test your revised program.

Objective 7.4 Converting Strings to Numbers and Numbers to Strings

Exercise 7.6 might have revealed a problem between numbers and strings. When entering a number in a text box, it must be entered as a string (because it is text). If you enter an alpha character (a character from the alphabet), Visual Basic generates a type mismatch error. Similarly, when displaying a number in a text box, or a label, the number must be displayed as a string. Assigning a number directly to a `Label` `Caption` property or a `TextBox` `Text` property generally doesn't generate an error because Visual Basic will convert it. It is good programming practice, however, to explicitly assign strings to object properties expecting them.

How can you convert one data type, such as a string, to another data type, such as a number? Visual Basic contains two broad functions to ensure that a string is represented as a number and a number represented as a string: the `Val()` and `Str()` functions. Visual Basic also contains numerous other functions for converting data types, which are covered in Chapter 18, "Numeric Functions," and Chapter 19, "String Functions." For now, you will use only the `Val()` and `Str()` functions.

7.4.1 The *Val()* Function

The `Val()` function is used to convert a string to a number. The syntax for this function is

```
Val(stringexpression)
```

The following is an example:

```
Val("600")
```

The `Val()` function evaluates string expressions from left to right, and it stops reading the string once it encounters an alpha character.

7.4.2 The *Str()* Function

The `Str()` function converts a number to a string. The syntax is

```
Str(number)
```

The following is an example:

```
Str(600)
```

When a number is converted to a string, a leading space is always reserved for its sign.

7.4.3 Revising the Money Exchange Example

The `Val()` and `Str()` functions should be used in the previous money exchange exercise. The revised `cmdDollar_Click()` procedure would be written as follows:

```
Private Sub cmdDollar_Click()
   Dim marks As Currency
   Dim dollars As Currency

   marks = Val(txtMarks.Text)
   dollars = marks * .6
   txtDollars.Text = Str(dollars)
End Sub
```

In this revised procedure, the marks entered into the program as text are converted to a number before they are assigned to the variable `marks`. After dollars are calculated, they must be displayed. In the revised procedure, the dollar amount is converted to a string before it is displayed in the text box. The `cmdMarks_Click()` procedure should be changed in a similar fashion.

Exercise 7.7 Using the *Val()* and *Str()* Functions

This exercise asks you to design a program to enter a value into a text box for either Fahrenheit or Celsius (see Figure 7.2). If Celsius is selected, the temperature is converted from degrees Fahrenheit to degrees Celsius. If Fahrenheit is selected, the temperature is converted from degrees Celsius to degrees Fahrenheit. Use a text box, `txtTemp`, to enter the temperature and a label, `lblTemp`, to display the output.

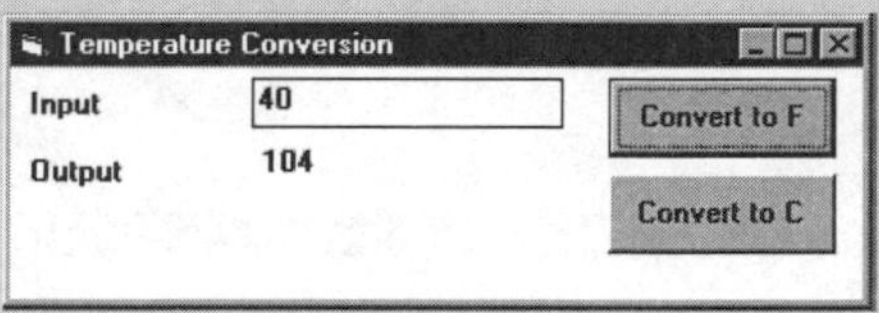

Figure 7.2

The Temperature Conversion application enables you to enter a temperature and convert it to degrees Fahrenheit or degrees Celsius.

1 Add the single text box, the three labels, and the two command buttons. Name the buttons **`cmdFahren`** and **`cmdCelsius`**. Name the form **Temperature Conversion**.

2 Write a `cmdFahren_Click()` procedure to convert degrees to Fahrenheit. Use the `Val()` and the `Str()` functions. The procedure is written as follows:

```
Private Sub cmdFahren_Click()
   'Convert to degrees Fahrenheit
   Dim DegreesF As Integer
   Dim DegreesC As Integer

   DegreesC = Val(txtTemp.Text)
   DegreesF = (DegreesC * 9/5) + 32
   lblTemp.Caption = Str(DegreesF)
End Sub
```

3 Write a `cmdCelsius_Click()` procedure to convert degrees to Celsius. Use the `Val()` and `Str()` functions. Write the procedure as follows:

```
Private Sub cmdCelsius_Click()
   'Convert to degrees Celsius
   Dim DegreesF As Integer
   Dim DegreesC As Integer
```

continues

```
    DegreesF = Val(txtTemp.Text)
    DegreesC = (DegreesF  - 32 ) *  5/9
    lblTemp.Caption = Str(DegreesC)
End Sub
```

4 Run and test the program. In testing, enter a letter instead of a number for the temperature.

Exercise 7.8 Displaying Your Taxes

This exercise also asks you to use the `Val()` and `Str()` functions. You are asked to enter your income and display the tax you owe. Figure 7.3 shows the interface you are asked to create.

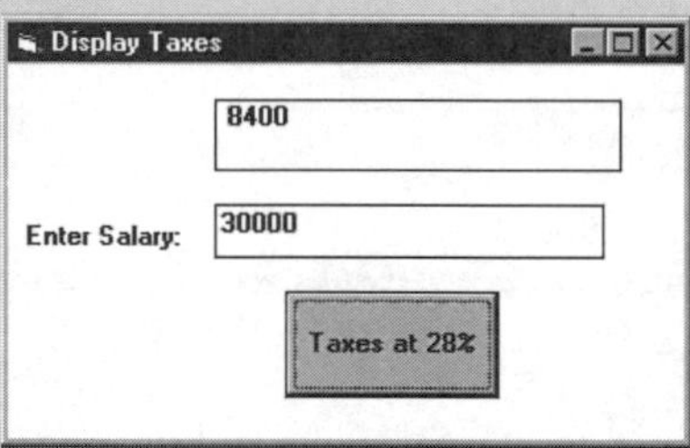

Figure 7.3

The Display Taxes form shows the amount of tax you will pay for a given income.

1 Add a text box (**`txtSalary`**), a label (**`lblTax`**), and a command button (**`cmdTaxes`**). Add the form caption **Display Taxes**. Add the caption **Taxes at 28%** to the command button.

2 Write the `cmdTaxes_Click()` procedure as follows:

```
Private Sub cmdTaxes_Click()
    Dim salary As Currency
    Dim tax As Currency
    tax = .28
    salary = Val(txtSalary.Text)
    lblTax.Caption = Str(tax * salary)
End Sub
```

3 Run and test your program. Enter a null value and click the `Taxes at 28%` button. Add `0` and click the button. With the `Val()` function, `lblTax` will display a zero when a null value or zero value is entered.

Chapter Summary

In Visual Basic, there are eight arithmetic operators when the negation operator is counted. These operators and their precedence are as follows:

Order	*Operator*	*Symbol*
1	Exponentiation	^
2	Negation	-
3	Multiplication and division	*, /
4	Integer division	\
5	Modulus	`Mod`
6	Addition and subtraction	+, -

Remember that an operator is the symbol used to designate an action to be performed on two operands (to form a binary linking), or on a single operand in the case of the negation operator. Also remember that integer division and the modulus operator can be used only on data of type `Integer` and type `Long` (long integer).

The order of precedence shown in the preceding table can be overruled through the use of parentheses, including nested parentheses.

An arithmetic expression is two or more operands connected by an arithmetic operator. This expression can be added directly to a `Print` statement, as follows:

```
Print (5 + (4 * 15)) /3
```

Or, it can be assigned to a variable:

```
y = (5 + (4 * 15)) /3
```

The unary operator requires a single operand, while the assignment operator is a binary operator. The assignment operator copies the value of the right operand to the left operand. The left operand can be a counter variable, such as the following:

```
count = count + 1
```

It also can be the property of an object.

Visual Basic contains two functions to ensure that a string and a number are represented properly. The `Val()` function converts a string to a number; the `Str()` function converts a number to a string. Because numbers are often entered in a program by using a text box (which enters a string), the string should be converted to a number before it is used in a program.

Skill-Building Exercises

1. The speed of light is 186,000 miles per second. Use four `Print` statements to display the following on a form titled LightSpeed:

```
The speed of light in miles per second is <enter computed value>
The speed of light in feet per second is <enter computed value>
The speed of light in miles per minute is <enter computed value>
The speed of light in miles per second squared is <enter computed value>
```

Write a `Form_Click()` procedure, and declare `light` as a variable of type `Long`.

2. Design a program to determine a person's total pay. You know the following:

- The rate of pay is $6 per hour.
- A person works 50 hours, and is paid 1.5 times the rate for hours greater than 40.

Design a `Click()` event to display the person's regular pay, overtime pay, and total pay.

3. Place three circles on a display, as follows:

- Radius 250, color red
- Radius 500, color blue
- Radius 750, color yellow

Control the radius using code. Add three labels to display the area of each circle. The formula is pi times the square of the radius. Pi is equal to 3.14159.

4. You receive a commission on sales according to the following formula:

- First $1,000: seven percent
- From $1,001 to $10,000: four percent
- From $10,001 to $50,000: three percent
- Over $50,000: two percent

Compute your commission for a sale of $150,000. Use five labels to show the sale amount, the commission earned at each level, and the total commission earned.

5. Design an application that converts meters into miles and miles into meters. From road races, you know that 10,000 meters is equal to 6.2 miles. Build your application to make it foolproof. When the user enters meters, set the text for miles equal to zero, the command button for miles visible, and the command button for meters invisible. Do the opposite if miles are entered by the user.

If...Then...Else Logic and the *Select Case* Statement

One of the most common logical structures in any computer language is the `If...Then...Else` structure. This structure allows a condition to be tested, with action taken based upon the evaluation of that condition. With simple `If...Then` logic, action is taken when a test condition evaluates to `True`. With `If...Then...Else` logic, one action is taken if the test condition is `True`; another is taken if the test condition is `False`.

Like the `If...Then` logical construct, the `Select Case` structure evaluates a test expression to determine if it is satisfied. However, `Select Case` functions in the same way as a set of `If` statements. Using `Select Case` is advantageous when you create a set of multiple-choice options. If the `Case` option matches the `Case` test expression, code is executed for that option.

In this chapter, you work with many `If...Then`, `If...Then...Else`, and nested `If...Then...ElseIf...Else` structures, as well as the `Select Case` structure. When you complete this chapter, you will be able to

- use relational operators,
- write an `If...Then...Else` logical construct,
- use logical operators,
- use the `ElseIf` statement,
- use the complete operator precedence,
- use the `Select Case` structure,
- write different types of `Select Case` test expressions.

Chapter Objectives

The objectives in this chapter are considerable. Besides working with `If...Then...Else` logic, you are expected to use both relational and logical operators. The objectives are as follows:

1. Understanding the relational operators
2. Knowing the different forms of the `If...Then...Else` syntax
3. Understanding the logical operators
4. Using nested `If...Then...Else` logic
5. Using `If...Then` logic to manipulate objects
6. Understanding and using the `Select Case` syntax

Objective 8.1 Understanding the Relational Operators

Before examining `If...Then...Else` structures, consider two important concepts: relational operators and the logic that follows from the use of this type of operator.

8.1.1 Relational Operators

The *relational operators* and their meanings are shown in Table 8.1.

Table 8.1 The Relational Operators

Operator	*Description*
=	Equality
>	Greater than
<	Less than
<>	Not equal to (no equality)
>=	Greater than or equal to
<=	Less than or equal to

Relational operators differ from arithmetic operators by returning a value for `True` (–1) or `False` (0). They are sometimes referred to as comparison operators because they are used to compare one value with another. For example, the following comparisons are all true:

```
5 < 6
7 = 7
1 <= 3
2 <> 4
```

Likewise, the following comparisons are false:

```
6 > 8
7 = 12
5 <= 2
2 <> 2
```

The syntax of these relational operators should be familiar. The only one that differs from standard mathematical notation is *not equal to,* shown as <>.

8.1.2 Relational Operators and Logic

Relational comparisons are used to express logic. For example, in English, you might state:

```
If (Is it raining?) Then
   "Open your umbrella"
Else
   "Carry your umbrella"
End If
```

This *structured English* (as it is called because it reads like a special form of English) begins with a question: Is it raining? If it is (the answer is *yes* or *true*), then do the first action specified: Open your umbrella. If it is not (the answer is *no* or *false*), then do the alternative action specified: Carry your umbrella. It is this form of English that is carried over to the Visual Basic syntax.

Objective 8.2 Knowing the Different Forms of the *If...Then...Else* Syntax

The `If...Then...Else` procedure always features a relational comparison as part of its test condition. However, the structure has three forms:

- **Form 1**—The first form can be placed on one line.

  ```
  If (test condition) Then (test if condition is True) [Else (test if condition is False)]
  ```

- **Form 2**—This second form is indented such that `If...Else...End If` are aligned. Statement blocks are always indented to make the structure easier to read.

  ```
  If test condition Then
    [statement block 1]
  [Else]
    [statement block-2]
  End If
  ```

- **Form 3**—The third form features a nested `If...ElseIf...Else...End If` alignment. Statement blocks are indented, as in the second form.

  ```
  If test condition 1 Then
     [statement block 1]
  [ElseIf condition 2 Then
     [statement block 2]]
  [Else]
     [statement block-n]
  End If
  ```

8.2.1 *If...Then* Syntax

The first form is the clearest. It avoids the use of the keywords `ElseIf` (one word) and `End If` (two words). As an example, consider the following code:

```
If (age >= 7) Then Print "Go to school"
```

With this form, the test condition is based on the value of the variable age. It is enclosed in parentheses, even though these are not required, to aid in reading. The code reads as follows: If a child's age is seven or more, the `Then` portion of the instruction is invoked, and the message `Go to school` is displayed. After the display, the next instruction in the program will be executed. If the child's age is less than seven, no message will appear. Instead, the program flow of control will immediately move to the next instruction in the program.

When using this form, the entire statement must be placed on one line. The following code would lead to an error:

```
If (age >= 7) Then
   Print "Go to school"
```

Fortunately, if you make this error, it will be easy to identify. The Visual Basic editor will inform you that it expects the `End If` keywords.

8.2.2 The *If...Then...Else* Format

The `If...Then...Else` syntax is more common to Visual Basic programs. With this form, the action that follows a `True` condition can be placed on a separate line, provided the `End If` statement is used. For example, the following code would not lead to an error:

```
If (age >= 7) Then
   Print "Go to school"
End If
```

Remember the following two rules when writing this code:

- The keyword `Then` must be on the same line as the keyword `If`.
- The `End If` statement is required.

With this syntax, it is possible to place more than one instruction between `If` and `End If`. Placing several instructions together forms what is called a *block of statements.* With a block, the `If` statement marks the beginning of the block, and the `End If` statement marks the end of the block. Suppose you had a computer-based, multiple-choice test, and you had declared the variable `Answer` for the requested user input and `Correct` as the variable to track total correct answers. Part of the procedure to track scoring might look like the following:

```
If Val(Answer) = 12 Then
   lblFeedBack.Caption = "Problem 15 is correct"
   Correct  = Correct + 1      ' increment the correct answer total
End If
```

In this example, if the value stored in the variable `Answer` is equal to 12 (the test is `True`), the block of statements placed between the `If` and `End If` keywords is executed.

The keyword `Else` is optional with this form; however, this keyword is required when an alternative response is needed. Consider this code:

```
If Val(Answer) = 12 Then
   lblFeedBack.Caption  = "Problem 15 is correct"
   Correct  = Correct + 1     ' increment the correct answer total
Else
   lblFeedBack.Caption = "Problem 15 is incorrect"
   Incorrect = Incorrect + 1    ' increment the incorrect answer total
End If
```

This code is read as follows:

> When the answer is not 12 (the test is `False`), skip the statements that follow `If`, and jump to the `Else` portion of the structure. Execute the instructions placed under the `Else` until reaching `End If`.

Program style is important when writing `If...Then...Else` instructions. The `Then` keyword must be placed to the right of the test condition. In addition, the block of instructions that follow should be indented to aid in their reading. The keywords `If`, `Else`, and `End If` are left justified to improve the readability and maintenance of the code.

Exercise 8.1 Using *If* Statements to Check the Value of a Number

This exercise asks you to enter a number from 1 to 10 and to use `If...Then` statements to determine whether the number is within this range. Figure 8.1 shows the interface you should create.

Figure 8.1

In this example, `If...Then` logic is used to determine whether a number falls within the range of 1 to 10.

1 Add two labels, one text box, and the **`cmdEnd`** command button. Change the caption of the form and write the `End` procedure.

2 Use a label to show the results. Name the label **`lblResult`**.

3 Use the text box to accept user input. Name the text box **`txtAnswer`**.

continues

4 Use a `LostFocus()` procedure for the text box. With a `LostFocus()` event, user input is evaluated after it is entered into the text box and focus has shifted to another control. The code is written as follows:

```
Private Sub txtAnswer_LostFocus()
   Dim Amt As Integer
   Amt = Val(txtAnswer.Text)
   If (Amt > 0) Then
      lblResult.Caption = "Number is OK"
   End If
   If (Amt > 10) Then
      lblResult.Caption = "Number is too high"
   End If
   If (Amt <= 0) Then
      lblResult.Caption = "Number is too low"
   End If
End Sub
```

Observe how indentation improves the readability of this code. Because there are no `Else` statements, each `If` statement could have been placed on a single line.

5 Run and test your code. Test values of –1, 0, 1, 9, 10, and 11.

8.2.3 The *If...Then...ElseIf...Else* Format

The third form of the `If...Then...Else` structure features what are known as *nested* or *multiple* `If...Then` tests. It reads as follows:

```
If (this is True) Then
   Do action 1
ElseIf (this is True) Then
   Do action 2
Else
   Do action 3
End If
```

With this form, it is possible to conduct more than one test to find a `True` condition. In this example, three actions are possible. But multiple `If...Then` tests are not limited to three. Multiple `ElseIf` conditional tests can follow an initial `If` test. This example could easily contain five conditional tests, which is called a nested `If...Then` structure. You will return to it after you are more familiar with the first two forms.

Objective 8.3 Understanding the Logical Operators

The logical operators and their meanings are shown in Table 8.2.

Table 8.2 The Logical Operators

Operator	*Description*
`And`	Conjunction
`Or`	Disjunction (inclusive `Or`)
`Xor`	Exclusive `Or`
`Not`	Complement
`Eqv`	Equivalence
`Imp`	Implication

These operators also return a `True` (–1) or `False` (0) value. They are sometimes referred to as *compound operators* because they are used to compare the results of two test conditions. Because these operators are more difficult to understand than the relational operators, consider each one carefully.

8.3.1 The *And* Logical Operator

Perhaps it would be best to start with an example. Consider the following code:

```
If (Money > 5000) And (Bills <= 5000) Then
   Print "I can actually pay off some of my debts"
End If
```

In this instance, both test conditions must be `True`: `Money` must be greater than `5000` and `Bills` less than or equal to `5000` for the condition to be `True`. When it is `True`, the message `I can actually pay off some of my debts` is displayed. If `Money` was greater than `5000`, but `Bills` was `5100`, the message would *not* be displayed. Likewise, if `Money` was less than `5000` and `Bills` was less than `5000`, the message would *not* be displayed.

The rule for the `And` logical operator is as follows: *Both test condition1 and test condition2 must be* `True` *for the expression as a whole to be* `True`.

8.3.2 The *Or* Logical Operator

The rule for the `Or` logical operator is as follows: *At least one test condition must be* `True` *for the expression as a whole to be* `True`.

The structure is as follows:

```
If (Money > 5000) Or (Bills <= 5000) Then
   Print "I can actually pay off some of my debts"
End If
```

With this code, if `Money` was greater than `5000` or `Bills` was `5000` or less, the message `I can actually pay off some of my debts` would be displayed. Note that this example would be troublesome in real life. If `Bills` was less than `5000`, but `Money` was `1000`, you could still pay off some of your debts.

Consider a more realistic example by returning to the test-answer code. Suppose that you asked the user for a number that could be evenly divided by four, and the possible numbers were in the range from 11 to 21. The correct answer could be 12, 16, or 20. To test for a correct answer, you would need to check these three possibilities. The `If...Then` statement then might look like the following:

```
If Text1.Text  = 12 Or Text1.text = 16 Or Text1.Text = 20 Then
   Answer = "The answer is correct"
   Correct = Correct + 1
Else
   Answer = "Sorry, the answer is incorrect"
   Incorrect = Incorrect + 1
End If
```

As indicated, the keyword `If` is used only once. If `Text1.Text = 12 Or If Text1.Text = 16, Or If Text1.Text = 20`, the statement as a whole would be correct. And, if the numbers are not 12, 16, or 20, the answer is incorrect.

8.3.3 The *Xor* Logical Operator

The rule for the `Xor` logical operator is as follows: *One test condition, and only one test condition, must be* `True` *for the expression as a whole to be* `True`.

Consider the same structure:

```
If (Money > 5000) Xor (Bills <= 5000) Then
   Print "I can actually pay off some of my debts"
End If
```

If `Money` was greater than `5000` and `Bills` was less than `5000`, the expression would be `False`. Why? Because more than one expression would be `True`. This is why the `Xor` operator is *exclusive* while the `Or` operator is *inclusive*. With `Or`, both sides can be `True` for the expression to be `True`. With `Xor`, only one side can be `True`—not both.

8.3.4 The *Not* Logical Operator

The `Not` operator can be confusing and is often discouraged when writing Visual Basic code. The rule for the `Not` operator is as follows: *If the expression is* `True`, *the result is* `False`. *If the expression is* `False`, *the result is* `True`.

The code is as follows:

```
If (Not (Money <= 5000)) And  (Bills <= 5000) Then
   Print "I can actually pay off some of my debts"
End If
```

This code reads as follows:

> If `Money` is not less than or equal to `5000` and `Bills` is less than or equal to `5000`, then print the message. Thus, if `Money` was greater than `5000` (not less than or equal to `5000`) and `Bills` was less than `5000`, the expression would evaluate to `True`.

8.3.5 The *Eqv* Logical Operator

The `Eqv` (equivalence) logical operator uses the following syntax:

```
result = expression1 Eqv expression2
```

The results table from this operator is as follows:

If Expression 1 Is	*And Expression 2 Is*	*The Result Is*
`True`	`True`	`True`
`True`	`False`	`False`
`False`	`True`	`False`
`False`	`False`	`True`

The rule for the `Eqv` operator is as follows: *Both test conditions must be* `True` *or both test conditions must be* `False` *for the expression as a whole to be* `True`.

This operator is not commonly used in Visual Basic; however, you need to know that it exists.

8.3.6 The *Imp* Logical Operator

The `Imp` (implication) operator is the most difficult of all. Fortunately, it is also little used. The syntax for this operator is as follows:

```
result = expression 1 Imp expression 2
```

The results table from this operator is as follows:

If Expression 1 Is	*And Expression 2 Is*	*The Result Is*
`True`	`True`	`True`
`True`	`False`	`False`
`True`	`Null`	`Null`
`False`	`True`	`True`
`False`	`False`	`True`
`False`	`Null`	`True`
`Null`	`True`	`True`
`Null`	`False`	`Null`
`Null`	`Null`	`Null`

In addition to `True` plus `True` is `True`, and `False` plus `False` is `True`, three other `True` results occur: when the first test is `False` and the second is `True` or `Null`, or when the first test is `Null` and the second is `True`.

8.3.7 The Expanded Operator Precedence

The addition of relational and logical operators expands the table of operator precedence. As shown in Table 8.3, there are 14 precedence levels in Visual Basic. Relational operators have a lower precedence than arithmetic operators; logical operators have a lower precedence than relational operators.

Table 8.3 The Order of Precedence of Operators

Level	*Operator*	*Name*
1	^	Exponentiation
2	-	Negation
3	*, /	Multiplication and division
4	\	Integer division
5	Mod	Modulo arithmetic
6	+, -	Addition and subtraction
7	&	String concatenation
8	+, >, <, <>, >=, <=	Relational operators
9	Not	Logical negation
10	And	Conjunctive
11	Or	Inclusive Or
12	Xor	Exclusive Or
13	Eqv	Equivalence
14	Imp	Implication

Precedence explains how such an expression as the following works:

```
If Money > 5000 And Bills <= 5000 Then
```

Before the logical `And` operator is used, the two relational operators are used to test the stated conditions. This sequence occurs because the relational operators have a higher precedence than the `And` operator. If the reverse were true, the expression would have to be rewritten with parentheses to override precedence, as follows:

```
If (Money > 5000) And (Bills <= 5000) Then
```

Objective 8.4 Using Nested *If...Then...Else* Logic

Visual Basic extends the usefulness of `If...Then...Else` logic by including the capability to test for multiple conditions with the `ElseIf` keyword. This enables you to write nested logic. Review the syntax of the third form:

```
If test condition1 Then
   [statement block 1]
[ElseIf condition 2 Then
   [statement block 2]]
[Else]
   [statement block-n]
End If
```

With this logic, more than one condition (`condition 1`, `2`, and `3`) can be tested to determine whether the result is `True` or `False`.

Suppose that you write a program that asks the user to quickly estimate in whole numbers (integers) the total of a seven-item grocery list. Further, suppose that you want to give immediate feedback on an estimate. You might have the person enter the amount in a text box and have a procedure compare the answer to different values and provide feedback accordingly. If the grocery list totaled $23.78, you might write the code for the event as follows:

```
If (Text1.Text = 23) Or (Text1.Text = 24) Then
   MsgBox "You're almost on the money; it's 23.78. Good estimate!"
ElseIf (Text1.Text =  21) Or (Text1.Text = 22) Then
   MsgBox "You're just a little low"
ElseIf (Text1.Text = 25) Or (Text1.Text = 26) Then
   MsgBox "You're just a little high"
Else
   MsgBox "Try again"
End If
```

Are you beginning to understand how nested logic works? If the first test is `False` (the answer is neither 23 nor 24), try the second test. If the second test is `False` (the answer is neither 21 nor 22), try the third test. If the third test is `False` (the answer is neither 25 nor 26), then move to the `Else` portion of the structure. This portion provides the default message when all tests are `False`.

Logical operators can be placed in condition statements to test for ranges of values, but here you must be careful because `Or` will not work the way you might expect. Use `And` instead. Visual Basic tests each condition from left to right, so ranges should be listed in either ascending or descending order, as in the following example:

```
If (Text1.Text >= 21)  And (Text1.Text < 23) Then
   MsgBox "You're just a little low"
ElseIf  (Text1.Text  >= 23)  And (Text1.Text <= 24) Then
   MsgBox "You're almost on the money; it's 23.78. Good estimate!"
ElseIf (Text1.Text > 24) And (Text1.Text <= 26) Then
   MsgBox "You're just a little high"
Else
   MsgBox "Try again"   'handles all values under 21 and over 26
End If
```

Do you understand why `Or` would not work in this code? If the user entered 12, the first condition would be `True` because the 12 is less than 23, and only one condition needs to be `True` for the test condition as a whole to be `True`.

Exercise 8.2 Using *If...Then...ElseIf* to Check the Value of a Number

This exercise uses the same interface as Exercise 8.1. Use `If...Then...ElseIf` to check the value of a number between 1 and 10.

1 Rewrite the `Lost_Focus()` procedure to read as follows:

```
Private Sub txtAnswer_LostFocus()
   Dim Amt As Integer
   Amt = Val(txtAnswer.Text)
   If (Amt > 10) Then
      lblResult.Caption = "Number is too high"
   ElseIf (Amt <= 0) Then
      lblResult.Caption = "Number is too low"
   Else
      lblResult.Caption = "Number is OK"
   End If
End Sub
```

2 Run and test this code. Use –1, 0, 1, 9, 10, and 11 as test values.

Objective 8.5 Using *If...Then* Logic to Manipulate Objects

The previous sections have been limited to programmed instructions, with little emphasis given to objects. In this section, which begins with changing the properties of command buttons, consider how `If...Then` logic is used with objects.

8.5.1 Using *If...Then* to Change the Properties of Command Buttons

Imagine a button that constantly shifts between two messages: One says `push me`, and the other says `push me again`. How might you design a form so that when the button is pushed, the second message appears requesting to be pushed again? When the button is pushed again, the first message reappears requesting to be pushed, and so on, in endless succession. You can accomplish this by changing the property of an object (such as a button) based on the results of a conditional test.

Exercise 8.3 Push My Button

This exercise asks you to design a form to toggle between two button captions. Figure 8.2 shows this simple interface.

Figure 8.2

`If...Then` *logic is used to create a toggle button.*

A program requirement that toggles between two states (`True` or `False`), suggests that a `Boolean` data type be used. In this instance, you are switching between two messages that are displayed as button captions. You can use `If...Then` logic to test which state you are in and execute code based on the results of the test.

1 Draw a button (**`cmdPushMe`**), and in the Property box, change its caption in plain bold font to **push me**.

2 Now write a procedure for the button so that its caption toggles back and forth. The code for the command button is as follows:

```
Private Sub cmdPushMe_Click()
Static urgent As Boolean

If urgent = False Then
  cmdPushMe.Font.Italic = True
  cmdPushMe.Caption = "push me again!"
  urgent = True
Else
  cmdPushMe.Font.Italic = False
  cmdPushMe.Caption = "push me"
  urgent = False
End If
End Sub
```

Notice that you want to retain information about the state you are in, so declare the `Boolean` variable with the keyword `Static`. This method of declaring variables is discussed in chapter 10, "Defining the Scope of Variables and Procedures."

3 Test and run the program; then, continue with the next exercise.

Exercise 8.4 Buttons to Push, Again

The last design won't win any computer game contests, but dress it up as much as you can. Suppose that you have a message after every sixth push that states, `Thanks, that feels much better!` (see Figure 8.3). It's still pretty boring, but at least you get some gratitude for slavishly pushing buttons.

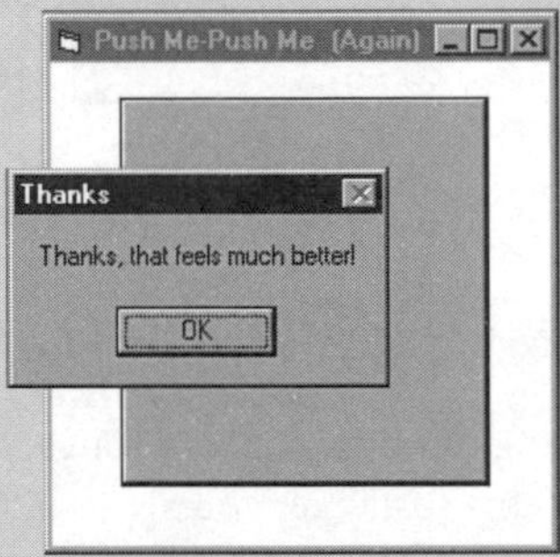

Figure 8.3

The button toggle exercise is extended to display a message using a message box.

1 Enter the following code into the `cmdPushMe_Click()` event:

```
Static urgent As Boolean
Static counter As Integer
Dim ans As Integer
Dim Msg As String

If urgent = False Then
   cmdPushMe.Font.Italic = True
   cmdPushMe.Caption = "push me again!"
   urgent = True
Else
   cmdPushMe.Font.Italic = False
   cmdPushMe.Caption = "push me"
   urgent = False
End If

Msg = "Thanks, that feels much better!"
counter = counter + 1
If counter = 6 Or counter = 12 Or counter = 18 Then
   ans = MsgBox(Msg, , "Thanks")
End If
```

Did you notice that you can have more than one `If...Then` structure in the same procedure? The message box function used in this exercise is discussed in Chapter 9, "Inputting Values and Formatting Output."

2 Run and test the program.

Exercise 8.5 Push Me: The Final Episode

Continue the last example. Suppose that you decide to be a bit obsessive—you really like clicking buttons. After 18 clicks, you decide that this has to stop. Here's a way to address the concerns of the "button obsessed." The procedure employs the `Mod` operator. The message appears after every six clicks for as long as the user keeps pressing the button. Now you can also use the `Not` operator to toggle the value of `Urgent` rather than directly assign the value in code.

1 At the `cmdPushMe_Click()` event, write the following:

```
Static urgent As Boolean
Static counter As Integer
Dim ans As Integer
Dim Msg As String

If urgent = False Then
   cmdPushMe.Font.Italic = True
   cmdPushMe.Caption = "push me again!"
Else
   cmdPushMe.Font.Italic = False
   cmdPushMe.Caption = "push me"
End If

Msg = "Thanks, that feels much better!"
counter = counter + 1
If counter Mod 6 = 0 Then
   ans = MsgBox(Msg, , "Thanks")
End If

urgent = Not urgent   ' this instruction toggles the value of urgent
```

2 Try it. What happened? A message is not displayed every time `cmdPushMe` is pushed.

Take a look at how the variable `Counter` is used. When the result of the test condition `counter Mod 6` is `0` (that is, after each sixth button click), control is passed to the message box to display the message. Also notice that because you are dealing with a `Boolean` data type, you can use the `Not` operator to toggle its value in the procedure's last instruction.

8.5.2 Using *If...Then* to Check Scroll Bar Values

`If...Then` structures can be used to change the properties of all types of controls. As this next exercise shows, `If...Then` can be used to check the position of a scroll bar and to display different messages depending on that position.

Exercise 8.6 Using *If...Then* to Control Messages for a Scroll Bar

Draw a three-inch horizontal scroll bar and a label of approximately the same width just above it (see Figure 8.4). Using `If...Then...Else` logic, have the label display the scroll bar position at the beginning, middle, or end. When the scroll button reaches the end, display a message box stating that you have reached it, and have the scroll button automatically reset itself to the beginning. At the starting position, have the label caption urge the user to move the scroll bar.

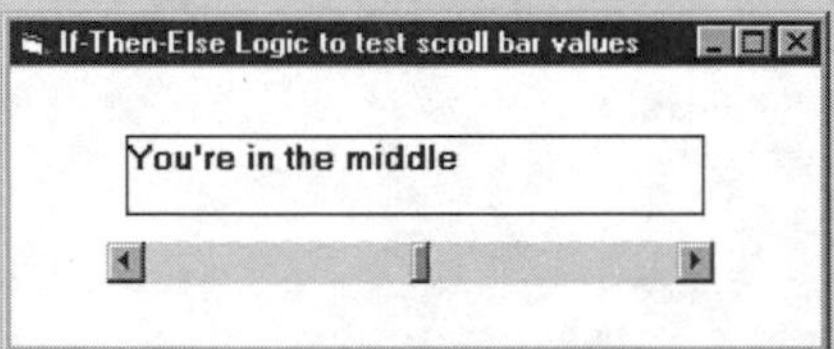

Figure 8.4

Use `If...ElseIf...Else...End If` statements to control messages displayed on a label.

1 Add the scroll bar and label, as shown in Figure 8.4. Set the `Min` value for the scroll bar to `0` and the `Max` value to `60`.

2 Write the following `HScroll1_Change()` procedure:

```
Private Sub HScroll1_Change()
   Dim Msg1 As String
   Msg1$ = "You've reached the end!"
   If Hscroll1.Value = 0 Then
      lblResult.Caption = "Move the Scroll Bar!"
   ElseIf Hscroll1.Value = 1 Then
      lblResult.Caption = "This is the beginning"
   ElseIf Hscroll1.Value = 30 Then
      lblResult.Caption = "You're in the middle"
   ElseIf Hscroll1.Value = 50 Then
      lblResult.Caption = "You're nearing the end"
   ElseIf Hscroll1.Value = 60 Then
      MsgBox Msg1$, , "You did it!"
      Hscroll1.Value = 0
   End If
End Sub
```

3 Write the same procedure for the `HScroll1_Scroll()` event. Remember that with a scroll bar, the `Scroll()` event is needed to control the scroll box.

4 Write the following `Hscroll1_GotFocus()` event:

```
Private Sub HScroll1_GotFocus()
   HScroll1.Value = 0
   lblResult.Caption = "Move the scroll bar!"
End Sub
```

> 5 Run and test your program. Why doesn't the `HScroll1_Scroll()` procedure work the same way as the `HScroll1_Change()` procedure?

Objective 8.6 Understanding and Using the *Select Case* Syntax

The `Select Case` structure derives its name from the selection of one case from a set of cases. This structure is advisable when there are several possible selections. It tends to be easier to read than a long series of `If...Then` expressions.

8.6.1 The *Select Case* Syntax

The syntax for the `Select Case` structure is less complex than it might initially appear. The words `Select Case` mark the beginning of the structure, and the words `End Select` mark the end. The complete syntax is as follows:

```
Select Case testexpression
   [Case expressionlist1
      [statementblock1]]
   [Case expressionlist2
      [statementblock2] ]
   Case Else
      [statementblock - n]]
End Select
```

Table 8.4 explains the components of the `Select Case` syntax:

Table 8.4 The Syntax of the *Select Case* Statement

Component	*Description*
`testexpression`	Any numeric or string expression
`expressionlist`	A comma delimited list of an expression, a range expression (using `To`), or a condition expression (using `Is`)

Consider the following example:

```
Name$ = "Ted"
Select Case Name$
   Case "Ted"
      Print "Hello Ted"
   Case "Jane"
      Print "Hello Jane"
   Case "B. J."
      Print "Hello B.J."
   Case Else
      Print "Hello!  Anyone home?"
End Select
```

In this example, the test expression must match one of the three literal string expressions: `"Ted"`, `"Jane"`, or `"B.J"`. If it does, then the appropriate message is printed. If no match is made, then the `Case Else` portion of the statement is executed, leading to the message, `"Hello! Anyone home?"` Because `"Ted"` matches the value assigned to the test expression `Name$`, the message `"Hello Ted"` is displayed.

As this example illustrates, the `Select Case` statement looks for a match between the `testexpression` and an `expressionlist`. When a match is found, the statement block of the `expressionlist` is executed. Following execution, control is passed to the statement immediately following `End Select`. If more than one of the `expressionlist` statements match the `testexpression`, then the statement block of the first `expressionlist` is executed, and all others are skipped.

`Case Else` indicates which statement block to execute if no match is found between the `testexpression` and the `expressionlist`. If there is no `Case Else`, and no `expressionlist` matches the `testexpression`, control passes to the statement following `End Select`. `Case Else` is advised when it is important to specify that no test expression match was made. For example, notice what happens if you rewrite the previous code to read as follows:

```
Name$ = "Roger"
Select Case Name$
   Case "Ted"  : Print "Hello Ted"
   Case "Jane" : Print "Hello Jane"
   Case "B. J." : Print "Hello B.J."
   Case Else : Print "Hello!  Anyone home?"
End Select
```

Because no match is made, the message `Hello! Anyone home?` is displayed. Notice that with the use of a colon, the statement block can be placed on the same line as the `expressionlist`.

8.6.2 Variations of *expressionlist* and *testexpression*

The `expressionlist` for a case or the `testexpression` itself can be written in several ways. This section discusses three possibilities.

Comma Delimited Multiple *expressionlist*

The `expressionlist` can contain comma delimited multiple expressions. For example, the following would be a valid `expressionlist`:

```
Case 0, 5, 8, 11, 13
```

With this test, a case is selected if the `testexpression` matches one of these values.

Range Delimited Expression

The `expressionlist` can also cover a range of values by using the keyword `To`, as in

```
Case 21 To 27
```

or by employing a comparison operator with the keyword `Is`, as in

```
Case Is >= 20
```

In this second example, the `Is` keyword used in `Case` statements is not the same as the `Is` comparison operator. In the preceding test expression, a `testexpression` value of `20` or greater would provide a match between the test and the expression.

Object Property *testexpression*

A `testexpression` can be the value of a control property, such as the value of a scroll bar. Consider the following code:

```
Select Case HScroll1.Value
```

The `testexpression` is the value of the scroll bar. It must search the various cases for a match of this value. In the next set of exercises, you draw heavily on this type of `testexpression`.

Many Visual Basic programmers prefer to use the `Select Case` statement in place of a complex `If...ElseIf...Else` statement because it is easier to read. It also tends to execute faster because multiple conditional tests need not be made. Exercise 8.7 demonstrates this concept. Compare this code to that written for Exercise 8.6.

Exercise 8.7 Moving the Scroll Bar

As in Exercise 8.6, place a horizontal scroll bar on a form in order to display different messages that depend on the position of the scroll box. Use Figure 8.4 once again for the visual layout.

1 Draw the interface shown in Figure 8.4, adding a scroll bar and a label. Set the scroll bar's `Min` value to 0 and the `Max` value to 45.

2 Write the `Hscroll1_Change()` and the `Hscroll1_Scroll()` event procedures as follows:

```
Private Sub HScroll1_Change()
   Dim ans As Integer
   Select Case HScroll1.Value
      Case 0
         Label1.Caption = "Move the scroll bar!"
      Case 1 To 10
         Label1.Caption = "You're near the beginning"
      Case 11 To 35
         Label1.Caption = "You're in the middle"
      Case 36 To 44
         Label1.Caption = "You're near the end"
      Case 45
         ans = MsgBox("You're at the end!", 0, "Time to Quit")
         HScroll1.Value = 0  ' reset the scroll bar value
   End Select
End Sub
```

continues

Notice that the test expression `HScroll1.Value` is matched by the values of the expression lists in the `Case` statements. If you excluded the `Case` statement for the value `45`, you could have entered `Case Is >= 35`. The keyword `Is` is demonstrated in Exercise 8.8.

3 Add the following code to display an opening message:

```
Sub Form_Load()
   Call HScroll1_Change ' displays opening message
   Call Hscroll1_Scroll
End Sub
```

Exercise 8.8 Extracting Salary Requirements and Giving the User Feedback

This exercise asks you to display a message box message following the entry of the salary a person thinks he or she deserves. Figure 8.5 shows this simple interface.

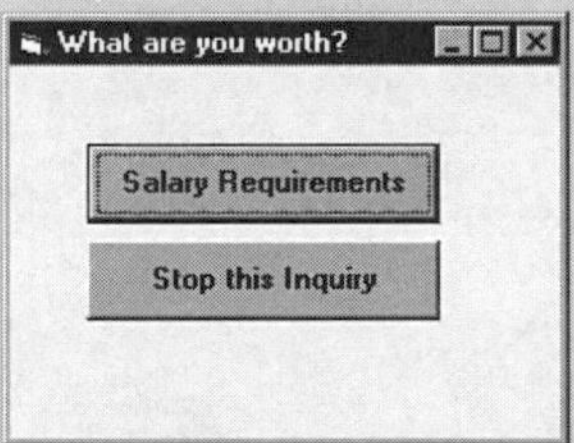

Figure 8.5

This design features a `Select Case` *structure to determine various salary requirements.*

1 Create the interface shown in Figure 8.5, adding the two command buttons **`cmdSalary`** and **`cmdEnd`**, their captions, and the form caption. Write the `cmdEnd_Click()` procedure.

2 In the `cmdSalary_Click()` event, enter the following code:

```
Private Sub cmdSalary_Click()
   Dim ans As Integer
   Dim Msg As String
   Msg$ = "Please enter your salary requirements in round dollar amounts
   ➥(no commas)"

   Select Case Val(InputBox(Msg$))
      Case Is < 10000
         ans = MsgBox("This is probably not the right position for you.",
         ➥0, "Sorry")
```

```
        Case 10000 To 15000
           ans = MsgBox("You're underestimating your value. We need positive
           ➥thinkers here!", 0, "Sorry")
        Case 15001 To 25000
           ans = MsgBox("You're an attractive candidate. Can we talk?", 0,
           ➥"We're Interested")
        Case 25001 To 40000
           ans = MsgBox("You're within range. Let's negotiate.", 0, "We're
           ➥Interested")
        Case 40001 To 50000
           ans = MsgBox("You obviously think well of yourself.  Let's talk
           ➥to see if you're worth it.", 0, "Perhaps")
        Case Is > 50000
           ans = MsgBox("We are not sure this position offers you enough
           ➥opportunity.", 0, "Sorry")
    End Select
End Sub
```

It is often difficult to determine just what your test expression should be. In this example, you take the numeric value of a string. With numeric values, you can cover all possible responses with numeric ranges. The test expression need not be complex. The simplest and perhaps most common `testexpression` tests for the index value of an object placed in a control array.

Exercise 8.9 Identifying World Capitals

This exercise uses the `List Index` property of a list box in combination with the `Select Case` statement (see Figure 8.6).

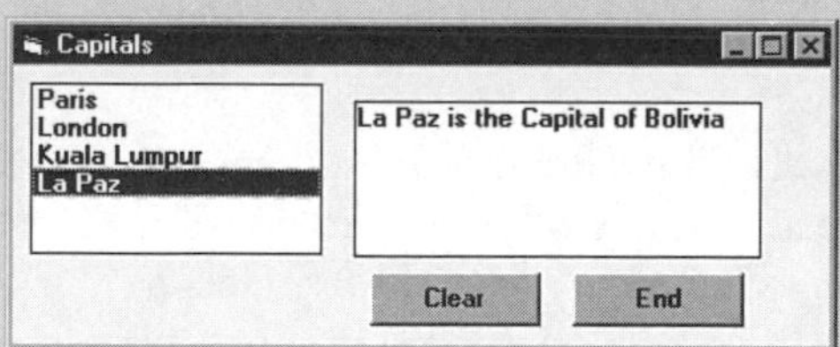

Figure 8.6

Click a capital city to identify its country.

1 Create the form shown in Figure 8.6. Add the necessary code to the `Form_Load()` event so that the capital cities appear when the program is run. Name the list **`lstCapital`**. A label named **`lblCountry`** occupies the space above the Clear and End buttons. Write the `cmdEnd_Click()` procedure.

continues

2 Set the `MultiSelect` property of `lstCapital` to `0` (None).

3 In the `lstCapital_Click()` event, enter the following code:

```
Private Sub lstCapital_Click()
   Select Case lstCapital.ListIndex
      Case 0
         lblCountry.Caption = "Paris is the capital of France"
      Case 1
         lblCountry.Caption = "London is the capital of the United Kingdom"
      Case 2
         lblCountry.Caption = "Kuala Lumpur is the capital of Malaysia"
      Case 3
         lblCountry.Caption = "La Paz is the capital of Bolivia"
   End Select
End Sub
```

Notice that list boxes (combo boxes as well) have built-in index properties. `Select Case` structures are used extensively when working with list and combo boxes.

4 In addition to the `ListIndex` property, list and combo boxes have a `ListCount` property that can be read at runtime. The `ListCount` property returns the number of items in the list. Because all array items in Visual Basic start with zero unless otherwise specified, the last `ListIndex` item is one fewer than `ListCount`. The total numbers of items in the list in the preceding example is 4, but the largest `ListIndex` number is 3. There are two ways to code the Clear button. One way is to simply reference the exact `ListIndex` numbers:

```
Private Sub cmdClear_Click()
   lblCountry.Caption = ""
   lstCapital.Selected(0) = False
   lstCapital.Selected(1) = False
   lstCapital.Selected(2) = False
   lstCapital.Selected(3) = False
End Sub
```

This works, but what if the number of list items changes? Suppose that the program has the user add items to the list? How do you program it so that the Clear button works, no matter what the user does to it? You can handle this by referencing the `ListCount` property within a loop and remembering that it is different from the largest `ListIndex` number.

```
Private Sub cmdClear_Click()
   Dim n As Integer
   lblCountry.Caption = ""
   For n = 0 To lstCapital.ListCount - 1
      lstCapital.Selected(n) = False
   Next n
End Sub
```

The loop in this code will probably not be clear to you until you read Chapter 11, "Working with Loops." It is used previously to illustrate that a loop is used when the number of items in a list is subject to change.

Exercise 8.10 Checking Text-Box Input for Correct Names

This exercise introduces the `Asc()` function to check whether users have put their names in the correct text box. It uses one `Select Case` structure embedded in another. Figure 8.7 shows the interface you are to create.

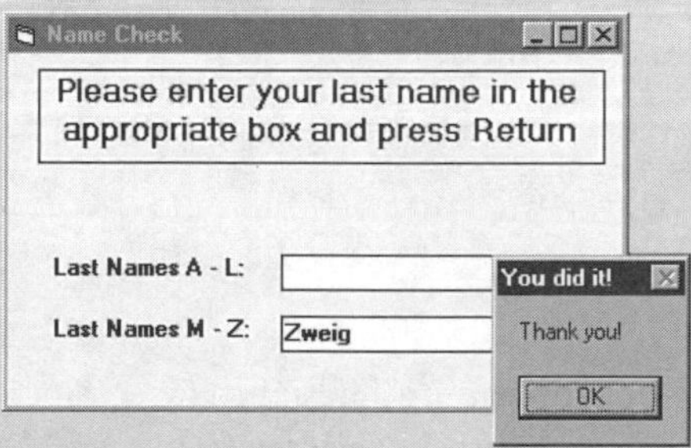

Figure 8.7

This design uses the `Select Case` *structure to determine whether the user has entered his or her last name in the correct text box.*

1 Draw the interface shown in Figure 8.7. There are two text boxes (**`txtNameAtoL`** and **`txtNameMtoZ`**). The instruction and alphabet range identifiers are labels.

2 Notice that you do not have a button on this form. Instead, use the `KeyDown()` event to execute your code. The user's natural tendency to press Return after entering input will be used to trigger code execution.

3 Enter the following code into the `txtNameAtoL_KeyDown()` event of the top text box:

```
Private Sub txtNameAtoL_KeyDown (Keycode As Integer, Shift As Integer)
   Select Case Keycode
      Case vbKeyReturn
         Select Case Asc(txtNameAtoL.Text)
            Case 65 To 76
               MsgBox "Thank you!",0,"You did it!"
            Case 97 To 108
               MsgBox "Please enter the first letter in uppercase",0,
               ➥"Comprehend?"
            Case Else
               MsgBox "Please enter your name in the appropriate
               ➥box",0,"Please"
               txtNameAtoL.Text  = ""
      End Select
   End Select
End Sub
```

continues

4 Enter the following code into the `KeyDown()` event of the bottom text box:

```
Private Sub txtNameMtoZ_KeyDown (Keycode As Integer, Shift As Integer)
   Select Case Keycode
      Case vbKeyReturn
         Select Case Asc(txtNameMtoZ.Text)
            Case 77 To 90
               MsgBox "Thank you!",0,"You did it!"
            Case 109 To 122
               MsgBox "Please enter the first letter in uppercase",0,
               ➥"Comprehend?"
            Case Else
               MsgBox "Please enter your name in the appropriate box", 0,
               ➥"Please"
               txtNameMtoZ.Text  = ""
      End Select
   End Select
End Sub
```

By referencing the built-in constant for the Return key, you can trigger code execution whenever the user presses the Return key after entering text. The keyword `To` is used in the `Case` expression list to set the ranges of the ANSI values (see Appendix A, "The ANSI Character Set"). These values are returned by the `Asc()` function applied to the text box input. For example, the number 109 returns the letter `m`. The `Select Case` structure is an efficient way to test for letters of the alphabet.

Chapter Summary

This chapter introduces the three forms of the `If...Then...Else` logical structure as well as relational and logical operators. Both types of operators are needed in writing useful `If...Then...Else` structures.

The three forms of the syntax should be more than familiar by now. In review, the complete syntax is as follows:

Form 1:

```
If (test condition) Then (test if condition is True) [Else (test if condition is
False)]
```

Form 2:

```
If test condition Then
   [statement block 1]
[Else]
   [statement block-2]
End If
```

Form 3:

```
If test condition 1 Then
   [statement block 1]
[ElseIf condition 2 Then
   [statement block 2]]
[Else]
   [statement block-n]
End If
```

In practice, the second and third forms are used most frequently.

Visual Basic has six relational operators and six logical operators. The relational operators all have the same precedence. All relational operators have a higher precedence than the logical operators.

Table 8.5 lists specific rules for the most frequently used logical operators.

Table 8.5 Rules for Logical Operators

Operator	*Rule*
`And`	Both test conditions must be `True` for the expression as a whole to be `True`.
`Or`	At least one test condition must be `True` for the expression as a whole to be `True`.
`XOr`	Only one test condition can be `True` for the expression as a whole to be `True`.
`Not`	If the expression is `True`, the result is `False`; if the expression is `False`, the result is `True`.

The `ElseIf` (one word) statement is used to write nested `If...Then` structures. By using `ElseIf`, a series of conditions can be tested to determine whether one evaluates to `True`.

`If...Then` logical structures are especially useful in changing the properties of objects based on test conditions. As illustrated in this chapter, `If...Then` can be used to change the properties of command buttons or to display different messages that depend on the value setting on a scroll bar.

The `Select Case` syntax, which is not complex, should be remembered. The syntax is as follows:

```
Select Case testexpression
   [Case expressionlist1
      [statementblock1]]
   [Case expressionlist2
      [statementblock2] ]
   Case Else
      [statementblock - n]]
End Select
```

The most difficult part of this syntax is the expression list, which can take several forms. These forms include an expression list, a range (using `To`), or a condition (using `Is`). The test condition can either be an assigned value or the value of an object's property.

Many Visual Basic programmers prefer to use `Select Case` rather than a series of `If...ElseIf` statements because of its readability and flexibility.

Skill-Building Exercises

1. Create an interface that contains a shape, a label, and a command button. Begin with a rectangle. When the user presses the command button the first time, display `I'm a Rectangle` in the label. When the user presses the command button the second time, change the shape and display `I'm a square`. Keep pressing the command button to change the shape and to display `I'm an oval`, `I'm a Circle`, `I'm a Rounded Rectangle`, and `I'm a Rounded Square`. After running out of shapes, display `No more shapes` on the last click. Use `If...Then` statements in your solution.

2. Create a limited expert system using `If...Then` statements. Ask the user whether the temperature is greater than, or equal to, 60 degrees and to click Yes or No. Use option buttons to ask the user whether it is raining and to click Yes or No. Have the user click a command button to analyze the responses to these questions. Display the evaluation of user input by using a label that displays the following possible messages:

Temp >= 60 degrees	*Raining*	*Message*
Yes	Yes	Bring umbrella
No	Yes	Bring coat and umbrella
Yes	No	No coat or umbrella is needed
No	No	Bring coat only

   If the user fails to check Yes or No to both questions, display the following message:

```
Need to check both conditions.
```

3. Write a program to compute a person's grade on an examination. Ask the user to enter a score and press the Return key. If the score is 90 to 100, assign an A; 80 to 89, assign a B; 70 to 79, assign a C; 60 to 69, assign a D; and below 60, tell the student to meet with you. Use a `Select Case` statement to display one of the following messages after the student enters a score:

```
Your grade is A. Well done!
Your grade is B. Good job!
Your grade is C. You passed!
Your grade is D. Do you need help?
Your performance was not acceptable. Please meet with me.
```

   If the person enters an invalid score (below 0 or above 100), display `Your score must be between 0 and 100.`

4. Place a set of option buttons inside a frame, naming the control array **optFood**. Label the buttons **Veggies**, **Dairy**, **Fruits**, **Meat**, **Junk Food**, and **Beverages**. Use a `Select Case` statement to display one of the following messages, depending on the user's selection:

```
Veggies are good for you.

Meat will make you beefy.

Dairy products will give you strong bones.

Junk food! Are you serious?

Fruits are easy to digest.

Beverages provide the body with fluid.
```

5. Here is the interface for the Pseudo Personality Test (see Figure 8.8). The shapes appear when the check box buttons are selected. Thus, more than one check box can be indicated. Based on the selections, the Evaluate button displays a personality evaluation of the user in the label.

Figure 8.8

The Pseudo Personality test interface asks you to click one or more of the shapes to reveal your personality type.

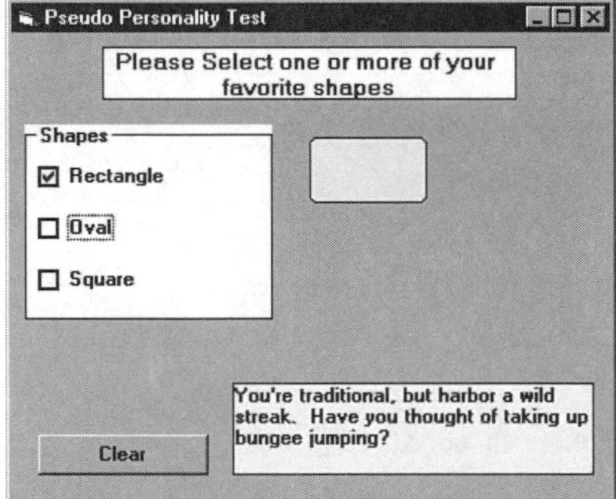

1. Draw the interface and then write the code for it. No shape should appear on the screen when beginning a test. When the user makes a selection, the shape will appear.

2. Cover all possible test selections by using `If...Then...Else` logic: A user can select a single shape or all three.

3. Make up your own evaluation messages for display in the label. There should be a total of eight combinations. Possible messages to display are as follows:

Shape Section	*Response*
Rectangle only	You're traditional, but harbor a wild streak.
Oval only	You have a tendency to be hedonistic, but inside are deeply spiritual and artistic.

Shape Section	*Response*
Square only	Ossified is the term that comes to mind.
Rectangle and oval	Harness your raging passions to a higher end, and you'll go far.
Rectangle and square	You're not totally rigid, but rigor mortis is a constant threat.
Oval and square	Your contradictions bedevil you.
Rectangle, oval, and square	Some might think you're well-rounded, but indecisive sums it up nicely.
No selection	Please restart by selecting one or more shapes.

4. Be sure that the test can be played more than once in succession. Add a Clear button to clear the shape selection check boxes, the shapes themselves, label output, and variable values so that a new test can be conducted.

5. After the Evaluate button is clicked, make it invisible and make the Clear button visible. After the Clear button is clicked, reverse this process.

9

Inputting Values and Formatting Output

This chapter takes a close look at the way in which you can use Visual Basic to enter data into processing, display it on the screen, and print it with a printer. Besides using text boxes, common ways of entering data into processing include the use of input and message boxes. The pop-up menu is another vehicle for enabling the user to make selections.

A second topic concerns how numbers, dates, and times can be formatted with Visual Basic. This chapter describes methods that you can use to display and print results in different ways. A typical design practice is to use one form to enter data and a second to display it. The results from the second form are then printed.

In this chapter, you learn several new tasks, including how to

- use the `InputBox()` function;
- use the `MessageBox()` function;
- design a pop-up menu;
- format numbers, dates, and time;
- use the `PrintForm` method.

Chapter Objectives

The objectives for this chapter focus on ways of entering data and displaying or printing processed results, as follows:

1. Understanding the uses of an input box
2. Understanding the uses of a message box
3. Learning how to add a pop-up menu
4. Learning how to format numbers, dates, and time
5. Understanding the `PrintForm` method

Objective 9.1 Understanding the Uses of an Input Box

An *input box* is a dialog box that opens and waits for the user to enter information or click a button. It differs from a message box, which only displays a message for the user to read. This chapter begins with the concept of the `InputBox()` function.

9.1.1 The *InputBox()* Function

A *function* differs from a *procedure.* A procedure does not return a value, while a function must return one and only one value. For example, in Chapter 6, "Variables and Constants," the `Date` function is used, which returns the current date from the system clock. If a value is to be returned, a function rather than a procedure *must* be written.

A function can be recognized by the way in which it is written. A function call is written as follows:

```
returnvalue = FunctionName (argumentlist)
```

On the other hand, the following example shows the way a procedure is written:

```
ProcedureName (argumentlist)
```

The `InputBox()` function is written as follows:

```
age = InputBox (argumentlist)
```

In this instruction, the `InputBox()` function returns one and only one value and assigns it to the variable `age`.

9.1.2 The Syntax of the *InputBox()* Function

The syntax for the `InputBox()` function is the following:

```
InputBox(prompt [,title][,default][,xpos] [,ypos][,helpfile,context])
```

Table 9.1 describes the components of the `InputBox()` function syntax.

Table 9.1 Syntax of the *InputBox()* Function

Component	*Description*
`prompt`	A string expression displayed as a message in a dialog box.
`title`	The title of the input box shown in the title bar.
`default`	The response if no other input is provided.
`xpos`	The horizontal distance between the left edge of the dialog box and the left edge of the display screen.
`ypos`	The vertical distance between the upper edge of the dialog box and the top of the display screen.

Component	*Description*
`helpfile`	A string expression that identifies the Help file that you use to provide context-sensitive help for the dialog box. If `helpfile` is provided, `context` must also be provided.
`context`	A numeric expression that is the Help context number that the Help author assigned to the appropriate Help topic. If `context` is provided, `helpfile` must also be provided.

The size of an input box cannot be changed; only its position on the screen can be altered. Nonetheless, you can add a considerable amount information to an input box, as Exercise 9.1 illustrates.

Exercise 9.1 Using an Input Box for Data Entry

Use an input box—rather than a form or a text box—to enter information for processing. Figure 9.1 shows the input box you are asked to create.

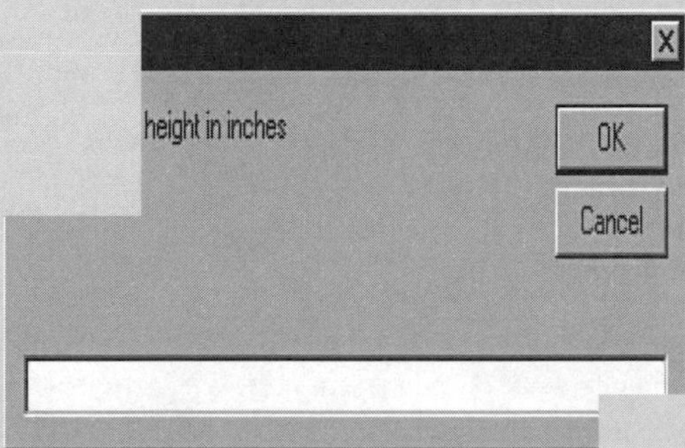

Figure 9.1

The input box enables you to collect additional information, such as the user's height in inches.

1. Open a new project and add the form caption **Input Box Data Entry**.
2. Add three labels and their captions. Add three text boxes named **`txtAge`**, **`txtHeight`**, and **`txtPounds`**.
3. Add the `Quit` command button and the `cmdQuit` procedure to end the program.
4. Add the following `Form_Load()` procedure:

```
form1.Show
txtAge.SetFocus
```

continues

If you try to set the focus without the `form1.Show` instruction, an error will result. The form must be seen before the focus can be set.

5 Add a `txtAge_GotFocus()` procedure. The procedure should get age, height, and weight from three input boxes—one value at a time. It should be written as follows:

```
Private Sub txtAge_GotFocus()
   'Program to get string values from the user
   'and convert them to integers
   Dim age As String
   Dim height As String
   Dim weight As String
   Dim answer As Integer

   'Get age and convert the string to a number
   age = InputBox("What is your age in years:", "Age", , 40, 2400)
   txtAge.Text = Val(age)

   'Get height and convert to a number
   height = InputBox("What is your height in inches", "Inches", , 40, 2400)
   txtHeight.Text = Val(height)

   'Get weight and convert to a number
   weight = InputBox("What is your weight in pounds", "Pounds", , 40, 2400)
   txtPounds.Text = Val(weight)
   answer = MsgBox("You're lying!!", 0, "Weight Detector")

   ' set focus to another text box to avoid regenerating the event
   txtHeight.SetFocus
End Sub
```

6 Test your completed program.

Notice that the input boxes appear as soon as the program is run because you set the focus to `txtAge` in the `Form_Load()` event (after showing the form) and have the code trigger the `txtAge_GotFocus()` event.

Objective 9.2 Understanding the Uses of a Message Box

In Exercise 9.1, you used a `MessageBox()` function to display an opinion about the truthfulness of the user. A `MessageBox()` function displays a message in a dialog box and waits for the user to respond. This response is limited to clicking the appropriate command button.

9.2.1 The Syntax of the *MessageBox()* Function

The syntax of the `MessageBox()` function is somewhat simpler and more limited than an input box. It consists of the following:

```
MsgBox(prompt[, buttons][, title][, helpfile, context])
```

Table 9.2 describes the components of the `MessageBox()` function syntax.

Table 9.2 Syntax of the *MessageBox()* Function

Component	*Description*
`prompt`	A string expression displayed as the message in the dialog box. The maximum length of `prompt` is approximately 1,024 characters, depending on the width of the characters used. If `prompt` consists of more than one line, you can separate the lines by using either a carriage return character (`Chr(13)`), a linefeed character (`Chr(10)`), or a combination of the two (`Chr(13) & Chr(10)`) between each line.
`buttons`	A numeric expression that is the sum of values specifying the number and type of buttons to display, the icon style to use, the identity of the default button, and the modality of the message box. If omitted, the default value for `buttons` is `0`.
`title`	A string expression displayed in the title bar of the dialog box. If you omit `title`, the application name is placed in the title bar.
`helpfile`	A string expression that identifies the Help file to use to provide context-sensitive help for the dialog box. If `helpfile` is provided, `context` also must be provided.
`context`	A numeric expression that is the Help context number the Help author assigned to the appropriate Help topic. If `context` is provided, `helpfile` also must be provided.

The `buttons` argument can lead to some complexity. Besides more than a dozen types, more than one type can be specified for a single message box. Consider the following code:

```
Private Sub Form_Click()
   Dim mbtype As Integer
   Dim Ans As Integer
   Dim Msg As String
   Dim Title As String
   Msg = "Are you OK?"
   Title = "Message Box Title"
   mbtype = vbYESNO + vbQuestion + vbDefaultButton1
   Ans = MsgBox(Msg, mbtype, Title)
End Sub
```

This procedure will display a message box with Yes and No as the choices. The default button will be Yes. A question mark icon will appear before the message "Are you OK?" because you are using the `vbQuestion` message box (see Figure 9.2).

There are many button argument types. Their built-in symbolic constants, integer values, and meanings are described in Table 9.3.

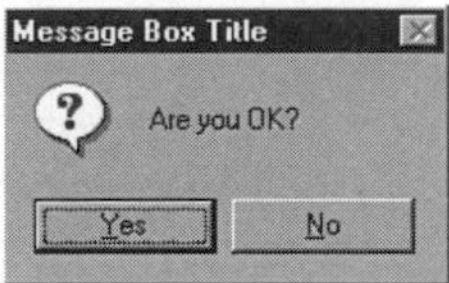

Figure 9.2
This message box displays a warning query and has Yes and No as choices, with Yes as the default.

Table 9.3 Button Arguments

Constant	*Value*	*Description*
`vbOKOnly`	0	Displays OK button only.
`VbOKCancel`	1	Displays OK and Cancel buttons.
`VbAbortRetryIgnore`	2	Displays Abort, Retry, and Ignore buttons.
`VbYesNoCancel`	3	Displays Yes, No, and Cancel buttons.
`vbYesNo`	4	Displays Yes and No buttons.
`VbRetryCancel`	5	Displays Retry and Cancel buttons.
`VbCritical`	16	Displays Critical Message icon.
`vbQuestion`	32	Displays Warning Query icon.
`VbExclamation`	48	Displays Warning Message icon.
`VbInformation`	64	Displays Information Message icon.
`VbDefaultButton1`	0	Makes the first button the default.
`vbDefaultButton2`	256	Makes the second button the default.
`VbDefaultButton3`	512	Makes the third button the default.
`VbDefaultButton4`	768	Makes the fourth button the default.
`VbApplicationModal`	0	Application modal: The user must respond to the message box before continuing work in the current application.
`VbSystemModal`	4096	System modal: All applications are suspended until the user responds to the message box.
`VbMsgBoxHelpButton`	16384	Adds Help button to the message box.
`VbMsgBoxSetForeground`	65536	Specifies the message box window as the foreground window.
`vbMsgBoxRight`	524288	Right justifies the text.
`vbMsgBoxRtlReading`	1048576	Specifies that text is to be read from right to left on Hebrew and Arabic systems.

Exercise 9.2 Exploring Different Types of Message Boxes

Figure 9.3 shows 12 different types of message boxes as button captions. Of the 12, the first column uses single message box types, while the second introduces message boxes with icons. The last column shows message boxes with combined types.

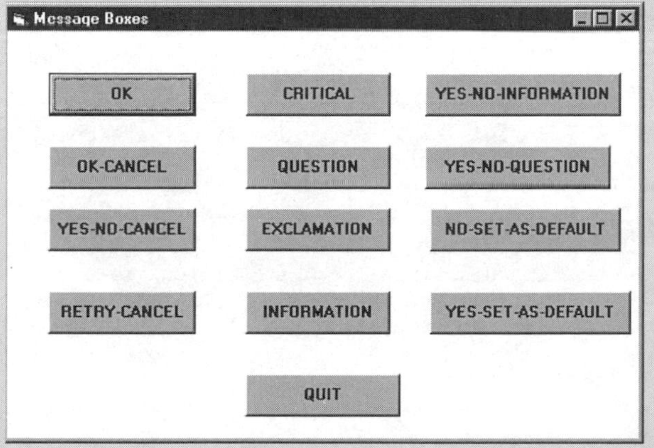

Figure 9.3

The different types of message boxes enable you to fulfill a variety of input and output needs in your programs.

1 Open a new project and change the caption to **Message Boxes.**

2 Place the command buttons on the form, and change the captions as shown in Figure 9.3. Add a `cmdQuit_Click()` procedure.

3 Declare the following variables at the form level:

```
Dim Ans As Integer
Dim title As String
Dim msg As String
```

4 Write the `cmdOK_Click()` procedure, as follows:

```
Private Sub cmdOK_Click()
   msg = "With this style the user can only click OK"
   title = "OK Message Box"
   Ans = MsgBox(msg, vbOKOnly, title)
End Sub
```

5 Test your design after adding this code.

6 Place the other command buttons (or as many as you find necessary) on the form. The messages (`msg`), titles (`title`), and answers (`Ans`) are as follows:

continues

- OK-Cancel message box

```
msg = "With this style the user can only click OK or Cancel"
title = "OK-CANCEL Message Box"
Ans = MsgBox(msg, vbOKCancel, title)
```

- Yes-No-Cancel message box

```
msg  = "With this style the user can only click Yes, No, or Cancel"
title = "Yes-No-Cancel Message Box"
Ans = MsgBox (msg, vbYesNoCancel, title)
```

- Retry-Cancel message box

```
msg  = "With this style Retry and Cancel are possible"
title = "Retry-Cancel Message Box"
Ans = MsgBox (msg, vbRetryCancel, title)
```

- Critical message box

```
msg  = "With this style the Critical icon is presented"
title = "Critical Message Box"
Ans = MsgBox (msg, vbCritical, title)
```

- Question message box

```
msg  = "With this style the Question icon is presented"
title = "Question Message Box"
Ans = MsgBox (msg, vbQuestion, title)
```

- Exclamation message box

```
msg  = "With this style the Exclamation icon is presented"
title = "Exclamation Message Box"
Ans = MsgBox (msg, vbExclamation, title)
```

- Information message box

```
msg  = "With this style the Information icon is presented"
title = "Information Message Box"
Ans = MsgBox (msg, vbInformation, title)
```

- Yes-No-Information message box

```
msg  = "The Information icon is presented with Yes-No Choices"
title = "Combination Message Box"
Ans = MsgBox (msg, vbYesNo + vbInformation, title)
```

- Yes-No-Question message box

```
msg  = "Do you want to continue?"
title = "Yes-No-Question Message Box"
Ans = MsgBox (msg, vbYesNo+ vbQuestion, title)
```

- No-Set-As-Default message box

```
msg  = "No is set as the default"
title = "Combination Message Box"
Ans = MsgBox (msg, vbYesNo + vbInformation + vbDefaultButton2, title)
```

- Yes-Set-As-Default message box

```
msg  = "Yes set as the default"
title = "Yes-No-Cancel Message Box"
Ans = MsgBox (msg, vbYesNoCancel +vbDefaultButton1, title)
```

7 Run and test all message boxes.

Message boxes are used to prompt the user to make choices or to provide the user with information that is critical to further processing of the program. Two of the button integer values you can use to customize a message box have no effect on buttons at all. These two values are `vbApplicationModal` (the default) or `vbSystemModal`. As a general rule, you should leave your message boxes to the default value of `vbApplicationModal`, and reserve the use of `vbSystemModal` for those situations in which there is a risk of your application destabilizing the current computer session.

9.2.2 The Return Values of the *MessageBox()* Function

The most critical part of employing message boxes is ascertaining user response so that the processing of your program can proceed according to the user's choices. In Exercise 9.2, all `MessageBox()` function return values are assigned to the integer variable `Ans`. Table 9.4 presents the values of possible user choices and their constant names.

Table 9.4 Return Values of the *MessageBox()* Function

Constant	*Value*	*Button*
vbOK	1	OK
vbCancel	2	Cancel
vbAbort	3	Abort
vbRetry	4	Retry
vbIgnore	5	Ignore
vbYes	6	Yes
vbNo	7	No

In Exercise 9.2, the value returned to `Ans` changes depending on a user's selection. You might want to extend the exercise by adding instructions so that the values of the variable `Ans` are printed on the form.

Objective 9.3 Learning How to Add a Pop-Up Menu

In Chapter 4, "Learning to Think Visually," you learned how to add a menu bar to a design. However, there are times when it is more convenient to open a pop-up menu, namely, one that is displayed on a form. The syntax for a pop-up menu is as follows:

```
object.PopupMenu menuname, flags, x, y, boldcommand
```

Table 9.5 explains the components of the pop-up menu syntax.

Table 9.5 Pop-Up Menu Syntax

Component	*Description*
`object`	Optional. An object expression that evaluates to an object in the Applies To list. If `object` is omitted, the form with the focus is assumed to be an object.
`menuname`	Required. The name of the pop-up menu to be displayed. The specified menu must have at least one submenu.
`flags`	Optional. A value or constant that specifies the location and behavior of a pop-up menu, as described in Settings.
`x`	Optional. Specifies the x-coordinate where the pop-up menu is displayed. If omitted, the mouse coordinate is used.
`y`	Optional. Specifies the y-coordinate where the pop-up menu is displayed. If omitted, the mouse coordinate is used.
`boldcommand`	Optional. Specifies the name of a menu control in the pop-up menu to display its caption in bold face. If omitted, no controls in the pop-up menu appear in bold.

The optional `flags` argument affects the location and behavior of the pop-up menu. This argument works only for applications running under Windows 95. The application will ignore this argument when running under 16-bit versions of Windows or Windows NT 3.51 and earlier. These settings are shown in Table 9.6.

Table 9.6 Settings for the *flags* Argument

Constant	*Value*	*Description*
		Location
`vbPopupMenuLeftAlign`	0	The default. Places the left side of the pop-up menu at x.
`vbPopupMenuCenterAlign`	4	Centers the pop-up menu at x.

Constant	Value	Description
		Location
`vbPopupMenuRightAlign`	8	Places the right side of the pop-up menu at x.
		Behavior
`vbPopupMenuLeftButton`	0	The default. Causes an item on the pop-up menu to react to a mouse click only when the left mouse button is used.
`vbPopupMenuRightButton`	2	Causes an item on the pop-up menu to react to a mouse click when either the right or the left mouse button is used.

As in the `MessageBox()` function, these constants can be combined to customize pop-up menu properties.

9.3.1 Building a Pop-Up Menu

Building a pop-up menu follows the same procedure as building a menu that appears on the menu bar, with one exception: the `Visible` property of the topmost menu name is turned off, making the menu invisible. An event, such as a mouse click, then makes the menu appear. Exercise 9.3 illustrates how this is done.

Exercise 9.3 Creating a Pop-Up Menu to Display Different Greetings

This exercise asks you to create a pop-up menu. When the user clicks the mouse on the form, the menu appears. From the menu, the user can select Greeting1, Greeting2, Greeting3, or End the program.

1 Create a new project, adding the caption **Pop-Up Menu** to the form.

2 Choose Tools, Menu Editor and build the following menu (leaving all items visible for the moment):

```
Caption             "Message"
Name                mnuMessage
......Caption       "Greeting1"
......Name          mnuGreeting1
......Caption       "Greeting2"
......Name          mnuGreeting2
```

continues

```
......Caption       "Greeting3"
......Name          mnuGreeting3
.........Caption    "End"
.........Name       mnuEnd
```

In this menu, mnuGreeting1, mnuGreeting2, and mnuGreeting3 are subordinate to mnuMessage; mnuEnd is subordinate to mnuGreeting3.

3 Write the following Form_Click() procedure:

```
Private Sub Form_Click()
   PopupMenu mnuMessage, 4, 1400, 1400
End Sub
```

4 Write the following menu procedures:

```
Private Sub mnuGreeting1_Click()
   Print "Good morning!"
End Sub

Private Sub mnuGreeting2_Click()
   Print "How are you today?"
End Sub

Private Sub mnuGreeting3_Click()
   Print "Feeling great, you say!"
End Sub

Private Sub mnuEnd_Click()
   Print "Sorry.  Have to run."
   End
End Sub
```

5 Now for the important point: Return to the menu builder, and check off the Visible check mark for mnuMessage.

6 Run your program. Click the form to make the pop-up menu appear.

9.3.2 Using a Pop-Up Menu

The question of when to use a pop-up menu is an interesting one. Some designers use pop-up menus when users need to refer to a menu often. A simple click of a command button or a mouse click opens the menu. One disadvantage of a pop-up menu is that only one can be displayed at a time; thus, calls to a pop-up menu are ignored if one is already displayed. Calls are also ignored if a pull-down menu is active.

In the newer version of Windows, context menus appear with the click of the right mouse button. This behavior is built in, for instance, with the Visual Basic Text Box control. If you want to build context menus for your applications, use the vbPopupMenuRightButton flag when building pop-up menus.

Objective 9.4 Learning How to Format Numbers, Dates, and Time

An important part of displaying or transmitting results is formatting them in a familiar form. This is especially important when converting numerical values to strings or to currency. Without formatting, the currency representing the numerical value might appear quite different from what the designer expected.

9.4.1 The *Format()* Function

The `Format()` function formats any expression according to a set of string instructions that you specify. Formatting, for instance, allows a string to appear as currency, with a specified number of leading or trailing zeros and with commas separating thousands. Format instructions can be either named or written in a user-defined format.

The syntax of the `Format()` function is as follows:

```
Format(expression[, format[, firstdayofweek[, firstweekofyear]]])
```

Table 9.7 explains the components of the `Format()` function syntax, while Table 9.8 and Table 9.9 show the settings of the function's `firstdayofweek` and `firstweekofyear` arguments, respectively.

Table 9.7 The *Format()* Function Syntax

Component	*Description*
`expression`	Any valid expression
`format`	A valid named or user-defined format expression
`firstdayofweek`	A constant that specifies the first day of the week (see Table 9.8 for the settings)
`firstweekofyear`	A constant that specifies the first week of the year (see Table 9.9 for the settings)

Table 9.8 The *firstdayofweek* Settings

Constant	*Value*	*Description*
`vbUseSystem`	`0`	Uses NLS API setting
`vbSunday`	`1`	Sunday (default)
`vbMonday`	`2`	Monday
`vbTuesday`	`3`	Tuesday
`vbWednesday`	`4`	Wednesday

continues

Table 9.8 continued

Constant	*Value*	*Description*
`vbThursday`	5	Thursday
`vbFriday`	6	Friday
`vbSaturday`	7	Saturday

Table 9.9 The *firstweekofyear* Settings

Constant	*Value*	*Description*
`vbUseSystem`	0	Uses NLS API setting
`vbFirstJan1`	1	Starts with the week in which January 1 occurs (default)
`VbFirstFourDays`	2	Starts with the first week that has at least four days in the year
`VbFirstFullWeek`	3	Starts with the first full week of the year

9.4.2 Using Named and User-Defined Formats with Numbers

The easiest way to format an expression is to use one of the named formats. Table 9.10 shows the available format names.

Table 9.10 Named Formats

Format	*Description*
`General Number`	Displays number with no thousands separator.
`Currency`	Displays number with thousands separator, if appropriate; displays two digits to the right of the decimal separator. Output is based on system locale settings.
`Fixed`	Displays at least one digit to the left and two digits to the right of the decimal separator.
`Standard`	Displays number with thousands separator, at least one digit to the left and two digits to the right of the decimal separator.
`Percent`	Displays number multiplied by 100 with a percent sign (%) appended to the right; always displays two digits to the right of the decimal separator.
`Scientific`	Uses standard scientific notation.

Format	Description
Yes/No	Displays No if number is 0; otherwise, displays Yes.
True/False	Displays False if number is 0; otherwise, displays True.
On/Off	Displays Off if number is 0; otherwise, displays On.

A full description of these named formats is available in Visual Basic's online Help. A standard number, for instance, displays the number with a thousands separator, at least one digit to the left and two digits to the right of the decimal separator. Therefore, the return from the `Format()` function call `Format(1234, "Standard")` would return `1,234.00`.

Examine the return from the `Format(8450, "Currency")` function call:

```
$8,450.00
```

In this example, using a named format simplifies the more difficult task of adding a dollar sign, pound signs, and leading and trailing zeros.

To create a user-defined numeric format, you can use one or more of the symbolic constants shown in Table 9.11.

Table 9.11 The User-Defined Numeric Format Constants

Symbol	Description
0	Prints a trailing or a leading zero in the format position if appropriate.
#	Never prints trailing or leading zeros.
.	Decimal place holder.
,	Thousands separator.
%	The expression is multiplied by 100. The percent sign (%) is inserted in the position where it appears in the format string.
-+$()	Literal character: displays a character as typed.

Examine the return values from the following `Format()` function calls:

Return	Format Function
$600.75	Format(600.75, "$###.##")
8,450	Format(8450, "#,#00")
$1	Format(0.5, "$#,##0")
300.00%	Format(3, "0.00%")
50E+03	Format(50000, "00E+00")

As you can see, using one of the named formats is much easier, but by using these symbolic constants, you can customize your output to appear exactly as you want. Remember that when writing format statements, you must place both named and user-defined formatting patterns in quotation marks.

Exercise 9.4 Entering and Displaying Numeric Data

This exercise asks you to enter payroll data on one form and display it on a second form. Figure 9.4 shows the forms you are asked to complete.

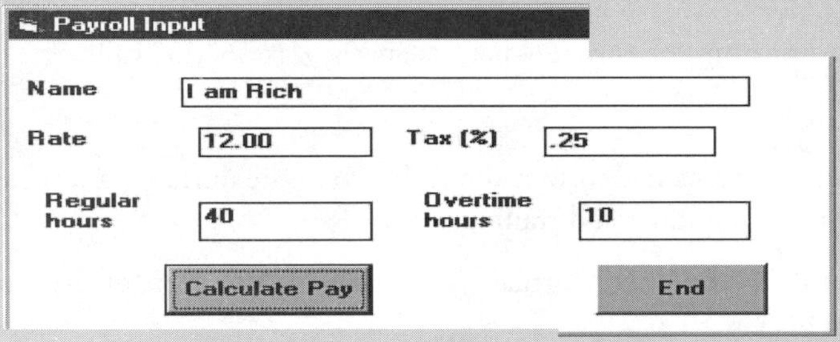

Figure 9.4

This data entry display asks you to enter the data needed to calculate and display gross pay, taxes paid, and net pay, using the format function.

1 Open a new project, and name the form **`frmInput`** and then add the caption **Payroll Input**.

2 Add the two command buttons and change the names to **`cmdCalculate`** and **`cmdEnd`**. Add the captions and write the `cmdEnd_Click()` procedure.

3 Add the five text boxes and the five labels. Name the text boxes as follows: **`txtName`**, **`txtRate`**, **`txtTax`**, **`txtRegHrs`**, and **`txtOverHrs`**. Add all label captions.

4 Add a second form and name the form **`frmRegister`**. Change the caption to **Payroll Register**.

5 Place a command button on this form and name it **`cmdReturn`**. *Return,* in this context, means return to the first form.

6 Write the event procedure for the `cmdReturn` button. The code should read as follows:

```
Private Sub cmdReturn_Click()
   frmInput.Show
   frmRegister.Hide
End Sub
```

7 Return to `frmInput`, and write the Calculate Pay procedure. Your code should read as follows:

```
Private Sub cmdCalculate_Click()
   Dim GrossPay As Currency
   Dim TaxPay As Currency

   Dim TaxRate As Single
   Dim PayRate As Currency
   Dim Regular As Integer
   Dim Overtime As Integer
   Dim NetPay As Currency

   PayRate = Val(frmInput.txtRate.Text)
   TaxRate = Val(frmInput.txtTax.Text)
   Regular = Val(frmInput.txtRegHrs.Text)
   Overtime = Val(frmInput.txtOverHrs.Text)

   GrossPay = (PayRate * Regular) + (1.5 * PayRate * Overtime)
   TaxPay = GrossPay * TaxRate
   NetPay = GrossPay - TaxPay

   'Show second form
   Load frmRegister
   frmRegister.Show

   frmRegister.Print
   frmRegister.Print
   frmRegister.Print "     Gross pay is :"; Format$(GrossPay, "Currency")
   frmRegister.Print "     Taxes are    :"; Format$(TaxPay, "Currency")
   frmRegister.Print "      Net pay is   :"; Format$(NetPay, "Currency")
End Sub
```

In this procedure, named formats are used in place of symbols.

8 Test your design with the following data: A person is paid $8.00 per hour, has a tax rate of 0.25, and works 40 regular hours and 10 overtime hours.

9.4.3 Using Named and User-Defined Date and Time Formats

The `Format()` function can also be used with dates and times. Named date and time formats are much easier to use than user-defined date and time formats. An extensive set of symbolic constants for dates and times can be employed in user-defined formats to obtain the exact output you want. The symbolic constants are available in Visual Basic's online Help. The named date and time formats are defined by the International section of the control panel. The format names and descriptions are shown in Table 9.12.

Table 9.12 Date and Time Formats

Format	*Description*
`General Date`	Displays a date and/or time. For real numbers, displays a date and time (for example, `4/3/98 05:34 PM`); if there is no fractional part, displays only a date (for example, `4/3/98`); if there is no integer part, displays the time only (for example, `05:34 PM`). The date display is determined by your system settings.
`Long Date`	Displays a date according to your system's `Long Date` format.
`Medium Date`	Displays a date using the `Medium Date` format appropriate for the language version of Visual Basic.
`Short Date`	Displays a date using your system's `Short Date` format.
`Long Time`	Displays a time using your system's `Long Time` format. It includes hours, minutes, and seconds.
`Medium Time`	Displays a time in 12-hour format using hours and minutes and the AM/PM designator.
`Short Time`	Displays a time in 24-hour format (for example, `17:45`).

Because the symbolic constants for user-defined dates and times are so extensive, they are not listed here. Push the F1 key and consult Visual Basic's online Help. Examples of user-defined dates and times are listed in the following table. In the examples, the slash (/) is used as the date separator and the colon (:) as the time separator.

Format Syntax	*Result*
`m/d/yy`	5/16/97
`dddd,mmmm dd,yyyy`	Monday, May 16, 1997
`d-mmmm`	16-May
`mmmm-yy`	May-97
`hh:mmm AM/PM`	09:30 AM
`h:mm:ss a/p`	9:30:12 a
`d-mmmm h:mm`	15-May 9:30

Exercise 9.5 Exploring the Use of Date Formats

This short exercise asks you to use named formats and user-defined formats. Figure 9.5 shows the output that you are asked to create.

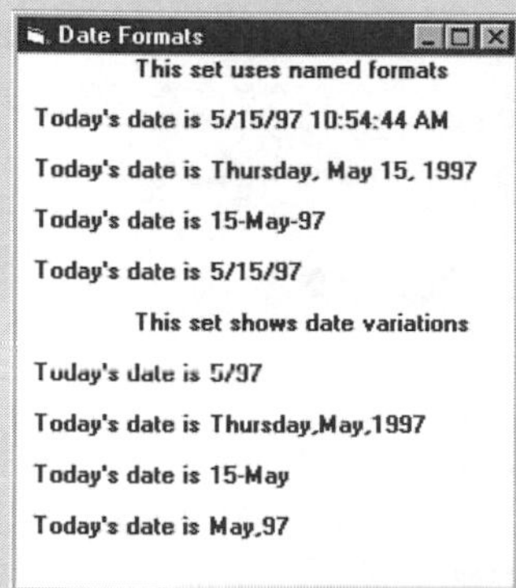

Figure 9.5

This display shows the four named date formats and four user-defined formats.

1 Open a project and change the form caption to **Date Formats**.

2 Write a `Form_Click()` procedure to format the `Now` function. Use all named date formats and user-defined formats. The procedure is written as follows:

```
Private Sub Form_Click()
   'Using named formats
   Print "                This set uses named formats"
   Print
   Print "  Today's date is "; Format(Now, "General Date")
   Print
   Print "  Today's date is "; Format(Now, "Long Date")
   Print
   Print "  Today's date is "; Format(Now, "Medium Date")
   Print
   Print "  Today's date is "; Format(Now, "Short Date")

   Print
   'User defined format symbols
   Print "                This set shows date variations"
   Print
   Print "  Today's date is "; Format(Now, "m/yy")
   Print
   Print "  Today's date is "; Format(Now, "dddd,mmmm,yyyy")
   Print
   Print "  Today's date is "; Format(Now, "d-mmmm")
   Print
   Print "  Today's date is "; Format(Now, "mmmm,yy")
End Sub
```

3 Run and test your program.

Exercise 9.6 Exploring the Use of Time Formats

This exercise is much like the last. It asks you to use both named time formats and special, symbolic formats. Figure 9.6 shows the output for this exercise.

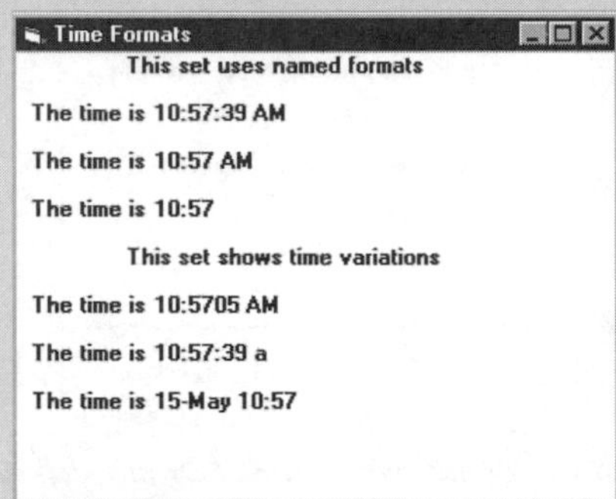

Figure 9.6

This display shows the three named time formats and three user-defined formats.

1 Begin a project, changing the form caption to **Time Formats**.

2 Write a `Form_Click()` procedure to format the `Now` function. Use all named time formats and user-defined formats. The `Form_Click()` procedure is written as follows:

```
Private Sub Form_Click()
   'Using named formats
   Print "               This set uses named formats"
   Print
   Print "  The time is "; Format(Now, "Long Time")
   Print
   Print "  The time is "; Format(Now, "Medium Time")
   Print
   Print "  The time is "; Format(Now, "Short Time")

   Print
   'User defined format symbols
   Print "               This set shows time variations"
   Print
   Print "  The time is "; Format(Now, "hh:mmmm AM/PM")
   Print
   Print "  The time is "; Format(Now, "h:mm:ss a/p")
   Print
   Print "  The time is "; Format(Now, "d-mmmm h:mm")
End Sub
```

3 Run and test your program.

Objective 9.5 Understanding the *PrintForm* Method

Besides displaying processed results on the computer screen, results can be printed by using a computer printer. There are two alternatives for sending information to a printer. The first uses the `PrintForm` method. The second uses `Printer` methods for which the `Printer` object is accessed with the keyword `Printer`.

9.5.1 The *PrintForm* Method

To print a form as it appears on the screen, insert the word `PrintForm` in an event procedure. This directs the computer to print, bit by bit, the image shown on the screen. The syntax for this method is as follows:

```
[form.]PrintForm
```

That's it!

Exercise 9.7 Printing a Form

This exercise asks you to design a simple interface and print it. Do the following:

1. Open a new project and change the form caption to **NotePad**.
2. Add a text box and two command buttons. For the text box, change the name to **`txtNotes`**. The text box should accept multiple lines.
3. Write the `cmdEnd_Click()` procedure.
4. Write the `cmdPrint_Click()` procedure as follows:

```
Private Sub cmdPrint_Click()
   PrintForm
End Sub
```

5. Test your design. Enter some text into the text box and click Print to print the text on the computer printer.

9.5.2 The *Printer* Object

Although the `PrintForm` method is simple, there are times when you do not want to print the entire form. For example, you might only want to print the text inside a text box. Additionally, you might want to change the font and configure the display page to better fit the printed page. The `Printer` object is required to accomplish such specific tasks.

The resolution of the `PrintForm` method is not nearly as fine as can be obtained by using the `Printer` object. To fully utilize the `Printer` object, however, you are required to use graphic methods.

Chapter Summary

This chapter considers several alternatives to using labels and text boxes for processing and displaying information.

The `InputBox()` function is used to display a dialog box for entering values into processing. The syntax is as follows:

```
InputBox(prompt [,title][,default][,xpos] [,ypos][,helpfile,context])
```

With an input box, a dialog box appears asking the user for information, and disappears after the user clicks OK.

The `MsgBox()` function is used to display a message or warning. The syntax is as follows:

```
MsgBox(prompt[, buttons][, title][, helpfile, context])
```

With a message box, a dialog box appears providing the user with information. It disappears after the user clicks an option, such as OK.

With a pop-up menu, a series of choices are presented to the user. These choices can range from opening an input screen to displaying a report. The advantage of a pop-up menu is that it can be opened quickly and placed on any position of the screen. It disappears after the user makes a choice.

The `Format()` function is used to change the appearance of numbers, dates, and times. Named formats exist for all three, and simplify the formatting process. Flexibility is added by allowing for the use of user-defined format symbols.

The `PrintForm` method directs the computer to print, bit by bit, the image shown on the display screen.

Skill-Building Exercises

1. Design a form with five labels to display the following:
   - Current hour and minutes (for example, `9:41`)
   - Morning or evening (for example, `AM`)
   - Month (for example, `January`)
   - Day (for example, `15`)
   - Year (for example, `19xx`)

   Add a `Click()` event to display `The time is` and `The date is`, formatted as stated here.

2. This exercise consists of the following two parts:
   - Design a form placing Date and End on the menu bar. When the user clicks Date, display an `InputBox()` to enable the user to enter the date if it differs from the default (use a `Short Date` as the default). Display the date using a label.
   - Using the same form, remove Date and End from the menu bar, placing both on a pop-up menu instead. Use the pop-up menu to allow the user to enter a date if it differs from the default, or to end the program.

3. Modify the payroll program in Exercise 9.4 by formatting the data to be displayed on the payroll register. When you have completed the exercise, `GrossPay`, `TaxPay`, and `NetPay` totals should be aligned as follows:

   $2,000.00

   $0,200.00

   $1,800.00

4. Modify the payroll program in Exercise 9.4 by adding a `CheckTax()` procedure. If the user enters a tax rate in excess of 1 (such as 15), call the procedure. Within the procedure, use a message box to tell the user `The tax rate must be less than 1`. Make `Check tax rate` the title of the message box.

   After displaying the message, set the focus to allow the user to enter an acceptable tax rate. Only when the rate is acceptable should the payroll register be displayed.

5. Modify Exercise 9.7 by printing the text inside the text box rather than the entire form. To complete this exercise, you will need to use the following methods with the `Printer` object:

`Printer.Print`	Prints the specified text or a blank line
`Printer.EndDoc`	Indicates the end of the document and releases the printer

To set the font, use the following statement:

```
Printer.FontSize = <set the font size here>
```

10

Defining the Scope of Variables and Procedures

Thus far you have worked with constants and numeric and string variables, giving little attention to how or where variables and constants are declared. In Visual Basic, the placement of the variable declaration together with keywords used to make the declaration determine the variable's *scope,* which describes how variables and constants can be seen by various parts of a program. Stated another way, *scope determines the life of a variable.*

In Visual Basic, variable scope is determined by where the variable is declared. A variable declaration can occur at the procedure, module, or global level. Variable scope is also determined by how the variable is declared with certain keywords. Variable declaration keywords include `Dim`, `Static`, and `Public`.

Procedures also have scope. Some procedures are available to the entire program, while others are only available to other procedures within the module in which they reside. Keywords that determine procedure scope include `Private`, `Public`, and `Static`.

The *scoping* of variables and procedures has particular importance in programs that contain more than one module, as most applications do. How can a variable be seen, for instance, between two forms? If data needs to be shared by two modules, how is this accomplished? Can data be shared without using variables that can be seen by the entire application? This chapter addresses these questions.

This chapter also introduces user-defined data types: a data type created by the grouping of other data types. User-defined data types are the beginning of objects, and make objects easier to understand.

In this chapter, you learn many terms and several tasks, including how to

- declare local-, module-, and global-level variables;
- create and use a code module;

- make a procedure call;
- pass data and objects from one procedure to another;
- define and use a user-defined data type;
- use more than one form in a program;
- call procedures on other forms;
- declare a static variable;
- write a `Sub Main()` procedure to initiate a program.

Chapter Objectives

The objectives for this chapter address the different ways of declaring variables in a Visual Basic program, as follows:

1. Understanding the differences in scope among local-, module-, and global-level variables
2. Understanding static scope
3. Understanding the scoping of procedures
4. Working with multiple forms in an application
5. Declaring and using a user-defined data type
6. Understanding procedures as object methods

Objective 10.1 Understanding the Differences in Scope Among Local, Module, and Global Variables

The scoping of variables determines how and where a variable is declared in a program and the other parts of a program that can read a value from that variable or assign a value to it. The three levels are defined as follows:

- **Local-level scope**—The variable is exposed only while a procedure is executing. Only code within the procedure itself can read or assign a value to the variable.
- **Module-level scope**—The variable is exposed to the entire module, whether the module is a form, a code .BAS file, or a class module. Any procedure within the module can read or assign a value to the variable.
- **Global-level scope**—The variable is exposed to the entire program. Any procedure within the program can read or assign a value to the variable.

Before moving to specific examples of variable scoping, it is important to understand how, why, when, and where variables are declared in a Visual Basic program. Within a procedure, form, or module, most variables are declared using the `Dim` statement, using the following syntax:

```
Dim variablename [As type]
```

In Visual Basic, variable declarations are not required before a variable is used and, in fact, are not required at all. Visual Basic automatically declares a variable with the name you use and assigns it type `Variant`, which is the default. This practice is not recommended, however, as it can slow down program execution speed, easily lead to errors, and make code extremely difficult to debug and maintain.

To avoid the problem of Visual Basic automatically assigning a type, place the following statement in the declaration section of a form or a standard module:

```
Option Explicit
```

This statement will recognize any variable that has not been declared in advance, and display an error whenever such a variable is found. Most Visual Basic programmers always work with `Option Explicit` turned on. This can be set by typing in `Option Explicit` yourself or by selecting the Require Variable Declaration check box in the Editor tab options after choosing Tools, Options (see Figure 10.1).

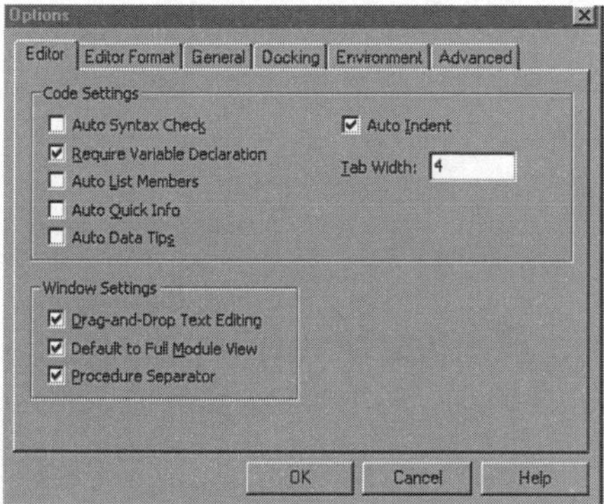

Figure 10.1

The second selection, Require Variable Declaration, requires you to explicitly declare all variables in your code.

10.1.1 Placement of the *Dim* Statement

`Dim` statements can be placed at the procedure or module level of a Visual Basic program. The module level includes those variables and procedures declared under the `(General)` object in the code window of a form, code, or class module.

When placing `Dim` inside a procedure, the declared variables are defined only within the procedure. This will become important when the local scope is examined and demonstrated (see section 10.1.4, "Local Scope and Local Variables"). A good programming practice is to place all `Dim` statements at the beginning of a procedure, rather than placing them throughout a procedure. This makes a program easier to read and helps to ensure that the variable has been defined before it is used.

10.1.2 Global Declaration with the *Public* Keyword

Another way of declaring a variable is by using the `Public` statement. The syntax is the same as `Dim` with the exception that the word `Public` replaces the word `Dim`, as follows:

```
Public variablename [As type]
```

The keyword `Public` means that the variable can be seen and used by every procedure in every module in a Visual Basic program. This knowledge will become important when you consider the concept of global scope (see section 10.1.6, "Global Scope and Global Variables"). The placement of global declarations of variables is especially important. If you want variables or constants to be seen by the entire application, they must be made in the declarations section of a standard module (or .BAS file). They cannot be made in procedures, or the general level of a form or class module.

Figure 10.2 shows the General Declarations section of a standard module. All standard modules have this General section. The Declarations drop-down menu is used to reveal all declarations that are general to the module. A standard module contains Visual Basic code and only code. A standard module is created by clicking the drop-down Module icon on the tool bar or by selecting Project, Add Module. This icon is the second from the left. A standard code module has a .BAS suffix.

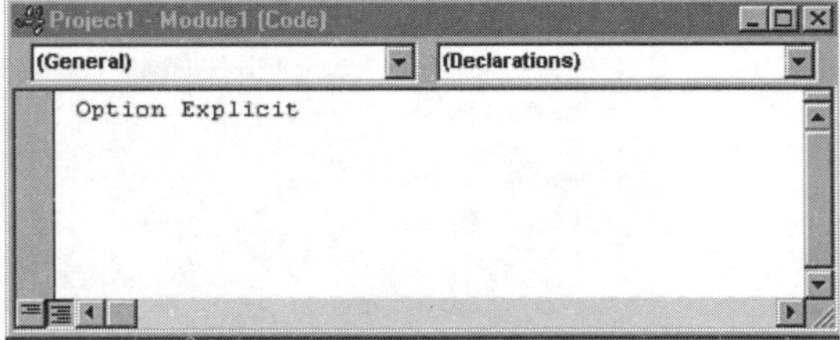

Figure 10.2

The General section of a standard module contains the `Option Explicit` *statement.*

10.1.3 The *Static* Declaration

A third way of declaring a variable is by using the `Static` statement. A `Static` variable can be used at the procedure level only. `Static` declares a variable that will retain its value when multiple calls are made to a procedure.

The syntax for a `Static` variable is similar to `Dim` and `Public`:

```
Static variablename [As type]
```

10.1.4 Local Scope and Local Variables

A local variable is recognized only within the procedure in which it is declared. The following instructions declare `Cat` as a local variable of type `String` within a procedure:

```
Private Sub cmdLocal_Click()
   Dim Cat As String
   Cat =  "Night Eyes"
   Print "The name of my cat is :"; Cat
End Sub
```

The variable, `Cat`, can only be seen and used by this `cmdLocal_Click()` procedure. When the procedure ends, the value assigned to `Cat` is cleared from memory. As a general rule, *always declare variables as locally as possible.* This reduces the chance that a variable will be assigned an errant value and reduces the size of your application's persistent memory footprint.

Exercise 10.1 Demonstrating Local Scope

This exercise demonstrates the concept of local scope. You are asked to write two procedures with the same variable name used in both. Because the life of the variable ends when the procedure ends, declaring two variables with the same name poses no problems. Figure 10.3 shows the interface you are asked to create. The following code segments show that the variable `Cat3` is declared twice—once within each procedure.

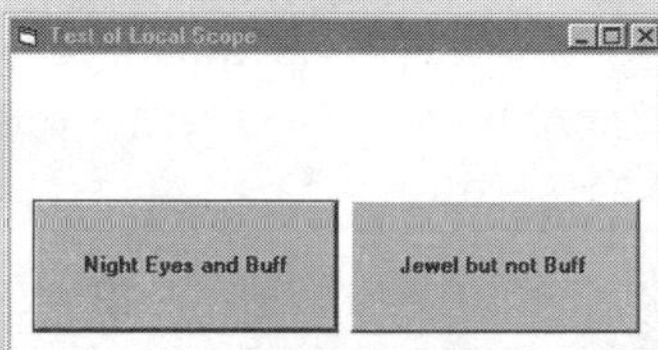

Figure 10.3

This design enables you to declare a variable with the same name in two different procedures to reveal different values.

1 Create the interface as shown in Figure 10.3. Name the first command button **`cmdCats1`** and the second command button **`cmdCats2`**. Add the caption **Test of Local Scope** to the form.

2 Write the `cmdCats1_Click()` procedure as follows:

```
Private Sub cmdCats1_Click()
   Dim Cat1 As String
   Dim Cat3 As String

   Cat1 = "Night Eyes"
   Print "The name of my first cat is "; Cat1
   Cat3 = "Buff"
   Print "The name of my third cat is "; Cat3
End Sub
```

3 Write the `cmdCats2 Click()` procedure as follows:

```
Private Sub cmdCats2_Click()
   'Cat1 is the same variable name as used in the
   'first procedure
   Dim Cat1 As String
   Dim Cat3 As String
   Cat1 = "Jewel"
   Print "The name of my second cat is "; Cat1
   'Display this message without assigning a value
   Print "The name of my third cat is "; Cat3
End Sub
```

4 Run and test your procedure. Observe what happens when you try to display the name of `Cat3` in the second procedure. The assignment in the first procedure is local to that procedure and does not carry over to the second procedure.

5 Save this project so that you can use it in the next exercises.

10.1.5 Module Scope and Module-Level Variables

Besides being used in a procedure, the `Dim` statement can be used in the General Declarations section of a module. The variable can then be accessed by any procedure written for that module. With module-level scope, the life of the variable ends when the module is removed from memory. For a form module, this would occur when the form is unloaded.

Exercise 10.2 Illustrating Module-Level Scope

This exercise is like the last and uses the same interface (refer to Figure 10.3). The difference lies in the placement of the variable declarations. The following code segments define the variable `Cat3` within the General section of the module level.

1 Modify the interface shown in Figure 10.3 to make the caption of the second button read **Jewel and Buff**. This time, `"Buff"` the cat will be displayed. Change the caption of the form to read **Test of Module Scope**.

2 Modify the `cmdCats1_Click()` procedure as follows:

```
Private Sub cmdCats1_Click()
   Dim Cat1 As String
   Cat1 = "Night Eyes"
   Print "The name of my first cat is "; Cat1
   Cat3 = "Buff"
   Print "The name of my third cat is "; Cat3
End Sub
```

Observe that `Cat3` is not declared.

3 Modify the `cmdCats2_Click()` as follows:

```
Private Sub cmdCats2_Click()
   'Cat1 is the same variable name as used in the first procedure
   Dim Cat1 As String
   Cat1 = "Jewel"
   Print "The name of my second cat is "; Cat1
   Print "The name of my third cat is "; Cat3
End Sub
```

Once again, `Cat3` is not declared.

4 Declare `Cat3` in the General section of the form module. This code reads as follows:

```
Option Explicit
Dim Cat3 As String
```

5 Run and test your program.

`"Buff"` the cat should be displayed by using either procedure. This behavior holds not only for form modules, but for standard modules as well (.BAS files). Declaring a variable with the keyword `Dim` in the General declarations section of a standard module makes that variable accessible to all procedures written for that standard module.

10.1.6 Global Scope and Global Variables

Global variables can only be declared in the General declarations section of a standard module. They are declared with the keyword `Public`. Once declared, global variables can be assigned or read by procedures in the entire application—including event procedures and all forms or general procedures written within all standard modules.

Before declaring a global variable, a wise practice is to add a standard module (.BAS file) to your project. This places all global variables in one place and makes it easy to check them later. To add such a file, click the Module icon on the tool bar, or choose Project, Add Module. A dialog box will appear with two tabs: one to assign an entirely new code module and one to add an existing code module. Select the tab to add a new code module. A module with the default file name `Module1.Bas` will be added to the Project Explorer Window. This new module has a `Name` property. As with a form, changing the name does not change the file name.

Exercise 10.3 Illustrating Global Scope

This exercise is similar to Exercise 10.2 and uses the same interface. The difference lies in the placement of the variable declarations and in the addition of a standard (.BAS) module, where `Cat3` is declared as a `Public` variable (see Figure 10.4).

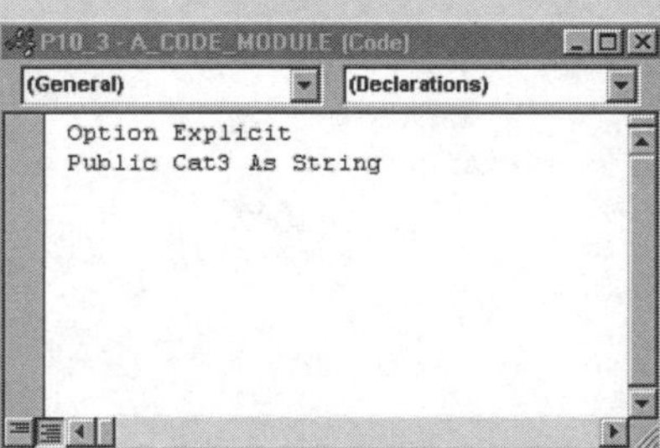

Figure 10.4

Declare global variables inside a standard code module.

1 Change the caption of the form to read **Using a Global Declaration**.

2 Neither the `cmdCats1 Click()` or `cmdCats2 Click()` procedure needs to be changed from Exercise 10.2. However, remove the declaration of `Cat3` from the General section of the form module.

3 Add a standard code module file, and declare `Cat3` as follows:

```
Option Explicit
Public Cat3 As String
```

4 Test your program. Change `Public` to `Dim` and determine what happens.

10.1.7 Overriding Scope Declarations

Local scope always takes priority over higher-level scope declarations. For example, if you declare a variable as global as in

```
Public Cat3 As String
```

followed by the same declaration in a procedure, written as

```
Dim Cat3 As String
```

the Local declaration will take priority, and no error will be reported.

Exercise 10.4 Overriding a Scope Declaration

Use the same interface as that in Exercise 10.3.

1 Change the form caption to read **Overriding a Global and Module Declaration**. Change the caption for `cmdCats2 Click()` to read **Jewel and Pansy**.

2 At the form module level, write the following:

```
Option Explicit
Dim Cat3 As String
```

This code should now appear in two places: in the Form1.frm module level and as a global variable in Module1.Bas.

3 Revise `cmdCats2_Click()` to read as follows:

```
Private Sub cmdCats2_Click()
   'Cat1 is the same variable name as used in the
   'first procedure
   Dim Cat1 As String
   Dim Cat3 As String
   Cat1 = "Jewel"
   Print "The name of my second cat is "; Cat1

   'Cat3 is a global variable in a .BAS file
   Cat3 = "Pansy"
   Print "The name of my third cat is "; Cat3
End Sub
```

4 Run and test your program. `Jewel and Pansy` should be displayed, not `Jewel and Buff`.

Objective 10.2 Understanding Static Scope

With static scope, the value of a locally declared variable is retained in memory after the procedure in which the variable is declared ends. So in this sense, the variable behaves like a module-level variable. Unlike module level variables, however, static variables expose themselves only to the local procedures in which they are declared. They persist as long as the module in which they are declared is in memory, but limit themselves to being assigned a value or incremented or decremented.

A classic use of a static variable is as a variable that serves to keep a count. Suppose you want to know how often an event procedure, such as a button click, is called. A static variable enables you to keep track of this number inside the procedure itself. The following code illustrates the use of such a *counter* variable:

```
Public Sub First()
   Static Num As Integer

   Num = Num + 1    'Increment Num by 1
End Sub
```

In this code, `Num` is incremented by `1` each time the procedure is processed. The first time the procedure is processed, the value of `Num` is set to `1`. The second time the procedure is processed, the value of `Num` is set to `2`. *Remember that static variables cannot be used at the module level.*

You can also use the `Static` keyword in a procedure declaration as in `Private Static Command1_Click()`. With this type of static declaration, all local variables within this button click procedure would retain their values between button clicks, even if declared with the `Dim` statement.

Objective 10.3 Understanding the Scoping of Procedures

So far we have discussed variable scope, a concept that determines which part of a program can assign values to, or read values from, a variable. In object-oriented terms, another way of describing scope is to say that objects *expose* themselves to other objects. They can expose themselves in either expansive or limited ways. A variable declared locally only exposes itself to other variables in that procedure. A variable declared at the module level exposes itself to all procedures in that module, whether it is a form module or a standard module (.BAS file). Globally declared variables are exposed to the entire application.

Procedures can also expose themselves. Because Visual Basic adds event procedures for you automatically inside the code editor, it is easy to forget that procedures, like variables, are declared. With variables, you use such keywords as `Dim` and `Public`. For procedures you use the keyword `Sub`. Functions are declared with the keyword `Function`. The `Sub` and `Function` statements declare the procedure or function, but they don't specify to what other objects the procedure or function should be exposed. This requires other keywords.

10.3.1 Adding and Calling a Procedure in a Form Module

By default, all built-in event procedures (such as `Form_MouseMove`) in form modules are declared so that they are exposed only to other procedures on the form. This scoping for standard event procedures is achieved by having the default keyword `Private` placed in the procedure declaration.

You can also add your own procedures to a form module. These procedures are sometimes referred to as *general procedures* to distinguish them from the event procedures associated with forms and controls. When you add your own procedures, you must decide how you want it exposed or scoped. By declaring a procedure with the `Private` keyword, you limit the exposure of the procedure to the module in which the procedure is declared.

To call a procedure, simply place the procedure name in code. The alternative is to use the optional `Call` statement. The following is the syntax for this statement:

```
[Call] name [(argumentlist)]
```

Table 10.1 describes the components of the `Call` statement.

Table 10.1 The Syntax of the *Call* Statement

Component	*Description*
`Call`	An optional keyword used to signify the transfer of program control to another procedure
`name`	The name of the procedure to pass control to
`argumentlist`	The variables or expression to pass to another procedure

In this book, you are always asked to use the keyword `Call` when calling another procedure. This makes `Call` statements easier to identify and procedures easier to follow.

To add your own procedure to a form module, select Tools, Add Procedure, and the dialog box you see in Figure 10.5 will appear. By default, the procedure you add is a `Sub` routine with `Public` scope.

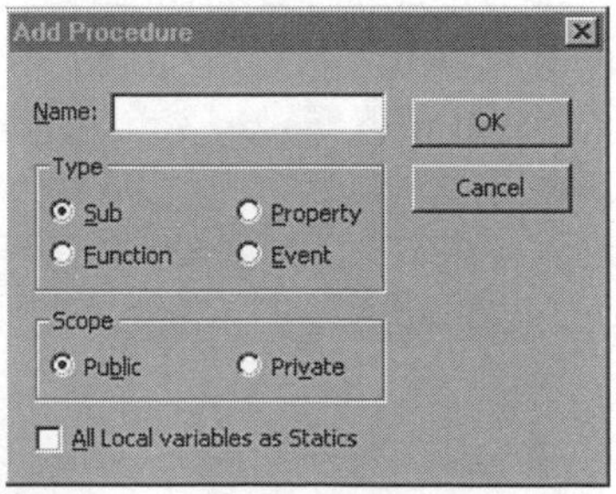

Figure 10.5

The Add Procedure dialog box is used to add a procedure (`Sub`) or a function to a project and to give it either `Public` or `Private` scope.

Exercise 10.5 Adding a General Procedure in a Form Module

In this exercise, you code an event procedure that calls a general procedure placed in a form module. Figure 10.6 illustrates the output for this exercise. The interface consists of two labels, `lblPrompt` and `lblDisplay`, and a text box, `txtInput`.

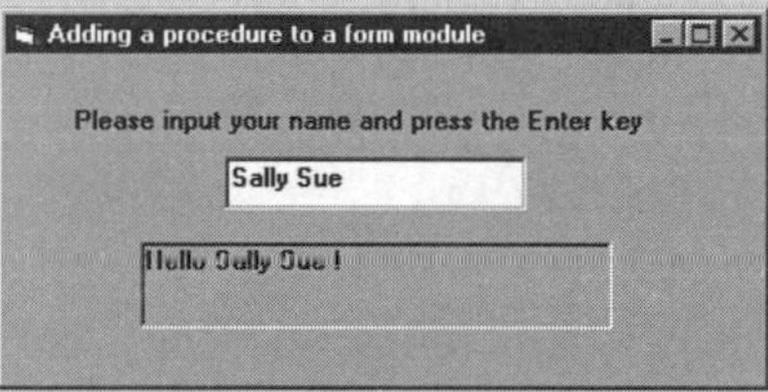

Figure 10.6

This interface tests the use of a general procedure added to a form module using the Add Procedure command.

1 Place the controls on the form as shown. The **`lblPrompt`** control is above **`txtInput`** and **`lblDisplay`** is below it. Change the caption of `lblPrompt` to **Please input your name and press the Enter key**. The `lblDisplay` control has its `BorderStyle` set to `Single`.

2 Add the following `txtInput_KeyPress()` procedure:

```
Private Sub txtInput_KeyPress(KeyAscii As Integer)
   If KeyAscii = vbKeyReturn Then
      Call DisplayName
   End If
End Sub
```

3 Add a new procedure and name it `DisplayName()`. With the code window open, choose Tools, Add Procedure, and set the scope to `Private`.

4 Write the following code:

```
Private Sub DisplayName()
   Dim username As String
   Dim msg As String
   username = txtInput.Text
   msg = "Hello " & username & " !"
   lblDisplay.Caption = msg
End Sub
```

5 Run and test your design.

The general procedure you added is called by the `KeyPress()` event procedure. As your programs become larger, code placed in the user event is often quite short. The code serves to call procedures in either a form or standard code module. We consider adding procedures to a standard code module next.

10.3.2 Adding and Calling a Procedure in a Standard Module

Adding a procedure to a standard code module is similar to adding a procedure to a form module. As before, choose Tools, Add Procedure from the menu; the dialog box will appear. The default scoping of the procedure is set to `Public`. This permits the procedure to be seen and called by any other module.

An important difference between a general procedure in a standard module and one in a form module is that a form must be explicitly loaded into memory before a public procedure in the form can be called by another module. Visual Basic handles the loading of standard code modules automatically. Any publicly scoped procedure in a standard code module can be called.

Another difference between general procedures placed in a form module and procedures placed in a standard code module is that control names need to be explicitly referenced. General procedures placed in a form module usually refer to controls on the form. Because standard code modules do not contain controls, any reference to them must include the form that contains the control. For example, instead of referring to a text box as simply `txtInput`, refer to it by its complete name, such as `Form1.txtInput`.

Exercise 10.6 Calling a Procedure in a Standard Code Module

This exercise is an extension of the earlier exercise. Instead of calling a general procedure on the form, however, you will add a procedure to a code module and call the procedure from a form.

1 Save the form and project created in Exercise 10.5 under new names (for example, **f10-6.frm** and **p10-6.vbp**). Keep the name of the form as `Form1`.

2 Add a standard code module to the project (choose Project, Add Module), and save the module as **m10-6.bas**. If you want, you may also name the module (for example, **M10_6**).

3 Important: Remove the `DisplayName()` procedure from the form. (You can use Edit, Cut, to place the code in the Windows Clipboard for use in Step 4.)

4 Add a new procedure to the code module. With the code window having the focus, choose Tools, Add Procedure, and set the scope of the procedure to `Public`. Add the code so that it reads as follows:

```
Public Sub DisplayName()
   Dim username As String
   Dim msg As String
   username = Form1.txtInput.Text
   msg = "Hello " & username & " !"
   Form1.lblDisplay.Caption = msg
End Sub
```

5 The code in the `txtInput_KeyPress()` event is the same as that in Exercise 10.5.

6 Run and test the program.

Notice the difference in the `DisplayName()` procedure written for the standard module. In the following statements, the references to the controls on `Form1` had to include the form reference:

```
username = Form1.txtInput.Text
Form1.lblDisplay.Caption = msg
```

Because the controls are in a separate module, the `DisplayName()` procedure must know which module contains `txtInput` and `lblDisplay`.

The coding of a module reference is unwieldy and inflexible. Let's assume you want to have a generic procedure that could display the string `Hello` on any form that contained a label named `lblDisplay`. Even better would be a procedure that could display any message on any label on any form. You learn how such procedures are written next.

10.3.3 Passing a Data Type to a Procedure

In Exercise 10.5 and Exercise 10.6, you executed code in general procedures, one in a form module and one in a standard module, respectively. Both procedures referred to specific controls on a specific form and displayed an identical message.

General procedures can be made more flexible if they are written to contain *arguments,* which are variables that are placed in the procedure declaration. These variable values are then *passed* to the procedure when the procedure is called. Reexamine the `DisplayName()` procedure in Exercise 10.5:

```
Private Sub DisplayName()
   Dim username As String
   Dim msg As String
   username = txtInput.Text
   msg = "Hello " & username & " !"
   lblDisplay.Caption = msg
End Sub
```

This procedure reads the value from the text box `txtInput`. A more flexible way of writing the procedure would be to have the procedure that called `DisplayName()`, the `txtInput_KeyPress()` event, pass the string expression in the `txtInput` text box to the `DisplayName()` procedure:

```
Private Sub DisplayName(username as string)
   Dim msg As String
   msg = "Hello " & username & " !"
   lblDisplay.Caption = msg
End Sub
```

```
Private Sub txtInput_KeyPress(KeyAscii As Integer)
   Dim person as string
   If KeyAscii = vbKeyReturn Then
      person = txtInput.text
      Call DisplayName(person)
   End If
End Sub
```

The declaration of `DisplayName()` now contains an argument. Whenever the `DisplayName()` procedure is called, it expects to receive a string expression. An error will now occur if the procedure is called without an argument. You may have observed that some of Visual Basic's built-in event procedure declarations contain arguments. Examine, for instance, the `txtInput_Keypress()` event. The procedure is designed to pass the ASCII value (an integer value) of the key that is pressed. A `MouseDown()` procedure contains even more arguments:

```
Private Sub Form_MouseDown(Button As Integer, Shift As Integer, X As Single,
➥Y As Single)
```

In this case, the procedure is passed the mouse button the user has clicked (`Button` argument), the status of the Shift, Alt, and Ctrl keys (the `Shift` argument), and the location of the mouse cursor (the `X` and `Y` arguments).

Exercise 10.7 Passing Primitive Data Types to a Procedure

In this exercise, you pass the `DisplayName()` procedure two primitive `Data` type arguments, a string expression, and an integer. Figure 10.7 illustrates the output for this exercise. The interface consists of three labels, `lblPrompt1`, `lblPrompt2`, and `lblDisplay`, two text boxes, `txtNameInput` and `txtAgeInput`, and a command button `cmdOK`.

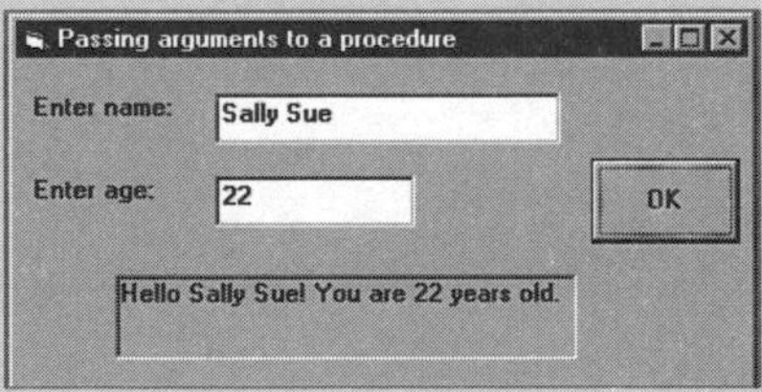

Figure 10.7

In this design, a person's name and age are passed to a procedure.

1 Place the controls on the form as shown in Figure 10.7. The **`lblPrompt1`** control is to the left of **`txtNameInput`** (caption of **Enter name:**) the **`lblPrompt2`** is to the left of **`txtAgeInput`** (caption of **Enter age:**). The **`lblDisplay`** control (at the bottom) has its `BorderStyle` set to `Single`.

2 Add the following `cmdOK_Click()` procedure:

```
Private Sub cmdOK_Click()
   Dim person As String
   Dim age As Integer
   person = txtNameInput.Text
   age = Val(txtAgeInput.Text)
   Call DisplayName(person, age)
End Sub
```

3 Add the `DisplayName()` procedure to the form module. With the code window open, choose Tools, Add Procedure and set the scope to `Private`. Write the following code:

```
Private Sub DisplayName(username As String, userage As Integer)
   Dim msg As String
   msg = "Hello " & username & "! You are " & userage & " years old."
   lblDisplay.Caption = msg
End Sub
```

4 Run and test your design.

The `cmdOK_Click()` procedure reads the values from the two text boxes, assigns them to the variables, and then passes the variables to the `DisplayName()` procedure. Observe the error that occurs if you remove one of the arguments from the instruction that calls `DisplayName()` (for example, `Call DisplayName(person)`).

When you call procedures and pass values, what really is being passed is the memory addresses of the passed variable. In Exercise 10.7, this method of passing values (called *passing by reference*) enables the `DisplayName()` procedure to access the values and create the displayed message. Visual Basic also enables you to pass copies of the variable rather than pass the address with the `By Val` statement. This difference in passing values to procedures and functions is covered in Chapter 22, "Modular Design."

10.3.4 Passing a Control Object to a Procedure

Visual Basic is made even more flexible with its capability to pass complex data types, such as objects, to procedures. You have worked with objects, mainly in the Visual Basic toolbox, when you've added visible controls to a form. You have manipulated these objects by changing their properties and invoking their methods in code.

For now, assume you want to write a generic procedure that receives values from passed arguments and then, after manipulating them, displays them in a specified label control. The label itself (really its address) can be passed to a procedure along with the primitive data types. Such a procedure can be placed in a standard code module and be made available to the entire application.

Exercise 10.8 Passing an Object to a Procedure

In this exercise, you extend Exercise 10.7 by adding the label as an argument in the `DisplayName()` procedure and place the `DisplayName()` procedure in a standard code module. The interface remains the same as that in the preceding exercise, but you now include a code module.

1 Open Exercise 10.7 and save the form under a new name (for example, save f10-7.frm as **f10-8.frm**). Save the project under a separate name as well (for example, save p10-7.vbp as **p10-8.vbp**). Rename the project in the project's `Name` property.

2 Add a code module to the project, and save it as **m10-8.bas**. Important: Remove the `DisplayName()` procedure from `Form1` (you may want to cut it for use below), and place the `DisplayName()` procedure in the `m8-8` code module. Change the scope of the procedure to `Public`, and add the third argument to the procedure. The code reads as follows:

```
Public Sub DisplayName(username As String, userage As Integer, dlabel As
➥Label)
   Dim msg As String
   msg = "Hello " & username & "! You are " & userage & " years old."
   dlabel.Caption = msg
End Sub
```

3 Change the `cmdOK_Click()` procedure as follows:

```
Private Sub cmdOK_Click()
   Dim person As String
   Dim age As Integer
   person = txtNameInput.Text
   age = Val(txtAgeInput.Text)
   Call DisplayName(person, age, Form1.lblDisplay)
End Sub
```

4 Run and test your design.

Notice the differences from the previous program. Because the scope of `DisplayName()` is now `Public`, it can be seen from outside the code module. The procedure declaration contains the declaration of a `Label` data type as an argument. This kind of declaration, a declaration of an object, is covered in detail in Chapter 12, "Object Types, Variables, and Collections." For now, you should know that this type of declaration and usage is common in Visual Basic.

The procedure that calls `DisplayName()`, the `cmdOK_Click()` event, passes the full reference of the Label control (`Form1.lblDisplay`) so that the code in the standard module can resolve where `lblDisplay` resides.

In Exercise 10.8, you passed a control to a procedure in a standard code module along with two primitive data types. By passing data via procedure arguments, data can be shared between two modules *without* using global variables. All of the variables in the exercise are local. Remember: Variables should always be declared as locally as possible. Using procedure arguments helps you keep variables local and restricts the use of global variables.

You can also pass data between form modules by using only local variables. Often this is accomplished by passing locally declared variables along with a form reference to code in a standard module. The standard code module, in turn, can show the value of the form variable. Demonstrating this, however, requires that you first learn about working with more than one form in a program. We turn to that topic next.

Objective 10.4 Working with Multiple Forms in an Application

Most Visual Basic applications require the use of multiple forms. With Visual Basic, you can have as many forms as you need for a project; however, only one form is loaded and displayed at project startup. Loading more than one form requires the `Load` statement and the `Show` method.

10.4.1 The *Load* Statement

The `Load` statement loads a form (or controls that are elements of a control array) into memory. At startup, Visual Basic only loads one form: the startup form that you designate. To examine which form is the startup form, select the Project Properties dialog box by choosing Project, [name of project], and selecting Properties. Under the General tab, you will see a combo box for the startup object (see Figure 10.8). The forms designed for the project and `Sub Main()` are displayed. You can select one and only one of these choices as the startup object. If you select `Sub Main()`, you must add a module that is called `Main()`. Visual Basic then starts from this procedure rather than from a form.

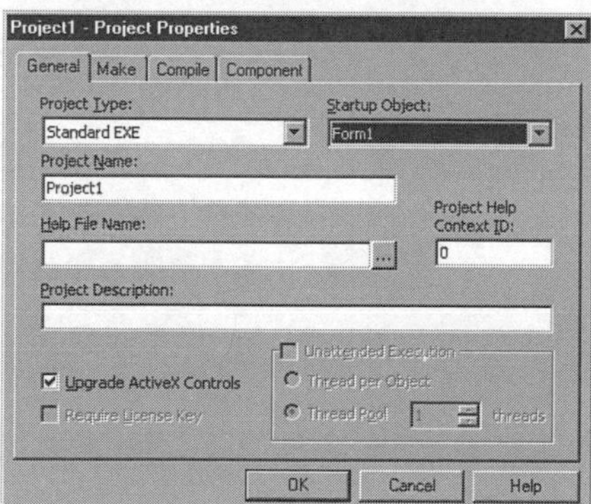

Figure 10.8

Specify the startup object using the General tab of the Project Properties dialog box.

The `Load` statement permits more than one form (or control) to be loaded at startup. The syntax is as follows:

```
Load object
```

For example, the following instruction loads the second form, provided `frmMain` is the startup form:

```
Load frmDisplay
```

10.4.2 The *Show* Method

The `Show` method displays a form. The following syntax is used to show a form stored in memory:

```
[form.]Show [style]
```

The following are the two style options for this method:

Modal (a style of 1)	A window or dialog box that requires the user to take an action before your application can proceed
Modeless (a style of 0)	A window or dialog box that does not require the user to take an action

What happens if you fail to load a form before you show it? During program execution, Visual Basic checks to determine whether the form has been loaded. If not, the `Show` method loads an unloaded form automatically. However, the reverse is not true. If you load a form into memory and fail to show it, it will remain hidden, but its `Public` procedures will be available to be called. Even so, you should normally load forms explicitly. This makes your code easier for you and others to read.

Exercise 10.9 Working with More than One Form

This exercise introduces working with more than one form. It asks you to receive input on one form, pass the data to a procedure in a standard code module, then display the data on another form. Figure 10.9 shows how your finished design should appear on-screen.

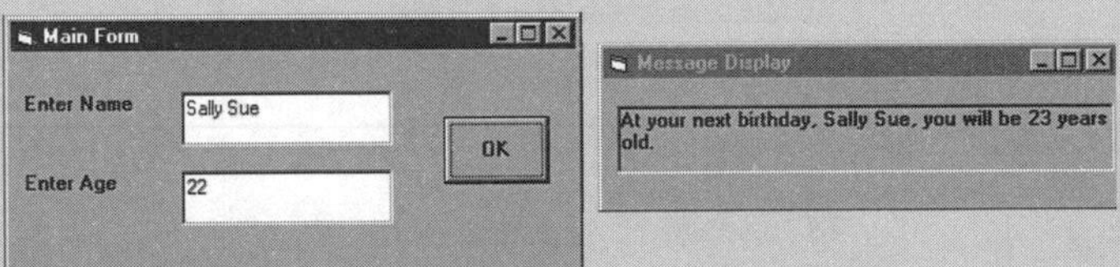

Figure 10.9

Working with more than one form requires using the `Load()` statement and the `Show` method.

1 Begin a new Standard EXE project. Save the default form as **f10-9A**, and name it **`frmMain`**. Add another form to the project (select Project, Add Form), save it as **f10-9B** and name it **`frmDisplay`**. Add a code module to the project and save it as **m10-9.bas**. Name the project **P10_9.vbp**, and save it to disk as **p10-9.vbp**.

2 Change the caption of `frmMain` to **Main Form**. As in the two previous exercises, the two text boxes are named **`txtNameInput`** and **`txtAgeInput`**, and the name of the command button is **`cmdOK`**.

3 The name of the label on **`frmDisplay`** is **`lblDisplay`**. Its `BorderStyle` is set to `Single`, and its `WordWrap` property is set to `True`.

4 Add the following `Form_Load()` procedure to `frmMain`:

```
Private Sub Form_Load()
   'Position startup form
   frmMain.Left = 2000
   frmMain.Top = 2000
End Sub
```

5 Add the following `cmdOK_Click()` procedure to `frmMain`:

```
Private Sub cmdOK_Click()
   Dim person As String
   Dim age As Integer
   person = txtNameInput.Text
   age = Val(txtAgeInput.Text)
   Call DisplayMessage(person, age)
End Sub
```

6 Choose Tools, Add Procedure to add a publicly scoped procedure named `DisplayMessage()` to the m10-9 standard code module:

```
Public Sub DisplayMessage(username As String, userage As Integer)
   Dim msg As String
   Dim nextBDay As Integer
   nextBDay = userage + 1
   msg = "At your next birthday, " & username & ", you will be "
   msg = msg & nextBDay & " years old."
   Load frmDisplay
   frmDisplay.Left = frmMain.Left + frmMain.Width + 100  ' position left
   frmDisplay.Top = frmMain.Top ' position top
   frmDisplay.lblDisplay.Caption = msg
   frmDisplay.Show
End Sub
```

7 Choose Project, P10_9 Properties to ensure that the startup object is `frmMain`.

8 Run and test your program.

Notice how the label again must be fully referenced inside the code module. You also could have passed a form reference and a label reference to the `DisplayMessage()` procedure. You are asked to do this in the "Skill-Building Exercises" section at the end of the chapter.

10.4.3 Using a *Main()* Procedure to Start an Application

Besides using a form at startup, you can specify that a procedure—always named `Main()` and placed in the standard code module—be the startup object. In most programs, a `Main()` procedure is used to check environmental parameters and to load program constants and other frequently called data into memory.

Exercise 10.10 *Sub Main():* Starting Your Program with Code Rather than a Form

This exercise asks you to use the same interface as Exercise 10.9 and make several minor adjustments to the code.

1 Save each of the elements from Exercise 10.9 as new elements in a new project. Save f10-9a.frm as **f10-10a.frm**, f10-9b.frm as **f10-10b.frm**, m10-9.bas as **m10-10.bas**, and the whole project as **p10-10.Vbp**. Rename the project **P10_10.vbp**.

2 By choosing Tools, Add Procedure, add a publicly scoped procedure named **Main** to the standard code module. Using the cut and paste features of the Windows Clipboard, remove the code from the `frmMain_Load()` event and place it in the `Sub Main()` procedure. Also move the relevant code from the `DisplayMessage()` procedure so that the code reads as follows:

```
Public Sub Main()
   Load frmMain
   frmMain.Left = 2000
   frmMain.Top = 2000
   Load frmDisplay
   frmDisplay.Left = frmMain.Left + frmMain.Width + 100
   frmDisplay.Top = frmMain.Top
   frmMain.Show
   frmDisplay.Show
End Sub

Public Sub DisplayMessage(username As String, userage As Integer)
   Dim msg As String
   Dim nextBDay As Integer
   nextBDay = userage + 1
   msg = "At your next birthday, " & username & ", you will be "
   msg = msg & nextBDay & " years old."
   frmDisplay.lblDisplay.Caption = msg
End Sub
```

3 Choose Project, P10_10 Properties, and make `Sub Main()` the startup object on the General tab of the Project Properties dialog box.

4 Run and test your code. Both forms should appear at project startup. The capability to pass values to the `DisplayMessage()` procedure should remain.

10.4.4 Benefits of *Sub Main()* Code and General Procedures in Standard Modules

Some Visual Basic programmers stress the importance of placing code in a `Main` procedure. A `Sub Main()` procedure enables an application to do error checking before launching the program. If you had a graphics-oriented application, for example, you might check the user's screen resolution and adjust the size of your forms and controls accordingly. A `Sub Main()` procedure is a good place to perform such system checking routines.

Should most of your procedures be placed on forms or in standard modules? Placing most of your procedures in standard modules makes some sense. The reasoning is quite straightforward. Code that is tied to a form is bound to the form. This does not permit code to be edited unless the form is opened. If the form is deleted, the code attached to it is also deleted. Code placed in a standard module is saved in a separate .BAS file. This code can be edited without opening a form. If a form is deleted, the code can be revised to reflect this occurrence; however, the code is not lost.

Often user interface events are small procedures that call lengthier procedures in standard module files. For example, when writing a `cmdButton_Click()` procedure, the code necessarily is tied to the button and the form. When the button is deleted, the code associated with this button should also be deleted. Suppose that the button called a procedure in an existing .BAS file. Deleting the button still allows another event, such as a menu-item click, to call the same code. Developing your programs in this way enables you to quickly make changes to the user interface without requiring a lot of new coding.

Objective 10.5 Declaring and Using a User-Defined Data Type

This chapter emphasizes scoping of variables. As you have seen, these variables can be either *primitive* (sometimes called simple) data types or *complex object* data types, such as components (including controls). Visual Basic includes an object type that can be considered halfway between a primitive data type and a full-fledge software component in its complexity: a *user-defined type*.

User-defined types combine variables of several different data types to form a single variable. Usually the variables making up the user defined type are primitive or simple data types. A user defined type is sometimes called a *structure* (in the C programming language), or a *record* (in the Pascal programming language). A user-defined type requires the `Type` statement.

10.5.1 Declaring a User-Defined Data Type

The syntax for a user-defined type begins with the keyword `Type` and ends with the keyword `End Type`. The complete syntax is as follows:

```
[Private | Public] Type UserType
    elementname [(subscripts)] As typename
   [elementname [(subscripts)] As typename
End Type
```

Table 10.2 describes the components of the syntax.

Table 10.2 The Syntax of the User-Defined Data Type

Component	*Description*
`Private` | `Public`	Indicates the scope of the user-defined type. If declared `Private`, it is only available to other procedures in the module in which it is declared. The default is `Public`.
`Type`	Indicates the beginning of a user-defined type.
`Usertype`	The name of the type, following standard naming conventions.
`elementname`	The name of an element of the user-defined type.
`typename`	One of the following: `Integer`, `Long`, `Single`, `Double`, `Currency`, `Boolean`, `String` (fixed-length or variable), `Date`, `Object` (reference), another user-defined type, `Variant`, or an object type (such as an Excel spreadsheet OLE object).
`End Type`	Indicates the ending of the definition.

The preceding syntax helps to explain what a user-defined type means. The type can consist of more than one element, and the elements can be of different types. A user-defined type can also contain another user-defined type as one of its types.

A user-defined type must be placed at the module level of an application. However, this placement does not lead to an allocation of memory: Memory is reserved once a user-defined type is declared. The declaration is similar to other type declarations:

```
Dim MyType As UserType
```

Consider the following example:

```
Public Type Asset
   Name As String * 20
   Location As Integer
   Value As Currency
End Type
```

This definition indicates that type `Asset` contains three members: one of type `String`, a second of type `Integer`, and a third of type `Currency`. The declaration using this type, then, might be the following (which declares the variable `Inventory` as being of type `Asset`):

```
Dim Inventory As Asset
```

After a user-defined type has been defined, it can be declared as a single variable or, more commonly, is placed into an array, which is a group of variables that share the same name. (Control arrays are introduced in Chapter 11 "Working with Loops," and variable arrays are addressed in Chapter 16, "Lists and Arrays," and Chapter 17, "Multidimensional Arrays, Arrays of User-Defined Types, and the Grid Control.") For now you will work with only single variables of user-defined types.

10.5.2 Using a User-Defined Data Type and Referencing Its Elements

Loading a structure with data requires a simple assignment statement, much like other variables. However, this is where the concept of a record begins to take on meaning. The following statement assigns the name `Laser Printer` to the first element of the user-defined type:

```
Inventory.Name = "Laser Printer"
```

Displaying the value stored in an element of a structure requires the structure variable name and the element to be displayed. The following statement prints the value stored in the first element of the user-defined type:

```
Print Inventory.Name
```

Exercise 10.11 Creating and Displaying Values Stored in a User-Defined Data Type

In this exercise, you define a product record, store the value for a single record, and display the contents of the record. Figure 10.10 shows the display you are asked to create.

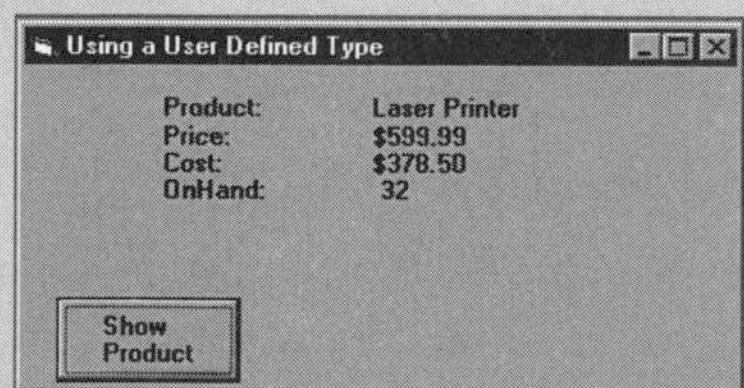

Figure 10.10

The user-defined type in this example is used to define a product record.

1 Add the command button (`cmdShow`) as shown in Figure 10.10, and change the form caption. Change the font for the form to bold.

2 Add a new standard (.BAS) code module, and write the following type definition:

```
Public Type Product
   Name As String * 20
   Price As Currency
   Cost As Currency
   OnHand As Integer
End Type
```

continues

3 Write the `Show Records` procedure as follows:

```
Private Sub cmdShow_Click()
   Dim inventory As Product
   inventory.Name = "Laser Printer"
   inventory.Price = 599.99
   inventory.Cost = 378.5
   inventory.OnHand = 32

   CurrentX = 1000
   CurrentY = 200
   Print "Product: ", inventory.Name
   CurrentX = 1000
   CurrentY = 400
   Print "Price:", Format(inventory.Price, "Currency")
   CurrentX = 1000
   CurrentY = 600
   Print "Cost:", Format(inventory.Cost, "Currency")
   CurrentX = 1000
   CurrentY = 800
   Print "OnHand:", inventory.OnHand
End Sub
```

4 Run and test your program.

This procedure initially declares `Inventory` as a variable of type `Product`. Following this, values for the four elements of the variable are assigned. Once assigned, all that remains is to display the contents of the variable.

10.5.3 Passing a User-Defined Data Type

The scoping of a user-defined data type declaration inside a standard code module is usually `Public`. This permits the data type to be used by any procedure in a program. As a data type, a user-defined data type can, of course, be passed to a procedure. However, the procedure must be scoped as `Private` to the module in which it is declared. The next exercise extends the previous exercise, and passes the user-defined type variable to a procedure that increases the price of the product by 4.5 percent.

Exercise 10.12 Passing a Variable of a User-Defined Data Type

This exercise changes Exercise 10.11 by adding a new procedure, which is passed a user-defined type variable of type `Product`.

1 Save the files from Exercise 10.11 under new names. For instance, save f10-11.frm as **f10-12.frm**, m10-11.bas as **m10-12.bas**, and p10-11.vbp as **p10-12.vbp**. Rename the project as **p10_12**.

2 Change the `cmdShow_Click()` event to read as follows:

```
Private Sub cmdShow_Click()
   Dim inventory As Product
   inventory.Name = "Laser Printer"
   inventory.Price = 599.99
   inventory.Cost = 378.5
   inventory.OnHand = 32
   Call AdjustPrice(inventory)
End Sub
```

3 Add the `AdjustPrice()` procedure, which increases the price of the item by 4.5 percent. This procedure is also at the form level. *Remember to set its scope to* `Private`.

```
Private Sub AdjustPrice(item As Product)
   ' raise price by 4.5%
   Dim newprice As Currency
   newprice = item.Price * 1.045
   CurrentX = 1000
   CurrentY = 200
   Print "Product: ", item.Name
   CurrentX = 1000
   CurrentY = 400
   Print "Price:", Format(newprice, "Currency")
   CurrentX = 1000
   CurrentY = 600
   Print "Cost:", Format(item.Cost, "Currency")
   CurrentX = 1000
   CurrentY = 800
   Print "OnHand:", item.OnHand
End Sub
```

4 Run and test your program.

Observe the syntax required to assign and reference an element of a user-defined type. When a user-defined type is passed, any element of that variable can be accessed.

Being able to aggregate data into a user-defined type is especially useful for storing data as records. In Exercise 10.11 and Exercise 10.12, you held information about a product, but the type could as easily be a person, building, or sales transaction.

Objective 10.6 Understanding Procedures as Object Methods

You might have noticed that a user-defined type can be employed to hold information about real things in the world, such as persons, places, or things. In short, user-defined types can hold information about any *object*. You can consider user-defined types as the beginning of object-oriented programming. You have seen how objects possess properties. Object properties are held in data structures that are very similar to user-defined type data structures.

Objects have two key differences from user-defined types, as follows:

- The scoping of their individual elements (members) can be either public or private. The public members become part of the object's *interface;* the private members are used only by the object itself.
- An object contains code that works on the data held by the object. This code is an object *method.*

You have already worked with object methods, such as the `Clear` method of a list box or the `Print` method of a form. Believe it or not, you have already created your own object methods. In Visual Basic, every time you add a procedure to a form module, such as an event procedure or a general procedure, you are declaring and coding an object method.

Procedures that are publicly declared on a form module can be called from another module, as long as the form is loaded into memory. When you call the procedure, you follow a syntax that is quite familiar:

```
object.method [argument]
```

By default, event procedures on forms, such as button clicks, are scoped to be `Private` to the form module in which they are declared. However, the scope of these event procedures can be changed to `Public`. They then can be called from any other module. Exercise 10.13 illustrates this principle.

Exercise 10.13 Creating and Calling a Public Procedure on a Form

In this exercise, you set the scope of a procedure on a form to `Public` and then call it from another form module.

1 Start a new project and add two forms. Name one form **`frmRequest`** and the other **`frmRespond`**. Save frmRequest as **f10-13a** and frmRespond as **f10-13b**. Name the project **p10_13**.

2 On `frmRequest`, add a command button named **`cmdRequest`**. On `frmRespond` add a command button named **`cmdRespond`** and a label named **`lblDisplay`** (see Figure 10.11).

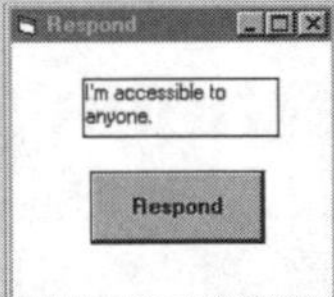

Figure 10.11

This design requires two forms such that a request on one form can be sent as a message to the second form.

3 In the `cmdRespond_Click()` event, set the scope by changing the keyword `Private` to `Public`, and add the following code:

```
Public Sub cmdRespond_Click()
   frmRespond.lblDisplay.Caption = "I'm accessible to anyone."
End Sub
```

4 Add the following code in the `cmdRequest_Click()` procedure:

```
Private Sub cmdRequest_Click()
   frmRespond.Top = frmRequest.Top
   frmRespond.Left = frmRequest.Left + frmRequest.Width + 200
   frmRespond.Show
   frmRespond.cmdRespond_Click
End Sub
```

5 Make `frmRequest` the startup object in the dialog box that appears when you choose Project, p10-13 Properties.

6 Run and test the program.

Notice the syntax used in this exercise. You dropped the use of the optional `Call` keyword. Because a `Click()` event has no arguments, you simply invoked a method of the `frmRespond` object, `frmRespond.cmdRespond_Click`.

A more accurate description of this exercise is to say that you invoked a method of the `frmRespond` object, which was an instance of the `frmRespond` class. A more complete explanation is provided in Chapter 12, "Object Types, Variables, and Collections," and Chapter 13, "Class Modules." For now, you should know that the procedures you write for forms are really object methods!

Chapter Summary

Defining scope is determined, primarily, by the placement of `Dim` statements, which can be placed at the procedure and module level of a program. `Dim` statements at the procedure level define scope, which is local to the procedure. `Dim` statements placed at the form module level define scope local to the form and all procedures written for that form. `Dim` statements placed at the module level define scope local to the module and all forms and procedures associated with that module.

A global declaration using the `Public` statement extends the scope of a variable to all modules, forms, and procedures.

A `Static` declaration is used at the procedure level only. Static scope extends throughout the life of a variable. It remains local, but the value of the variable is retained in memory between procedure calls. A common static variable is a counter variable.

Higher scope declarations can be overridden by lower scope declarations. If a variable is defined as a global variable, the same variable name can be used in a declaration statement in a procedure. The local declaration takes priority over the global declaration.

The concept of scope can be difficult to master. As a general rule, *keep the scope of variables as local as possible by using as few module levels as possible and avoiding global variables.* Why this rule? Because the life of a local variable ends when a procedure ends; it cannot create errors in several places throughout a program. This is in sharp contrast to a global variable, which can affect each procedure in a program.

Procedures also have scope. They can be private to the module in which they are declared, or they can be made public. Public procedures in standard code modules are available to the entire application. Public procedures in form modules are available only if the form is loaded into memory.

The `Call` statement is used to call another procedure. To call a procedure, the name of the procedure and the arguments to be passed, if any, must be indicated.

Primitive (simple) and complex variables and objects can be passed to procedures. By default, the address of these variables are passed—not their actual values.

Variable scope becomes especially important in multi-module projects. In applications with more than one form, the `Load` statement and the `Show` method are used to load forms into memory and to make them visible. Because only one form or module is loaded at project startup, additional forms must be loaded by using the `Load` statement and must be shown by using the `Show` method.

A `Sub Main()` procedure can be used at startup in place of a form. A main benefit of a `Sub Main()` procedure is that code that must be executed before loading a form can be executed in this procedure. Placing most of your code in standard modules allows it to be saved as a file, independent of forms that it might require in processing.

Visual Basic enables the creation of user-defined types: unique data types that are a grouping of other data types, most often primitive data types. The variables that make up a user-defined type are called members. User-defined data types can be passed as variables to procedures that have `Private` scope.

Skill-Building Exercises

1. This project asks you to use four forms and a standard code module. Each form should be designed to display a different color. Name the forms **`frmColor1`**, **`frmColor2`**, **`frmColor3`**, and **`frmColor4`**, but change no other properties.

   Use a standard code module to declare and write a `Main()` procedure to call four procedures: `color1`, `color2`, `color3`, and `color4`. Within each procedure, load and show the form, change the form caption, and change the form background color.

2. This project also asks you to use four forms, named **`frmBlue`**, **`frmRed`**, **`frmYellow`**, and **`frmGreen`**. When the program runs, only `frmBlue` should be visible. Click `frmBlue` to show `frmRed`. Click `frmRed` to show `frmYellow`. Click `frmYellow` to show `frmGreen`. Click `frmGreen` to print text on each of the forms as indicated by the following table:

Text	*Form*
THIS	`frmBlue`
IS	`frmRed`
A GREAT	`frmYellow`
DAY	`frmGreen`

3. Extend Exercise 10.8. Request the user's city of residence, pass the variable along to the `DisplayName()` procedure, and add the city name to the message displayed.

4. Extend Exercise 10.9. Pass the form on which the message is to be displayed as well as the label that should display the message to the `DisplayName()` procedure. All form and label references in the procedure should then use the passed object arguments.

5. Write a program that defines an employee user-defined data type. Include the employee number, first name, last name, pay rate, and date hired as elements in the data type definition. Create a button-click procedure that assigns values to this data type and then displays the values on the form.

6. Extend the preceding exercise. Add a form module general procedure that assigns values to the elements of the user-defined type and then passes the variable to a procedure that increases the pay rate of the employee by 5 percent. Display the results of processing on the form.

11

Working with Loops

Now that you understand variable and procedure scope and the `If...Then...Else` construct (having worked through Chapter 6, "Variables and Constants," and Chapter 8, "`If...Then...Else` Logic and the `Select Case` Statement," respectively) it is time to move to a more fascinating topic: *loops.* The loop logical structures provide the Visual Basic designer with enormous flexibility.

Loops enable code to be repeatedly executed under a variety of conditions. This is why loops are associated with the term *iteration* (literally, *repetition*). This chapter first addresses three types of loops. The first type is known as a `For-Next` loop (which has a syntax that looks more difficult than it really is). It executes a fixed number of times until a condition is met. The second and third types are `Do` and `While` loops, respectively. These loops continue to execute while or until a specified condition is satisfied.

You are also introduced to the concept of a *control array,* which is nothing more than a set of like objects with the same name. Following this, you are introduced to the `Do` loop syntax—which, as you will discover, takes multiple forms—and the easier-to-understand `While` loop syntax.

You will learn several tasks in this chapter, including how to

- code a simple `For-Next` loop structure,
- use a control array,
- increment and decrement a loop,
- code a nested `For-Next` loop,
- use the `DoEvents` statement,
- move shapes with a loop,
- use the `Do-Loop Until` and `Do-Loop While` structures,
- use a nested `Do-Loop Until` structure,

- set the Timer control,
- use a `While-Wend` structure.

Chapter Objectives

The six objectives for this chapter focus on learning to write programs using the variety of loop structures in Visual Basic 5, as follows:

1. Understanding the `For-Next` loop syntax
2. Using a control array in a program
3. Using a nested `For-Next` loop in a program
4. Understanding how `Do` loops work
5. Understanding how `While-Wend` loops work
6. Adding the Timer control to a project

Objective 11.1 Understanding the *For-Next* Loop Syntax

With loops, code executes in cycles. It starts at the top of the loop clause, goes through the body of the loop, and continues to cycle until a condition is met. Certain loops can cycle indefinitely, or until the user decides to interrupt them. As an example, the snooze button on your alarm clock can be used to demonstrate a loop structure. Consider the following pseudocode:

```
Beginning of Loop:     When the alarm goes off, then

     Body of Loop:     Press the snooze button.
                       Sleep a few more minutes.

Ending of Loop:        Return to the beginning of the loop.
```

As this loop indicates, you can sleep forever as long as you continue to press the snooze button when the alarm goes off.

11.1.1 The *For-Next* Loop Syntax

A `For-Next` loop works the same way as the alarm clock example except that the maximum number of times you intend to press the snooze button is set. For example, you will press it exactly five times. When executing a known (designer designated) number of cycles, the logical structure recommended is the `For-Next` structure. The structure sets the maximum number of cycles executed by using a counter variable, which changes with each iteration of the cycle. The syntax of a `For-Next` loop is as follows:

```
For counter = startvalue To endvalue [Step incrementvalue]
   [statement block 1]
     [Exit For]
   [statement block 2]
Next [counter]
```

Table 11.1 explains the components of the `For-Next` loop syntax.

Table 11.1 The Syntax of the *For-Next* Loop

Component	*Description*
`For`	The beginning of the `For-Loop` structure. The word `For` must appear at the start of the loop.
`counter`	A numeric variable of type `Integer` or type `Long`.
`startvalue`	The initial value of `counter`.
`To`	The keyword to separate the `start` from the `end` value.
`endvalue`	The final value of `counter`.
`Step`	The keyword to indicate that an increment (or decrement) value will be given.
`incrementvalue`	The value by which the loop `counter` is incremented (or decremented). The default is `+1`.
`statement block`	The instructions placed within the loop. It is also referred to as the *loop body.*
`Exit For`	An alternative way to exit the loop.
`Next`	The end of the `For-Next` loop structure. It causes the `Step` increment to be added to, or subtracted from, `counter`.

Does this syntax seem complex? It is until you work through some examples. Consider the following code fragment:

```
For SumIt = 10 to 20 Step 5
   Print SumIt
Next
```

Can you determine how the values will be displayed? The values will be the following:

```
10    15    20
```

Consider what happens. The first time through the loop, the variable `SumIt` is set to `10`. This value is printed. The second time through the loop, `SumIt` is incremented by the `Step` increment, which changes `10` to `15`. This value is printed. The final time through the loop, `SumIt` is incremented again by `5`, which is added to `15`. The value `20` is printed.

As a beginning rule, *call the counter variable something recognizable and relevant to the procedure.* For instance, if you want to sum a set of numbers, you might well call the counter variable `SumIt`, as in the preceding code fragment. Avoid such names as `X`, `Y`, or `Foo`.

Another important rule is to *indent the body of the loop and align the* `For` *with the* `Next` *statement*, as shown in the following code:

```
For SumIt = 10 to 20 Step 5
   Print SumIt
Next
```

As you will soon appreciate, indenting becomes very important when the loop body contains many instructions, or when one loop is placed inside another loop.

Exercise 11.1 Displaying the Iterations of a Loop

In this simple example, you explore how the option `Step` works in `For-Next` logical structures. Figure 11.1 shows the form you are asked to design.

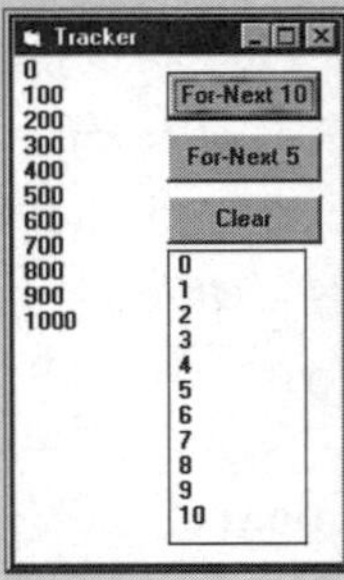

Figure 11.1

These `For-Next` *loops run from 0 to 100 in steps of 10 and steps of 5.*

1 On a tall, narrow form, draw three buttons near the upper-right corner, one underneath the other. Add a Picture Box control to this form. Add the caption **Tracker**.

2 Add the captions **For-Next 10**, **For-Next 5**, and **Clear**.

3 On the top button's `Click()` event, enter the following code:

```
Private Sub cmdForNext10_Click()
  Dim Tracker As Integer
  For Tracker = 0 To 100 Step 10
    Form1.Print Tracker * 10
    Picture1.Print Tracker / 10
  Next Tracker
End Sub
```

4 On the second button, enter the same code, but change the `Step` value to `5` so that the loop condition reads as follows:

```
For Tracker = 0 To 100 Step 5.
```

5 On the last button's `Click()` event, enter `Form1.Cls` and `Picture1.Cls` to clear the form and the picture. This code reads as follows:

```
Private Sub cmdClear_Click()
  Form1.Cls
  Picture1.Cls
End Sub
```

6 Run and test the program. Figure 11.1 shows the results of clicking the topmost button.

How does this program work? With the `Step` value of the top button set at `10`, the procedure takes every tenth value of the range `Tracker` and multiplies it by 10 (displaying it on the form) and divides it by 10 (displaying it in the picture box). With the `Step` value set to `5`, the middle button's procedure takes every fifth value of the `Tracker` range and multiplies and divides it by 5. The middle button's procedure cycles twice as many times as the first, and displays twice as many return values.

Adding `Step` to a `For-Next` procedure is optional. If it isn't added, the loop's default increment is `1`.

11.1.2 *Step* Increment and Decrement

The `Step` increment in a `For-Next` loop can be positive or negative. Consider the following code fragment:

```
For num = 100 To 5 Step -5
   Print num
Next num
```

Can you guess what will be printed? The first value will be `100`, the second `95`, and so on. Looking in a backward direction is useful for creating interesting patterns. When the `Step` increment is negative, the term decrement takes on importance. Remember that when the `Step` increment is *positive,* you *increment* the counter; when it is *negative,* you *decrement* the counter. To decrement, you must use the `Step` keyword, even if you are decrementing by one.

Objective 11.2 Using a Control Array in a Program

A control array is a group of like objects with the same name. This might seem confusing at first, but it's not difficult once you write a few programs. Consider the following code fragment:

```
Private Sub cmdMoney_Click()
  Dim arraynum as integer
  Dim NewPrice as Currency
```

```
  For arraynum = 0 To 4
     NewPrice = Val((txtPrice(arraynum).Text) * 1.075)
     txtPrice(arraynum).Text = NewPrice   ' a control array
  Next                                    ' next loop iteration
End Sub
```

This code is somewhat different from the code you have seen before. In this example, five text boxes are specified, with each one given the name `txtPrice`. The first object of the array is `txtPrice(0)`, while the last is `txtPrice(4)`. A control array is advised when like objects have similar code events. They are distinguished from one another by their index numbers (`0`, `1`, `2`, `3`, and `4`).

It is easy to make a control array. First, design the first object and assign it a name. Then, select and copy the object (Ctrl+C or choose Edit, Copy), and paste it (Ctrl+V or choose Edit, Paste). When you paste the new control, a dialog box asks whether you want to create a control array (see Figure 11.2). If you do, select Yes. The `Index` property of the second object is changed from `0` to `1`, thus providing the second object with a unique identifier, but with the same object name as the first object.

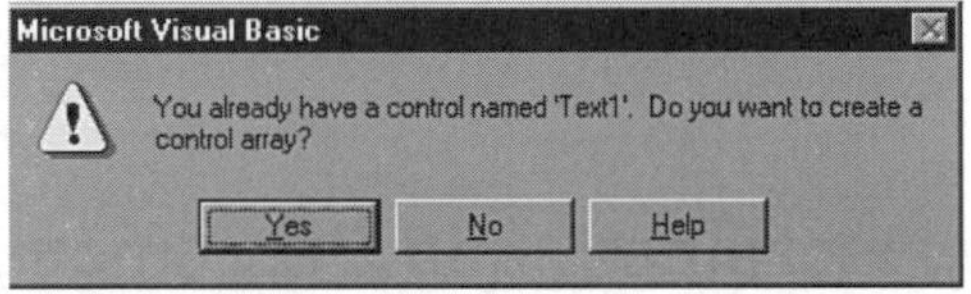

Figure 11.2

When you copy and paste a control on a form, you are asked whether you want to create a control array.

11.2.1 The Counter Variable and Control Arrays

The counter variable is important in several ways in a loop structure. In the preceding `For-Next` example, `arraynum` is declared as the counter variable. It is then used twice to refer to the index of the `txtPrice` control array. The linking the first time through the loop can be read as follows (with `0` indicating the value of `arraynum`):

```
NewPrice = Val(txtPrice(0).Text) * 1.075
txtPrice(0).Text = NewPrice
```

The second time through the loop, the index value is changed from `0` to `1`. The new code reads as follows:

```
NewPrice = Val(txtPrice1(1).Text) * 1.075
txtPrice(1).Text = NewPrice
```

11.2.2 Why Use Control Arrays?

Now do you have a better understanding about the value of control arrays? The preceding code could have been designed with five different text boxes, such as `txtPrice1`, `txtPrice2`, and so on; however, that design would not have allowed you to use a loop to change prices. It would have required five separate sets of coded instructions. Five might not seem like a large number, but imagine an online wholesale ordering system for which individually ordered items are displayed in labels and then marked up by the order taker. Suppose that 80 items have been ordered. With a control array and a loop, all those item prices could be changed with the same code.

Exercise 11.2 Showing Fill Styles

Loops are very useful for dealing with lists of values, including object property values. For instance, the `FillStyle` property of a shape has eight styles (`0` through `7`). Using a loop, it is easy to display these eight fill styles for Visual Basic Shape controls. Figure 11.3 shows the design you are asked to create.

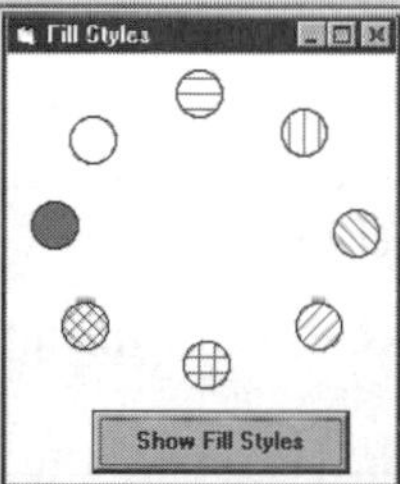

Figure 11.3

This display features a control array of eight shapes, each assigned a different fill style.

1 Draw a shape (a circle) on a form and set its `Visible` property to `False`. Set the form background color and the shape fill color to contrasting colors.

2 Copy and paste the shape, responding Yes to whether you want to create a control array.

3 Create an eight-member array of identical shapes on the form.

4 Add a button and the caption **Show Fill Styles**, and add the following code for the button `Click()` event:

```
Private Sub cmdShow_Click()
  Dim FillNum As Integer
  For FillNum = 0 To 7
      Shape1(FillNum).FillStyle = FillNum  'FillNum is used twice here
      Shape1(FillNum).Visible = True
  Next
End Sub
```

5 Run and test the program. As one test, change the loop from 1 to 8, and observe the problem.

Objective 11.3 Using a Nested *For-Next* Loop in a Program

Multiple `For-Next` loops, with one set of `For-Next` statements placed inside a second, leads to a nested `For-Next` loop. With a nested loop, the innermost loop must finish its execution before the outside loop is incremented or decremented.

11.3.1 The Nested Loop Syntax

The following is the syntax for a nested For-Next loop, with one For-Next loop placed inside an outer For-Next loop:

```
For counter1 = startvalue To endvalue [Step incrementvalue]
  [statement block 1]
     For counter2  = startvalue To endvalue [Step incrementvalue]
       [statement block 2]
     Next [counter2]
  [statement block 1 continued]
Next [counter1]
```

This indentation used in writing a nested For-Next loop is important. It enables the eye to separate the outer loop from the inner loop.

To illustrate how a nested loop works, consider the following code fragment:

```
Dim outer, inner As Integer
For outer = 10 to 20 Step 5
  For inner = 1 to 5
     Print inner * outer
  Next inner
Next outer
```

Can you determine how the values will be displayed? The following is the first set of values displayed:

```
1 * 10 = 10      2 * 10 = 20      3 * 10 = 30      4 * 10 = 40      5 * 10 = 50
```

This completes the processing of the inner loop, at which point the outer loop is incremented by five. The following is the second set of values:

```
1 * 15 = 15      2 * 15 = 30      3 * 15 = 45      4 * 15 = 60      5 * 15 = 75
```

This completes the processing of the inner loop once again. Control is passed by the instruction `Next outer`, which increments the outer loop by five one more time. The following is the third set of values:

```
1 * 20 = 20      2 * 20 = 40      3 * 20 = 60      4 * 20 = 80      5 * 20 = 100
```

11.3.2 The *DoEvents* Statement

The DoEvents statement passes control to Windows. In other words, it forces Visual Basic to yield execution to Windows so that Windows can process all messages in the operating system message queue before being asked to process additional code.

As illustrated by Exercise 11.3, numeric calculations in the processor execute far more rapidly than the video system can repaint the screen. Without the DoEvents statement, most computer graphic systems simply freeze until all the numeric calculations are completed. Following this, the results are displayed. By using DoEvents, you ensure that screen redraws are completed and calculated values displayed in the list box before beginning another inner loop cycle.

You will use `DoEvents` in programs in which it is essential to control the sequencing of events, especially as it affects graphical output.

Exercise 11.3 Using Nested Loops and List Boxes

This exercise asks you to use a nested loop and to place the results of processing into one of five list boxes. It also contains an inner loop with a `DoEvents` statement. This statement ensures that a number is fully displayed before the next display instruction is reached. Figure 11.4 shows the display that you are asked to create.

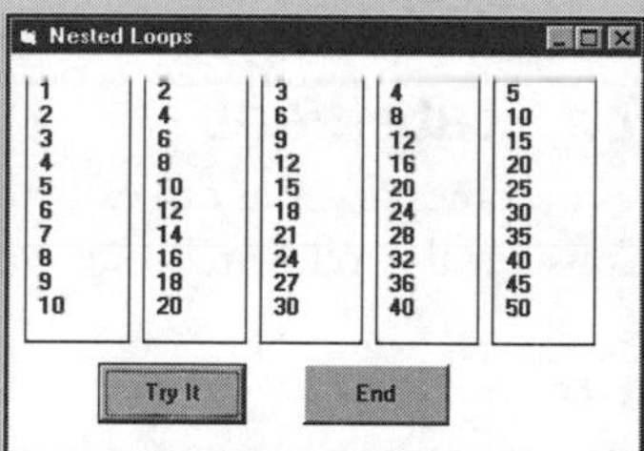

Figure 11.4

This display shows the results of using an outer loop to designate list boxes and an inner loop to populate each list box.

1 Create a control array of five list boxes, add the two command buttons, and add the captions to the command buttons and to the form. Use a lowercase *L* (that is, *l*) in naming the list boxes **`lstNum()`**. Make sure that it is not the numeral 1, which looks similar to an *l*, but is illegal as the first letter of an object name. Make the list boxes different colors to clearly separate the boxes on the screen.

2 Write the `cmdTryIt_Click()` procedure as follows:

```
Private Sub cmdTryIt_Click()
  Dim inner As Integer
  Dim outer As Integer
  Dim item As String
  Dim sleep As Long

  For outer = 0 To 4
     lstNum(outer).Visible = True
     For inner = 1 To 10
       DoEvents
       item = Str((outer + 1) * inner)
       lstNum(outer).AddItem(item)
         For sleep = 1 To 100000
         Next sleep
     Next inner
  Next outer
End Sub
```

continues

Within the second loop (loop `inner`), the value of `outer` times `inner` (adding 1 where necessary) is computed. The result is assigned to the variable `item`. The `AddItem` method is used, making the assignment of the `item` to the list. Following this, a third nested loop is indicated. This loop acts to delay the display of numbers on the display screen.

3 Run and test your program. As one of your tests, comment out the `DoEvents` statement. This will help you to better understand the function provided by this statement.

11.3.3 The *Exit For* Statement

With nested loops, it sometimes becomes necessary to exit from a loop structure before the loop completes all iterations. Consider the following code sample:

```
For sleep = 1 To 100000
  If sleep > 35000 Then Exit For
Next sleep
```

This code reduces the number of iterations from 100,000 to 35,000. When 35,000 is reached, the innermost loop ends; however, the entire nested loop does not end. Instead, control is transferred to the statement that immediately follows the `Next sleep` instruction, which in Exercise 11.3 was the `Next inner` instruction. To escape this loop, a second `Exit For` statement is required.

11.3.4 Moving Shapes with *For-Next* Loops

`For-Next` loops are especially useful for manipulating shapes and other types of visual objects by relying on the Visual Basic coordinate system. As an example, consider the following code fragment:

```
For Position = 1 To 50
  Shape1.Left = Shape1.Left + 60
Next Position
```

In this example, `Shape1.Left` refers to the movement of the shape to the left. `Shape1.Left + 60` indicates that each iteration of the loop moves `Shape1` 60 twips to the right. This is equivalent to 1/24 of an inch with each loop iteration.

Exercise 11.4 Moving a Shape to Make the Moon Rise

This example asks you to move a shape by making the moon rise in the sky. Figure 11.5 shows the form you are asked to design, which places the moon at its highest position in the sky. At that point, a message appears.

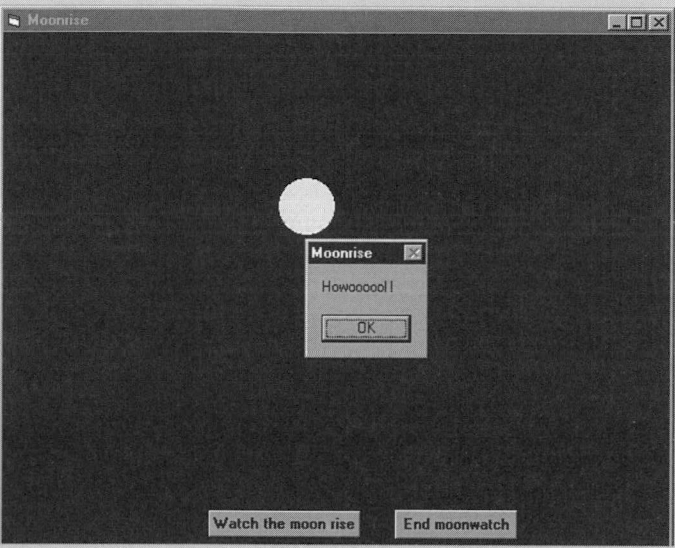

Figure 11.5

A `For-Next` *loop is used to make the moon rise from the lower-left corner of the form to the position illustrated.*

1 On the form, set the `WindowState` property to `Maximize (2)` and the `BackGroundColor` property to a dark blue. Title the form **Moonrise**.

2 Place a Shape control in the bottom-left corner, and change its properties so that it is a one-inch yellow circle.

3 Set the circle's `Left` property to `250` and its `Top` property to `5800`. (This assumes a 640 by 480 resolution, and that the form `ScaleMode` is in its default mode of `Twips`).

4 At the bottom of the page, draw a command button wide enough to hold the caption **Watch the moon rise.** This caption will change to **Watch the moon fall** when the dialog box appears. Add a second command button, **End moonwatch**. Write the `End` procedure.

5 Add the following code to the `cmdRise_Click()` command button:

```
Private Sub cmdRise_Click()
  Dim Position As Integer
  Dim ans As Integer

  For Position = 1 To 50
     Shape1.Left = Shape1.Left + 60
     Shape1.Top = Shape1.Top - 80   ' ascend
  Next
```

continues

```
    ans = MsgBox("Howoooool!", 0, "Moonrise")

    Shape1.Left = 250
    Shape1.Top = 5800
  End Sub
```

Note that the coordinates of a Visual Basic form begin with 0,0 (left, top). To make the moon rise, you need to subtract from the `Top` property value. After the moon reaches the highest point, the message box appears.

Exercise 11.5 The Moon Descending

Suppose that you wanted to watch the moon also descend from the night sky. It would be a simple matter of having the moon begin to descend when it reaches the uppermost position in its path. To make the moon descend, however, you must add to the `Top` property value. When the moon reaches its highest point, you will change the caption on the command button to **Watch the moon fall**. When the message box button is clicked, the moon will descend.

1 Change the `cmdRise()` procedure to work as follows:

```
Private Sub cmdRise_Click()
  Dim Position As Integer
  Dim ans As Integer

  For Position = 0 To 50
    Shape1.Left = Shape1.Left + 60
    Shape1.Top = Shape1.Top - 80   ' ascend
  Next Position

  cmdRise.Caption = "Watch the moon fall"
  ans = MsgBox("Howoooool!", 0, "Moonrise")

  For Position = 0 To 50
    Shape1.Left = Shape1.Left + 60
    Shape1.Top = Shape1.Top + 80  'descend
  Next Position

  Shape1.Left = 250
  Shape1.Top = 5800
  cmdRise.Caption = "Watch the moon rise"
End Sub
```

In the second loop, the moon continues to travel to the right of the screen while it descends. Instead of subtracting, twips must be added for this movement to take place.

2 Run the modified program. Because the message box suspends operation, clicking it starts execution again, and the moon descends.

By changing the counter variable and the size of the movement of `Shape1.Left` and `Shape1.Top` in each cycle iteration in Exercise 11.4 and Exercise 11.5, you can adjust the speed and distance of the moon's movement. Experiment to get the results you like.

Objective 11.4 Understanding How *Do* Loops Work

`Do` loops continue to do something (to loop) while a condition is `True` or until a condition becomes `True`. They are especially useful when dealing with situations in which the exact number of iterations required (loops) cannot be determined in advance. In designing a `Do` loop, a conditional test is used. The test returns a value of `True` or `False`, and works in the same way as the `If...Then` conditional test.

11.4.1 The *Do-Loop* Syntax

The syntax of the `Do-Loop` is complex. As shown in the following examples, you can place the conditional test of the loop either at the beginning or the end of the loop construct. The two options are as follows:

- **Option 1**—In the following code, the conditional test is at the beginning of the `Do-Loop` structure:

  ```
  Do [{While | Until} condition]
     [statement block 1]
     [Exit Do]
     [statement block 2]
  Loop
  ```

- **Option 2**—In the following code, the conditional test is at the end of the `Do-Loop` structure:

  ```
  Do
     [statement block 1]
     [Exit Do]
     [statement block 2]
  Loop [{While | Until} condition]
  ```

Both the Option 1 syntax and Option 2 syntax permit the following four types of `Do` loops to be written:

- The `Do-While Loop` structure:

  ```
  Do While(this condition is True)
     [statement block]
  Loop
  ```

- The `Do-Until Loop` structure:

  ```
  Do Until(this condition is True)
     [statement block]
  Loop
  ```

- The `Do-Loop While` structure:

```
Do
   [statement block]
Loop While(this condition is True)
```

- The `Do-Loop Until` structure:

```
Do
   [statement block]
Loop Until(this condition is This)
```

In this syntax, every keyword is optional except the keywords `Do` and `Loop`. Even so, be careful! If you make a construct with only `Do` and `Loop` (without `While` or `Until`) with no conditional test, you can easily create an infinite loop and suspend your computer's operation within an unending cycle. You might then have to exit Visual Basic, lose all unsaved work, and launch the program again. You will read about this in detail later in section 11.4.4, "Overflow Errors and Infinite Loops," but first look at the different `Do-Loop` structures.

11.4.2 Working with the Four *Do-Loop* Styles

In this section, you further examine the four `Do-Loop` styles. You should remember that with `Do-While Loop` and `Do-Until Loop`, it is possible to completely avoid the loop body. With `Do-Loop While` and `Do-Loop Until`, you must execute the loop body at least once.

Do-While Loop

`Do-While Loop`, the first of these four forms, performs a relational test at the top of the loop and tests for a positive condition. The loop continues to execute while the relational test remains `True`. Consider the following pseudocode example:

```
Do While("You are hungry")
   Eat
Loop
```

In this loop, you will continue to eat while you are hungry.

Do-Loop While

`Do-Loop While` is similar to the `Do-While Loop` except that the test condition is placed at the bottom of the loop. This ensures that the loop executes at least once. This is particularly useful when you want a user to enter a value and test after data entry. Continuing the last example:

```
Do
   Eat
Loop While("You are hungry")
```

In this loop, you would eat first before checking to determine whether you are hungry. (It's easy to gain weight with this practice, isn't it?)

Do-Until Loop

The `Do-Until Loop` acts like the `Do-While Loop` except that it tests for the negation of the `While` condition. Consider how the previous example needs to be reworded with `Do-Until`:

```
Do Until("You are Not hungry.")
   Eat
Loop
```

As stated, you would continue to eat until you *are not* hungry, which is the negation of you *are* hungry.

Do-Loop Until

`Do-Loop Until` also tests for the negation of the `While` condition, after executing the loop body at least once. The following syntax continues with the example:

```
Do
   Eat
Loop Until("You are Not hungry.")
```

In this instance, you would eat at least once before deciding that you were not hungry.

Exercise 11.6 Using *Do-Loop Until*

To survey the different ways `Do` loops can be constructed, use the form shown in Figure 11.6. The code of the `cmdDoLoop_Click()` event (top button) will display the products of 4 from 1 to 60.

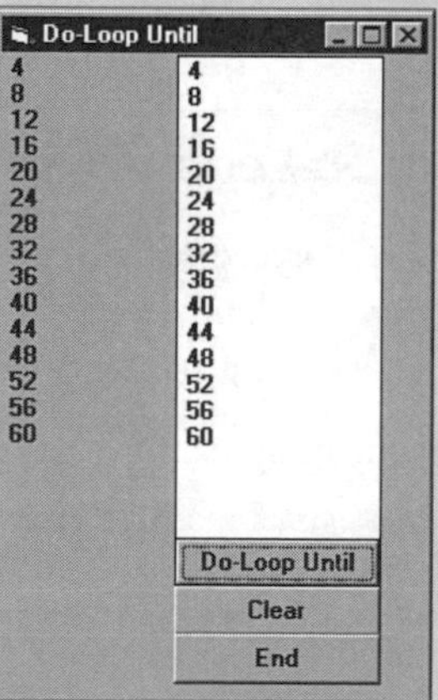

Figure 11.6

This example is similar to Exercise 11.1, but uses `Do` *loops rather than* `For-Next` *loops.*

continues

1 Draw the picture box and three buttons (**cmdDoLoop**, **cmdClear**, and **cmdEnd**) on a form as shown, and add the appropriate captions.

2 For the Clear button, enter `Form1.Cls` and `Picture1.Cls` instructions.

3 For the Do-Loop Until button, enter the following code:

```
Private Sub cmdDoLoop_Click()
  'program to display products of 4 from 1 through 60

  Dim increment As Integer
  Dim result As Integer
  increment = 1
  Do
    'multiply the increment variable by 4
    result = increment * 4
    Form1.Print result
    Picture1.Print result
    'increment the variable
    increment = increment + 1
  Loop Until result >= 60
End Sub
```

4 Run the program.

Similar to the tracker exercise using `For-Next` loops (see Exercise 11.1), this code displays an increment of the products of four. The counter variable (entitled `increment`), however, must be incremented in the body of the loop. Alternatively, you could move the conditional clause to the top of the construct so that the code would read `Do Until result >= 60`, and the bottom of the construct would simply read `Loop`.

Exercise 11.7 Using *Do-Loop While*

How would you need to change the preceding code if you used `While` rather than `Until`? In this case, you would need to change the relational operator (from >= to <), so that the code fragment would read `Loop While result < 60`.

1 Use the same form as the one designed for Exercise 11.6, or create a new one.

2 Insert the following `Do-Loop While` instructions:

```
Do
  result = increment  *  4    ' multiply the increment variable by 4
  Form1.Print result
  Picture1.Print result
  increment = increment + 1      ' increment the variable
Loop While result < 60
```

3 Run the program.

Like the `Do-Until Loop`, the `While` keyword can be placed at the top of the construct with the single word `Loop` placed at the bottom of the loop structure.

11.4.3 The *Exit-Do* Statement

The `Exit-Do` statement is an optional statement. It provides another way to exit a `Do` loop. The syntax for this statement is as follows:

```
Exit {Do | For | Function | Sub}
```

When this statement is reached in a `Do` loop, control is transferred from the loop body to the statement that follows the `Loop` statement.

As a general rule, `Exit Do` always should be used with an `If` statement, as in the following example:

```
If Value >= 50000 Then
  Exit Do
End If
```

If the test condition in this example is `True` and the value is greater than `50000`, it leads to an exit from the `Do` loop.

If `Do` loops are nested, as in the following structure, then the `Exit Do` statement in the innermost `Do` loop transfers control to the statement `y = y + 1`, which follows `Loop Until x > 10`. `Exit Do` does not terminate the entire nested loop.

```
Do                             'Beginning of outer loop
  Do                         'Beginning of inner loop
    Exit Do                'Exit from inner loop
  Loop Until x > 10          'End of inner loop
  y = y + 1
Loop Until y > 15              'End of outer loop
```

`Exit Do` can be used as a statement without qualification, or it can be made the `Then` branch of an `If...Then` statement. As an example, the code you have been writing can be changed to read:

```
Do
  result = increment  *  4  'multiply the increment variable by 4
  Form1.Print result
  Picture1.Print result
  increment = increment + 1 'increment the variable
  If result >= 60 Then
     Exit Do
  End If
Loop
```

Although this code works, be careful. Think of what would happen if you changed the code to read as follows:

```
Do
  If result <= 60 Then
     Exit Do
  End If
  result = increment * 4  ' multiply the increment variable by 4
  Form1.Print result
  Picture1.Print result
  increment = increment + 1 ' increment the variable
Loop
```

Because no result has been determined, the code does not know how to make the relational test. Nothing would be displayed, and the loop would end.

11.4.4 Overflow Errors and Infinite Loops

When working with loops, an incorrect setting of the test condition can lead to *overflow errors.* An uncontrolled loop in which the conditional test cannot be met will eventually reach some limit. For instance, an integer counter variable might run past 32,767, and you will encounter an overflow error. Suppose that you changed the loop in Exercise 11.7 to read as follows:

```
Do
  result = increment * 4    ' multiply the increment variable by 4
  Form1.Print result
  Picture1.Print result
  increment = increment + 1   ' increment the variable
Loop While result >=  0
```

With this code, an overflow message is produced because the loop would eventually run past 32,767. With other data types, you might encounter a stack overflow error because you simply run out of memory. Depending on the size of your computer's memory and the speed of its processor, the time will vary before you get the overflow message.

Remember the following rule when working with `Do` loops: *Make sure that you have an out. In writing the loop, write a statement that moves a condition closer and closer to meeting the requirements of the test condition.*

Stack and variable overflows are one type of problem; an *infinite loop* is another. While writing loop constructs, you can write an infinite loop for which there is no exit other than using Ctrl+Break keys on the keyboard to escape. At times, you might even have to reboot your computer and lose unsaved work. So be certain that your loop contains a statement that will satisfy the test condition. For example, suppose that you thought you would use a loop to test repeatedly whether an inputted value is correct. You might, at first, write a procedure like the following in which `Reply` represents the input:

```
Do
  Reply = Val(Text1.text)
  If Reply > 1 And Reply < 10 Then
     MsgBox "Thank you"
     Exit Do
  Else MsgBox "That's incorrect"
  End If
Loop
```

What happens? If `Reply` is correct (the return value is `True`), then there is no problem. However, if the user enters an incorrect answer, you get the message box with the message `That's incorrect`. And you get it again, and again, and again. Eventually, you might need to reboot your computer because you can't escape the loop. As a rule, just stick with an `If...Then...Else` conditional test, rather than a `Do` loop.

11.4.5 Converting *For-Next* Loops to *Do* Loops

`For-Next` loops are good points of departure for practicing coding `Do` loops. Most `For-Next` loops can readily be converted to a `Do-Loop` structure; however, the counter variable found in the `For-Next` structure will have to be rewritten. The `For-Next` code used earlier in Exercise 11.2 looks like the following:

```
Private Sub cmdShow_Click()
  Dim FillNum As Integer

  For FillNum = 0 To 7
     Shape1(FillNum).FillStyle = FillNum
     Shape1(FillNum).Visible = True
  Next FillNum
End Sub
```

To convert this code to a `Do-Loop` structure, you must add your own increment instruction and place this instruction inside the loop, as follows:

```
Private Sub cmdShow_Click()
  Dim FillNum As Integer
  FillNum = 0
  Do
     Shape1(FillNum).FillStyle = FillNUm
     Shape1(FillNum).Visible = True
     FillNum = FillNum + 1      'increment the variable
  Loop While FillNum <= 7
End Sub
```

The result is the same. Try it.

Exercise 11.8 Changing a *For-Next* Structure to a *Do-Loop* Structure

In this exercise, you convert a `For-Next` structure to a `Do-Loop` structure.

1. Add the array of shapes as shown in Figure 11.7.
2. Add the command buttons (**`cmdShow`**, **`cmdEnd`**), and add the captions shown in Figure 11.7. Write the `cmdEnd_Click()` event procedure.
3. Write the `cmdShow_Click()` procedure:

```
Private Sub cmdShow_Click()
   Dim FillNum As Integer
      FillNum = 0
   Do
      Shape1(FillNum).FillStyle = FillNUm
      Shape1(FillNum).Visible = True
      FillNum = FillNum + 1      'increment the variable
   Loop While FillNum <= 7
End Sub
```

continues

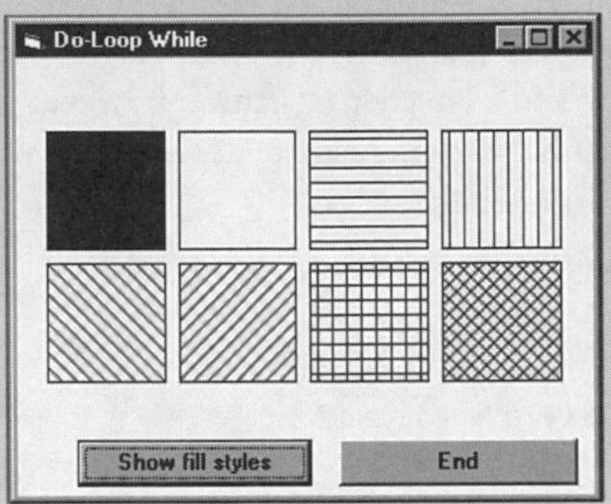

Figure 11.7

Display fill styles with a `Do-Loop`.

4 Run and test your design.

What is different between this code and the previous code? The counter variable is incremented inside the loop body. With a `For-Next` loop, the counter variable is incremented when it reaches the `Next FillNum` statement.

Exercise 11.9 Using Nested *Do* Loops to Create the Mercury 1 Commemorative Launch

The first American in space, Alan Shepard, didn't orbit Earth. He just went up into space and splashed down into the Caribbean. To commemorate the flight, code an image of a rocket going into space and then floating down with a parachute attached. Use two `Do` loops with the inner loop nested inside the other. Figure 11.8 shows the form you are asked to create.

1 Use Paint, the bitmap editing software bundled with Windows. Under the Image menu, set the Attributes so that the image is 0.4 inches wide and one inch long.

2 Draw a rocket with the nose touching the top of the image frame at the center. Set the background color to a light blue.

3 Save the image twice in your Visual Basic subdirectory: once as rocket.bmp and again as para.bmp. Now edit para.bmp to add a parachute canopy with the parachute cords converging at the bottom of the image at the center. Use the same background color as rocket.bmp. Save the revised image.

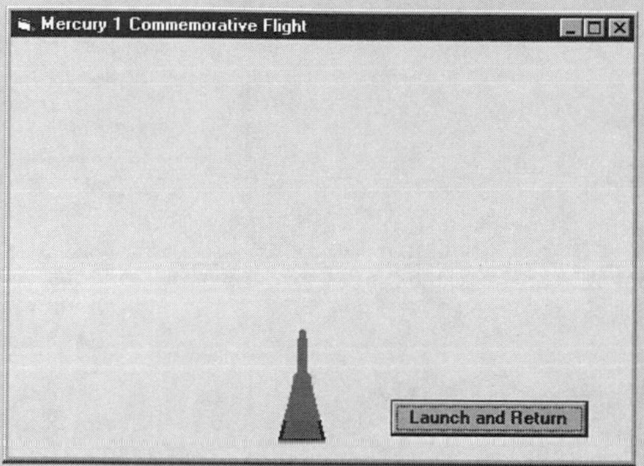

Figure 11.8

The Mercury 1 Commemorative Launch contains a rocket ready to launch and a parachute (hidden) that will return the rocket to the Earth.

4 Return to Visual Basic, start a new project, and set the form properties to `Form1.Top = 0`, `Form1.Height = 7260`, and `Form1.Width = 6200`. Set the form background color the same as in your .BMP images.

5 Place two image controls on the form, each with a height of `1440` and a width of `570`. Name one **`imgRocket`** and the other **`imgChute`**.

6 In `imgRocket`, load the rocket.bmp image. (Double-click the `Picture` property on the property window.)

7 In `imgChute`, load the para.bmp image.

8 Set the `imgRocket.Left` property to `2760` and the `imgRocket.Top` property to `5520`.

9 Set the `imgChute.Left` property also to `2760` and its `Top` property to `-8800`. (It will disappear from the form; but don't worry, it's still there.) Set the `imgChute.Visible` property to `False`.

10 Draw a button to the right of the rocket image, and add the caption **Launch and Return**.

11 Write the following code for the `cmdLaunch_Click()` event:

```
Private Sub cmdLaunch_Click()
  Dim counter As Integer
  cmdLaunch.Enabled = False     ' prevent multiple launches
  counter = 0
  Do
     Do
       imgRocket.Top = imgRocket.Top - 70 ' ascend at rapid rate
```

continues

```
      Loop Until imgRocket.Top <= -7360      ' top of imgRocket is at the
                                             ' bottom of imgChute
                                             ' -8800(imgChute.Top) +
                                             ' 1440(height of images)

      imgChute.Visible = True                ' show the parachute

      Do
       imgChute.Top = imgChute.Top + 25
       imgRocket.Top = imgRocket.Top + 25     ' descend at a slower rate
      Loop Until imgRocket.Top >= 5520        ' stop when rocket lands at
                                              ' original position

      imgChute.Visible = False             ' make parachute disappear
      imgChute.Top = -8800                 ' move it back off of form
      counter = counter + 1                ' increment the counter
    Loop Until counter = 1                 ' limit loop to 1 iteration,
                                           ' reset if you like

    cmdLaunch.Enabled = True               ' permit another launch
  End Sub
```

Let's review what you coded. This is very similar to the Moonrise, except that this time you have two `Do-Loop` inner structures nested within an outer `Do-Loop` structure. When `cmdLaunch` is clicked, you disable it. This prevents the user from clicking multiple times and relaunching the code. The first loop is executed, and `imgRocket` ascends at a rate of 70 twips per cycle. The loop is exited, and the `Visible` property of `imgChute` switched to `True`.

You enter the second loop, at which point `imgRocket` and `imgChute` descend at a slow rate because with every cycle they only descend by 25 twips. Once the rocket lands, the `imgChute Visible` property is switched to `False` and reset to its original position. The command button, `cmdLaunch`, is enabled to permit the user to relaunch.

Both inner loops are placed within a larger loop with a counter to provide for several launchings of the rocket. If you only wanted Alan Shepard to go into space once, you would remove this outer loop.

12 Run the program. Depending on your computer's processor and graphics system, the rocket will ascend or descend at different speeds. Adjust the code to suit your machine.

13 Save the program because you will come back to it in Exercise 11.11.

Objective 11.5 Understanding How *While-Wend* Loops Work

Besides the four forms of the `Do` loop, Visual Basic supports a `While-Wend` statement. The syntax for this statement is as follows:

```
While condition
   [statement block]
Wend
```

This format is sometimes favored in place of the `Do-While Loop` statement; however, the latter does provide more flexibility. There is no corresponding `Exit While` to permit an immediate exit from the loop. As an example of this syntax, consider the following code:

```
count = 0
While count < 20
   Print "Hello! How are you today?"
   count = count + 1
Wend
```

Like `Do-While Loop` structures, the counter variable should be initialized prior to entry into the loop body. Within the loop body, the counter variable is incremented (or decremented) in order to eventually satisfy the test condition.

Objective 11.6 Adding the Timer Control to a Project

Visual Basic has a built-in control that cycles repetitively named the Timer control, which can sometimes be used instead of a loop logical structure, especially when you want to control precisely the timing of the executed code.

The important thing to remember about the Timer control is that it is periodic. The code placed in the `Timer()` event will continue to execute until the control is disabled. The Timer control doesn't appear during runtime, so don't worry about its placement on the form.

11.6.1 The *Interval* Property

If you inspect the properties of the Timer control, you will discover that there are very few. You will generally adjust the `Interval` property, for which the syntax is as follows:

```
[form.]timer.Interval [= milliseconds]
```

Table 11.2 shows the settings for `milliseconds`.

Table 11.2 The *milliseconds* Settings

Setting	*Description*
`0`	Disables the `Timer` (the default).
`1 to 65535`	Sets the number of milliseconds between calls to the `Timer`. 10,000 milliseconds equals 10 seconds.

Thus, the following instruction tells the `Timer` to fire every 10 seconds:

```
Timer.Interval = 10000
```

The Timer continues to fire until it is disabled, as discussed next.

11.6.2 The *Enabled* Property

The second property of great importance to the Timer control is the `Enabled` property. The `Timer` is disabled by the setting:

```
Timer.Enabled = False
```

This action suspends the countdown set by the `Interval` property. Changing the setting to `True` resumes the countdown set by the `Interval` property, as follows:

```
Timer.Enabled = True
```

Exercise 11.10 Setting the Timer

Before adding a timer to the Mercury 1 Commemorative Launch, use the Timer control. Figure 11.9 shows the form you are asked to design.

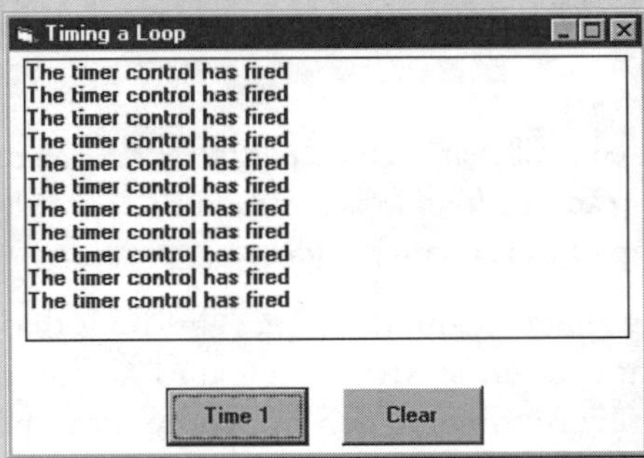

Figure 11.9

For this exercise, the timer will keep running until it is disabled.

1. Draw a form with a picture box and the two command buttons (**`cmdTime`** and **`cmdClear`**). Add a caption to the form and to all buttons.
2. Double-click the Timer control to add the timer to the form.
3. Go to the `Timer1` property box. Set the `Enabled` property to `False`, which turns the timer off.
4. The `Interval` property determines at what frequency the code in the `Timer()` event will be executed. The interval is divided into thousandths of a second. To have the `Timer()` event execute the code every second, set the `Interval` property to `1000`.
5. Enter the following in the `Click()` event for the `cmdClear` button:

```
Private Sub cmdClear_Click()
  Picture1.Cls
  '  timer1.Enabled = False  ' be sure this line is commented out at first
End Sub
```

6 Enter the following in the `Click()` event for the `cmdTime` button:

```
Private Sub cmdTime_Click()
  timer1.Enabled = True
End Sub
```

In this code the timer is turned on. It remains on until it is disabled.

7 Double-click the Timer control, and the `Timer1_Timer()` event will appear. Enter the following:

```
Private Sub Timer1_Timer()
  Picture1.Print "The timer control has fired"
End Sub
```

8 Run the program.

9 Now change the `Clear` button's `Click()` event code to disable the timer. The revised code is as follows:

```
Private Sub cmdClear_Click()
  Picture1.Cls
  timer1.Enabled = False   ' instruction is no longer commented out
End Sub
```

10 Run the program again. You should now have a greater feeling of control.

When using the Timer control, you will usually set its `Enabled` property to `False` during design time, and enable it or disable it with code.

Remember that conditional loops do not affect underlying property assignments. Consider the following code:

```
Private Sub cmdTime_Click()
  Dim count as integer
  count = 1
  While count < 20
     timer1.Enabled = True
     count = count + 1
  Wend
End Sub
```

When you first look at this code, you might think that the `Enabled` property for the timer will be `True` only while the `count` variable is `< 20`. But object properties are only changed by direct assignments, such as the assignment `timer1.Enabled = False` in the `cmdClear_Click()` event.

Exercise 11.11 Adding a Timer Countdown to the Mercury 1 Commemorative Launch

Now use a `Timer()` event so Mercury 1 can get a proper countdown and blastoff. Figure 11.10 shows the form you are asked to design.

continues

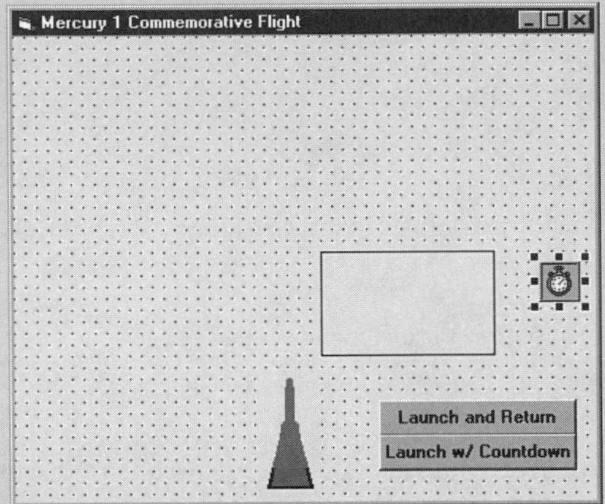

Figure 11.10

This revised Mercury 1 launch design contains a Timer control that displays a countdown before launching.

1 Add a Timer control to the form, and add a second button below the Launch and Return button with the caption **Launch w/ Countdown**.
2 Add a label about 1.5 inches wide above the buttons. This area will display the countdown messages. Name the label **`lblCountDown`**.
3 Place a border around the label, and give it the same background color as the form.
4 Because the label display will be controlled by the `Timer()` event, set its `Visible` property to `False`. You will make it visible when you need to display the countdown, and make it invisible when you don't.
5 Set the `Enabled` property of the Timer control to `False`. Set the `Interval` property to one second (enter **`1000`**).
6 You will need an additional variable. Call it `Countdown`, and declare it at the module level. Enter the following in the General declarations window of the form module:

```
Dim Countdown as Integer
```

7 Enter the following code in the `Timer1_Timer()` event:

```
Private Sub Timer1_Timer()
  lblCountDown.Visible = True      ' make countdown display visible
  lblCountDown.Caption = 10 - Countdown

  If Countdown = 10 Then
     lblCountDown.Caption = 10 - Countdown & " WE HAVE IGNITION!"
  ElseIf Countdown = 11 Then
     lblCountDown.Caption = "Blastoff!"
```

```
    ElseIf Countdown = 12 Then
       lblCountDown.Visible = False   ' make countdown display invisible
       lblCountDown.Caption = ""      ' clear display for subsequent
                                      ' launches
       Timer1.Enabled = False     ' turn off the timer event
       Call cmdLaunch_Click()       ' call the procedure to launch rocket
    End If

    Countdown = Countdown + 1           ' increment variable
  End Sub
```

Examine these instructions. Once per second the `Timer1_Timer()` event occurs, and `lblCountDown` displays the difference of `10 - Countdown`. Once per second the `Countdown` variable value is also increased by `1`. When `Countdown` reaches `10`, `lblCountDown` displays `0 (10 - Countdown)` and adds an encouraging message. When `Countdown` reaches `11`, `Blastoff` is displayed (instead of `-1`). When `Countdown` reaches `12`, `lblCountDown` is made invisible and the caption cleared. `Timer1` is disabled and the launch procedure is initiated by calling the `cmdLaunch_Click()` event.

8 So what do you need to add to the Countdown button `Click()` event? Only two instructions:

```
Private Sub cmdCount_Click()
   Countdown = 0
   Timer1.Enabled = True
End Sub
```

Remember that after the first launch, the `Countdown` variable now equals `12`. By setting it to `0` every time the Countdown button is pushed, you ensure that the `Timer1_Timer()` event will start the countdown at `10`.

9 Next, for consistency's sake, disable both launch buttons when the Countdown button is clicked, and enable them both after the rocket has landed. Modify the `cmdCount_Click()` procedure to read as follows:

```
Private Sub cmdCount_Click()
  cmdLaunch.Enabled = False
  cmdCount.Enabled = False
  Countdown = 0
  Timer1.Enabled = True
End Sub
```

10 The following code should be added to the `cmdLaunch_Click()` event right below where `cmdLaunch` itself is enabled again:

```
cmdCount.Enabled = True
```

Examine the earlier code to determine this position.

Chapter Summary

This chapter addresses several topics critical to the study of Visual Basic. It discusses the meaning of a loop and demonstrates how code can be written to execute in cycles. To perform this type of processing, the `For-Next` syntax must be understood. In review, this syntax is as follows:

```
For counter = startvalue To endvalue [Step incrementvalue]
  [statement block 1]
   [Exit For]
  [statement block 2]
Next [counter]
```

Besides a `counter` variable, the syntax must include `startvalue` and `endvalue` and the keyword `Next`, which signals the end of the loop. Observe that the entire statement block is optional. When you wrote the sleep loop to slow the program, no statements appeared within the body of the loop. Remember that a loop counter can be incremented or decremented. To decrement a counter, begin with the highest value, and subtract from this value using a negative `Step increment`.

The concept of a control array is perhaps the most difficult conceptual topic presented. Keep this concept in the back of your mind: *A control array is nothing more than a group of objects with the same name.* To help you, Visual Basic provides the index number for each object within the group. You can place more than one control array on a design. For example, you might have a control array of text boxes and another control array of list boxes.

A nested `For-Next` loop is nothing more than one `For-Next` loop placed inside another. Even so, remember how iteration works. The innermost loop must complete its entire execution before control is passed to the next outer loop. There is no limit on the number of nested loops that you can use; however, going beyond three might make your code difficult to read.

The `DoEvents` statement enables an event to be completed before the next event is started. The `DoEvents` statement passes control to Windows and enables windows messages to be processed before executing the next line of code.

The `Exit-For` statement is used to exit a loop before all iterations have been completed. Most often, a test condition is written to determine if it is time to exit a loop. The exit itself is for the immediate loop only.

The coordinate system in Visual Basic defines *x* and *y* positions on the display screen, with x = 0 and y = 0 set as the defaults in the upper-left corner of the display screen. Objects are moved using a unit of measurement called a *twip* (1,440 twips equals one inch). The `Left` and `Top` properties are used to move an object.

The two syntax options for `Do` loops are as follows:

- **Option 1**—In the following code, the conditional test is at the beginning of the `Do-Loop` structure:

```
Do [{While | Until} condition]
   [statement block 1]
   [Exit Do]
   [statement block 2]
Loop
```

- **Option 2**—In the following code, the conditional test is at the end of the `Do-Loop` structure:

```
Do
   [statement block 1]
   [Exit Do]
   [statement block 2]
Loop [{While | Until} condition]
```

With the first option, the body of the loop need not be executed at all if the test condition is `False` to begin with (using the `Do-While Loop`), or if the test condition is `True` to begin with (using the `Do-Until Loop`). With `Do-Loop While` or `Do-Loop Until`, the body of the loop is executed at least once.

The `While-Wend` loop is another type of loop structure. The loop body of this structure continues to be executed until the test condition becomes `False`.

The Timer control is sometimes used instead of a loop structure. The Timer control is periodic: It fires during a specified interval and is turned on and off by the `Timer.Enabled` property or by setting the `Interval` property to `0`. The `Timer_Timer()` procedure (see Exercise 11.10) is often used to control the actions taken each time the timer is fired.

Skill-Building Exercises

1. Write a program to rotate the fill shapes that are shown earlier in this chapter in Figure 11.3. Make three rounds in all, visiting each fill style three times. Do the following:

1. Make each shape style visible, and keep it visible using a go-to-sleep loop.

2. Make each shape style invisible, and keep it invisible using a go-to-sleep loop.

3. Use `DoEvents` to enable completion of each nested loop.

2. Modify Exercise 11.3 to display the following pattern:

```
10      8       6       4       2
10      8       6       4
10      8       6
10      8
10
```

In your design, use five list boxes. Use nested loops in your solution. Add a `DoEvents` instruction.

3. Write a program using `Do` loops that displays numbers from 1 to 10 raised to the second power. The output should look as follows:

```
Number raised
to the 2nd power is

1                                               2
2                                               4
```

Continue this pattern to number 10. Display the numbers on the form and the results in a picture box.

4. Change the code in the Moonrise program so that it uses a `Do-Loop` structure rather than a `For-Next` structure.

5. Using a combination of `Do` loops and `If...Then` logic, add two scroll bar controls to the Mercury 1 Commemorative Launch program that enable the users to control the speed of ascent and descent. Also add a check box or option button that enables them to control how many times the rocket will launch. (*Hint:* You will need to add several extra variables.)

12

Object Types, Variables, and Collections

When building a Visual Basic application, you always work with objects. These objects include forms, controls, and software components made available to you by other applications or software components you create yourself. As described in Chapter 1, "Running a Visual Basic Program," Visual Basic 5 is much more object-oriented than earlier versions.

Because forms and controls are visible, it is difficult to think of them as data types. However, they are data types of a special kind. They are *object* data types. They differ from user-defined types because they hold both data (properties) and code (methods), and they receive events.

`Object` variables are declared as object data types, just as you would declare a variable as a primitive data type or as a user-defined data type. Whenever you add a form to a project, Visual Basic automatically declares it as a global form, using the name assigned. This gives it an `Object` variable declaration.

Using `Object` variables, much like string and numeric variables, it becomes possible to manipulate forms, controls, and other complex data structures. As an example, it is possible to create new instances of forms during runtime. Using a control array, it is possible to create new instances of Visual Basic controls. Finally, you can create your own objects in Visual Basic. These objects contain code to be used by your own application or exposed and used by other applications. In advanced Visual Basic, you can even create your own controls.

The tasks you will learn in this chapter include the runtime use of built-in Visual Basic objects as well as the creation of new instances of forms and controls. Although the practical, real-world use of this chapter's exercises might not be readily apparent, the exercises introduce the syntax for manipulation of objects that will be useful as you progress through this book—especially the syntax necessary for working with objects that you create yourself. In this chapter, you will

- declare an `Object` variable;
- learn the scope and life of `Object` variables;
- use the `TypeName` function;
- use a specific object type;
- use the `Set`, `Nothing`, and `New` keywords;
- change the property of an object created at runtime;
- use the forms and controls collections with the `For Each...Next` structure;
- dynamically load elements of control arrays.

Chapter Objectives

All of the tasks for this chapter are condensed into five objectives, as follows:

1. Understanding the makeup of `Object` variables
2. Knowing the difference between generic and specific object types
3. Working with multiple instances of forms
4. Understanding forms and controls collection
5. Dynamically loading a control array

Objective 12.1 Understanding the Makeup of *Object* Variables

An `Object` variable is declared similarly to a string or numeric variable: Use `Dim`, `Static`, or `Public` keywords to declare an `Object` variable. There are, however, several critical differences between an `Object` variable and a variable of a primitive data type. First and foremost, objects are instances of *classes*. Second, `Object` variables are *references* to objects, not the objects themselves.

When you declare an `Object` variable, you are creating a reference to that object. The object itself cannot be manipulated until the object is bound to the variable using the `Set` keyword. How objects are bound to `Object` variables is addressed in detail in this section, but first you need to review classes and object references so that these concepts are clear.

12.1.1 Objects as Class Instances

When you add a control to a form, you are in fact creating an instance of that specific control's class. Think of a class as a cookie cutter or a template. Each instance of the class gets its shape and behavior from the cookie cutter. The class defines the properties of the object (data), the methods of the object (the code), and the events to which the class will respond. When you add a control to a form, it "knows" the events to which it might be asked to respond.

Using a cookie cutter or a template is the essence of encapsulation. You, as a programmer, can use control objects (software components) and not have to be aware of their inner workings. You work with the objects' public *interface:* the properties, methods, and events the designer of the control has made available to you. However, you never are required to examine the computer code used in the creation of the class.

Visual Basic provides a function that can be used to determine an `Object` variable's class: the `TypeName` function. Although the function also works with simple data types, its main use is to return a string expression describing the class membership of a variable. The following is the syntax for the `TypeName` function:

```
TypeName(varname)
```

The `varname` argument contains any variable except a user-defined type variable.

Exercise 12.1 Using the *TypeName* Function to Read Object Classes

This initial exercise shows the use of the `TypeName` function, and unveils the classes behind the Visual Basic controls and objects with which you have already worked.

1. Create the interface shown in Figure 12.1. Name the form **`frmTest`** and add five controls: **`Image1`**, **`Label1`**, **`Command1`**, **`Shape1`**, and **`Vscroll1`**. Add the list box and name it **`lstType`**.

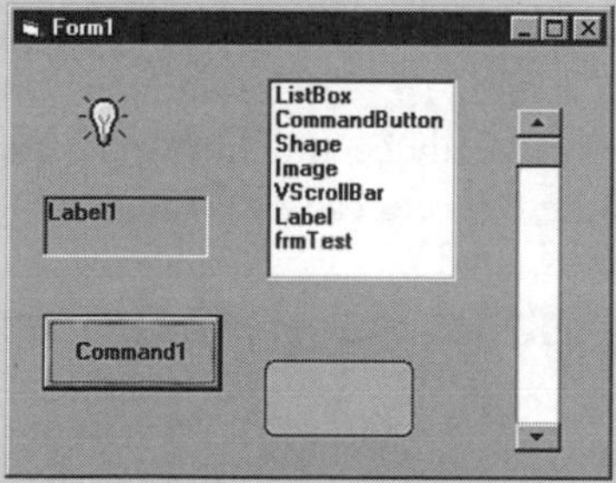

Figure 12.1

This interface will enable you to use the `TypeName` *function to determine class membership of each control placed on the form.*

continues

2 In the `Command1_Click()` event, enter the following code:

```
lstType.AddItem TypeName(lstType)
lstType.AddItem TypeName(Command1)
lstType.AddItem TypeName(Shape1)
lstType.AddItem TypeName(Image1)
lstType.AddItem TypeName(VScroll1)
lstType.AddItem TypeName(Label1)
lstType.AddItem TypeName(frmTest)
```

3 Run and test the program. You should see the class names of all the control objects, as shown in Figure 12.1.

In section 12.4.2, "Using the `For Each...Next` Structure to Work with Collections," you are introduced to the `For Each...Next` syntax, which allows you to loop through all of the controls on the form, and display their type names and hence their class membership. As the output in the list box illustrates, each time you add a control to a form you are creating an instance of its class. For example, if you place two labels on this form, you create two instances of the label class.

Notice the difference for the form named `frmTest`. Its class name is `frmTest`, not Form. When you add a form module to a Visual Basic project and add code to it, you are *not* adding an object to your program; you are defining a class. When you run the project and load the form, you create an instance (object) of that class definition.

How is this possible? Think again about what makes an object: properties, methods, and events that are defined for the *class*. Every time you add an event procedure or a general procedure to the form, you are defining a class method. Forms have pre-defined properties that make up their class properties. Even so, Visual Basic enables you to add your own properties to a form. In advanced Visual Basic, you can even create your own events for a form.

Visual Basic provides a specific module designed for creating your own classes: the *class module* (which is addressed in Chapter 13, "Class Modules"). Whether you realize it or not, by starting with a form, you have already been defining classes. Once a class is defined, multiple instances (objects) of these classes can be created.

12.1.2 *Object* Variables as References to Objects

`Object` variables refer to an object, but they are not the object itself. Any object occupies more than four bytes in memory, but if you look up the size of `Object` variables, you will see that they occupy four bytes. With eight bits to a byte, that makes 32 bits, which not coincidentally is the size of a memory register in a 32-bit operating system, such as Windows. `Object` variables hold the *memory address* of the object. In other computer languages, such a variable is called a *pointer variable*. In Visual Basic, `Object` variables have some properties of pointers, but not all of them.

Let's examine some instructions that bind an `Object` variable to an object you recognize: a command button. Consider the following code for a form that contains a command button named `Command1`:

```
Dim objectref as CommandButton
Set objectref = Command1
objectref.caption = "I have a new caption"
```

The first instruction declares `objectref` as an object of type `CommandButton`. The second uses the `Set` statement to *bind* the `Object` variable to the `Command1` command button. The syntax for this statement is as follows:

```
Set objectvar = {[New] objectexpression ¦ Nothing}
```

The keywords `New` and `Nothing` are optional. The third instruction changes the caption of `Command1`.

Instead of using a reference to a command button, you could, of course, simply assign a new value to the command button's caption property; in most instances you would do that because a command button is a software component defined for you as a control. As a control, an instance of a command button is created for you when you add it to a form.

Visual Basic also allows you to use software components other than controls. These objects are created during runtime and are defined in class modules. To *create* an object during runtime, you create an instance of the class. This is done by declaring an `Object` variable and then binding the variable with the `Set` keyword. There are several ways of making these `Object` variable declarations. We turn to this subject next.

Objective 12.2 Knowing the Difference Between Generic and Specific Object Types

In the preceding example, we used a specific object data type, a `CommandButton`, to declare an `Object` variable. This variable was then bound to a Visual Basic component: a Command Button control. It is possible, however, to use the generic word `Object` as our `Object` variable type. We could thus make an `Object` variable declaration for a command button with the following instruction:

```
Dim mycmd as Object
```

The *generic object types* are as follows:

- **`Object`**—Any object
- **`Form`**—Any form in an application
- **`Control`**—Any control in an application

There are many more *specific object types* built into Visual Basic—mainly the software components contained in the toolbox. These include the following:

`CheckBox`	`Frame`	`OLE`
`ComboBox`	`Grid`	`OptionButton`
`CommandButton`	`HScrollBar`	`PictureBox`
`CommonDialog`	`Image`	`Shape`
`Data`	`Label`	`TextBox`
`DirListBox`	`Line`	`Timer`
`DriveListBox`	`ListBox`	`VScrollBar`
`FileListBox`	`Menu`	

As a rule, use a specific object type whenever possible. This makes it easier for Visual Basic to determine how to resolve references to objects and makes your application run much faster. It also makes your Visual Basic code much easier to read.

The difference in speed depends on whether `Object` variables in your code are *early-bound* (the fastest) or *late-bound* (slowest). A variable is bound when Visual Basic can resolve which kind of object a variable refers to. If Visual Basic can resolve the `Object` variable when your application is compiled—which specific `Object` variable declarations allow it to do—then Visual Basic knows where to retrieve information about the class when it encounters the variable in code.

However, if the `Object` variable declaration is generic, Visual Basic must first make inquiries during runtime to resolve which class the object belongs to. It then must retrieve the class information. This adds considerable overhead. Thus, as a general rule, remember to always declare `Object` variables as specifically as possible.

Exercise 12.2 Using Generic and Specific Object Declarations

This short exercise illustrates how a generic type object declaration differs from a specific type object declaration.

1 Create the interface shown in Figure 12.2.

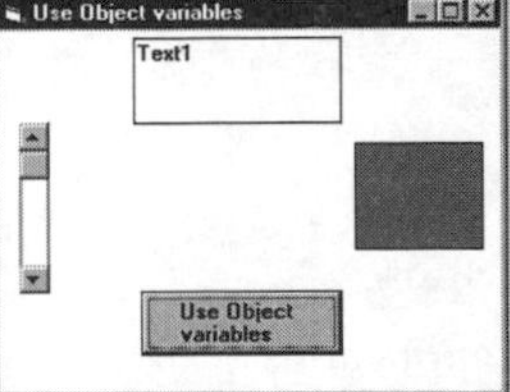

Figure 12.2

This interface shows how generic and specific object variables can be bound to controls.

2 In the `Command1_click()` event, enter the following code:

```
Private Sub Command1_Click()
    Dim genobj As Control           'A generic declaration
    Dim sobj As VScrollBar          'A specific declaration

    Set genobj = Text1              'Set first generic variable
    genobj.Left = 200

    Set genobj = Shape1             'Set second generic variable
    genobj.Left = 200

    Set genobj = Command1           'Set third generic variable
    genobj.Left = 200

    Set sobj = VScroll1             'Set specific Object variable
    sobj.Left = Form1.ScaleWidth - sobj.Width
End Sub
```

3 Run and test the program.

In this exercise, you use the generic object type `Control` and the specific object type `VScrollBar`. You successively bind controls to the variable `genobj` with the `Set` statement. The specific `Object` variable, `sobj`, is bound to `Vscroll1`. Observe the results if you attempt to bind the `Object` variable `sobj` to one of the other controls, such as `Command1`.

Objective 12.3 Working with Multiple Instances of Forms

Designing a form in Visual Basic is actually the creation of a new class. As a class instance, individual forms are *instantiated* when you use the `Load` statement. Usually you only create one instance of a form. But it is possible to create many instances of a single form. Sound strange? Exercise 12.3 will demonstrate this concept.

12.3.1 The *New* Keyword

To create multiple instances of objects from classes, you can use the `New` keyword. In this book, we use the `New` keyword to create multiple instances of the same form, multiple instances of components we create with class modules, and software components exposed to our applications by other applications.

The following instruction declares the variable `NextForm` as a new instance of `Form1`:

```
Dim NextForm As New Form1
```

Initially, the new instance is an exact copy of the original form as it exists at design time. The shape and contents of the new form are the same as the original form.

To inspect the new form, write the following:

```
NextForm.Show
```

12.3.2 The *Move* Method

When you create a new instance of a form, it becomes an exact copy and completely covers the original. To separate the new instance from the original, it must be moved. The `Move` method does this with the following the syntax:

```
[object.]Move left[,top[,width[,height]]]
```

Table 12.1 explains the components of the `Move` syntax.

Table 12.1 The Syntax of the *Move* Method

Component	*Description*
`object`	The form or control to move
`left`	The value for the left edge of the `object`
`top`	The value for the top edge of the `object`
`width`	The new width of the `object`
`height`	The new height of the `object`

Even though the `left` argument is the only argument required (`right` does not appear in the syntax), you cannot specify `width` or `height` unless previous arguments are used. For example, `width` cannot be used without specifying `left` and `top`. `Height` cannot be used without specifying `left`, `top`, and `width`.

Exercise 12.3 Creating a New Instance of a Form

This brief exercise asks you to create multiple instances of a form. Figure 12.3 illustrates the form you are asked to create and how multiple instances appear.

Figure 12.3

These forms are all identical and represent instances of a class defined by a form.

1 Create the visible form shown in Figure 12.3, adding a text box (**txtMessage**) and two command buttons (**cmdNew** and **cmdEnd**). Write the cmdEnd procedure.

2 Write the cmdNew_Click() procedure as follows:

```
Private Sub cmdNew_Click()
   'Declare a new form object
   Dim F As New form1

   F.Show
   F.Move Left + (Width \ 10), Top + (Height \ 10)
End Sub
```

3 Run and test your program. Observe that the F.Move method is needed to separate the new form from the original.

The optional keyword New always ensures that a new instance of a form is created at runtime. However, you cannot use the keyword New with a generic form class. It can be used only with a specific form class, such as Form1 or frmMain, not simply Form. For example, the following statement is not allowed because Form is generic:

```
Dim MyForm As New Form
```

The New keyword cannot be used to create runtime instances of controls. Controls must be created during design time. Visual Basic does permit you to dynamically add and remove controls from control arrays. You will dynamically load and unload control arrays in Objective 12.5, "Dynamically Loading a Control Array," later in this chapter. The following instruction is not allowed, for example, because New cannot be used with a control:

```
Dim MyText As New Text1
```

12.3.3 *Set* and *New* with the Keyword *Nothing*

In Exercise 12.3, you used the keyword New in the declaration of the Object variable itself. This binds the Object variable to the object in one statement. You then used the Show Method to display the new form instance. In practice, the New keyword is generally used in combination with the Set keyword. The following is the syntax for the declaration of the Object variable and its binding:

```
Set objectvar = New objecttype
```

The statements:

```
Dim MyForm As frmMain
Set MyForm = New frmMain
```

appear equivalent to the statement:

```
Dim MyForm As New frmMain
```

They are not quite the same, however. When the statement

```
Set MyForm = New frmMain
```

is used, you can use the keyword `Nothing` to make the following assignment:

```
Set MyForm = Nothing
```

This allows the `Object` variable to be *dereferenced,* indicating that an `Object` variable does not refer or point to any object. When an `Object` variable is set to `Nothing`, Visual Basic will not attempt to create another class instance the next time you attempt to use the variable. Consider the code:

```
Dim MyForm As frmMain
Set MyForm = New frmMain
MyForm.Show
Set MyForm = Nothing
If MyForm Is Nothing Then
   [do the following because Nothing will always be true]
End If
```

Using the `Nothing` keyword allows the memory referenced by the `Object` variable, `MyForm`, to be reclaimed.

12.3.4 Changing Properties of Form Instances

When you use the `New` keyword with a form, Visual Basic creates a new instance of a form. Each instance originally retains the properties of the original class definition, making it an exact clone. However, you can change the properties of a new instance of a form at runtime. For example, you can change the caption, messages placed in text boxes, labels, colors, and so forth, as the next exercise demonstrates.

Exercise 12.4 Changing *Object* Variable Properties

This exercise asks you to use the same form as the last exercise, but to change the caption of the new form instance, the caption on the command button, and the message in the text box. In addition, you are asked to change the message in the text box on the original form and to disable the New Instance command buttons on both the original and the copy of the original form. Figure 12.4 shows how the new form instance will appear.

1 Write the `cmdNew_Click()` procedure as follows:

```
Private Sub cmdNew_Click()
   Dim F As Form1
   Set F = New Form1
   F.Show
   F.Move Left + (Width \ 10), Top + (Height \ 10)
   F.Caption = "My Caption my Caption"
```

```
    'New notation to separate object from property
    F.txtMessage.Text = "Finally, a new message"
    F.cmdNew.Caption = "Don't press"
    F.cmdNew.Enabled = False
    Form1.cmdNew.Enabled = False
    Form1.txtMessage.Text = "That's all folks!"
End Sub
```

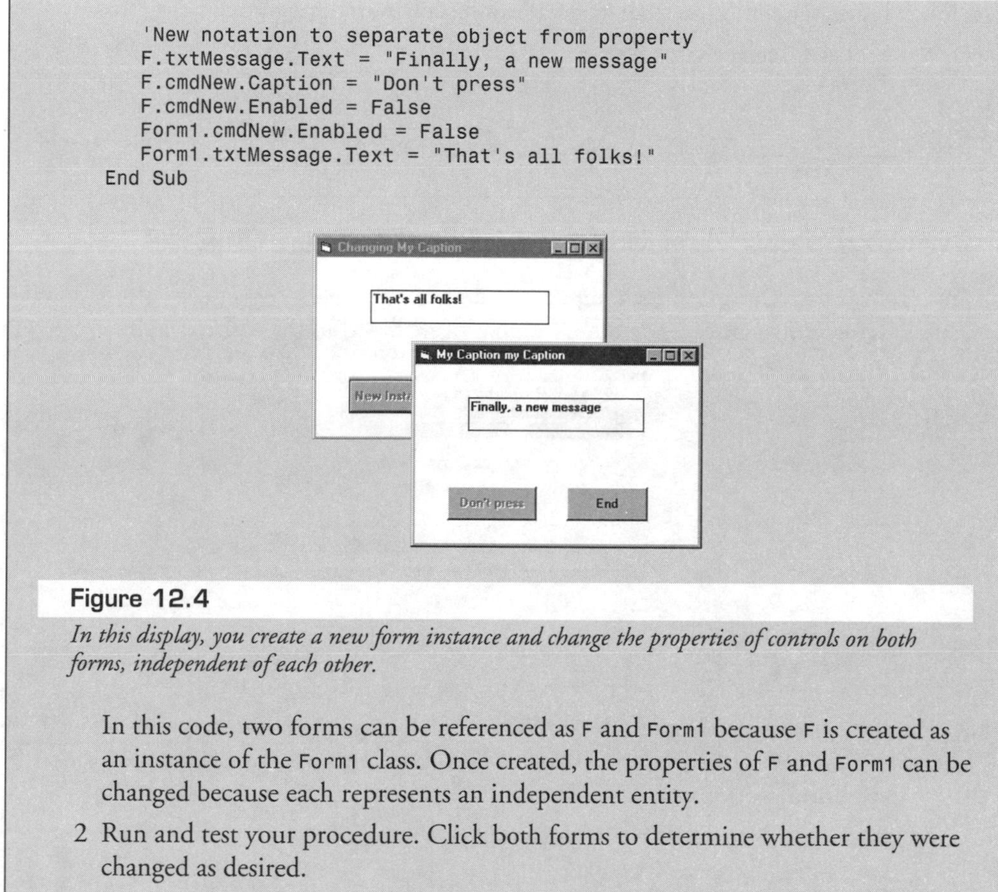

Figure 12.4

In this display, you create a new form instance and change the properties of controls on both forms, independent of each other.

In this code, two forms can be referenced as `F` and `Form1` because `F` is created as an instance of the `Form1` class. Once created, the properties of `F` and `Form1` can be changed because each represents an independent entity.

2 Run and test your procedure. Click both forms to determine whether they were changed as desired.

Objective 12.4 Understanding Forms and Controls Collection

Visual Basic provides collections to make it possible to access all the forms in an application or the controls placed on a form. For each application, there is a forms and a controls collection. These collections are declared globally with the pre-defined keywords `Forms` and `Controls`. Visual Basic contains several types of collections. In advanced Visual Basic, you can even create your own objects and your own object collections.

12.4.1 Collection Syntax

The following is the syntax for a forms collection in which `index` is a numerical expression:

```
Forms(index)
```

The following statement unloads all loaded form instances in a program:

```
For i = 0 to Forms.Count -1
   Unload Forms(i)
Next i
```

Observe that the index of the forms collections begins with `0`.

Exercise 12.5 Unloading an Instance of an Object

This exercise asks you to load and unload instances of an object. As shown in Figure 12.5, a Remove Instance command button has been added to the earlier design.

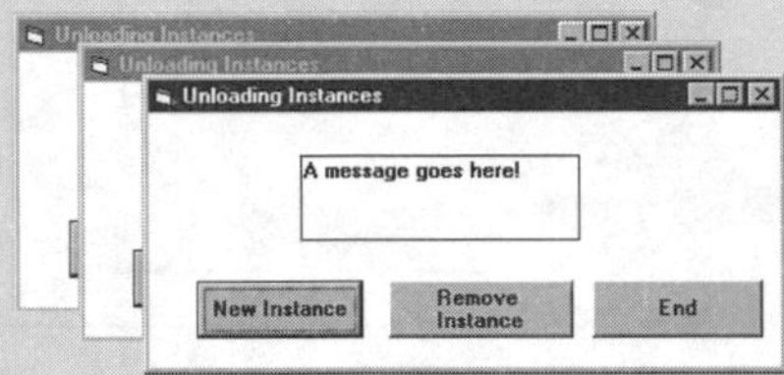

Figure 12.5

Besides adding a form instance, any instance created can also be removed using coded statements.

1 Create a new project and save the form from Exercise 12.3 as a new form named **f12-5.frm**. Add the **Remove Instance** command button, and change the form and button captions.

2 Add the following code to the `cmdRemove_Click()` event:

```
Private Sub cmdRemove_Click()
   Dim i As Integer
   If (forms.Count > 1) Then
        Unload forms(forms.Count - 1)
   Else
         cmdRemove.Enabled = False
   End If
   forms(forms.Count - 1).Print forms.Count
End Sub
```

Observe that all forms will be unloaded until only one form remains because the form index begins with `0`. To reference the last member of the forms collection, you can use the expression `forms(forms.Count - 1)`. In the `cmdRemove_Click()` event, as each form is unloaded, the count of the forms collection is printed on the last member of the collection until `Count` equals `1`.

3 Add the `cmdNew_Click()` code:

```
Private Sub cmdNew_Click()
   Dim F As New Form1
   F.Show
```

```
    F.Move Left + (Width \ 10), Top + (Height \ 10)
    F.Print forms.Count
End Sub
```

4 Run and test your program.

12.4.2 Using the *For Each...Next* Structure to Work with Collections

To work with collections, use the `For Each...Next` structure. The syntax for this loop structure is as follows:

```
For Each element In group
   [statements]
   [Exit For]
   [statements]
Next [element]
```

Table 12.2 explains the components of the `For Each...Next` structure.

Table 12.2 The *For Each...Next* Syntax

Component	*Description*
`element`	A variable used to iterate through the elements of the collection. For collections, `element` can only be of type `Variant`, a generic object, or any specific OLE automation object variable.
`group`	The name of an object collection.
`statements`	One or more instructions that are executed on each item in the group.

As an example, when working with the forms collection, the statements read as follows:

```
Dim frmNew as Form
For Each frmNew In Forms
   [statements]
   [Exit For]
   [statements]
Next [element]
```

Observe that `Form` is singular and `Forms` is plural. `Forms` is a collection, whereas `Form` is an element of the collection.

Exercise 12.6 Iterating Through the Forms and Controls Collections

This exercise iterates through the controls collection of two identical forms and illustrates use of the `For Each...Next` statement to list each control. Figure 12.6 shows the interface you are asked to create.

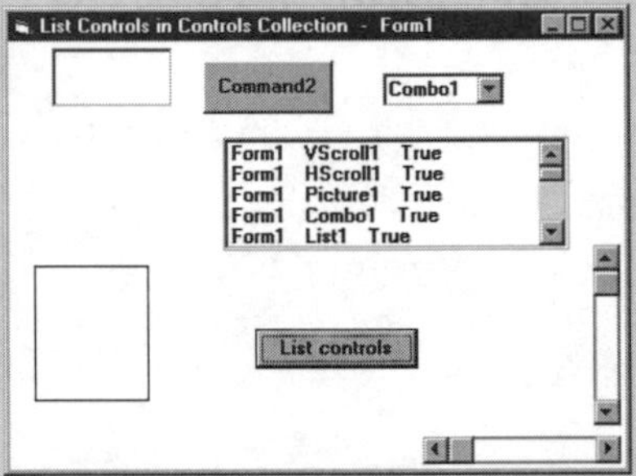

Figure 12.6

Place several controls on a form, add a list box to show the results, and use the `For Each...Next` *statement to list all controls assigned to each instance of a form.*

1 Build the two forms. The forms contain a list box, picture box, OLE control, frame, combo box, command button, and scroll bars. Name one **`Form1`** and the other **`Form2`**. Save the forms as **f12-6A.frm** and **f12-6B.frm**. Save the project as **p12-6.vbp**.

2 Make `Form1` the startup object, and put the following code in its `Form_Load()` event:

```
Load Form2
Caption = Caption & "  -  " & Name
```

This places both forms in memory.

3 Set the `Visible` property of one or two of the controls to `False`.

4 Enter the following code for the command button on `Form1` named `cmdTest` with the caption **List controls**:

```
Private Sub cmdTest_Click()
  Dim ctrl As Control
  Dim tstForm As Form
  Dim item As String
  For Each tstForm In Forms
    For Each ctrl In Controls
     item = tstForm.Name & "    " & ctrl.Name & "    " & ctrl.Visible
     List1.AddItem item
    Next
  Next
End Sub
```

Each element in the `For Each...Next` syntax must be an object of the same type as that in the group. In this instance, you declare the variable `ctrl` as a generic `Control` object and `tstForm` as a generic `Form` object. The nested loops iterate through the collection hierarchy. Each element of the `Forms` collection is first selected, and all controls contained within that form (`Form1`) are listed. Following this, all the controls contained on the second element in the `Forms` collection (`Form2`) are listed.

Also observe that `Form2` is loaded into memory, but it is not shown. Even so, the controls placed on this form can be listed. If the form had not been loaded into memory, it would not have been considered part of the forms and controls collections.

Objective 12.5 Dynamically Loading a Control Array

Although elements of a control array are not treated as `Object` variables, they can be loaded and unloaded at runtime. In this manner, they behave like `Object` variables. By adding an element to a control array, a new "instance" can be created.

Two properties, `LBound` and `UBound`, make working with dynamic control arrays easy. Read at runtime, they return the lowest and highest elements, respectively, in the control array. This enables you to write a single loop procedure to process values in the control array event as its size is dynamically changed.

Suppose you developed an application for online ordering. As users made selections, an invoice could be created, adding one label at a time. Once the order was completed, the total could be calculated with code similar to the following:

```
Private Sub Calculate()
   Dim  as Integer
   Dim RunTotal as Currency
   RunTotal=0
   For  n = lblOrderitem.LBound to  lblOrderItem.UBound
      RunTotal = Runtotal + Val(lblOrderItem(n).Caption)
   Next
End Sub
```

Exercise 12.7 Using a Dynamic Control Array

This exercise asks you to dynamically load a control array of shapes on the display screen. The interface is very simple (see Figure 12.7).

continues

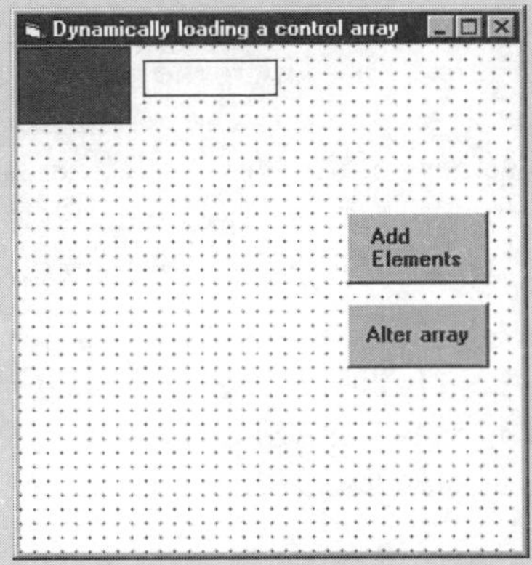

Figure 12.7

This application demonstrates creating instances of shapes and labels and changing their properties during runtime.

1 Build the interface shown in Figure 12.7. It contains two unmarked controls: a shape and a label. Each has its `Index` property set to `0`. Set their `Visible` properties to `False`. The display also contains two command buttons (**`cmdNew`** and **`cmdAlter`**).

2 Write the `cmdNew_Click()` procedure.

```
Private Sub cmdNew_Click()
    Static  n As Integer
    Dim hgt As Integer

    hgt = CInt(Shape1(0).Height / 2)
    Label1(0).BorderStyle = 0

    If Label1.UBound <> 6 Then
       For n = 1 To 6
         Load Shape1(n)
         Load Label1(n)
         Shape1(n).Shape = n - 1
         Label1(n).Caption = Shape1(n).Index
         Shape1(n).Top = Shape1(n - 1).Top + Shape1(n - 1).Height
         Label1(n).Top = Shape1(n).Top + hgt
         Shape1(n).Visible = True
         Label1(n).Visible = True
       Next
    Else
       Exit Sub
    End If
End Sub
```

By changing the `BorderStyle` property of the original control, `Label1(0)`, before loading new control array elements, the appearance of each subsequent array member is changed. Each individual array member can now be referenced and have its properties changed.

3 Add a new command button to the form and name it **cmdAlter**. Write the following code:

```
Private Sub cmdAlter_Click()
    Dim n As Integer
    For n = Label1.LBound To Label1.UBound
        Label1(n).BorderStyle = 1
        Label1(n).Font.Size = 10
    Next
End Sub
```

By employing the `LBound` and `UBound` properties of the `Label1` control array, you can change every element of the array without knowing how many elements were contained within it.

4 On your own, add a procedure that unloads the related Shape control element every time a Label control is clicked.

Chapter Summary

This chapter introduced the use of `Object` variables. `Object` variables can be declared like other variables. Their scope is also similar to other variables. They differ, however, in that they are references to objects. These object references are not assigned to an object until the `Set` statement is used. You can declare an `Object` variable by using the generic objects of `Object`, `Form`, or `Control`; or, preferably, by using specific object references. For controls, these specific object references could be included in such declarations as the following:

```
Dim X as TextBox

Dim Y as Label

Dim Z as CommandButton.
```

`Object` variables that refer to specific forms and objects can be declared by using the `New` keyword. If declared in conjunction with the `Set` statement, object references to forms can be cleared from memory using the `Nothing` keyword. The properties of each instance of the form can be changed, as well as the properties of the controls that the form contains.

Visual Basic maintains a forms collection for the entire application and a controls collection for each form. You can iterate through these collections by using the `For Each...Next` structure. For form and control collections, the `Object` variable that refers to each collection member can be declared with generic variables, as shown in the following example:

```
Dim f as Form
For Each f in Forms
   [Add statements here]
Next
```

```
Dim fControl as Control
For Each fcontrol in Controls
   [Add statements here]
Next
```

You can dynamically load and unload controls, provided they are elements of a control array. The `LBound` and `UBound` runtime properties of the arrays refer to the last and first elements, respectively, of the control array. These are altered dynamically as the size of the array changes.

Skill-Building Exercises

1. Add a check box and a command button to a form. In the command button's `Click()` event, enter code that declares a specific `Object` variable for a check box, binds the variable to the check box on the form, then changes the variables `Caption`, `BackColor`, `Left`, and `Top` properties.

2. Using the `Forms` collection `Count` property, modify Exercise 12.5 so that the width of newly created forms is incrementally increased by ten percent. *Hint:* The most recently created form can be referenced as `Forms(Forms.Count - 1)`.

3. Place a command button and a shape control on a form. During design time, make the shape a blue circle. Write a program so that in the button `Click()` event, a specific `Object` variable is declared, the shape control is bound to this variable, and the variable is then passed to a form-level procedure.

   The form-level procedure should change the shape's `BackColor` to red and increase its `Height` and `Width` properties by a factor of 3. In the procedure, assign the shape's `Name` to the shape's `Tag` property and then print the new `Tag` property value on the form.

4. Add a label, a picture box, an option button, a check box, a shape control (with `BackStyle` set to `Opaque`), a list box, and a command button on a form. Set them randomly on the form. In the command button, write a procedure that, using the controls collection, passes the controls and their controls collection index to a procedure. In the procedure, align the controls across the top of the form, change their `BackColor` properties to red, and assign the control's name to the control's `Tag` property.

   Before exiting the button `Click()` event, list the controls `Tag` properties in the list box. *Hint:* To align the controls, test the index number. If it isn't zero, then you can reference the previous index number to set the control's position.

5. Write a program that enables the user to enter price data into a text box. Every time the user presses the Return key, the price is displayed in a label that is dynamically loaded and then shown. The program should also include a means of summing the prices and displaying the results. *Hint:* Use the `KeyPress()` event, and check if `KeyAscii = vbKeyReturn`. Also, use the control array `LBound` and `UBound` properties. Remember, if you add a new control, `Load lblPrice(lblPrice.Ubound + 1)`, it becomes `lblPrice(lblPrice.Ubound)`.

13

Class Modules

This chapter introduces you to building your own objects in Visual Basic by defining classes. You have defined classes, methods, and properties already. Every time you add a procedure to a form, you define a class method. Every time you declare a `Public` variable at the module level you, in essence, create a form property. Visual Basic permits you to explicitly define your own form properties with special procedures called `Property` procedures.

Visual Basic forms can be used as software components by being deployed in multiple Visual Basic projects. Developers more typically, however, build software components with class modules, which enable the developer to define classes explicitly. They encourage a deliberate object-oriented approach to application development. Like forms, class modules have their own properties, methods, and events. Class modules differ from forms, principally by not having a visual interface.

Class modules contain only code, but the code differs from that held in standard code modules. Code in a class module defines the properties, methods, and events for objects created from that class. Class modules hold their data and code together. Another way of describing this is to say that classes *encapsulate* data and code in a way that makes objects created from the class easy to manipulate. Standard code modules, by contrast, contain procedures, which vary widely in their functions; standard code modules do not explicitly hold data and code together. As important, code in class modules can be compiled to become an ActiveX DLL, ActiveX EXE, and, in advanced Visual Basic, an ActiveX control. Other applications can then create objects from the class that you have defined.

Class modules were built into Visual Basic to allow for the design of robust, reusable components. Components that are to be used by many applications need to contain code that is well-tested and as free of errors as possible. This chapter addresses some of the fundamental ways in which class modules are designed to make code more robust. Creating procedures for class modules is a more involved process than writing procedures for standard code modules or forms. It takes forethought and careful analysis before the first line of code is written. The tradeoff is code that tends to be reliable and easy to maintain.

In this chapter, you learn to

- define your own property for a form,
- add `Property Let` and `Property Get` procedures to a form,
- add a class module to a project,
- define class properties and class methods,
- instantiate objects from the class you create,
- use class `Initialize()` and `Terminate()` events,
- add objects you have defined to an instance of the `Collection` class.

Chapter Objectives

Because this chapter focuses on the definition and use of classes, it first reviews the fundamental features of objects and introduces the mechanics of creating several simple classes. It reviews the way instances of classes (objects) can be placed in collections. This chapter concludes with some guidelines on creating class modules. The objectives of this chapter are as follows:

1. Reviewing object features
2. Adding a class module to a project
3. Adding objects to a collection
4. Employing good programming practices when creating class modules

Objective 13.1 Reviewing Object Features

Objects are instances of classes. Objects possess properties, contain methods, and can handle events. Properties hold data that describe the object; methods tell the object what to do; events are actions the object undertakes when it receives a particular message. Properties and methods of the class are referred to as *class members*. Let's review each of these features of objects in more detail.

13.1.1 Object Properties

Properties hold data that describe an object. You have already worked with the properties of forms as well as some of the properties of Visual Basic's system objects, such as the `Screen` or `App` object. The `Caption` property of a form, for instance, describes the string expression displayed on the form's title bar. This property can be read as

```
strVariable$ = Form1.Caption
```

or it can be written to

```
Form1.Caption = "I have a new caption!"
```

Some properties are read-only. The `ListCount` property of a list box, for example, is a read-only property.

The properties of an object that can be read or written to are part of the object's *interface*. That is, code that seeks to manipulate or read object properties "sees" only those properties the object exposes. It does not see data that the object does not expose.

How does this work in practice? Because you have already worked with forms, let's start with form properties.

Exercise 13.1 Public and Private Module Level Declarations

In this exercise, you will experiment with the scoping of module-level variables and review concepts introduced in Chapter 10, "Defining the Scope of Variables and Procedures." You work with two forms (see Figure 13.1). One form launches a second form. The first form assigns a value to, and reads a value from, variables declared at the module level of the second form. One of the variables is declared as `Public` and the other is declared as `Private`.

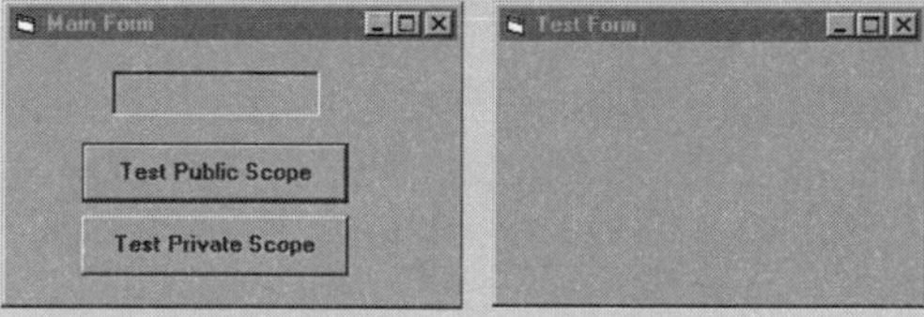

Figure 13.1

Create two forms to test private and public scope.

1 Create two forms, as illustrated in Figure 13.1. The form on the left is named **`frmMain`**, and the form on the right is named **`frmTest`**. Change the captions as shown. Save these two forms to disk, respectively, as **f-13A.frm** and **f-13B.frm**. Save the project as **p13_1.vbp**.

2 On `frmMain`, add two command buttons named **`cmdTestPublic`** (upper button) and **`cmdTestPrivate`** (lower button). The label control is named **`lblDisplay`** and has its `BorderStyle` property set to `1` (fixed single).

3 On `frmTest`, make the two following declarations at the module level and add the `Form_Load()` event:

```
Public PublicProperty As Integer
Private PrivateProperty As String
```

continues

```
Private Sub Form_Load()
    PublicProperty = 67
    PrivateProperty = "I am a string"
End Sub
```

4 On `frmMain`, add the following code to the button `Click()` events and the `Form_Load()` event:

```
Private Sub cmdTestPublic_Click()
   'This can be displayed since PublicProperty is declared
   ' as type public
   lblDisplay.Caption = Str$(frmTest.PublicProperty)
End Sub

Private Sub cmdTestPrivate_Click()
   'This cannot be displayed since PrivateProperty is declared
   ' as type private
   cmdTestPrivate.Caption = frmTest.PrivateProperty
End Sub

Private Sub Form_Load()
   frmTest.Top = frmMain.Top     ' position frmTest
   frmTest.Left = frmMain.Left + frmMain.Width + 200
   frmTest.Show
End Sub
```

5 Specify `frmMain` as the startup object, then run and test the program.

When you click the top button, the `cmdTestPublic_Click()` event assigns a value to the publicly declared variable on `frmTest` and then reads that value and displays it on the label `lblDisplay`. The `cmdTestPrivate_Click()` event, however, should generate an error message like that shown in Figure 13.2.

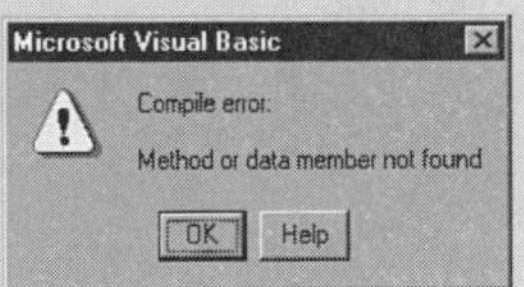

Figure 13.2

This dialog box displays a compile error stating that a method or data member was not found.

The message generated in Exercise 13.1 states, "Method or data member not found." As far as Visual Basic is concerned, a publicly declared variable is a *data member* or property of the form. Because the `PrivateProperty` variable was declared with the `Private` statement, its scope was limited to the `frmTest` module. It was not part of the public *interface* for the `frmTest` class. The variable `PublicProperty`, however, did become part of the class interface. It could be assigned a value and that value could be read. Also observe how the variable `PublicProperty` was read in the `cmdTestPublic_Click()` event. It was referred to with the same `object.property` notation you have used with other objects.

The PublicProperty variable in frmTest can be assigned any integer value. If assigned a string expression, an error occurs. Suppose you wanted to create a variable that could only take positive integer values. A publicly declared integer variable at the module level might be assigned an errant value and introduce errors into processing. Visual Basic provides special procedures to cover this contingency. They are called Property procedures, and are executed when an object property is assigned a value or read. Property procedures act as a buffer between the direct assignment or reading of object property values.

13.1.2 *Property* Procedures and Data Hiding

Although a publicly declared variable in a module acts like a property of a class, and is assigned new values and read with the object.property syntax, class properties are properly defined with Property procedures. Property procedures allow variables that are defined as Private in a module to be assigned new values and read from outside the object. They promote one of the central tenets of object-oriented programming: *the encapsulation of data into hidden data members.* These terms will become clear as you work with Property procedures. Creating a class property in Visual Basic that can be both assigned new values and read is usually a three step process:

1. Declare a Private variable at the *module level* of a form or class module.
2. Create a Property Let procedure to assign new values to the object property.
3. Use a Property Get procedure to read the object property value.

A read-only property has only a Property Get procedure. Values returned by the Property Get procedure can be assigned or determined only from within the object. The following exercise takes you through the mechanics of creating a Property procedure.

Exercise 13.2 Hiding Class Data

Like Exercise 13.1, this exercise contains two forms: frmMain and frmTest (see Figure 13.3). The second form, frmTest, contains the variable declarations and the Property procedures. This exercise demonstrates the way data members of a class stay hidden and are accessed only with Property procedures.

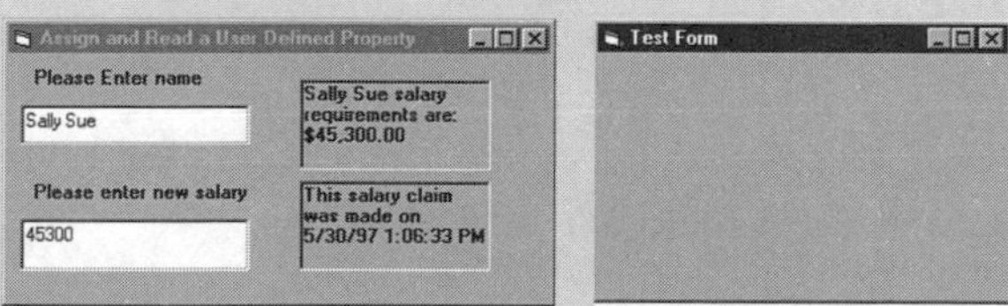

Figure 13.3

Create two forms to test the hiding of class data.

continues

1 Create the two forms that you see illustrated in Figure 13.3. The first form, **frmMain**, has two text boxes on the left: **txtName** in the upper position and **txtSalary** in the lower position. Above each text box are labels that prompt the user to enter data. Set the TabIndex property for txtName to 0 and the TabIndex for txtSalary to 1.

2 On the right side of frmMain are two larger labels with their BorderStyle properties set to 1 (fixed single). The upper label is named **lblSalary**, and the lower label is named **lblTimeStamp**. Both of these labels have their WordWrap properties set to True.

3 Change the captions for the forms as shown.

4 Save frmMain to disk as **13-2A.frm** and frmTest as **13-2B.frm**. Save the project as **p13_2.vbp**.

5 In the frmMain Form_Load() event, enter the following code to position frmTest and make it visible:

```
Private Sub Form_Load()
   frmTest.Top = frmMain.Top
   frmTest.Left = frmMain.Left + frmMain.Width + 200
   frmTest.Show
End Sub
```

6 Add the following procedures to the txtName_KeyPress() and TxtSalary_KeyPress() events:

```
Private Sub txtName_KeyPress(KeyAscii As Integer)
   If KeyAscii = vbKeyReturn Then
     frmTest.EmployeeName = txtName.Text
     txtSalary.SetFocus
   End If
End Sub

Private Sub txtSalary_KeyPress(KeyAscii As Integer)
   If KeyAscii = vbKeyReturn Then
      frmTest.Salary = Val(txtSalary.Text)
      lblSalary.Caption = frmTest.EmployeeName &
      ➥" salary requirements are: "& Format$(frmTest.Salary, "Currency")

      lblTimeStamp.Caption = "This salary claim was made on " &
      ➥frmTest.TimeStamp
   End If
End Sub
```

7 For safety's sake, add the two lines of code to the txtName_LostFocus() event:

```
Private Sub txtName_LostFocus()
  frmTest.EmployeeName = txtName.Text
  txtSalary.SetFocus
End Sub
```

8 At the module level of `frmTest`, declare the following variables to complete the first step of the three-step class `Property` procedure:

```
Option Explicit
'The lowercase m notation stands for a private data class member
Private m_EmployeeName As String
Private m_Salary As Double
Private m_TimeStamp As Date
```

These module-level variables are declared, and will have their values read and assigned by `Property` procedures.

9 Add code to the `Form_Load()` event to the `frmTest` form to initialize these variables:

```
Private Sub Form_Load()
   ' initialize object property values
   m_Salary = 0
   m_TimeStamp = Now
   m_EmployeeName = "Nog"
End Sub
```

10 Add the `Let` and `Get` procedures for the `Salary` property to complete the final two steps of the three-step class `Property` procedure:

1 With the code window of `frmTest` having the focus, choose Tools, Add Procedure from the Visual Basic menu bar. You should see the dialog box in Figure 13.4. Select the Property option button, make sure the scope of the procedure is `Public`, and name the procedure **`Salary`**.

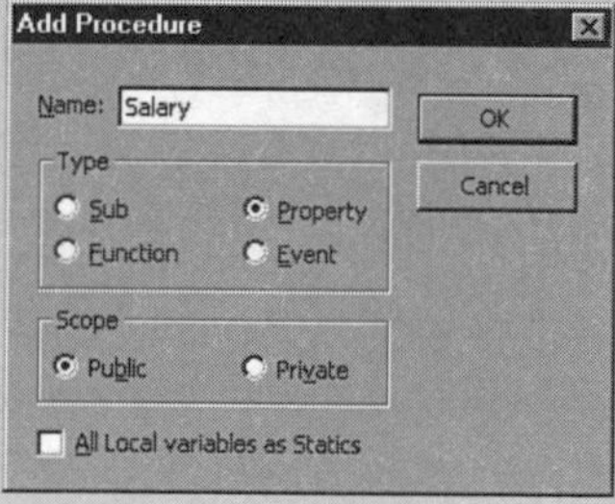

Figure 13.4

The Add Procedure dialog box provides space for a name and allows you to set the type and the scope.

2 Click OK in the Add Procedure dialog box, and you will immediately see two procedure declarations added to the code window:

```
Public Property Get Salary() As Variant
End Property

Public Property Let Salary(ByVal vNewValue As Variant)
End Property
```

continues

Visual Basic adds these procedures automatically to the code window. The `Let` procedure assigns values to a variable; the `Get` procedure reads values. By default, the object property is a `Variant` data type. Data types for `Property Let` and `Property Get` procedures must match. Because the `Salary` property will return a `Double` numeric value, change the procedure declarations to read as follows:

```
Public Property Get Salary() As Double
End Property

Public Property Let Salary(ByVal vNewValue As Double)
End Property
```

3 Add the following code for the salary `Let` and `Get` procedures:

```
Public Property Let Salary(ByVal vNewValue As Double)
   If m_EmployeeName <> "Nog" Then
      Select Case vNewValue
       Case Is <= 0
          MsgBox "Sorry, salary must be a positive value."
       Case 0 To 10000
          MsgBox "You are worth more than that!"
          m_Salary = vNewValue
       Case Else
          m_Salary = vNewValue
      End Select
          m_TimeStamp = Now
      Else
          MsgBox "Employee must be identified before salary can be
          ➥assigned."
      End If

End Property

Public Property Get Salary() As Double
   Salary = m_Salary
End Property
```

Observe that before a new value can be assigned to the `Salary` property, an employee must first be identified. The `m_TimeStamp` variable is also assigned a value when a new salary is assigned.

11 Add a second class `Property` procedure to `frmTest` for the `EmployeeName` property (remembering to change the data types to `String`!):

```
Public Property Get EmployeeName() As String
     ' make sure assignment has been made to the property
     If m_EmployeeName <> "Nog" Then
        EmployeeName = m_EmployeeName
     Else
        EmployeeName = ""
     End If
End Property

Public Property Let EmployeeName(ByVal vNewValue As String)
   m_EmployeeName = vNewValue
End Property
```

12 Add a third TimeStamp class property, which is a read-only property of type Date. (Delete the Property Let procedure when it is inserted by the Add Procedure dialog box.)

```
Public Property Get TimeStamp() As Date
   TimeStamp = m_TimeStamp
End Property
```

13 Make sure that frmMain is the startup object, then run and test the program.

This is a lot of code to accomplish what could be accomplished more easily by standard coding procedures. But observe the increased flexibility of encapsulating data into class properties. Visual Basic, for instance, provides no signed numeric data types. It is therefore impossible to restrict Integer, Single, or Double numeric data types to positive values unless input is checked each time, such as in a text box KeyPress() event.

By placing this value as an object property, the value can be checked in the Property Let procedure of a class. This was done in the case of the Salary property. The Let Salary procedure also could check to see that an employee had been identified before a salary value could be assigned.

Encapsulation of data into class properties allows multiple forms to use the same data-validation procedure and can easily be updated as conditions change. Although data placed into class properties is more time-consuming than simply using publicly declared variables, class data promotes code reuse—shortening software development projects in the long term.

Observe also that the data held by the object is now hidden. All of the newly added data members of frmTest were declared privately. The only way to alter their values or read them is through the Property Let and Property Get procedures. This *data hiding* is central to robust, object-oriented programming. These kinds of procedures are sometimes referred to as class *mutator* methods (a Property Let procedure) or *accessor* methods (a Property Get procedure).

13.1.3 Object Methods

In Exercise 13.1 and Exercise 13.2, you extended the class definition of a form named frmTest. Just as data can be assigned to class properties, code to operate on that data can be placed in class methods. Methods can be private to the object or made part of the object's interface by declaring the procedure as type Public.

Consider, for instance, a ListBox class. One of the public procedures of the ListBox class is Clear. This method is now available to you as a class method. How the ListBox class implements the Clear method is hidden from you. All you know is that it works. The ListBox class therefore *encapsulates* the Clear method. The ListBox class also has privately scoped procedures that are not accessible. For example, how does the ListBox class reorder

ItemIndex numbers when an item is deleted? It presumably invokes a method internally, but this method is not made available to you through the ListBox interface.

When adding class methods, you need to determine whether the method will become part of the class interface or hidden from the user.

Exercise 13.3 Adding a Class Method

This exercise extends Exercise 13.2 by adding a method that calculates a 4.7 percent increase above the new salary. (We're getting greedy!) The user can now select an option button (see Figure 13.5) to calculate such a raise. Because this option works solely on the calculation of the Salary property itself and is not invoked separately as an object calculation, the method need not be Public. It can simply be used internally by the object to calculate the requested value.

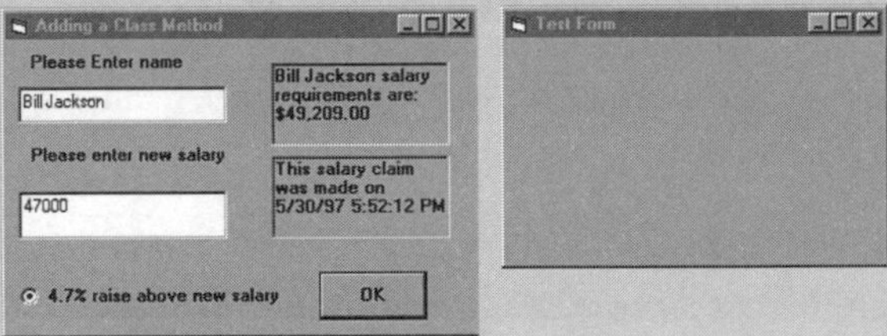

Figure 13.5

The raise button returns a value of True *or* False*, and determines whether the method to increase a person's salary is to be executed.*

1 Save the files from the previous exercise under new names (such as, f13-2A.frm as **f13-3A.frm**. f13-2B.frm as **f13-3B.frm**, and p13-2.vbp as **p13-3.vbp**). Remember to change the Name property of the project to **p13_3**.

2 Add the option button **optRaise** and the command button **cmdOK** to frmMain. Position frmMain and frmTest, as illustrated in Figure 13.5.

3 Add the following code to the optRaise_Click() event:

```
Private Sub optRaise_Click()
   frmTest.Raise = optRaise.Value
End Sub
```

4 Erase (or cut) the code from the txtSalary_KeyPress() event, and place or paste the following code in the cmdOK_Click() event:

```
Private Sub cmdOK_Click()
   frmTest.salary = Val(txtSalary.Text)
   lblSalary.Caption = frmTest.EmployeeName &
   ➥" salary requirements are: " & Format$(frmTest.salary, "Currency")
   lblTimeStamp.Caption = "This salary claim was made on " &
   ➥frmTest.TimeStamp
End Sub
```

5 Add an additional declaration for the `Raise` property in the General declarations section of the `frmTest` module:

```
Private m_Raise As Boolean
```

The module now contains four class property variables: `m_EmployeeName`, `m_Salary`, `m_TimeStamp`, and `m_Raise`.

6 As with the other class properties, add the accessor and mutator methods for the `Raise` object property:

```
Public Property Get Raise() As Boolean
   Raise = m_Raise
End Property

Public Property Let Raise(ByVal vNewValue As Boolean)
   m_Raise = vNewValue
End Property
```

7 Add the following line of code to the `frmTest_Load()` event to initialize this new data member:

```
m_Raise = False
```

8 Add the following method to the `frmTest` module, which calculates the new salary with a 4.7 percent increase:

```
Private Function AddRaise(salary As Double)
   AddRaise = (salary * 0.047)
End Function
```

9 Finally, change the `Property Let Salary` procedure so that the user can calculate a new salary with or without a 4.7 percent increase:

```
Public Property Let salary(ByVal vNewValue As Double)
   If m_EmployeeName <> "Nog" Then
     Select Case vNewValue
       Case Is <= 0
         MsgBox "Sorry, salary must be a positive value."
       Case 0 To 10000
         MsgBox "You are worth more than that!"
         If m_Raise = True Then
            'Call the AddRaise function
            m_Salary = vNewValue + AddRaise(vNewValue)
         Else
            m_Salary = vNewValue
         End If
       Case Else
         If m_Raise = True Then
            'Call the AddRaise function
            m_Salary = vNewValue + AddRaise(vNewValue)
         Else
            m_Salary = vNewValue
         End If
     End Select
     m_TimeStamp = Now
```

continues

```
    Else
      MsgBox "Employee must be identified before salary can be assigned."
    End If
End Property
```

10 Run and test the program.

13.1.4 Object Events

Objects can also contain events. Most of the code you have written so far are procedures that respond to user events, such as a button click or the press of a key on the keyboard. These events are triggered when the object receives an operating system message. When you click a mouse button on a form, for instance, an operating system message is sent to the object (a window); and if the form's `Click()` event has code in it, the code is fired.

Visual Basic allows you to create your own events for objects, but it is rare that you will need to add them. The visual interface for your program can almost always be handled by the events provided by Visual Basic, and the standard and custom controls provided in the different Visual Basic versions. In addition, a rich market of third-party controls exists for Visual Basic.

The capability to design events for objects is used primarily in advanced Visual Basic when creating ActiveX controls. This usually requires extensive knowledge and use of the Windows API, a topic beyond the scope of this book. The software components that we create will not require custom events. Most will be invisible components created from class modules. We turn to that objective next.

Objective 13.2 Adding a Class Module to a Project

In the preceding exercises, you added your own properties and methods to forms. You, in effect, extended the definition of classes named `frmTest`. Custom properties for forms were introduced first because you are already familiar with forms. But the `frmTest` classes were not especially efficient. They contained a lot of extra data and a visual interface that was not needed to work with salary calculations for employees.

The `frmTest Property` procedures that you added in the preceding two exercises are part of the code that you might use to define an object called `Employee`. This object would contain data about an `Employee` (object properties) and make certain calculations with that data (object methods). As you can see, a full-blown `Employee` class definition might contain quite a bit of code.

Data that we might want to hold about the employee include the employee's name, social security number, company ID number, date of birth, date of hiring, pay grade, and emergency information. Methods to operate on that data might include not only payroll calculations (including local taxes, state taxes, federal taxes, payroll savings matching plans, and automatic charitable contribution deductions), but also benefit calculations based upon years of service.

Think of another, simpler, object: a customer order entry. An order entry usually includes the product name, the quantity ordered, the unit price of the item, and the total amount of the order entry. Although in most online ordering applications you would want to have this information displayed, you can think of an order entry item as a discrete object. Such an object can be described in a class module.

13.2.1 Inserting Class Modules into a Project

Adding a class module to a project is very similar to adding a form module or standard code module to a project. Choose Project, Add Class Module from the Visual Basic menu bar. A standard dialog box will appear with tabs named New and Existing. You can choose to insert a new module or an existing module.

Like forms and standard modules, class modules have a `Name` property (press F4). It is important that you do not accept the default name provided. Instead, immediately give the class module a meaningful name. Like form and standard code modules, class modules are saved to disk under a separate file name. Class modules saved to disk have a .CLS file extension.

When you examine a class module, the code window is nearly identical to that of a standard module. There is a General Declaration section and a Class object (click the left drop-down combo box). The Class object has two events listed: Initialize and Terminate (click the right drop-down combo box). The first event is easy to understand. It fires whenever an object of that class is instantiated with the `Set` statement, or declared with the `New` keyword and then referenced in code. The `Terminate()` event is less obvious. It executes when the *last variable reference* to the object leaves memory. This will become clear when you write code for the `Initialize()` and `Terminate()` events in sections 13.2.3, "The `Initialize()` Event," and 13.2.4, "The `Terminate()` Event," respectively.

13.2.2 Creating an *OrderItem* Class

Creating a new class with a class module should seem straightforward to you. You have already worked with `Property Let` and `Property Get` procedures, and coded general methods for form classes. The main difference is that when it comes time to create (instantiate) an object from the class, you must bind an object variable with the `Set` statement or declare an object variable with the `New` keyword.

Exercise 13.4 Creating and Using an *OrderItem* Class

In this exercise, you will use a class module to create an `OrderItem` class. You will instantiate the class as needed as the user makes order selections. In this case, imagine that you have a limited list of tools from which the user can select (see Figure 13.6). The exercise consists of a single form module, named `frmOrder`, and a single class module, named `OrderItem`.

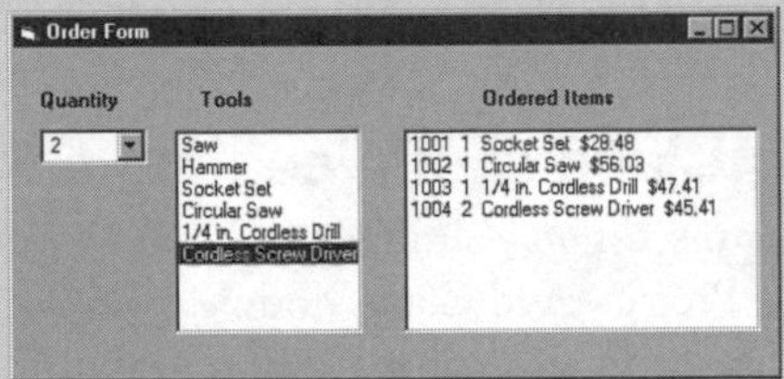

Figure 13.6

This order form allows you to select the quantity of an item and display the ordered items by product number.

1 Start a project and create the form shown in Figure 13.6. Name the form **`frmOrder`** and save it to disk as **f13-4.frm**. Name the project **p13_4** and save it to disk as **p13-4.vbp**.

2 On the form is a combo box (**`cboQuantity`**), and two list boxes: **`lstProducts`** (in the middle) and **`lstOrder`** (on the right). Change the form caption, as shown, and add the labels identifying the controls.

3 In the `Form_Load()` event, add code to populate the `cboQuantity` and `lstProduct` with data:

```
Private Sub Form_Load()
   Dim i As Integer

   ' populate the combo box with quantity choices
    For i = 1 To 10
      cboQuantity.AddItem Str$(i)
    Next
    ' display the first item
    cboQuantity.ListIndex = 0

   ' populate the list box with products
   With lstProducts
     .AddItem "Saw"
     .AddItem "Hammer"
     .AddItem "Socket Set"
     .AddItem "Circular Saw"
     .AddItem "1/4 in. Cordless Drill"
     .AddItem "Cordless Screw Driver"
   End With
End Sub
```

The design enables the user to make a quantity selection and then, by clicking the `lstProducts` list box, have the order line item displayed in the `lstOrder` list box. You need to define the `OrderItem` class before you can code the `lstProducts_Click()` event.

4 Add a class module to your project. Press F4 and change the `Name` property of the class module to **`OrderItem`**. Save it to disk as **cls13-4.cls**.

In this imaginary store, tools are priced by taking a markup from cost. The standard markup is 45 percent. (In retailing, markup is measured from the retail price. So an item bought wholesale for $50.00 and sold at retail for $100.00 is said to have a 50 percent markup. A $50.00 item sold for $90.00 has a 40 percent markup: $50.00 + (.8 * 50).) This calculation requires that you know the unit cost and the quantity ordered. In addition, the `OrderItem` class holds data on the product name, a line item number, and the item total.

5 Declare the following private data members at the General Declaration level of the `OrderItem` class:

```
Option Explicit

Private m_OrderNum As Integer
Private m_ProductName As String
Private m_Quantity As Double
Private m_UnitCost As Currency
Private m_ItemTotal As Currency
Private Const Markup = 0.9
```

6 Add the mutator and accessor methods (property procedures) for the class data members:

```
Public Property Get orderNum() As Integer
  orderNum = m_OrderNum
End Property

Public Property Let orderNum(ByVal vNewValue As Integer)
   m_OrderNum = vNewValue
End Property

Public Property Get ProductName() As String
   ProductName = m_ProductName
End Property

Public Property Let ProductName(ByVal vNewValue As String)
  m_ProductName = vNewValue
End Property

Public Property Get quantity() As Double
   quantity = m_Quantity
End Property

Public Property Let quantity(ByVal vNewValue As Double)
   m_Quantity = vNewValue
End Property
```

continues

```
Public Property Get UnitCost() As Currency
   UnitCost = m_UnitCost
End Property

Public Property Let UnitCost(ByVal vNewValue As Currency)
  m_UnitCost = vNewValue
End Property
```

We have yet to add all the data validation checking that normally would be present in the `Property Let` methods.

7 Add a read-only property that returns the order item total:

```
Public Property Get ItemTotal() As Currency
    m_ItemTotal = (m_UnitCost + (m_UnitCost * Markup)) * m_Quantity
    ItemTotal = m_ItemTotal
End Property
```

8 On `frmOrder`, add the following code to the `lstProducts_Click()` event:

```
Private Sub lstProducts_Click()
  Dim toolorder As OrderItem
  Static ordernumber As Integer
  Dim quantity As Integer
  Dim product As String
  Dim itemtotal As Currency
  Dim strDisplay As String

  Set toolorder = New OrderItem
  product = lstProducts.Text
  ordernumber = ordernumber + 1
  quantity = Val(cboQuantity.Text)

' ascertain the unit cost
 Select Case product
    Case "Saw"
       toolorder.UnitCost = 7.99
    Case "Hammer"
       toolorder.UnitCost = 6.99
    Case "Socket Set"
       toolorder.UnitCost = 14.99
    Case "Circular Saw"
       toolorder.UnitCost = 29.49
    Case "1/4 in. Cordless Drill"
       toolorder.UnitCost = 24.95
    Case "Cordless Screw Driver"
       toolorder.UnitCost = 11.95
 End Select

 With toolorder
    .orderNum = 1000 + ordernumber
    .ProductName = product
    .quantity = quantity
 End With
 'display the order

  strDisplay = toolorder.orderNum & "  "
  strDisplay = strDisplay & toolorder.quantity & "  "
```

```
    strDisplay = strDisplay & toolorder.ProductName & "  "
    strDisplay = strDisplay & Format(toolorder.ItemTotal, "Currency")

   lstOrder.AddItem strDisplay
  End Sub
```

9 Run and test the program.

The `lstProducts_Click()` event merits further explanation. Every time the user clicks the `lstProduct` list box, a new instance of the `OrderItem` class is created by the instruction `Set toolorder as New OrderItem`. Unlike a form, which is instantiated when a form instance is created with the `Load` statement, classes defined in class modules need to be instantiated with object variables. The object variable `toolorder` is now ready to have values assigned to its properties.

Values are read from the combo box for quantity of the order line item and from the list box for the product name. The unit cost is determined by the product name and assigned. The `orderNum` property is determined by arbitrarily incrementing a `Static` variable. Finally, when all of the properties are set, the order line item can be displayed in the `lstOrder` list box with the `ItemTotal` property read from the `OrderItem` class.

Should the calculation of the line item total be encapsulated in the read-only `ItemTotal` property or should it, instead, have been placed in a class method, perhaps a `CalculateLineItem` method? Making these kinds of determinations is sometimes a close call—or even a matter of taste.

13.2.3 The *Initialize()* Event

In Exercise 13.4, you could count on the fact that the application using the `OrderItem` class assigned values to the `Product Name`, `Quantity`, and `Unit Cost` properties before the `ItemTotal` property was read. But this is a dangerous assumption, particularly if the class is to be used in multiple applications. It is the responsibility of the class module to ensure that interdependent properties are managed so as to avoid errors, not the responsibility of other applications or applications that instantiate objects from the class.

Suppose you tracked average order item sizes. This calculation would require a division operation, and you would have to ensure that the class did not attempt to divide by zero. A class member with a `Date` data type that had to be within a certain range would need data validation code added to the mutator (`Property Let`) method. A `ShipDate` property, for instance, might not be allowed to have a date prior to the `OrderDate` property.

You can minimize these kinds of errors by assigning default values to class properties every time the class is instantiated. This will not prevent all errors, but will ensure that default values are assigned and that the user is presented with non-null values when accessing an object property.

As discussed earlier, Visual Basic class modules have two built-in events: the `Initialize()` event and the `Terminate()` event. The `Initialize()` event fires every time a class instance (object) is created. In this event, all data members of the class should be initialized and assigned default values. You can complete the code for Exercise 13.4 by adding the following code to the `OrderItem` class module:

```
Private Sub Class_Initialize()
   m_ProductName = "Product"
   m_UnitCost = 0
   m_Quantity = 0
   m_OrderNum = 1001
End Sub
```

Remember that you can find the `Initialize()` event by selecting the implicit Class object under the General Declarations item in the left-hand combo box in the class module code window.

13.2.4 The *Terminate()* Event

`Terminate()` events are used to execute cleanup code. Such cleanup code usually involves memory-management tasks, such as setting object variable references to `Nothing` or reclaiming memory allocated to large arrays. (Arrays are addressed in Chapter 16, "Lists and Arrays.")

The `Terminate()` event fires when the *last reference* to the class instance is removed from memory by falling out of scope or by having all of the variables that refer to that object set to `Nothing`. It is possible to have multiple object variables refer to the same class instance. It is only when all of these variable references are removed from memory that the `Terminate()` event fires.

Exercise 13.5 Demonstrating the *Terminate()* Event

In this short exercise, you learn how multiple object variables can refer to the same object and the effect this has on the `Terminate()` event. Figure 13.7 shows the form you are asked to create.

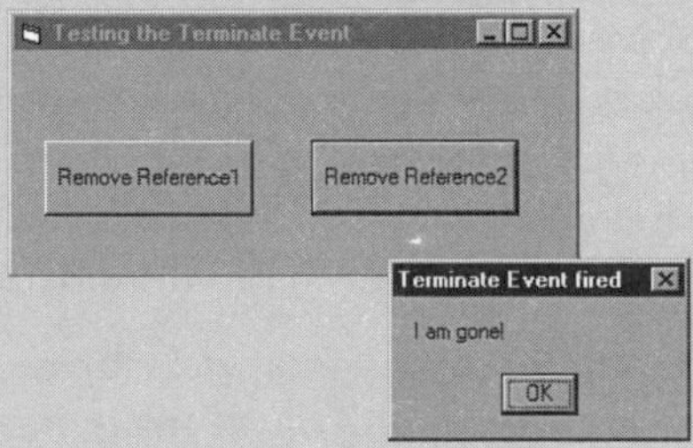

Figure 13.7

This design displays a dialog box to verify that the `Terminate()` *event has executed properly.*

1 Save cls13-4.cls as **cls13-5.cls** on disk.

2 Create a new project, and save it as **p13-5.vbp**. Name the project **p13_5**.

3 Add cls13-5.cls to the project.

4 Name the form **`frmTestTerminate`**, and save it to disk as **f13-5.frm**. Name the buttons **`cmdRemove1`** and **`cmdRemove2`**. Change the captions of the buttons and the form as illustrated.

5 Add the following code in the `Terminate()` event of the `OrderItem` class:

```
Private Sub Class_Terminate()
  MsgBox "I am gone!", 0, "Terminate Event fired"
End Sub
```

6 Make the following object variable declarations at the module level of `frmTerminate`:

```
Dim Reference1 As OrderItem
Dim Reference2 As OrderItem
```

7 In the `Form_Load()` event, bind the `Reference1` variable to a new class instance and then set the `Reference2` variable to the same object:

```
Private Sub Form_Load()
  Set Reference1 = New OrderItem
  Set Reference2 = Reference1
End Sub
```

8 Finally, add code to set these object variables to `Nothing` in the button click procedures:

```
Private Sub cmdRemove1_Click()
  Set Reference1 = Nothing
End Sub

Private Sub cmdRemove2_Click()
  Set Reference2 = Nothing
End Sub
```

9 Run and test the program. First click Remove Reference 1 and then click Remove Reference 2. The `Terminate()` event does not fire until the last reference to the object is removed from memory.

13.2.5 Examining the *OrderItem* Class with the Class Browser

In Chapter 1, "Running a Visual Basic Program," you were briefly introduced to the Object Browser. Just as you can use the Object Browser to observe classes in the Visual Basic libraries, the Object Browser permits you to examine classes that you have defined for individual projects.

You have a choice in how broadly you want to expose classes defined in class libraries. If they are confined to an individual project, the code is compiled into the .EXE, and the class is not exposed for others to use. If, however, you compile classes defined in class modules into an ActiveX .DLL or ActiveX .EXE, the class can be used by other applications. We examine these choices in Chapter 23, "Creating an ActiveX DLL."

To observe the members of your project's classes, you need only press the F2 key. This launches Object Browser. Alternatively, you can choose View, Object Browser from the Visual Basic menu bar. For the `OrderItem` class defined in Project p13_5, Object Browser should look like that illustrated in Figure 13.8.

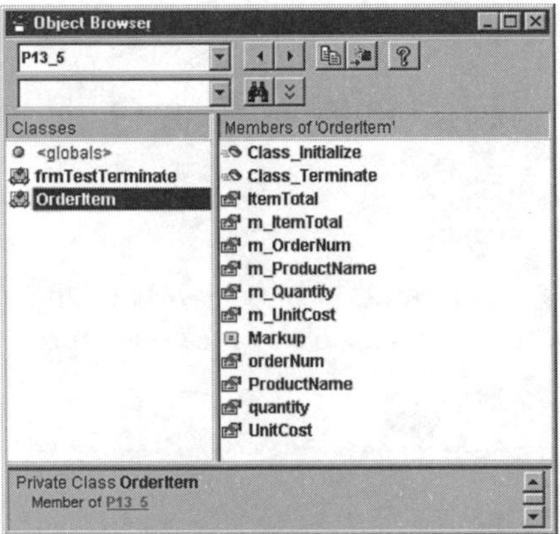

Figure 13.8

Object Browser displays the members and events of a class.

By selecting any one of the class members in the list box on the right side of the Object Browser, information about that class member is revealed at the bottom of the Object Browser. The data types for data members are displayed. Arguments for class methods can be displayed, as can the exposure of class properties. For example, select the ItemTotal property for the OrderItem class. The Object Browser indicates it is read-only. Select the Markup constant. The Object Browser displays its numeric value. Double-click the Markup constant to view the code.

As the classes you create become larger and as you begin to share class modules among projects, you will become quite familiar with the Object Browser. It is a handy way to remind yourself of the private data members, as well as the public interface the object provides for use by other modules.

Objective 13.3 Adding Objects to a Collection

As illustrated by the `OrderItem` class, an individual object is created every time the user places an order for an individual item at a certain quantity. It would be convenient if you could group these objects together. Fortunately, Visual Basic provides you with a generic `Collection` object for just this purpose. Although the objects need not be of the same class, you will find it convenient to place members of the same class into a `Collection`.

You will recall that you worked with the built-in forms and controls collection in Chapter 12, "Object Types, Variables, and Collections." You iterated through the collection by using `For Each...Next` syntax. Now that you can create you own objects, you look at the `Collection` object with a little more specificity.

As an object, the `Collection` must first be declared and bound with an object variable. The following instructions declare the `Tools` object variable as a `Collection` and bind it to a new class instance:

```
Dim Tools as Collection
Set Tools = New Collection
```

After a collection has been created, objects can be added to it. An important difference between the `Collection` object and the `Forms` and `Controls` collections is that instances of the `Collection` object follow a *one-based* index. That means the index value of the first member of the collection is `1`, not `0`, as with the `Forms` and `Controls` collection. Iteration through the collection can be accomplished in two ways:

- Using a standard `For...Next` loop you could write the following instructions:

  ```
  Dim i as integer
  For i = 1 to Tools.Count
        ' execute code here
  Next
  ```

- If the members of the `Collection` were all of the `OrderItem` class, you could write the instructions as follows:

  ```
  Dim counter as OrderItem
  For Each counter in Tools
        ' execute code here
  Next
  ```

As an object, the `Collection` object has its own properties and methods. Fortunately, these are very few, but very useful. The `Collection` object has a single property: `Count`. This is a read-only property that returns the number of objects in a collection. The `Collection` object has three methods, as explained in Table 13.1.

Table 13.1 The *Collection* Object Methods

Method	*Description*
`Add`	Adds an object to a collection
`Remove`	Removes an item from a collection
`Item`	Returns an item from a collection by its index or by its key

Let's examine each of these methods in turn.

13.3.1 The *Add* Method of the *Collection* Object

You add objects to a collection with the `Add` method. When objects are added, they can be also given a numeric index or a unique string expression to identify them. This allows them to be more easily retrieved with the `Item` method. The `key`, however, is optional. The complete syntax for the method is as follows:

```
object.Add item, key, before, after
```

Table 13.2 explains the components of the `Add` method syntax.

Table 13.2 The Syntax of the *Add* Method

Component	*Description*
`object`	Required. A `Collection` object that has been instantiated.
`item`	Required. An expression of any type that specifies the member to add to the collection.
`key`	Optional. A unique string expression that specifies a key string that can be used, instead of a position index, to access a member of the collection.
`before`	Optional. An expression that specifies a relative position in the collection. The member to be added is placed in the collection *before* the member identified by the `before` argument. For a numeric expression, `before` must be a number from 1 to the value of the collection's `Count` property. For a string expression, `before` must correspond to the `key` specified when the member being referred to was added to the collection. You can specify a `before` position or an `after` position, but not both.
`after`	Optional. An expression that specifies a relative position in the collection. The member to be added is placed in the collection *after* the member identified by the `after` argument. If numeric, `after` must be a number from 1 to the value of the collection's `Count` property. If a string, `after` must correspond to the key specified when the member referred to was added to the collection. You can specify a `before` position or an `after` position, but not both.

Keys, whether numeric or a string key, cannot be duplicated. Suppose you had an object variable, `toolorder`, that represented an instance of the `OrderItem` class. The following instruction would add the object to the collection:

```
Tools.Add toolorder
```

The following instruction would add the item with the `ProductName` as the unique string identifier:

```
Tools.Add toolorder, toolorder.ProductName
```

13.3.2 The *Remove* Method of the *Collection* Object

The `Remove` method removes items from the collection. You identify the numeric or string index expression when the item was first added to the collection. The syntax is straightforward:

```
object.Remove index
```

Table 13.3 explains the components of the `Remove` method syntax.

Table 13.3 The Syntax of the *Remove* Method

Component	*Description*
`object`	Required. A `Collection` object.
`index`	Required. An expression that specifies the position of a member of the collection. If a numeric expression, `index` must be a number from 1 to the value of the collection's `Count` property. If a string expression, `index` must correspond to the `key` argument specified when the member referred to was added to the collection.

13.3.3 The *Item* Method of the *Collection* Object

The `Item` method returns an item from a collection by the `index` or `key` used when the object was added to the `Collection`. The syntax for the method is as follows:

```
object.Item(index)
```

Table 13.4 explains the components of the `Item` method syntax.

Table 13.4 The Syntax of the *Item* Method

Component	*Description*
`object`	A `Collection` object.
`index`	Required. An expression that specifies the position of a member of the collection. If a numeric expression, `index` must be a number from 1 to the value of the collection's `Count` property. If a string expression, `index` must correspond to the key argument specified when the member referred to was added to the collection.

Exercise 13.6 Using a *Collection* Object

In this somewhat lengthy exercise, you substitute individual labels in control arrays for the `lstOrder` list box used in Exercise 13.4. This allows order items to be aligned (see Figure 13.9). You also add a `Collection` object to your program that you use to calculate an ongoing order total.

1 Open Exercise 13-4 and save f13-4.frm to disk as **f13-6.frm**, save cls13-4.cls as **cls13-6.cls**, and save p13-4.vbp as **p13-6.vbp**. Change the Name property of the project to **p13_6**.

2 Comment out the `Terminate()` event code for the `OrderItem` class.

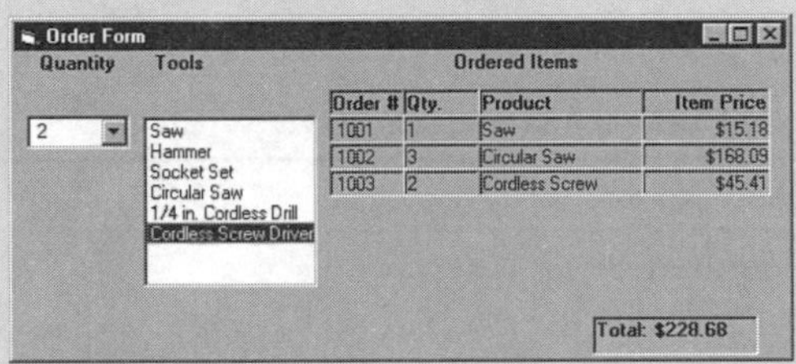

Figure 13.9

This application expands the previous exercise by creating and displaying label control arrays.

3 Alter `frmOrder` so that it appears as displayed in Figure 13.9. It will need to be made wider to accommodate the labels that will substitute for `lstOrder`. Remove `lstOrder` from the form, and add the four labels displayed. These labels are named **`lblOrderNum`**, **`lblQuantity`**, **`lblProductName`**, and **`lblPrice`**. The `BorderStyle` for each of these labels is set to `1 (Fixed Single)`. The `FontBold` properties are set to `True`. Set the `Alignment` property of `lblPrice` to `1` (right align). Change the captions of the labels as shown.

4 Important: change the `Index` property for each of these four labels to `0`. This will make them each the first element in a Label control array.

5 In the bottom right-hand corner of the form, add a label named `lblTotal`, and remove any text from the `Caption` property.

6 At the module level of `frmOrder`, make the following declaration:

```
Dim purchase As Collection
```

7 In the `Form_Load()` event, add a line of code that instantiates this object:

```
Set purchase = New Collection
```

8 Change the `lstProducts_Click()` event so that it will create a new `OrderItem` object, assign it data, display the data in newly loaded elements of the control arrays, add the object to the purchase collection, and iterate through the collection to ascertain the order total.

```
Private Sub lstProducts_Click()
  Dim toolorder As OrderItem
  Dim Counter As OrderItem
  Dim product As String
  Static ordernumber As Integer
  Dim quantity As Integer, i As Integer, topindex As Integer
  Dim total As Currency
  Dim strDisplay As String
  Dim hgt As Integer, tposition As Integer

  Set toolorder = New OrderItem

  product = lstProducts.Text
  ordernumber = ordernumber + 1
  quantity = Val(cboQuantity.Text)

  Select Case product
    Case "Saw"
       toolorder.UnitCost = 7.99
    Case "Hammer"
       toolorder.UnitCost = 6.99
    Case "Socket Set"
       toolorder.UnitCost = 14.99
    Case "Circular Saw"
       toolorder.UnitCost = 29.49
    Case "1/4 in. Cordless Drill"
       toolorder.UnitCost = 24.95
    Case "Cordless Screw Driver"
       toolorder.UnitCost = 11.95
 End Select

  With toolorder
      .orderNum = 1000 + ordernumber
      .ProductName = product
      .quantity = quantity
  End With

  'display the ordered item by loading
  'and positioning the new labels
  'get top index
  topindex = lblOrderNUM.UBound

  'get height and position of labels
  hgt = lblOrderNUM(topindex).Height
  tposition = lblOrderNUM(topindex).Top

  'load new labels
  Load lblOrderNUM(topindex + 1)
  Load lblProductName(topindex + 1)
  Load lblQuantity(topindex + 1)
  Load lblPrice(topindex + 1)

  'position new labels
  lblOrderNUM(topindex + 1).Top = tposition + hgt
  lblProductName(topindex + 1).Top = tposition + hgt
  lblQuantity(topindex + 1).Top = tposition + hgt
  lblPrice(topindex + 1).Top = tposition + hgt
```

continues

```
        'make them visible
        lblOrderNUM(topindex + 1).Visible = True
        lblProductName(topindex + 1).Visible = True
        lblQuantity(topindex + 1).Visible = True
        lblPrice(topindex + 1).Visible = True

        'make fonts plain
        lblOrderNUM(topindex + 1).FontBold = False
        lblProductName(topindex + 1).FontBold = False
        lblQuantity(topindex + 1).FontBold = False
        lblPrice(topindex + 1).FontBold = False

        'display values
        lblOrderNUM(topindex + 1).Caption = Str$(toolorder.orderNum)
        lblProductName(topindex + 1).Caption = toolorder.ProductName
        lblQuantity(topindex + 1).Caption = toolorder.quantity
        lblPrice(topindex + 1).Caption = Format(toolorder.ItemTotal, "Currency")

        purchase.Add toolorder    ' add item to collection

        total = 0                 ' initialize variable

        For Each Counter In purchase          ' calculate total
          total = total + Counter.ItemTotal
        Next

        'display total
        lblTotal.Caption = "Total: " & Format$(total, "Currency")
      End Sub
```

9 Run and test the program.

A lot is going on in this procedure. Most of the code loads and positions the Label control arrays. The upper boundary of the control array for one of the labels (`lblOrderNum`) is determined. This is then incremented to load new labels and position them on the form. The main new code is near the end of the procedure. The `toolorder` object is added to the purchase collection and then the collection is iterated through to calculate a running order total.

No changes were made to the `OrderItem` class itself other than commenting out the `Terminate()` event code. You are beginning to see code reuse!

Notice that we did not use an `index` or `key` when adding the `OrderItem` objects to the purchase collection. Because this code only adds and totals ordered items, we did not need to retrieve individual `Collection` members with the `Item` method, so we did not use the `index` or `key` arguments.

Let's say that you wanted to extend the functionality of the program in Exercise 13.6 so that users could remove ordered items before finally committing to a final purchase, or at least change the quantity of an individual order item. It might also be nice to have sales tax automatically calculated. In addition, you might need to have differential pricing: one price for retail customers and another for wholesale customers. Which of these functions should be handled by the `OrderItem` class and which should be handled by a hypothetical `Order` class?

As you work with classes and extend their functionality, issues about which class should handle which function quickly become central concerns of the developer. In fact, you should not begin to create a class until you have a very good idea about the function of the class and its relationship to other classes.

The `Collection` object provides a lot of the functionality central to an `Order` class, but not all of it. How, for instance, should the individual order numbers for each line item be renumbered if an item is deleted? This would presumably be a function of an `Order` class, but it could not be handled by a `Collection` object alone. In fact, wrapping a `Collection` object inside a class makes a lot of sense.

Objective 13.4 Employing Good Programming Practices when Creating Class Modules

You now have been exposed to the object-oriented nature of Visual Basic. If your head is swimming, don't worry. The longer you work with class modules, the more natural "object oriented thinking" becomes. Although there is much still to learn about class modules, you have been exposed to the heart of it. The following are some critical issues to remember:

- Data members of class modules should (almost) always be private to the module rather than declared publicly. Even if you think it isn't possible that an errant value could be assigned to the variable, as the class definition evolves, you will see a need to put a data-validation procedure in place. Yes, all those `Property Let` and `Property Get` procedures can become tedious, but it is worth it in the long run.
- Use a naming convention for class data members and their associated `Property` procedures that is easy to work with. Using an `m_` prefix for class data members is one standard practice. The associated `Property Let` and `Property Get` procedures have the prefix removed.
- All data members should be initialized with default values.
- When designing a class, remember that you are aiming for robust, reusable code.
- Properties are often described as an attribute of an object, and methods as actions that work on an object. Often it is difficult to distinguish between what is properly a property and what is a method. The distinction is less important than making sure the code is reliable; and, when made part of the class interface, readily understandable to other users of the class.
- Collections created from the `Collection` class are one-based, not zero-based.
- As your classes begin to take on significant size, don't forget the Object Browser. It's a handy tool!

Chapter Summary

This chapter introduced you to class modules. Like form and standard code modules, class modules can be added to the project, named, and saved to disk as a file. Class modules are designed to promote object-oriented design in Visual Basic programs.

Class modules promote the encapsulation of data by allowing privately scoped module-level variables to be accessed through `Property Let` and `Property Get` procedures. Other names for these procedures are *mutator* (`Property Let`) methods and *accessor* (`Property Get`) methods. Methods for class modules can be created by creating publicly scoped procedures in the class module.

Objects are instantiated from class modules by the use of object variables. Object variables can be declared with the `New` keyword, as in `Dim toolorder as New OrderItem`. The object will be instantiated when the first reference is made to the new variable in code, as in `toolorder.ProductName = Saw`. Alternatively, the object variable can be simply declared and then instantiated with the `Set` statement and the `New` keyword, as in the following example:

```
Dim toolorder as OrderItem
Set toolorder = New OrderItem
```

Class modules have two built-in events: `Initialize()` and `Terminate()`. The `Initialize()` event can be used to assign default values to class data members. The `Terminate()` event is used to execute cleanup code. The `Terminate()` event fires when the last variable reference to the object is removed from memory by the `Nothing` statement or by falling out of scope.

Objects can be placed into collections by the use of the `Collection` object. The lowest index number of a collection created from the `Collection` object is `1`. `Collection` objects have three main methods: `Add`, `Remove`, and `Item`. Either a numeric index or a string key can be used when adding an object to the collection. The index is used to retrieve the object with the `Item` method or delete the object with the `Remove` method.

Visual Basic's object-oriented emphasis places special demand on designing classes well. Key concepts to remember when designing classes are as follows:

- Keep data members private.
- Use an easily workable naming convention.
- Initialize all data members.

Skill-Building Exercises

1. Extend Exercise 13.4 by adding a `TimeStamp` property to the `OrderItem` class. Display the property in the `lstOrder` list box when it is selected. Be sure that all data members of the class are properly initialized.

2. Using the program you created in the preceding exercise, add a `WholeSale` property to the `OrderItem` class. Set this as a `Boolean` value. Change the `Property Get ItemTotal` procedure to accommodate this new property. Wholesale prices have a 25 percent markup. Change the interface for `frmOrder` to accommodate the changes you have made. *Hint:* use a check box.

3. Create a small class of your own called **`Vehicle`**, which holds information on the make and model of the vehicle, the year it was manufactured, and its useful life. For cars manufactured before 1991, the useful life is seven years; for vehicles manufactured after 1991, the useful life is eight years. *Hint:* Use the `Year` function to determine the current year. Make sure the class uses **`CarYear`** for the year of manufacture, not `Year` to avoid name conflicts.

4. Create an interface to use the class that you created in the preceding exercise. On the left-hand side of a form, place a combo box to let the user select a year between 1980 and 1998. Include a list box or combo box so the user can select one of four manufacturers: Ford, General Motors, Chrysler, or Honda. Supply a text box so the user can input model type. Finally, add a command button captioned **OK**, which will instantiate the class object and assign it property values and display those values in another list box.

5. Extend the preceding exercise by adding a class method (function) called `DisplayUsefulLife`. This method will return the useful life that the vehicle still has left from the current year, or how many years past its useful life the vehicle has enjoyed. Display this information in the second list box. As users enter data on the form and press the command button, add these objects to a collection. In a separate label, display the average year of the vehicles in the collection as automobiles are added.

14

Programming User Events

Thus far, you have used a small number of user events, mostly the clicking of buttons, selecting from list box and combo box items, and selecting from menu commands. Visual Basic makes available an assortment of other user events—including keyboard actions and mouse events, such as dragging and dropping icons. Many users prefer highly graphical environments and the visual cues such environments provide. This chapter explores how to better exploit this environment by introducing several new types of user events.

The following sections concentrate on new keyboard and mouse events—the principal actions for receiving user input. New keyboard events include `KeyUp()`, `KeyDown()`, and `KeyPress()`; new mouse events include `MouseDown()`, `MouseUp()`, `MouseMove()`, `DragOver()`, and `DragDrop()`. These keyboard and mouse events are shared by many controls, and their differences are discussed here.

The tasks in this chapter involve employing various user events in your programming. By the end of this chapter, you will

- know how to trap individual keys using the `KeyUp()`, `KeyDown()`, and `KeyPress()` events;
- know how to trap key combinations, such as Ctrl+F12 or Alt+Y;
- understand the different built-in mouse procedure arguments;
- know how to use `MouseDown()`, `MouseUp()`, and `MouseMove()` events;
- understand and use the `Drag` method;
- understand and use the `DragOver()` event;
- understand and use the `DragDrop()` event.

Chapter Objectives

The objectives for this chapter address the three sets of user events: keyboard events and two types of mouse events.

Common mouse events involve clicking the mouse button and moving the mouse, which lead to `MouseDown()`, `MouseUp()`, and `MouseMove()`, respectively. More advanced and interesting mouse events include `DragOver()` when you drag the mouse over an object and `DragDrop()` when you drag the mouse to an object and release the mouse button. These and related issues are addressed in the three objectives for this chapter:

1. Understanding `KeyPress()`, `KeyUp()`, and `KeyDown()` events
2. Working with mouse events
3. Understanding `Drag()` and `DragDrop()` events

Objective 14.1 Understanding *KeyPress()*, *KeyUp()*, and *KeyDown()* Events

The keyboard events in Visual Basic enable you to capture just about any single key or key combination. In the Windows development environment, it is standard design practice to provide the user with a variety of methods for performing the same action. Frequently used menu commands should be provided with keyboard actions known as shortcut keys. Some applications should also provide the user with a toolbar of icons to perform frequently used functions. This section shows you how to capture keyboard events that are not provided by shortcut keys designated in the Menu Editor.

14.1.1 The Form *KeyPreview* Property

The different key events can be confusing at first, but they share one thing in common: *They work only on the object that has the current focus.* The object of focus for keyboard events is usually a control, such as a text box or a combo box. A form, in general, can only have the focus if it is either blank or all controls are disabled. If you want to add a key event for a form, you need to set the form's `KeyPreview` property to `True`. The form will then receive keyboard events before any control on the form control does. This is especially useful if you want to have certain keys perform the same action, no matter which control has the current focus. Such keys might be F10 for saving a file, or the key combination Alt+X for exiting a program. However, this does not apply if you have used these keys as shortcut keys in the Menu Editor window.

14.1.2 *KeyPress()* versus *KeyDown()* and *KeyUp()* Events

Two types of events that are similar are `KeyPress()` and `KeyDown()` (or `KeyUp()`). A `KeyPress()` occurs when a key corresponding to an ASCII character is pressed. Use Visual Basic Help to inspect this character set. `KeyDown()` and `KeyUp()` events occur as any key on the keyboard is pressed and released. The difference between these two types of events is subtle, but important:

- A `KeyPress()` event supplies a generated character when a key is pressed, and an event is triggered when that character is recognized in code.

- `KeyDown()` and `KeyUp()` events are triggered by the physical key on the keyboard itself, not by an ASCII generated character. To interpret the event, you must supply the code to do it.

`KeyDown()` and `KeyUp()` events have two arguments: `KeyCode` and `Shift`. If you press an uppercase *K* or a lowercase *k*, the `KeyDown()` event would get the same `KeyCode`. The event indicates that it was the physical key itself that was pressed. The argument `Shift` provides the Shift key state. By examining it, the computer can determine whether the key-entered letter was uppercase or lowercase.

`KeyUp()` and `KeyDown()` can capture more keys of the keyboard than a `KeyPress()` event because the ASCII code is limited by a 0 to 127 character set. By using the built-in `KeyCode` constants, however, there is little that you cannot capture. There are over 90 `KeyCode` constants, with the full list of constants provided in Visual Basic's online Help. `KeyCode` constants are provided for special keys on the keyboard, all numeric keys, all letters of the alphabet, all keys on the numeric keypad, and all function keys. Table 14.1 shows examples of `KeyCode` constants.

Table 14.1 Examples of *KeyCode* Constants

Constant	*Value*	*Description*
`vbKeyBack`	8	Backspace key
`vbKeyTab`	9	Tab key
`vbKeyClear`	12	Clear key
`vbKeyReturn`	13	Enter key
`vbKeyShift`	16	Shift key
`vbKeyCapital`	20	Caps Lock key
`vbKeyEscape`	27	Esc key
`vbKeySpace`	32	Spacebar key
`vbKeyDelete`	46	Del key
`vbKeyA`	65	The letter A key (same as ASCII)
`vbKey0`	48	The number 0 key (same as ASCII)
`vbKey1`	49	The number 1 key (same as ASCII)
`vbKeyNumpad4`	100	The 4 key on the numeric pad
`vbKeyMultiply`	106	Multiplication (*) key on the numeric pad
`vbKeyF1`	112	F1 key
`vbKeyF2`	113	F2 key

The `KeyPress()` event has only one argument: `KeyAscii`. For example, the code for a command button `KeyPress()` event would read as follows:

```
Command1_KeyPress(KeyAscii As Integer)
```

It treats an uppercase *K* and a lowercase *k* as two separate characters. It can handle all number and letter keys, punctuation keys, and a few special keys, such as the or Esc key (unless a command button has the `Cancel` property set to `True`), the Backspace key, and the Enter key (unless there is a command button with `Default` set to `True`).

Which event should you use? The obvious answer is to use `KeyDown()` and `KeyUp()` for handling keys the `KeyPress()` event can't. These include arrow (directional) keys, Page Up and Page Down, numeric keypad keys, and function keys not used by the menu commands. You can also use `KeyDown()` and `KeyUp()` if you want to know when a key is being held down and when it is released, such as in certain graphics programs.

One way to become familiar with the difference between the two events is to simply enter the following code in the `KeyPress()` event and experiment with different keys:

```
Print Chr(KeyAscii)
```

Next, comment out the code in the `KeyPress()` event, and enter the following code in the `KeyDown()` event, and continue the experiment:

```
Print Chr(KeyCode)
```

Compare how upper- and lowercase keys are handled by the two events.

Exercise 14.1 Using the Return Key

The Return key has the ASCII value 13 and is used to trigger events in text-based controls after the user has completed input. Filling out an online data-entry form, for example, frequently requires this type of event. A benefit of using the `vbKeyReturn` event is that it avoids the need for a command button to trigger an event.

1 Place a text box and a label on a form (see Figure 14.1). Name the text box **`txtKeyPress`** and the label **`lblKeyPress`**. Set the text `MultiLine` value to `True`.

2 In the `txtKeyPress_KeyPress()` event, enter the following code:

```
If KeyAscii = vbKeyReturn Then
   lblKeyPress.Caption = txtKeyPress.Text
End If
```

3 Place some sample text in the text box, and press the Enter key.

4 Add a command button to the form and set its `Default` property to `True`. Write no code for this control.

5 Run the program by typing in text and pressing the Enter key. Surprised? The text will not be displayed. When the button is set to `True`, it becomes the default Command button.

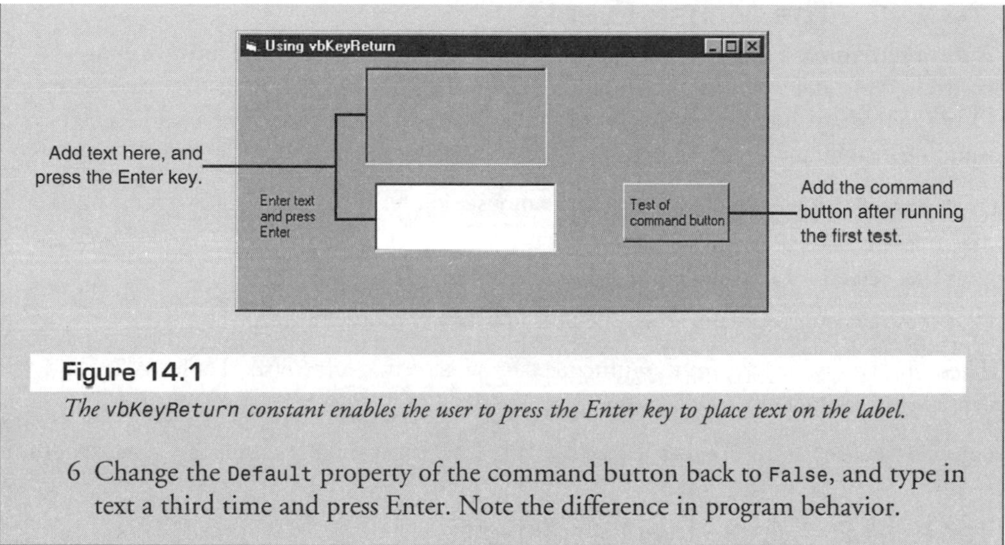

Figure 14.1

The `vbKeyReturn` *constant enables the user to press the Enter key to place text on the label.*

6 Change the `Default` property of the command button back to `False`, and type in text a third time and press Enter. Note the difference in program behavior.

Exercise 14.2 Using a Function Key as a Menu Shortcut Key

This exercise is similar to the preceding one, but demonstrates the use of a function key coupled with the `KeyDown()` event.

1 With the form from Exercise 14.1 still on-screen, change the `KeyPreview` property of the form to `True`.

2 In the `Form_KeyDown()` event, enter the following code:

```
Private Sub Form_KeyDown (Keycode As Integer, Shift As Integer)
   If Keycode = vbKeyF12 Then End
End Sub
```

3 Start the program, enter some text into the text box, and press Enter. Then, press F12. You exit the program.

4 In Design mode, change the form's `KeyPreview` property back to `False`, and try it. This time it doesn't work because `txtKeyPress` has the focus, and the program can't read F12.

Objective 14.2 Working with Mouse Events

Mouse events provide the Visual Basic developer enormous flexibility in making an application easy to use. Forms and certain controls, such as picture boxes and images, can detect the position and state of the mouse button.

14.2.1 The Mouse Events

A form recognizes a mouse event when the pointer is over a part of the form with no controls. Most controls can recognize a mouse event when the pointer is over the control. Other than drag and drop events (which are addressed later in Objective 14.3, "Understanding `Drag()` and `DragDrop()` Events"), the mouse has three events:

- **`MouseMove()`**—Occurs whenever the mouse pointer is moved to a new point on the screen
- **`MouseUp()`**—Occurs when the user releases any mouse button
- **`MouseDown()`**—Occurs when the user presses any mouse button

These mouse events have four arguments: `Button`, `Shift`, `X`, and `Y` (see Table 14.2). The syntax reads as follows:

```
Command1_MouseDown(Button As Integer, Shift As Integer, X As Single, Y As Single)
```

Table 14.2 Mouse Event Arguments

Component	*Description*
`Button`	Refers to the specific mouse button pressed (left, right, or middle)
`Shift`	Refers to the state of the Shift, Alt, and Ctrl keys
`X` and `Y`	Refer to the coordinates of the form or control to which the mouse pointer is pointing

A `MouseMove()` event over a control does not trigger a simultaneous `MouseMove()` event on the underlying form.

Exercise 14.3 Using the *Move* Method with a *MouseDown()* Event

In this exercise, you move an image placed on a form, using the mouse `Move` method. Figure 14.2 shows the interface.

1. Embed an icon, such as that of a small house, in an image control placed on a form. Name the control **`imgIcon`**.
2. Enter the following code into the `Form_MouseDown()` event:

```
Private Sub Form_MouseDown(Button As Integer, Shift As Integer, X As
➥Single, Y As Single)
     Dim xadjust, yadjust As Integer
     xadjust = imgIcon.Width \ 2
     yadjust = imgIcon.Height \ 2
     imgIcon.Move X - xadjust, Y - yadjust
End Sub
```

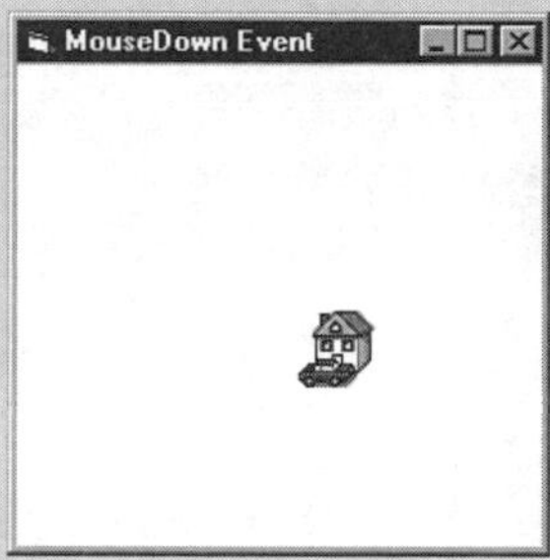

Figure 14.2

An icon image is placed on the form and moved from one location to another using the mouse.

3 Run the program. By this simple reference to the mouse pointer's x- and y-coordinates, you can position a control anywhere you want on the form. Because the `Move` method places the upper-left corner of the control at the x- and y-coordinates, you can center the image by employing the adjustment variables you see in the procedure.

Exercise 14.4 Using the *MouseMove()* Event

This exercise asks you to create a design that continues to trigger a continuous event as the mouse is moved. Figure 14.3 shows the design interface.

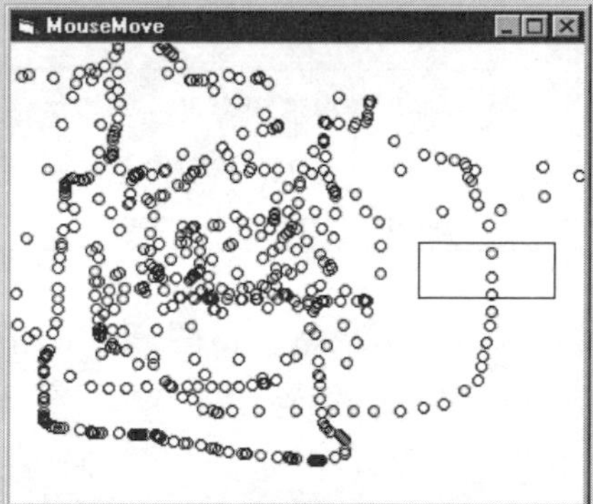

Figure 14.3

The `MouseMove()` *event enables the user to draw on the form.*

continues

1 Place a shape on the form, and enter the following code into the `Form_MouseMove()` event:

```
Private Sub Form_MouseMove (Button As Integer, Shift As Integer, X As
➥Single, Y  As Single)
   shp1.Left = X + 500
   shp1.Top = Y - 500
End Sub
```

2 Run the program to watch the `MouseMove()` event operate continuously as the mouse is moved. This means that the computer is recognizing many separate events in quick succession. It is prudent to avoid this event for something that requires a great deal of computing time.

3 Add the following line of code to the `Form_MouseMove()` event above:

```
Form1.Circle (X,Y), 40
```

This code fragment employs the graphic method `Circle`, and illustrates the frequency with which circles are drawn by the `MouseMove()` event. Experiment by moving the mouse pointer both quickly and slowly.

14.2.2 The *Button* Argument with Mouse Events

As shown earlier in Table 14.2, the `Button` argument in mouse events refers to the left, middle, or right mouse buttons. These buttons can be identified by referencing the built-in constants in the `Button` argument shown in Table 14.3.

Table 14.3 *Button* Argument Constants

Constant	*Value*	*Description*
`vbKeyLButton`	`0×1`	Left mouse button
`vbKeyRButton`	`0×2`	Right mouse button
`vbKeyMButton`	`0×4`	Middle mouse button

In `MouseUp()` and `MouseDown()` events, you can only test for one button at a time. In `MouseMove()` events, however, you can test to see whether more than one button is being used.

Exercise 14.5 Coding a *MouseUp()* Event

This short exercise asks you to experiment with the `MouseUp()` event.

1 On a new form, enter the following code for `Form_MouseUp()` event:

```
Select Case Button
   Case vbKeyLButton
     Print "You released the left mouse button"
   Case vbKeyRButton
     Print "You released the right mouse button"
End Select
```

2 Run the program.

3 If you press more than one button, Visual Basic interprets it as two events. Try it.

Continue with the next exercise.

Exercise 14.6 Coding a *MouseMove()* Event

Use the same interface as in Exercise 14.5, but replace the `MouseUp()` event with a `MouseMove()` event. The three built in constants for the mouse buttons are as before:

Left mouse button = 1

Right mouse button = 2

Middle mouse button = 4

If you want to test for mouse buttons depressed simultaneously, you can add these constants. If the left and right buttons were depressed at the same time, for instance, you could test for the button value 3.

1 Comment out the code in the `MouseUp()` event provided for the last exercise.

2 Enter the following code into the `Form_MouseMove()` event:

```
Private Sub Form_MouseMove(Button As Integer, Shift As Integer, X As
➥Single, Y As Single)
   Select Case Button
      Case vbKeyLButton
         Circle (X, Y), 40
      Case vbKeyRButton
         Circle (X, Y), 80
      Case vbKeyLButton + vbKeyRButton
         Circle (X, Y), 120
   End Select
End Sub
```

3 Run the program. Unlike `MouseUp()` and `MouseDown()` events, pressing the left and right buttons together causes a separate event.

Objective 14.3 Understanding *Drag()* and *DragDrop()* Events

In program design, controls in Visual Basic can be *dragged.* That is, the user can hold the mouse button down and move (drag) a control. When the mouse button is released, the control is *dropped.*

Drag and drop events have both a *source* and a *target.* The source is the control that is being dragged. The target is the form (or control) over which the source is being dragged or upon which the source is being dropped.

14.3.1 Drag and Drop Elements

Properties, events, and methods control drag and drop operations, as shown in Table 14.4.

Table 14.4 Drag and Drop Elements

Element	*Description*
	Properties
`DragMode`	Enables automatic or manual dragging of an icon
`DragIcon`	Identifies which icon is displayed when the control is dragged
	Events
`DragDrop()`	Recognizes when a control is dropped onto an object (a target event)
`DragOver()`	Recognizes when a control is dragged over an object (a target event)
	Method
`Drag`	Starts or stops manual dragging (typically a `Source` method)

The easiest way to enable dragging is to set the Source control's `DragMode` property to Automatic in the property manager window. This indicates that the Source control can always be dragged. However, it does not mean that the control's position will be permanently moved: A move has to be coded separately.

14.3.2 The *DragOver()* Event

The `DragOver()` event is a *target* event. It is coded for objects over which you expect to be dragging other objects. The `DragOver()` event has the following syntax:

```
Command1_DragOver(Source As Control, X As Single, Y As Single, State As Integer)
```

Table 14.5 explains the components of the `DragOver()` event syntax.

Table 14.5 The *DragOver()* Event Syntax

Component	*Description*
`Control`	The generic object reference used by the `Source` argument. Visual Basic passes the control that is being dragged over the target object.
`State`	Refers to the status of the source object.
`X` and `Y`	Refer to the current mouse pointer position over the target object.

The `State` argument has the values shown in Table 14.6.

Table 14.6 *State* Argument Values

Constant	*Value*	*Description*
`vbEnter`	`0`	Source control dragged into target
`vbLeave`	`1`	Source control dragged out of target
`vbOver`	`2`	Source control dragged from one position in target to another

Exercise 14.7 Coding a *DragOver()* Event

This exercise and the next ask you to use the same display form (see Figure 14.4). Objects on the form include three image controls and a command button. Their names are **`imgCloud`**, **`imgCat`**, **`imgSun`**, and **`cmdReset`**, respectively.

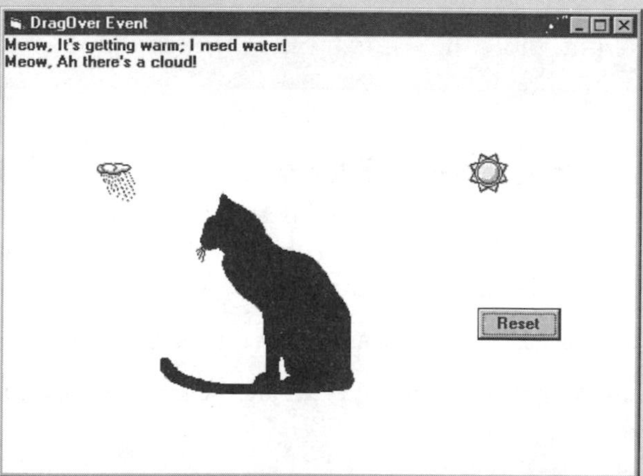

Figure 14.4

Drag the cloud or the sun over the cat to reveal a message.

continues

1 In the `cmdReset_Click()` event, enter the following code:

```
imgCat.Visible = True
imgCloud.Visible = True
imgSun.Visible = True
Form1.Cls
```

2 Set the `DragMode` property in `imgSun` and `imgCloud` to Automatic.

3 In the `imgCat_DragOver()` event, enter the following code:

```
Private Sub imgCat_DragOver(Source As Control, X As Single, Y As Single,
➥State As Integer)
  Select Case Source
   Case imgCloud
      If State = vbEnter Then
         Print "Meow, Yuk, It's raining!"
         'Make sun disappear
          imgSun.Visible = False
      ElseIf State = vbLeave Then
           'Make sun reappear"
            imgSun.Visible = True
            Print "Meow, Ah there's the sun!"
      End If
   Case imgSun
      If State = vbEnter Then
           Print "Meow, It's getting warm; I need water!"
           'Make cloud disappear"
           imgCloud.Visible = False
      ElseIf State = vbLeave Then
         'Make cloud reappear"
          imgCloud.Visible = True
          Print "Meow, Ah there's a cloud!"
      End If
   End Select
End Sub
```

4 Run and test the program. Save it; it will be used in the next exercise.

14.3.3 The *DragDrop()* Event

The `DragDrop()` event is also a target event and is generated when one object is *dropped* on another. It is triggered when the user drags an object with the mouse button depressed and then drops it by releasing the mouse button. The syntax is as follows:

```
Image1_DragDrop(Source As Control, X As Single, Y As Single)
```

The `Source` and `X` and `Y` arguments for this event are the same as those for the `DragOver()` event (see Table 14.5).

Exercise 14.8 Coding a *DragDrop()* Event

In this exercise you are asked to drag an object over another object and release the mouse button to give the impression that the object has been dropped into a container of some type. The container object might be a trash can icon, a file icon, or a mail box icon. Use the same design interface as in Exercise 14.7.

1 In the `imgCat_DragDrop()` event, enter the following code:

```
Select Case Source
   Case imgCloud
      Print "I'm wet, this cat's going to scat"
      imgCat.Visible = False
   Case imgSun
      Print "Ouch, That's hot! This cat's going to scat"
      imgCat.Visible = False
End Select
```

2 Run and test the program.

14.3.4 Using the *Drag* Method

To have more precise control over when a control can be dragged, you can change the `DragMode` property setting to Manual. Dragging the control then needs to be explicitly enabled by employing the `Drag` method. The syntax for the method is as follows:

```
object.Drag action
```

`Object` is a control that can be dragged, and `action` is a method argument that specifies the action to perform. If action is omitted, the default is to begin dragging the object. The `Drag action` settings are described in Table 14.7.

Table 14.7 *Drag Action* Settings

Constant	*Value*	*Description*
vbCancel	0	Cancels drag operation
vbBeginDrag	1	Begins dragging control
vbEndDrag	2	Drops control

Exercise 14.9 Using the *Drag* Method

This exercise illustrates use of the `Drag` method. Create the interface shown in Figure 14.5. The interface includes a list box named **`lstDocs`**, an image with a trash can icon embedded in it named **`imgTrash`**, and a label aligned with the bottom of the form named **`lblStatus`**.

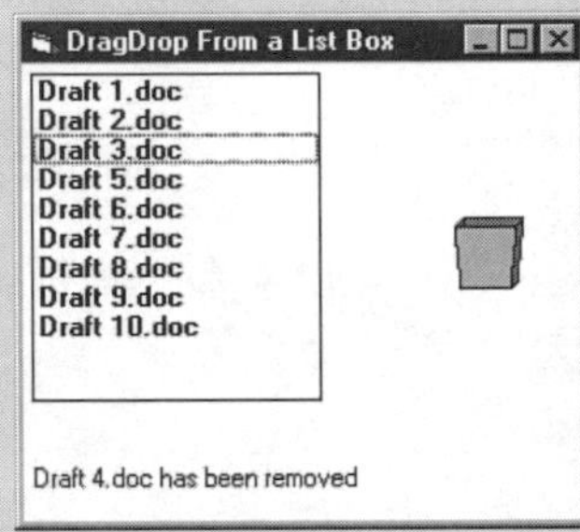

Figure 14.5

This exercise enables you to drag an item from the list box and drop it into the trash can.

1 In the `Drag` icon property for `lstDocs`, embed an image similar to that shown in Figure 14.6.

Figure 14.6

The Drag icon image appears when you drag an item to the trash can.

2 Set the `DragMode` property of `lstDocs` to Manual.

3 In the `Form_Load()` event, add the following code:

```
Private Sub Form_Load()
   Dim n As Integer
   For n = 1 To 10
      lstDocs.AddItem "Draft" & Str$(n) & ".doc"
   Next n
End Sub
```

4 In the `lstDocs_MouseMove()` event, enter the following code:

```
Private Sub lstDocs_MouseMove(Button As Integer, Shift As Integer, X As
➥Single, Y As Single)
   Select Case Button
     Case vbKeyLButton
       If X > (lstDocs.Width - 100) Then lstDocs.Drag vbBeginDrag
   End Select
End Sub
```

5 In the `lstDocs_MouseUp()` event, enter the following code:

```
lstDocs.Drag vbEndDrag
```

6 In the `imgTrash_DragDrop()` event, enter the following code:

```
Private Sub imgTrash_DragDrop(Source As Control, X As Single, Y As Single)
   If TypeOf Source Is ListBox Then
      lblStatus.Caption = Source.List(Source.ListIndex) & " has been
      ➥removed"
      Source.RemoveItem (Source.ListIndex)
   Else
      lblStatus.Caption = "Unrecognized control."
   End If
   lstDocs.Drag vbEndDrag
End Sub
```

7 Run and test the program.

Two sections of this code merit special comment. In the `Mouse_Move()` event for `lstDocs`, the code tests to see whether the left mouse button is depressed. If this test evaluates to `True` and the mouse pointer comes within 100 twips of the right edge of the list box, dragging is enabled.

In the `imgTrash_DragDrop()` event, the generic object reference `Control` is used. Visual Basic passes to the `DragDrop()` event the control that is being dropped. The `TypeOf` argument in an `If...Then` statement tests for object type.

Chapter Summary

In this chapter, you worked with a broader range of user events than you have in earlier chapters. These included capturing individual keys with `KeyPress()`, `KeyUp()`, and `KeyDown()` events and exploiting the rich variety of mouse events built into Visual Basic.

The `KeyPress()` event captures ASCII generated characters, while `KeyDown()` and `KeyUp()` events capture the physical key itself. `KeyUp()` and `KeyDown()` events capture more keyboard keys than the `KeyPress()` event, which is limited to keyboard keys that generate ASCII characters. To add a shortcut key to your application that is not on the form menu bar, set the form's `KeyPreview` property to `True`. On data entry forms, you can capture the user pressing the Return key by using the `KeyPress()` event and the `vbKeyReturn` key constant. When doing this, make sure that no command button on the form has its `Default` property set to `True`.

Mouse events in Visual Basic include `MouseDown()`, `MouseUp()`, and `MouseMove()`. These events have four built-in arguments: `Button`, `Shift`, `X`, and `Y`. They enable a wide range of mouse actions to be coded, including mouse button clicks in combination with `Shift` arguments (Shift, Ctrl, and Alt keys) and mouse clicks in defined areas using `X` and `Y` arguments.

Drag events include the `DragOver()` and `DragDrop()` events. These events include both a source and a target and also have built-in arguments. The `DragOver()` event arguments include `Source`, `X`, `Y`, and `State`. These enable you to detect which specific control is being dragged over a target, the location of the mouse pointer as the control is being dragged, and the point at which the mouse pointer enters the target, is over the target, and is leaving the target.

The `DragDrop()` event's three arguments—`Source`, `X`, and `Y`—enable detection of which control is being dropped upon a target and the position of the mouse pointer as a control is dropped. To have explicit control over when Source controls can be dragged, the program designer can set the Source control's `DragMode` properties to Manual and then explicitly enable and disable dragging with the `Drag` method. When a Source control that has been enabled with the `Drag` method is released, it generates a `DragDrop()` event on the target.

Skill-Building Exercises

1. Write a procedure that asks whether the user wants to exit the program when the Esc key is pressed. If the answer is yes, end the program. Otherwise, continue.
2. Add a Line control to a form and set its index property to 0 and make it invisible. On the form's `Mouse_Down()` event, code a procedure that adds a new Line control to the array, positions its `X1,Y1` coordinates at the `X2,Y2` coordinates of the line that preceded it, and makes the line visible. In this manner, the user can build a complex polygon with mouse clicks. Your procedure will need to accommodate the first mouse click with subsequent clicks displaying lines.
3. Add a 3,000×3,000 twip picture box to a form and a label immediately below it. In the picture box's `Mouse_Move()` event, code a procedure that alerts the user through four individual messages displayed in the label when the mouse pointer is less than 80 twips from the left, right, top, or bottom of the picture box.
4. Use the `TypeOf` statement to test controls in a `DragDrop()` event. Add a picture box on the right-hand side of a form and a label, an option button, a check box, and an image on the left. Code a drag and drop procedure using a TypeOf statement so that when the control on the left is dropped on the picture box, a message is displayed on the picture box identifying what type of control was dropped (see Figure 14.7).

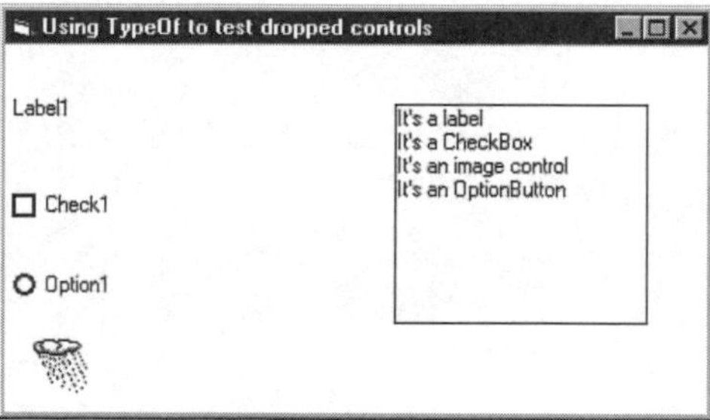

Figure 14.7

All four controls on this display can be dragged and dropped into the image box.

5. Alter the preceding exercise so that when a control is dragged over the picture box, its type is displayed in the picture box. Have the message displayed once, and when the dragged control leaves the picture box, have the message erased.

15

Introduction to Debugging

Debugging is the process of diagnosing and fixing errors in your code. Visual Basic provides a set of tools to assist you in debugging, which this chapter introduces. As you write code to handle a greater variety of tasks and your applications become larger, Visual Basic's debugging capabilities will prove increasingly useful.

The debugging tools in Visual Basic represent the capabilities of the `Debug` system object. This object is essentially a spy on the computer's memory. It retrieves values from memory that you request. The `Debug` object also will execute coded instructions if the objects and variables of the instructions are within the scope of the procedure being examined.

One of the most important facts to remember about Visual Basic's debugging tools is that they can be used for much more than just fixing mistakes: They also let you observe how good code works. Observing variables change value during a procedure or seeing program control switch from one procedure to another, for example, teaches you a lot about how a program has been put together. Visual Basic's debugging tools enable these kinds of observations and many more. In this chapter, we examine Visual Basic's debugging tools by observing how they are used with functioning code. When you complete this chapter, you will be able to

- discern some of the different types of errors that you can encounter,
- set a breakpoint in your code and perform both `Step Into` and `Step Over` code execution while in break mode,
- employ the debugging tools of the Visual Basic Debug toolbar and menu,
- add a watch expression,
- use Quick Watch,
- use the `Debug` object's Locals window,
- set a breakpoint on expressions when they become `True`,
- query the `Debug` object with the Immediate window,
- query the `Debug` object using the `Print` method during runtime.

Chapter Objectives

The objectives of this chapter teach you about different types of programming errors, the use of the debugging commands, and the use of the `Debug` object and its `Print` method, as follows:

1. Understanding the different types of programming errors
2. Working with Visual Basic's third mode: break
3. Working with watch expressions
4. Understanding the `Debug` object and its `Print` method

Objective 15.1 Understanding the Different Types of Programming Errors

Programming errors generally break down into two types: syntax errors and logic errors. *Syntax errors* can generally be prevented by choosing Tools, Options and checking the Auto Syntax Check box on the Editor Tab. During design time, if you enter code with syntax errors, you are prompted with a message informing you of the nature of the error. Color coding provided by the Visual Basic editor also helps you to catch this type of error.

Logic errors are more troublesome. With this type, code seems to execute correctly, but produces incorrect results. Visual Basic's analytic debugging tools can assist you in tracking down errors of logic by letting you observe code as it executes step by step.

There is another general category of errors called *runtime errors.* These occur if code attempts to execute an instruction that is impossible to carry out. For example, if you try to save a file to a disk that is not formatted or attempt to divide by zero, a runtime error message will be generated. Many of these errors occur when you first attempt to run code and can be corrected immediately, such as by correcting a variable type mismatch error. They most often represent a logic error in that you forgot to do something or wrote code that could not be executed.

You can trap these runtime errors and handle them with a message to the user. You will work more with error trapping in later chapters, which deal with file input/output and database applications, beginning with Chapter 20, "File-Processing Controls and Sequential File Processing."

Objective 15.2 Working with Visual Basic's Third Mode: Break

So far you have worked in two modes in Visual Basic: design time and runtime. Visual Basic provides another mode known as *break mode.* During break mode, Visual Basic's debugging tools are available to help analyze and fix your code. You can easily tell which mode you are in by looking at the designation in the brackets to the right of "Microsoft Visual Basic" on the Visual Basic title bar.

You might have noticed, and already used, the buttons on the Visual Basic Standard toolbar for testing and stopping execution of your code during design time (see Figure 15.1). They look like the play, pause, and stop controls on a compact disc player.

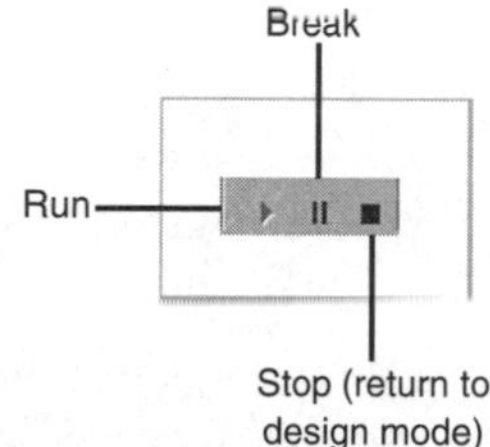

Figure 15.1

The buttons on the Standard toolbar run a program, end execution, or place the program in break mode.

These same buttons are available on the Debug toolbar, which is shown in Figure 15.2. To make the Debug toolbar visible, choose View, Toolbars, and select Debug. Like other toolbars, the Debug toolbar can be added to the toolbars underneath the Visual Basic title and menu bar. It can also be made "free floating" by dragging it onto the desktop.

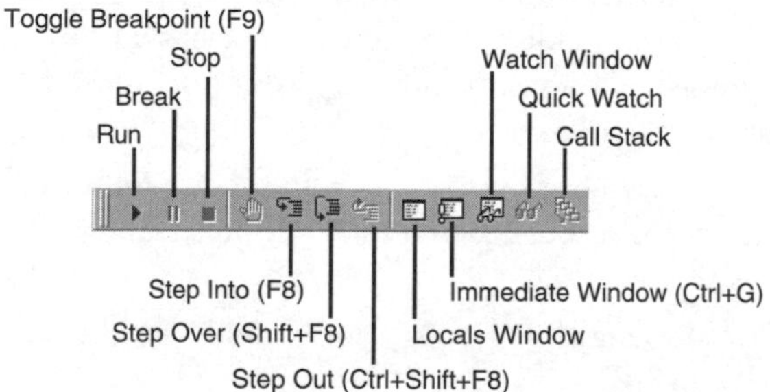

Figure 15.2

The buttons on the Debug toolbar allow you to run, break, and stop program execution, to step through code, and to watch code values using special debug windows.

During code execution, pressing the Break button on the Debug toolbar puts a program into break mode. When you press it, the Start and End debugging buttons immediately to the right on the toolbar are enabled, as are the Step Over and Step Into buttons, and the Locals, Immediate, and Watch Windows. The other debugging buttons are disabled. These buttons and the available corresponding keyboard shortcuts are illustrated in Figure 15.2.

The first step in debugging is to set a breakpoint in code. Although you can press the Break button next to the Run button, typically you set the breakpoint by selecting a code statement where you would like program execution to stop. While in the code window with your cursor on the line where you would like the break to begin, simply press F9 or click the Toggle Breakpoint button. When the code is next executed, operation is suspended immediately before the breakpoint. After a breakpoint is reached, you can step through your code. Let's begin with a simple exercise.

Exercise 15.1 Setting a Breakpoint and Stepping Through a Procedure

For this exercise, you will use a modified version of Exercise 11.4, "Moving a Shape to Make the Moon Rise," from Chapter 11, "Working with Loops." For all exercises, place the floating Debug toolbar on the screen.

1 Open project p15-1.vbp, which is stored on the disk packaged with the book.

2 Press the Run button. Once in runtime mode, press the Break button immediately to the right of the Run button. You should see that the Visual Basic title bar indicates you are in break mode.

3 Press the Immediate Window button (or press Ctrl+G). The Immediate window should now appear and have focus. This window is the one you will work with most frequently in debugging operations. It provides an immediate "view" into memory. In the Immediate window, type **`Print Form1.Caption`**. You should see the caption of the form displayed. Next, type **`Print Form1.Left`**. These two instructions invoke the `Print` method of the `Debug` object.

4 Change the focus back to the Moon Rise window, and press the button captioned "Watch the moon rise." Nothing happens, right? That's because you placed the program in break mode, suspending it's operation before the command button was pressed. To continue program execution, press the Run button again (or press F5).

5 In general, it is better to set a breakpoint within the code window itself, so you know exactly where code is suspended and can continue debugging from there. You do that next.

In the `cmdRise_Click()` event of p15-1.vbp, you should have the following code:

```
Private Sub cmdRise_Click()

  Do While shpMoon.Top >= 200
    shpMoon.Left = shpMoon.Left + 60
    shpMoon.Top = shpMoon.Top - 80   ' ascend
  Loop

  MsgBox "Howwoooooooooool!", 0, "MoonRise"

  Do
    shpMoon.Left = shpMoon.Left + 60
    shpMoon.Top = shpMoon.Top + 80  'descend
  Loop While shpMoon.Top <= 4200

  shpMoon.Left = 120
  shpMoon.Top = 4300
  cmdRise.Caption = "Watch the moon rise"
End Sub
```

Set your cursor on the line of the code fragment `Do While shpMoon.Top >= 200` at the beginning of the procedure. To set a breakpoint in the code window, you have three options: simply press F9; press the Toggle Breakpoint button (the hand icon) on the Debug toolbar; or select Debug, Toggle Breakpoint. The entire line of code should be highlighted with a color. (The color can be changed in the Editor Format dialog box by choosing Tools, Options.)

6 Run the program, and press the Watch the Moon Rise button. Immediately, the code window appears with a rectangle around the code you have selected as the breakpoint. You are now in break mode. The rectangle represents the code that will be executed next as you step through the procedure.

7 To step through the code, simply press F8 or click the button with the single footstep (the Step Into button) on the Debug toolbar. Arrange the different windows on your screen so you can see the code window as well as the Moon Rise application. As you step through each iteration of the ascending loop, you should see the moon's position change slightly.

Exercise 15.2 Stepping Over a Procedure

You used the single step procedure (`Step Into`) in Exercise 15.1 by pressing the F8 key. This time, you will use `Step Over` to avoid looking at code stored in a procedure. `Step Over` code is identical to `Step Into` code except when the code contains a call to a procedure. A single step will step into the called procedure. `Step Over` executes the called procedure and then stops at the next statement immediately following the procedure call. Sound confusing? This exercise should make it clear.

1 Open p15-2.vbp on your disk.

2 The code is almost identical to the previous exercise except the loop that controls the moon's descent is contained in a separate procedure (`Descend`). In addition, two variables—`msg` and `loopcounter`—have been added. The `cmdRise_Click()` event and the `Descend()` procedures are as follows:

```
Private Sub cmdRise_Click()
   Dim msg As String
   Dim loopcounter As Integer

   msg = "Howwoooooooooool!"
   Do While shpMoon.Top >= 200
     shpMoon.Left = shpMoon.Left + 60
     shpMoon.Top = shpMoon.Top - 80   ' ascend
     loopcounter = loopcounter + 1
   Loop
```

continues

```
    cmdRise.Caption = "Watch the moon fall"
    MsgBox msg, 0, "Moonrise"
    loopcounter = 0
    Call Descend

    shpMoon.Left = 120
    shpMoon.Top = 4300
    cmdRise.Caption = "Watch the moon rise"
End Sub

Private Sub Descend()
    Dim loopcounter As Integer
    Do
      shpMoon.Left = shpMoon.Left + 60
      shpMoon.Top = shpMoon.Top + 80   'descend
      loopcounter = loopcounter + 1
    Loop While shpMoon.Top <= 4200
End Sub
```

3 Put a breakpoint on the line (shown in bold) `Call Descend`.

4 Run the program. When you get to the breakpoint, try single stepping through the code first. Notice how you move into the Descend procedure?

5 Exit the program and restart. When you get to the breakpoint, click the Step Over button or press Shift+F8. This time, the called procedure (`Descend`) should execute and the step rectangle should proceed to the next line of code in `cmdRise_Click(): shpMoon.Left = 120`.

15.2.1 Clearing Breakpoints

A breakpoint remains in your code until you clear it. You clear it either by pressing F9 or the Toggle Breakpoint button when the cursor is on the breakpoint. Alternatively, you can clear one or more breakpoints in your code by selecting Debug, Clear All Breakpoints or by pressing Ctrl+Shift+F9.

Exercise 15.3 Stepping Out of a Procedure

If you discover you are inside a called procedure and want the called procedure to continue executing, you can *step out* of a procedure. Execution will continue until the next line of code immediately following the call to the procedure. This exercise illustrates how this debugging feature works.

1 Open p15-2.vbp on your disk, or if it is open already, make sure that all breakpoints have been cleared.

2 In the `Descend` procedure, place a breakpoint on the single word instruction `Do` at the beginning of the loop (shown in bold):

```
Private Sub Descend()
   Dim loopcounter As Integer
   Do
     shpMoon.Left = shpMoon.Left + 60
     shpMoon.Top = shpMoon.Top + 80    'descend
     loopcounter = loopcounter + 1
   Loop While shpMoon.Top <= 4200
End Sub
```

3 Run the program. When the program hits the breakpoint, single step through several iterations of the loop. Then, press the Step Out button on the Debug toolbar (or press Ctrl+Shift+F8). The program will continue executing until the `Descend` procedure is complete and program control has returned to the `cmdRise_Click()` event. You should see the code execution rectangle now over the instruction `shpMoon.Left = 120` in that event.

15.2.2 The Locals Window

Besides an Immediate window, the `Debug` object includes a Locals window, which shows the values of all locally declared variables within the procedure being examined by the `Debug` object. These values can be examined and even changed from the Locals window. The Locals window can greatly accelerate diagnosing errors in your code. A quick scan of variable values often indicates the source of a problem.

Exercise 15.4 Using the Locals Window

In this exercise, you will employ the Locals window to examine the value of locally declared variables and then change the value of the variable.

1 Open p15-2.vbp on your disk, or if it is open already, make sure all breakpoints have been cleared.

2 In the `cmdRise_Click()` procedure, place a breakpoint at the instruction `MsgBox msg, 0, "Moonrise"` (shown in bold):

```
Private Sub cmdRise_Click()
   Dim msg As String
   Dim loopcounter As Integer

   msg = "Howwoooooooooool!"
   Do While shpMoon.Top >= 200
     shpMoon.Left = shpMoon.Left + 60
     shpMoon.Top = shpMoon.Top - 80   ' ascend
     loopcounter = loopcounter + 1
   Loop
```

continues

```
    cmdRise.Caption = "Watch the moon fall"
    MsgBox msg, 0, "Moonrise"
    loopcounter = 0
    Call Descend

    shpMoon.Left = 120
    shpMoon.Top = 4300
    cmdRise.Caption = "Watch the moon rise"
End Sub
```

3 Run the program until you hit the breakpoint. Click the Locals Window button on the Debug toolbar. You should see the window, as illustrated in Figure 15.3.

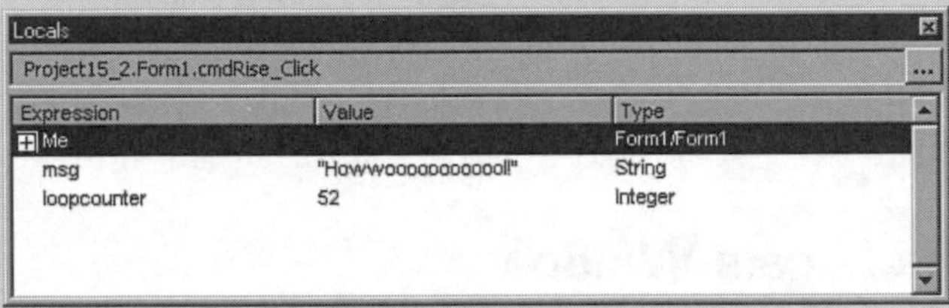

Figure 15.3

The Locals window displays the values of local variables and the objects of the Form module.

The Locals window shows the two variables—`msg` and `loopcounter`—and the implicit variable `Me`, referring to the form module.

4 In the Locals window, select the value for `msg`. Once the value column has focus, a cursor should appear. Change the value of `msg` from `"Howwoooooooooool!"` to some other value.

5 Press the Step Into button or press F8. You should see the `MsgBox` statement display the newly assigned value.

Objective 15.3 Working with Watch Expressions

Setting a breakpoint and stepping through code is one pillar of debugging in Visual Basic. The other is setting a watch expression for values. With a watch expression, you can monitor an expression or variable as it changes or reaches a value that you define. This is especially useful, for instance, if something goes wrong within a long loop procedure when values reach a certain range. By defining the range for the watch expression, you can avoid tedious repetitions of the loop and focus on where you think the problem begins.

Another advantage of watch expressions is that you can add them at design time or in break mode. You can even instruct Visual Basic to enter break mode when an expression reaches a certain value.

Exercise 15.5 Adding a Watch Expression at Design Time and Using Quick Watch

This exercise asks you to set two types of watch expressions: a regular watch and a quick watch. A *regular watch* enables you to watch how the value of an expression changes when you are in break mode. Regular watch selections appear in a Watches window. A *quick watch* also enables you to check the value of a variable while in break mode; however, the watch is not retained in the Watches window.

1 Open p15-2.vbp and make sure that all breakpoints have been cleared.

2 Select Debug, Add Watch. (Alternatively, right-click within the code window and then select Add Watch). You will see the dialog box for adding a watch expression. Change the dialog box so that it conforms to the one presented in Figure 15.4. You will watch the `loopcounter` variable.

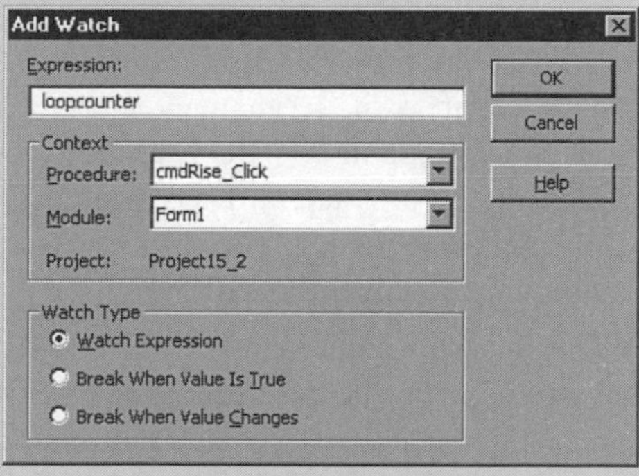

Figure 15.4

The Add Watch dialog box is used to define the variable or expression that you want to watch.

3 Notice that you have set the context to the procedure `cmdRise_Click()` placed on `Form1` for Project p15_2. The context sets the scope of variables watched in the expression. Because the variable `loopcounter` is declared locally, you set its context to the procedure in which it is declared.

The context setting would be set to All Procedures for a variable declared at the module level. When you press the OK button on the Add Watch dialog box, the Watches window appears with the variable value that you intend to watch (see Figure 15.5)

continues

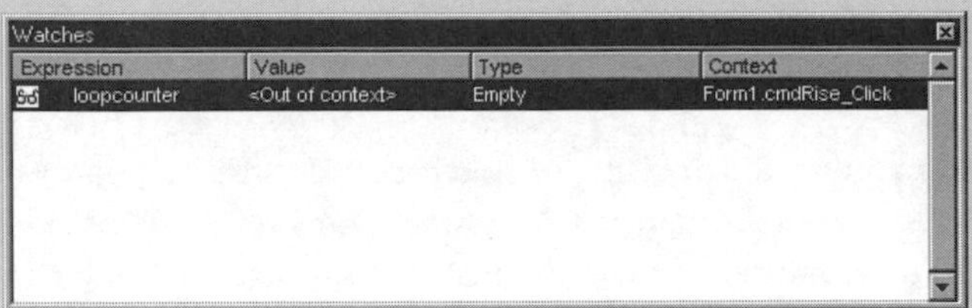

Figure 15.5

The Watches window shows the value, type, and context for each selected expression.

4 Add another watch expression for the `loopcounter` variable in the `Descend` procedure.

5 Set a breakpoint at the following instruction:

```
MsgBox msg, 0, "Moonrise"
```

6 Run the program.

7 When the program enters break mode, change the focus to the Watches window. The watch pane should display the value of `loopcounter` for the `cmdRise_Click()` procedure. Now step into (F8) the next line of code. After pressing the OK button on the message box, continue single stepping through the `Descend` procedure for several loop iterations.

You should see the value of the `loopcounter` for `cmdRise_Click()` remain at zero. The `loopcounter` variable for the `Descend` procedure, while at first out of context, will begin to increment in value with each loop iteration.

8 Stay within the `Descend` procedure, and use your pointer to select the expression `shpMoon.Top`. Press the Quick Watch button on the toolbar (the eyeglasses icon), or press Shift+F9.

9 A dialog box displays the integer value of `shpMoon.Top` and provides a choice of Add Watch or Cancel. Note the value of `shpMoon.Top` and then select Cancel.

10 Step through another two iterations of the `Descend` loop and select `shpMoon.Top` again. Press the Quick Watch button or press Shift+F9. How has the value changed?

One feature of Visual Basic's debugging tools is that there is nearly always more than one way of accomplishing the same thing. All of these values could have been examined in the Locals window as well as in the Watches window. You could also have selected the Immediate window and entered the following:

```
Print shpMoon.Top
```

Exercise 15.6 Setting a Break When a Watch Expression Becomes *True*

An added advantage of watch expressions is that you can use them to set breaks. By assigning a value or using a comparison operator in an expression, you can have the code break when a variable or expression reaches a certain value.

1 Go back to design mode. Clear all breakpoints by selecting Debug, Clear All Breakpoints (or Ctrl+Shift+F9). Remove all watch expressions by selecting them in the Watches window and pressing the Delete key.

2 Select Debug, Add Watch. The dialog box in Figure 15.4 appears.

3 In the expression box, enter **`shpMoon.Left >= 2000`**. Be sure the context is set to the `cmdRise_Click()` procedure. Under Watch Type, select the Break When Value Is True option button (see Figure 15.6).

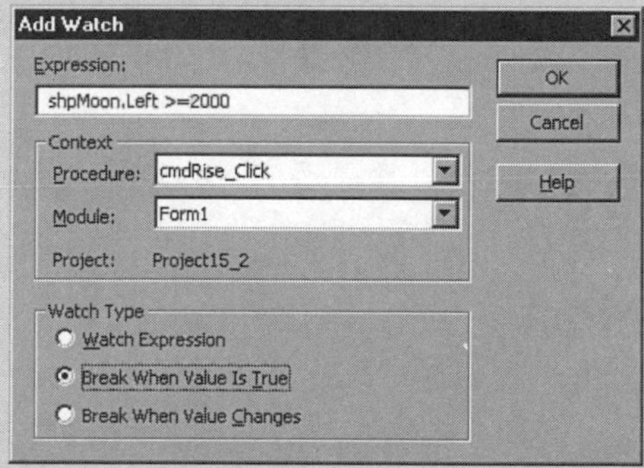

Figure 15.6

The Add Watch window offers you three types of watches: a regular watch expression, create a breakpoint when the value is true, and create a breakpoint when the value changes.

4 Run the program. When it breaks, change the focus to the Watches window. The watch pane tells you the expression is now `True`.

5 To get the precise value of `shpMoon.Left`, select it and take a quick watch (Shift+F9) of its value, or examine its value in the Locals window.

When adding and editing watch expressions, you must select the right expression to evaluate. If you choose to watch an entire assignment, you will often receive a message in the watch pane that evaluates to `False`. For example, if you select the entire expression `shpMoon.Top = shpMoon.Top - 80`, the Debug window will work, but consistently return a value of `False`.

continues

What you might really want to know is the value of `shpMoon1.Top` as it goes through the loop. Simply set `shpMoon.Top` as the expression to evaluate, or use a comparison operator so the program will break where you want, as in the following:

```
shpMoon.Top >= 2000
```

Objective 15.4 Understanding the *Debug* Object and Its *Print* Method

Visual Basic treats the Debug window as a separate object named, appropriately enough, `Debug`. You have already used its one method—`Print`—which can prove extremely useful. With the `Debug.Print` method, the Immediate window can display values of expressions and variables at points you designate in your code. You can also query the immediate pane itself. You can even enter a `Debug.Print` request directly into your code and then remove it later when you are ready to make an executable file.

Exercise 15.7 Using the *Debug.Print* Method in Code

This exercise asks you to insert `Debug.Print` expressions in your code and to observe the results in the Immediate window.

1 Go back to p15-2.vbp and clear all watch expressions and breakpoints.

2 Insert a `Debug.Print` instruction at the sixth line (which is set in bold) in the `CmdRise_Click()` event:

```
Private Sub cmdRise_Click()
   Dim msg As String
   Dim loopcounter As Integer

   msg = "Howwoooooooooool!"
    Do While shpMoon.Top >= 200
      Debug.Print shpMoon.Left; shpMoon.Top; loopcounter
      shpMoon.Left = shpMoon.Left + 60
      shpMoon.Top = shpMoon.Top - 80  ' ascend
      loopcounter = loopcounter + 1
    Loop

     cmdRise.Caption = "Watch the moon fall"
     MsgBox msg, 0, "Moonrise"
     loopcounter = 0
    Call Descend
    shpMoon.Left = 120
    shpMoon.Top = 4300
    cmdRise.Caption = "Watch the moon rise"
End Sub
```

3 Run the program, but, before clicking Watch the Moon Rise, position the Immediate window to the side so you can observe it. Click the command button.

Observe the Immediate window. Visual Basic's performance slows as it prints the values for each iteration of the ascending loop, but it's a good way to observe how code is working! Using the `Print` method in code is especially useful with very troublesome procedures for which you are having difficulty determining the problem.

Chapter Summary

In this chapter, you became familiar with Visual Basic's debugging tools. Debugging tools are especially helpful in large, complex projects in which the source of undesired program behavior is not readily apparent. Debugging tools can also help examine the functionality of unfamiliar code. Visual Basic's debugging tools are not difficult to use, but they are so flexible that it can be difficult to decide which tool is best for examining your code. There are always several ways of accomplishing the same thing.

One pillar of debugging is setting a breakpoint and stepping through code. You can single step through code—one instruction at a time. You can also step over procedures, which is useful for ignoring procedure calls you are confident work properly. As you step through code, you can examine expression values by using the Immediate Watch window.

The other pillar of debugging is setting and observing watch expressions. When regular watch expressions are added, they can be seen in the Watches window. Watch expressions can be added either during design time or in break mode. A quick watch enables you to check the value of a variable while in break mode.

These two pillars of debugging are combined when a watch expression is added that causes the program to break when an expression becomes `True` or a value changes. Using these powerful features, you can have your program break when your code approaches an expression value that is proving troublesome. From there, you can step through your code and examine possible sources of difficulty using the Immediate Watch window.

The `Debug` object has one method: `Print`. This method can be used on a query basis in the immediate pane when the program is in break mode. You can also place `Debug.Print` instructions in your code so that expression values will be outputted to the Immediate window for examination after the code has stopped executing.

Skill-Building Exercises

The exercises in this section ask you to modify programs that you wrote in Chapter 11, "Working with Loops."

1. The debugging tools are most useful with complex applications with multiple forms in which values are passed from one procedure to another. You will use the debugging tools later as you develop more complex applications. To review the debugging tools, it is best to start by setting breakpoints and watch expressions on code you are familiar with. Take Exercise 11.11, "Adding a Timer Countdown to the Mercury 1 Commemorative Launch," and practice using Visual Basic's debugging tools on it.
2. Return to Exercise 11.3, "Using Nested Loops and List Boxes," in Chapter 11, which introduced nested loops. Set a breakpoint so that the program breaks every time you are about to exit the inner loop and return to the outer loop.
3. In Exercise 11.3, set a breakpoint that places the program in break mode when the variable `item` becomes equal to, or greater than, 26.
4. In Exercise 11.3, set a breakpoint that places the program in break mode when the variable `sleep` becomes equal to, or greater than, 99999. Then, single step through the program until you reach the `sleep` loop again.
5. Change the code in Exercise 11.3 so that the value of `sleep` on every 10,000 iterations of the `sleep` loop is displayed in the Immediate window. *Hint:* Use a `Mod` operator.

16

Lists and Arrays

This chapter introduces you to the concept of the *variable array.* In Chapter 11, "Working with Loops," you learned about the concept of the control array. With a control array, like objects are given the same name. A variable array works in a similar way: It is a list of more than one variable with the same name and the same data type. This chapter begins with the subject of *lists* because lists help illustrate a one-dimensional array. We also provide more practice in working with list and combo boxes.

Several new concepts are presented in this chapter, while other concepts that were touched upon only briefly in previous chapters are discussed in detail. When you complete this chapter, you will be able to

- select and assign an item from a list,
- add an item to a list during runtime,
- remove an item from a list during runtime,
- move items from one list to a second list,
- declare a variable array in more than one way,
- use both a control and a variable array in an instruction,
- use the array functions `LBound` and `Ubound` to determine the range of an array.

Chapter Objectives

The first three objectives for this chapter tackle list boxes and combo boxes. The last two then deal with declaring and working with one-dimensional arrays. The objectives are as follows:

1. Selecting an element from a list box
2. Selecting an element from a combo box
3. Selecting multiple items from a list
4. Understanding one-dimensional variable arrays
5. Working with control and variable arrays

Objective 16.1 Selecting an Element from a List Box

A list is an array, provided the variables are of the same data type. With Visual Basic, lists are commonly associated with List Box and Combo Box controls. Both List Box and Combo Box controls feature one-dimensional arrays because all values are placed in a single column.

In Chapter 4, "Learning to Think Visually," you were asked to design a list box, but were not asked to select an element from the list for use elsewhere. In this section, you gain additional experience in working with list boxes. You will work more extensively with list box and combo box runtime properties.

Before a list box can be utilized by a program, it must be loaded. This is accomplished with the `AddItem` method. In review, the syntax for the `AddItem` method when loading a list is as follows:

```
listname.AddItem item [,index]
```

Table 16.1 explains the components of the `AddItem` method.

Table 16.1 The Syntax of the *AddItem* Method

Component	*Description*
`listname`	The name of the control
`item`	The string expression to add to the control
`index`	The position of the item in the control when the index begins with 0

If an index is omitted and you want the list to be sorted, set the control `Sorted` property to `True`.

16.1.1 The *Selected* Property

Once items are placed in a list, a specific item can be selected. Although selection is usually made by the user, a selection can be made within Visual Basic using the `Selected` property. The syntax for this property is as follows:

```
[form.]{filelistbox|listbox}.Selected(index)[= {True|False}]
```

The `Selected` property settings are Boolean (`True` or `False`). When an item is selected, the `Selected` property becomes `True`. For other items in the list, the `Selected` property remains `False`, unless multiple selections are permitted. What might not be apparent is that the `Selected` property is not available to you in design time. It can only be read or assigned during runtime.

16.1.2 The *ListCount* and *ListIndex* Properties

Determining the selected item from a list requires knowledge of two additional properties:

- `ListCount`—Determines the number of items in the list box.
- `ListIndex`—Determines the index of the selected item in a list.

This information is known after a list has been loaded at runtime, making both properties runtime properties. The `ListIndex` settings are shown in Table 16.2.

Table 16.2 *ListIndex* Settings

Setting	*Description*
`-1`	No item is currently selected.
`N`	The *n*th item in the list has been selected. `N` varies from 0 to the last element in the list (the number of elements minus one).

The following expression returns the string of the selected item:

```
list1.List (ListIndex)
```

Exercise 16.1 Selecting Items from a List

This silly little exercise asks you to create a list of fruit (you can add nuts as well if you like), select one from the list, and display the selected fruit using a label. Figure 16.1 shows how the form is to be designed.

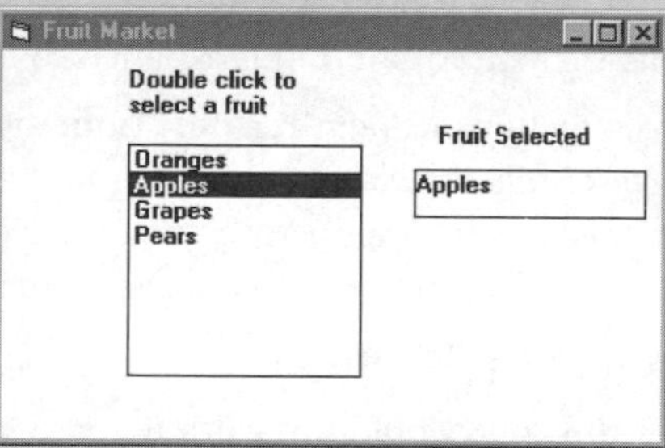

Figure 16.1

Double-clicking an item in the list on the left places it on the label.

1 Design the form as shown, adding a list box (`lstFruit`) and three labels (`lblFruit` and two descriptive labels). Set the caption of this form to **Fruit Market**.

continues

2 Write a `Form_Load()` procedure to load the list at runtime. Use `AddItem` to add the strings to the list, as follows:

```
Private Sub Form_Load()
   lstFruit.AddItem "Oranges"
   lstFruit.AddItem "Apples"
   lstFruit.AddItem "Grapes"
   lstFruit.AddItem "Pears"
End Sub
```

3 Write a `lstFruit_DblClick()` procedure to take a list box selection and display it as a caption. Write the following:

```
Private Sub lstFruit_DblClick()
   Dim item As Integer
   'Search the entire list to determine which item is selected

   For item = 0 To (lstFruit.ListCount - 1)
      If lstFruit.Selected(item) = True Then
        lblFruit.Caption = lstFruit.List(item)
        Exit For
      End If
   Next item
End Sub
```

The tricky parts of this procedure involve the `item` counter variable, the upper boundary of which is only known at runtime. You determine how many iterations of the loop should be made by making reference to the `ListCount` property. In this instance, `ListCount` equals `4` because four fruit are contained in the list.

Because items in list boxes are part of arrays that begin at 0, subtract 1 from `ListCount` to determine the number of loop iterations to execute. The item counter variable runs from 0 to 3 (`ListCount - 1`). The loop checks to determine which of the list items has been selected—the `Selected(item)` property is `True`. When the `True` element is found, it is assigned to the label caption.

Because you are selecting only one item at a time from the list box, it is possible to reference the `ListIndex` property without using a loop. Every time an item is selected by the user, the `ListIndex` property changes, so you could write the following single line of code for the `DblClick()` event, and get the same behavior:

```
lblFruit.Caption = lstFruit.Text
```

Note, however, that this code would not work if you wanted to select multiple items before updating the label. This is illustrated in Exercise 16.4.

16.1.3 The *AddItem* and *RemoveItem* Methods

Once items have been added to a list, the list need not remain unchanged. Additional items can be added to a list in a procedure using the `AddItem` method. Items can also be removed from a list using the `RemoveItem` method. The syntax for `RemoveItem` is as follows:

```
listname.RemoveItem index
```

The component `index` represents the position of the item within the list, and must be specified with this method.

Exercise 16.2 Adding and Removing Items from a List

This exercise expands Exercise 16.1. As shown in Figure 16.2, you are asked to add command buttons for adding an item to the list (**Add Element**) and removing an item from the list (**Delete Element**). In addition, you are asked to keep track of the number of elements in the list (**No. of Elements**).

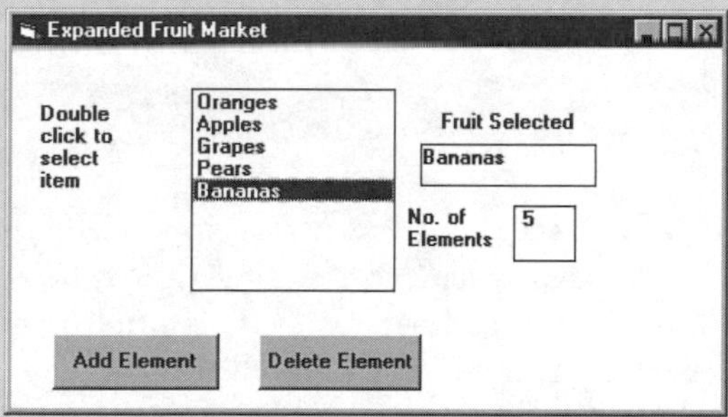

Figure 16.2

With this design, you can select fruit from the list, add fruit, or delete fruit from the list.

1 Add the additional command buttons (**`cmdAdd`** and **`cmdRemove`**) and two labels (**`lblNum`** and one descriptive label), as shown in Figure 16.2.

2 Modify the `Form_Load()` procedure to display the number of items in the list following the loading of all items. The new procedure should read as follows:

```
Private Sub Form_Load()
   lstFruit.AddItem "Oranges"
   lstFruit.AddItem "Apples"
   lstFruit.AddItem "Grapes"
   lstFruit.AddItem "Pears"
   lblNum.Caption = Str$(lstFruit.ListCount)
End Sub
```

In this procedure, the `ListCount` property determines the number of items in the list (which is four).

3 Write a `cmdAdd_Click()` procedure to add an item to the list. Add the fruit **Bananas**. Remember to revise the number of items in the text box. The code reads as follows:

```
Private Sub cmdAdd_Click()
   lstFruit.AddItem "Bananas"
   lblNum.Caption = Str$(lstFruit.ListCount)
End Sub
```

continues

Observe how the `AddItem` method is used in this instance. Besides the `Form_Load()` procedure, `AddItem` can be placed in a command click procedure.

4 Write a second `cmdRemove_Click()` procedure to remove an item from the list. After a list item is selected, it can be removed. Do the following:

```
Private Sub cmdRemove_Click()
   Dim item As Integer

   For item =(lstFruit.ListCount - 1) To 0 Step -1
      If lstFruit.Selected(item) = True Then
         lstFruit.RemoveItem item
            Exit For
      End If
   Next item

   lblNum.Caption = Str$(lstFruit.ListCount )
End Sub
```

In this procedure, the `ListCount` property is again used. The `ListCount` property changes as items are added or removed from the list. By decrementing from the end of the list to the beginning when removing multiple items, you avoid generating index errors. You can refresh the `lblNum.Caption` by reassigning the `ListCount` property after each addition or deletion.

You could also reference the `ListIndex` property to remove items. The `cmdRemove_Click()` procedure could be written as follows:

```
lstFruit.RemoveItem(lstFruit.ListIndex)
lblNum.Caption = Str$(lstFruit.ListCount)
```

But again, this would work only if you want to remove one item at a time from a list.

Objective 16.2 Selecting an Element from a Combo Box

Working with lists is complicated by the requirement of placing a loop within a procedure to determine which list element or elements have been selected. With a combo box, this requirement does not hold.

Transferring a selection from a Combo Box control is much easier than using a List Box control. After a combo box is filled with data, using a `Form_Load()` procedure, the combo box selection can be assigned to another control using the following syntax:

```
control.Text = combo1.Text
```

Because a selected item appears in the text box, a search using loops is not required to identify the item.

Exercise 16.3 Transferring Information from Combo Boxes to a Text Box

This short exercise asks you to select a pet and ship the pet to a city. Figure 16.3 shows two combo boxes and a text box: The selection in the first combo box is combined with the city in the second combo box.

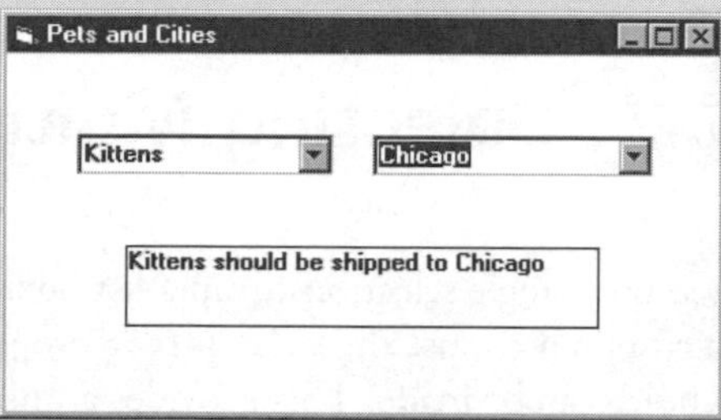

Figure 16.3

Pull down on the first combo box to select a pet and the second to select a city.

1 Create the design showing two combo boxes (**`cboPets`** and **`cboCities`**) and one label with borders (**`lblOutput`**). Change the name of the form to **Pets and Cities**.

2 Write a `Form_Load()` procedure to add **Fish**, **Kittens**, **Snakes**, and **Mice** to the first combo box, and **Boston**, **Chicago**, **New York**, and **Atlanta** to the second combo box. The procedure is as follows:

```
Private Sub Form_Load()
   cboPets.AddItem "Fish"
   cboPets.AddItem "Kittens"
   cboPets.AddItem "Snakes"
   cboPets.AddItem "Mice"

   cboCities.AddItem "Boston"
   cboCities.AddItem "Chicago"
   cboCities.AddItem "New York"
   cboCities.AddItem "Atlanta"
End Sub
```

3 Write a `cboPets_Click()` procedure to assign the pet selection to `lblOutput` caption.

```
Private Sub cboPets_Click()
   lblOutput.Caption = cboPets.Text
End Sub
```

Observe how simple this procedure is compared to the procedure for the list box.

continues

4 Write a `cboCities_Click()` procedure to assign the city selection and to complete the sentence.

```
Private Sub cboCities_Click()
   lblOutput.Caption = cboPets.Text & " should be shipped to " &
   ➥cboCities.Text
End Sub
```

Objective 16.3 Selecting Multiple Items from a List

Thus far, we have considered only single selections from a list box. To work with multiple selections from a list (not a combo box), use the `MultiSelect` property. With multiple selections, more than one choice can be made. This provides a unique advantage to the use of lists over combo boxes.

The syntax for the `MultiSelect` property is as follows:

```
[form.]{filelistbox | listbox}.MultiSelect = Setting
```

Either a `filelistbox` or a `listbox` are control candidates. Table 16.3 explains the `MultiSelect` settings.

Table 16.3 The *MultiSelect* Settings

Setting	*Description*
0	Multiple selection is not allowed (the default).
1	Simple multiple selection. Clicking the mouse or pressing the Spacebar selects an item in the list.
2	Extended multiple selection. Shift+click or Shift+arrow key extends the selection from the previous selected item to the current item.

Exercise 16.4 Making Multiple Selections

This exercise asks you to make several selections from a list of birthday supplies and assign them to a second list. Figure 16.4 shows the completed design.

Figure 16.4

Selecting from the Available list box places items in the Selected list box.

1 Add the two list boxes (**`lstAvailable`** and **`lstSelected`**), the labels, and the command button (**`cmdOrder`**) shown in Figure 16.4. Label the form **Birthday Supplies**.

2 Set the `MultiSelect` property to `Simple` for the list on the left-hand side (`lstAvailable`).

3 Write the `Form_Load()` procedure to add the items shown in the `lstAvailable`, as follows:

```
Private Sub Form_Load()
   lstAvailable.AddItem "Balloons"
   lstAvailable.AddItem "Streamers"
   lstAvailable.AddItem "Punch"
   lstAvailable.AddItem "Cookies"
   lstAvailable.AddItem "Clown"
   lstAvailable.AddItem "Cake"
   lstAvailable.AddItem "Candles"
End Sub
```

4 Write the `cmdOrder_Click()` procedure to find the highlighted items on the Available list and assign them to the Selected list. In the assignment, use the `AddItem` method and the `ListIndex` property.

```
Private Sub cmdOrder_Click()
   Dim item As Integer
   'Loop searches entire list to determine which
   'items to move to the second list
   For item  = 0 To (lstAvailable.ListCount - 1)
      If lstAvailable.Selected(item) = True Then
         lstSelected.AddItem lstAvailable.List(item)
      End If
   Next item
End Sub
```

continues

> If an element in the first list is selected (is `True`), it is added to the second list. In multiple selections, a loop is mandatory to iterate through the list to determine which list elements have been selected. You cannot simply reference the current `ListIndex` or text properties as you can with single selection situations.

Objective 16.4 Understanding One-Dimensional Variable Arrays

With a better understanding of lists and combo box selections firm in your mind, it's time to move to the subject of one-dimensional variable arrays. A *one-dimensional variable array* is a list of more than one variable with the same name. (We consider multidimensional arrays in Chapter 17, "Multidimensional Arrays, Tables, Arrays of User-Defined Types, and the MSFlexGrid Control.")

16.4.1 The Purpose of Arrays

Arrays are used in programming to avoid the cumbersome ordeal of having to invent unique names for every variable in a program. As an example, suppose you have five friends and want to assign them their names. You could use a single variable for each friend, as follows:

```
Dim friend1 As String
Dim friend2 As String
Dim friend3 As String
Dim friend4 As String
Dim friend5 As String
```

However, an easier way of writing this is to use a one-dimensional variable array. This reduces the preceding individual statements to the following:

```
Dim friend(1 to 5) As String
```

The `friend` component is the name of the variable array, and `1 to 5` specifies the number of array elements.

16.4.2 Declaring Array Variables

Although not difficult, the declaring of an array is troublesome in that arrays, like other variables, have scope. They can be declared in several ways: at the procedure level, at the module level, and globally. The scope rules for declaring an array should be familiar (see Chapter 10, "Defining the Scope of Variables and Procedures"):

- **Global array**—Place the `Public` statement in the Declarations section of a standard code module to declare the array. No further declarations are needed.
- **Module-level array**—Place the `DIM` statement in the General Declarations section of a form module or standard module. No further declarations are needed.

- **Array with local scope**—Use one of the following within a procedure, preferably at the beginning:

  `Static` scope for the array

  `Static` scope for the entire procedure and `DIM` for the array

  `ReDim` for the array

As these rules indicate, changing the scope of an array variable changes the way in which it is declared. The first item of "Skill-Building Exercises" at the end of this chapter asks you to write several procedures to test the different ways of declaring an array. Before you do this, however, you need to consider the differences between the keywords `Dim` and `ReDim`, and between array elements and array subscripts.

Dim and *ReDim*

When working with variable arrays, the difference between `Dim` and `ReDim` becomes important. The `Dim` statement declares variables and allocates storage space. However, when you write a statement at the procedure level, such as the following, you cannot resize the array at a later time:

```
Dim friends(1 to 5) As String
```

Instead, use `ReDim` as in the following statement:

```
ReDim friends(1 to 5) As String
```

With `ReDim`, you can dynamically control or *redimension* the size of the array. For example, let's assume that the amount of your friends increases from five to eight. You can write the following code to dynamically increase the size of the array from five to eight array elements:

```
Size = 8
ReDim friends(1 to Size)
```

Dynamic allocation is important for several reasons:

- It enables the size of an array to be changed as items are added or deleted from a list. In many programs, the size of a list will not be known in advance. `ReDim` allows the size of a list to be kept flexible.
- `ReDim` enables an array to be declared at the procedure level. As a rule, define all variables, including array variables, at the lowest level or scope possible. `ReDim` provides this capability.
- It is possible to release the memory allocated for an array. By reducing the size of an array to zero, all memory is released, as in the following instruction:

  ```
  ReDim Friends(0)
  ```

You can also declare an array as follows:

```
Dim Codes() As Integer
```

The number of `Codes()` is not specified. This permits use of the following statement later in the program for dynamic allocation of memory:

```
ReDim Codes(5) As Integer
```

Variable Array Elements and Subscripts

One of the most confusing concepts of working with arrays is distinguishing between the number of array elements and the use of array subscripts. Consider the following declaration:

```
ReDim days(6) As String
```

This declaration indicates that the variable `days` actually contains seven elements because, by default, all arrays in Visual Basic are zero-based. The numeric value in the array declaration specifies the upper index value of the array. What it does not tell you is how to reference a specific day, such as the first of the six days. A specific day is referenced as follows: `days(0)`, `days(1)`, `days(2)`, `days(3)`, `days(4)`, `days(5)`, `days(6)`. The values `0`, `1`, `2`, `3`, `4`, `5`, and `6` are called subscripts. To determine the value of the first element of the array, a subscript of `0` is used.

You might ask, Can array elements and subscripts be numbered beginning with the number 1? Fortunately, there are two ways to accomplish this. The first is to insert an `Option Base` statement in the Declarations section of a module. Consider the following statement:

```
Option Base 1
```

This specifies a subscript of `1` rather than `0` for the first element.

A second way—the one favored by the authors—is to use the keyword `To` in the declaration of an array. The following statement specifies the beginning and the ending subscript numbers:

```
ReDim days(1 To 7) As String
```

16.4.3 Assigning a Value to an Array Element

Given this initial understanding of how arrays are declared, you might question next how values are assigned to array elements. Assigning a value to an array element is accomplished using an assignment statement, such as the following:

```
friend(0) = "B. J. McGathering"
friend(1) = "M. H. Pickel"
```

Continue until your assignments are finished. The expression on the right-hand side is assigned to the array variable. A loop represents another way of assigning an expression to an array element, as in the following example:

```
For Inside = 1 to 10
    num(Inside) = Inside
Next Inside
```

In this example, `num(1)` to `num(10)` contain the values of `1` through `10`.

16.4.4 Displaying the Contents of an Array

The quickest way of displaying the contents of an array is to use a `For...Next` loop when the number of array elements is known. Consider the following code:

```
For OfMine = 0 To 4 Step 1
   Print Spc(25) ; Friend(OfMine)
Next OfMine
```

In this example, the first value displayed would be the value stored in `Friend(0)`.

Exercise 16.5 Assigning Values to an Array

This simple exercise asks you to assign the days of the week to an array named `days` and to print out the contents of the array once all assignments have been made. Figure 16.5 shows the output of this design.

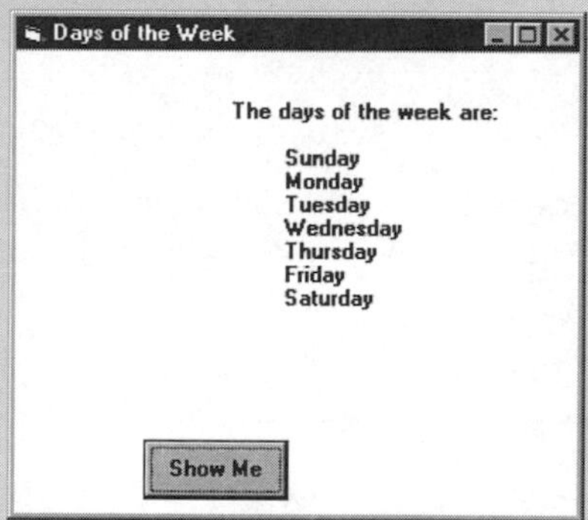

Figure 16.5

This form displays the contents of the variable array named `days(1 to 7)`.

1 Place a single command button (**`cmdShowMe`**) on the form, and change the form title to **Days of the Week**.

2 Write a `cmdShowMe_Click()` procedure as follows:

```
Sub cmdShowMe_Click()
   ReDim days(1 To 7)  As String
   Dim index As Integer

   days(1) = "Sunday"
   days(2) = "Monday"
   days(3) = "Tuesday"
   days(4) = "Wednesday"
   days(5) = "Thursday"
   days(6) = "Friday"
   days(7) = "Saturday"
```

continues

```
    Print
    Print
    Print Spc(20); "The days of the week are:"
    Print

    For index = 1 To 7
       Print Spc(25); days(index)
    Next index
End Sub
```

This array could be declared in one of four ways. The first is shown in the preceding code: It uses `ReDim`. The four ways are as follows:

a.
```
Sub cmdShowMe_Click()
   ReDim days(1 To 7)  As String
```

b.
```
Sub cmdShowMe_Click()
   Dim days(1 To 7)  As String
```

c.
```
Sub cmdShowMe_Click()
   Static days(1 To 7)  As String
```

d.
```
Static Sub cmdShowMe_Click()
   Dim days(1 To 7)  As String
```

With the fourth alternative, the entire procedure is declared as type `Static`. This means that the contents of the array `days` will not be lost if the procedure is called more than once.

3 Run and test your program, trying all four ways of declaring the array.

16.4.5 Using the *LBound* and *UBound* Functions

The `LBound` and `UBound` functions enable you to test the upper and lower boundaries of an array when they are unknown to you. The syntax is as follows:

```
LBound(arrayname[, dimension])
UBound(arrayname[, dimension])
```

The components of the syntax are explained in Table 16.4.

Table 16.4 The Syntax of the *LBound* and *Ubound* Functions

Component	*Description*
`arrayname`	Name of the array variable. It follows standard variable naming conventions.
`dimension`	A whole number indicating which dimension's lower or upper boundary is returned. (We will turn to multidimensional arrays in Chapter 17, "Multidimensional Arrays, Tables, Arrays of User-Defined Types, and the MSFlexGrid Control.")

Exercise 16.6 Using the *LBound* and *Ubound* Functions

This brief exercise shows how to test for the upper and lower boundaries of an array. Figure 16.6 shows the interface that you are to create.

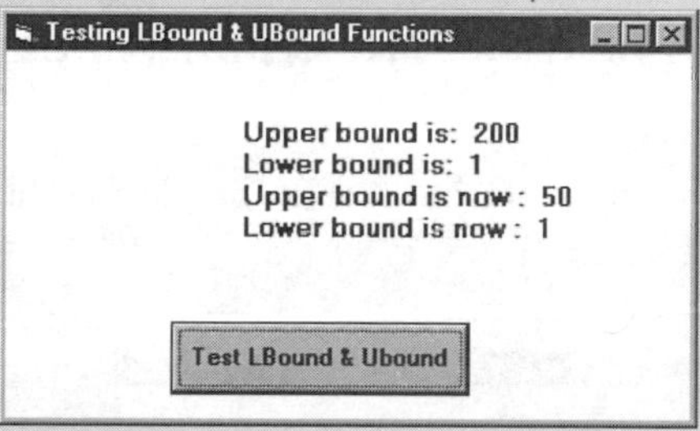

Figure 16.6

This display indicates that the upper and lower boundaries of the array are changed by the program.

1. Start a new project, and add a command button named **`cmdTest`**.
2. In the `cmdTest_Click()` event, enter the following code:

```
Private Sub cmdTest_Click()
   ReDim arraynum(1 To 200) As Integer
   Cls
   Print
   Print
   Print Spc(15); "Upper bound is: " & Str$(UBound(arraynum))
   Print Spc(15); "Lower bound is: " & Str$(LBound(arraynum))

   ReDim arraynum(1 To 50)
   Print Spc(15); "Upper bound is now : " & Str$(UBound(arraynum))
   Print Spc(15); "Lower bound is now : " & Str$(LBound(arraynum))
End Sub
```

The values returned from the `UBound` and `LBound` functions can be used in `For...Next` loops, which iterate through arrays that are dynamically changed during runtime.

Objective 16.5 Working with Control and Variable Arrays

Sometimes it is necessary to combine controls with variable arrays. Consider the following statement:

```
txtText(element).Text = arrayVariable(element)
```

This assignment statement assigns the value of the variable array element to a text box within an array of text boxes. Using both control arrays and variable arrays in a single instruction is common and useful. Exercise 16.7 and Exercise 16.8 demonstrate this technique.

Exercise 16.7 Working with a Control and a Variable Array

This exercise asks you to modify the last exercise by placing the names of the week into seven labels (see Figure 16.7).

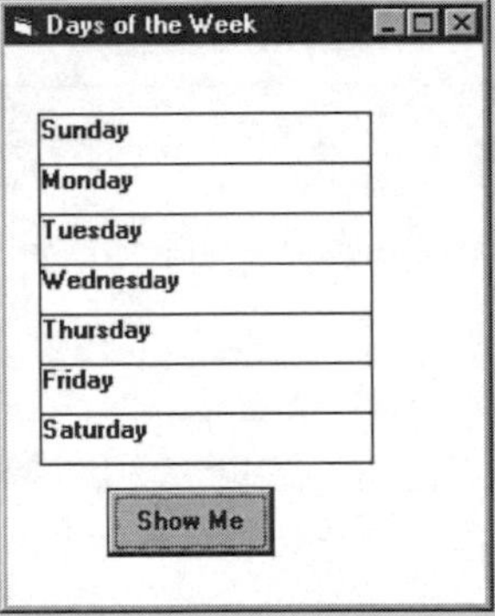

Figure 16.7
The names of the week are presented by a control array of seven labels.

1 Add the control array of seven labels to your design, using the name **lblDays**.
2 Rewrite the `cmdShowMe_Click()` procedure to assign the seven elements of the variable array `days` to the seven text boxes, as follows:

```
Private Sub cmdShowMe_Click()
   ReDim days(0 To 6)  As String  ' need to change because control arrays
                                  ' are zero based
   Dim index As Integer

   days(0) = "Sunday"
   days(1) = "Monday"
   days(2) = "Tuesday"
   days(3) = "Wednesday"
   days(4) = "Thursday"
   days(5) = "Friday"
   days(6) = "Saturday"
   For index = 0 To 6
      lblDays(index).Caption = days(index)
   Next index
End Sub
```

In this code, the index references each member in the control array and each element in the variable array. This single line of code should help clarify the similarity between a control and a variable array. Both require an index or subscript. More importantly, the value of the elements of a variable array can be assigned as a property value to a member of a control array.

Exercise 16.8 Creating an Application with Variable Arrays

Suppose that you are responsible for informing people on vacation about possible side trips and their costs. A side trip can be chosen from a list, causing the price of the trip to appear. After all side trips are scheduled, a total price is calculated. When customers change their minds, the add-on trip listing can be cleared, and the customer can start again. Figure 16.8 shows a layout for this design. You are working with the following prices:

```
Tyrolean Weekend            $455.55
Rhine River Cruise           630.75
Personal Louvre Guide        120.35
Milan Fashion Tour           330.33
Schwartzwald Spa Weekend     765.45
```

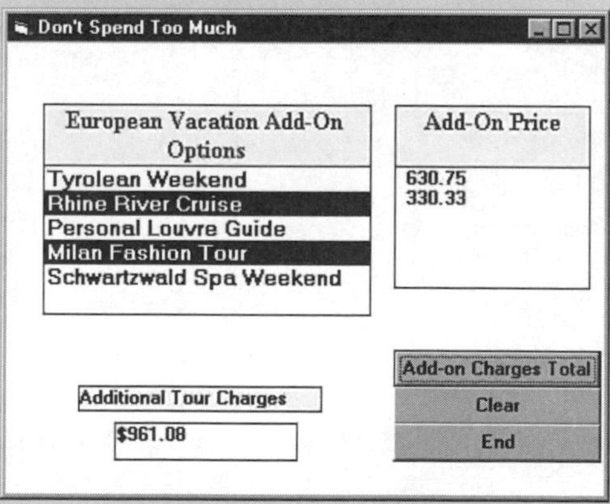

Figure 16.8

The Don't Spend Too Much application enables you to determine the cost of each side trip and the total cost of your selections.

continues

1 Add the two list boxes (**lstOptions** and **lstPrice**), the three command buttons (**cmdAddPrice**, **cmdClear**, and **cmdEnd**), and the label (**lblTotal**), as shown in Figure 16.8. The Additional tour charge box is a Label control, not a Text Box control. Write the end procedure, and change the form caption to **Don't Spend Too Much**. An alternative title is **Go Ahead; Spend. You Only Live Once!**

2 Write the `Form_Load()` procedure to add the five list items to `lstOptions`.

3 Write the `lstOptions_Click()` multi-select procedure. If an item is selected, add the item to the second `lstPrice`.

```
Private Sub lstOptions_Click()
   Select Case lstOptions.ListIndex
      Case 0
         lstPrice.AddItem "455.55"
      Case 1
         lstPrice.AddItem "630.75"
      Case 2
         lstPrice.AddItem "120.25"
      Case 3
         lstPrice.AddItem "330.33"
      Case 4
         lstPrice.AddItem "765.45"
   End Select
End Sub
```

4 Write the `cmdAddPrice_Click()` procedure to compute the total charge. Assign all prices to a variable array named `AddPrice()`. Use a `For...Next` loop to determine whether an item has been selected. If it has, add the price to the total. Display the total price as a caption under the label Additional Tour Charges.

```
Private Sub cmdAddPrice_Click()
   Static AddPrice(0 To 4)  As Currency
   Dim Choice As Integer
   Dim RunTotal As Currency

   'Add prices to array
   AddPrice(0) = 455.55
   AddPrice(1) = 630.75
   AddPrice(2) = 120.25
   AddPrice(3) = 330.33
   AddPrice(4) = 765.45

  lstPrice.Clear  ' clears listbox of multiple selections of the same
                  ' option

 'Loop to get choices
   For Choice = 0 To 4
      If lstOptions.Selected(Choice) = True Then
         lstPrice.AddItem Str$(AddPrice(Choice))
         RunTotal = RunTotal + AddPrice(Choice)
      End If
   Next Choice

   lblTotal.Caption = Format (Str$(Runtotal), "Currency")
End Sub
```

5 Write the `cmdClear_Click()` procedure to clear the list. Use a `For...Next` loop to iterate through the list box.

```
Private Sub cmdClear_Click()
   Dim n As Integer
   'Turn off previous choices
   For n = 0 To 4
     lstOptions.Selected(n) = False
   Next n
   'Clear price list
   lstPrice.Clear
   'Clear totals box
   lblTotal.Caption = ""
End Sub
```

16.5.1 Dynamically Sizing Global Arrays

The capability to dynamically size arrays is one of Visual Basic's greatest strengths. Frequently, a global declaration of an array is made with no elements defined, such as the declaration in a standard module:

```
Public friend() As String
```

This array can then be dynamically resized within procedures, as in the following instruction:

```
ReDim friend (1 to 5) As String
```

Once done working with an array, you can clear memory by simply redimensioning it, as follows:

```
ReDim friend(0) As String
```

Alternatively, you can use the `Erase` statement, which deallocates the memory:

```
Erase friend
```

16.5.2 The *Preserve* Keyword

How do you manage an array that has had values assigned that you want to preserve as it is resized? In this case, use the keyword `Preserve`. Suppose you had first made the declaration `ReDim friends(1 to 3) As String` and assigned the values `"Bob"`, `"Jean"`, and `"Bill"` to `friends(1)`, `friends(2)`, and `friends(3)`, respectively. You could preserve those values when resizing the array with the declaration:

```
ReDim Preserve friends(1 to 4)
```

A value could be assigned to the new array element `friends(4)`, and previous assignments to other array elements would be preserved. If you had resized the array with the following statement, all previous assigned values would be lost:

```
ReDim friends (1 to 4)
```

Exercise 16.9 Demonstrating the *Preserve* Keyword

This exercise shows how values can be retained by using the keyword `Preserve` when an array is resized. You are asked to create the display shown in Figure 16.9.

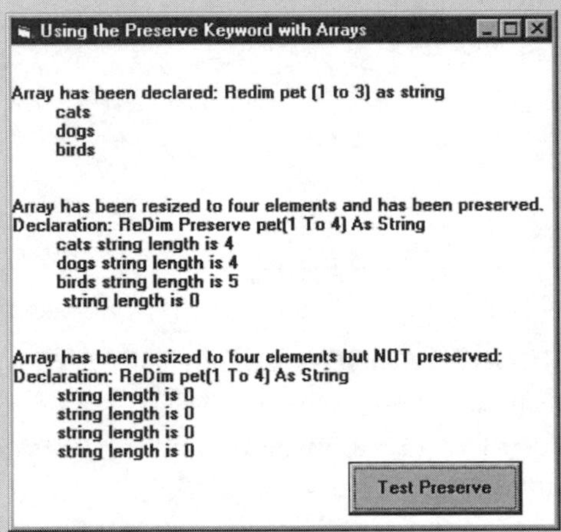

Figure 16.9

This display shows the effect of both using and not using the keyword `Preserve`.

1 Create a new project, and add a single command button to the form named **`cmdTestPreserve`**.

2 In the `cmdTestPreserve_Click()` event, enter the following code:

```
Private Sub cmdTestPreserve_Click()
   Font.Bold = True
   Dim i As Integer
   ReDim pet(1 To 3) As String   ' dimension array

   pet(1) = "cats"         ' assign values
   pet(2) = "dogs"
   pet(3) = "birds"

   Print
   Print
   'display them
   Print "Array has been declared: ReDim pet (1 to 3) As String"

   For i = LBound(pet) To UBound(pet)
      Print Spc(5); pet(i)
   Next
   Print
   Print

   ReDim Preserve pet(1 To 4) As String    ' redimension array and
                                           ' preserve previous values
```

```
        ' display them and the string length of array elements
        Print "Array has been resized to four elements and has been preserved."
        Print "Declaration: ReDim Preserve pet(1 To 4) As String "

        For i = LBound(pet) To UBound(pet)
           Print Spc(5); pet(i) & " string length is " & Len(pet(i))
        Next
        Print
        Print

        ReDim pet(1 To 4) As String   ' redimension array without preserving
                                      ' previously assigned values

        ' display them and the string length of array elements
        Print "Array has been resized to four elements but NOT preserved:"
        Print "Declaration: ReDim pet(1 To 4) As String"

        For i = LBound(pet) To UBound(pet)
           Print Spc(5); pet(i) & "string length is " & Len(pet(i))
        Next
     End Sub
```

Remember to use the `Preserve` keyword in order to save values to arrays if you resize them dynamically.

Chapter Summary

This chapter begins with lists and combo boxes as examples of one-dimensional lists of items. When only one item can be selected from a list, one `Selected` property will be set to `True`; for all other items, the `Selected` property will be set to `False`. When multiple items can be selected from a list, each item must be tested (using a loop) to determine which selected properties are set to `True`.

With a combo box, determining which item has been selected is much easier because it always appears in the combo text box.

The largest section of this chapter—Objective 16.4, "Understanding One-Dimensional Variable Arrays"—addresses the description and use of one-dimensional variable arrays. A one-dimensional array is a list of more than one variable with the same name.

Variable arrays can be declared in several ways. The preferred way is generally the use of `ReDim`, as in the following example:

```
ReDim friend(5) As String
```

`ReDim` permits the size of an array to be dynamically controlled, keeps the variable declaration at the procedure level, and permits memory to be released by reducing the size of the array to zero.

Besides ReDim, it is possible to declare a global array, to declare an array at the form level, or to create a local array with Static scope (for the array or for all variables in the procedure).

Array subscripts often begin with zero. In the declaration above, the array friends() contains six elements: The subscript of the first item is (0), and the subscript of the sixth item is (5). To begin subscripts and elements with the number 1, declare an array using Option Base 1 or the To keyword, as in the following example:

```
ReDim friend (1 To 6) As String
```

Values are assigned to array variables in the same manner as assignments are made to non-array variables. For example, the following is a valid assignment:

```
friend(1) = "Hank"
```

The manipulation of an array is performed using loops. Displaying the contents of the array declared above is done as follows:

```
For OfMine = 1 to 6
   Print friend(OfMine)
Next OfMine
```

Variable and control arrays are often combined. The following instruction assigns the value stored in a variable array to a property of a control array:

```
lblFriends(index).Caption = friend(index)
```

The upper and lower boundaries of arrays can be tested with the UBound and LBound functions. To ensure that values previously assigned to an array are not lost when the array is redimensioned, use the Preserve keyword. If you had made an original declaration of ReDim cities(1 to 5) As String and then assigned the values "Seattle", "Portland", "Chicago", "Boston", and "St. Louis" to the array, you could retain these values when you redimension the array to add "Phoenix" with the following declaration:

```
ReDim Preserve cities (1 to 6) As String.
```

Skill-Building Exercises

1. This project consists of several parts. It asks you to use different rules for declaring an array. Figure 16.10 illustrates an array testing program that displays the test results in a text box. See section 16.4.2, "Declaring Array Variables," earlier in this chapter for the rules for declaring arrays in this exercise.

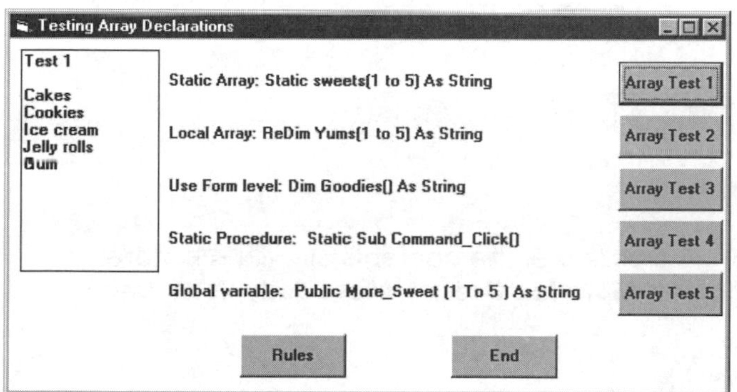

Figure 16.10

Five array tests are used to show the different ways of dimensioning an array.

1. Declare `sweets()` at the procedure level. The array contains: **Cakes**, **Cookies**, **Ice Cream**, **Jelly Rolls**, and **Gum**. Complete the following procedure:

```
Private Sub cmdTest1_Click()
   'Declare sweets as a static array of type string
   Static sweets(1 to 5) As String
   Dim i As Integer
   'Clear the results box
   'Load the array
   'Clear the text box
   'Display the array
End Sub
```

2. Declare `yums()` at the procedure level. The array contains: **Cakes**, **Cookies**, **Ice Cream**, **Jelly Rolls**, and **Gum** (the same as the last array). Complete the following procedure:

```
Private Sub cmdTest2_Click()
   'This procedure declares yums()
   'as array of type string
   ReDim yums(1 To 5)  As String
   Dim I As Integer
   'Load the array
   (Use the code from Step 1 to complete this procedure.)
End Sub
```

3. Declare `goodies()` at the form module level to allow the same array to contain different things.

```
Option Explicit
```

Then, write a procedure to call two other procedures, named `some()` and `more()`. Complete the following procedure:

```
Private Sub cmdTest3_Click()
   Dim i As Integer
   'Goodies is dimensioned at the form level
   'Call the two procedures
End Sub
```

Complete the `some()` procedure. `Goodies` contains the same items as before: **Cakes**, **Cookies**, **Ice Cream**, **Jelly Rolls**, and **Gum**.

```
Private Sub some()
   Dim i As Integer
   'Clear the text box
   'Load the Goodies array
   'Display the array
End Sub
```

Complete the `more()` procedure. The contents of `Goodies` is **More Cakes**, **More Cookies**, **More Ice Cream**, **More Jelly Rolls**, and **More Gum**:

```
Private Sub more()
   Dim i As Integer
   'Load the Goodies array
   'Clear the text box
   'Display the array
End Sub
```

4. Declare the procedure as type `Static`. `More_sweets` contains the same items as `Goodies`. Complete the following procedure:

```
Private Static Sub cmdTest4_Click()
   'The word static is placed in the definition of the procedure
   Dim i As Integer
   Dim More_sweets(1 To 5)  As String
   'Load the More_sweets array
   (add code to complete)
End Sub
```

5. Declare the global array `More_Sweets` within a standard code module (a .BAS file):

```
Option Explicit
```

Complete the procedure:

```
Private Sub cmdTest5_Click()
   Dim i As Integer
     (add code to complete)
End Sub
```

What conclusions can you draw from these five tests?

2. Write a procedure that declares an array that can contains the names of ten of your favorite cities or towns, assigns values to the array, and displays them in a list box with a loop that uses the `LBound` and `UBound` functions. Extend this exercise so that when one of the cities is clicked, the message "(city name) is a great place to visit!" is displayed.

3. Write a program that lists ten of your favorite movies in a list box. Using a text box or other means (for example, the `InputBox` function), permit the user to add to the list. Also provide a way for the user to remove items from the list.

4. Extend the preceding exercise by globally declaring an unassigned array of your ten favorite movies. Change the selection property of the list box so that the user can make multiple selections. Write a procedure so that, depending upon user selections, the global array is redimensioned to the number of user selections, selected movies are assigned to the global array, and the array (using `LBound` and `UBound` functions) is printed on a display label. At the end of the procedure, clear the array from memory.

5. Experiment with the prceding exercise using the `Preserve` keyword in the array declaration and removing code that clears the array from memory. What changes do you notice in program behavior?

17

Multidimensional Arrays, Tables, Arrays of User-Defined Types, and the MSFlexGrid Control

Besides one-dimensional arrays and lists, Visual Basic provides several ways of creating two-dimensional arrays or tables, as well as even larger dimensioned arrays. In this chapter, we begin with the concept of multidimensional arrays, followed by the MSFlexGrid custom control and user-defined data types. This chapter concludes with a more complex type of array, known as an array of structures.

This chapter presents several new and important concepts. When you complete this chapter, you will be able to

- declare a multidimensional array,
- use a table look-up procedure,
- transfer items from a multi-column list,
- use the MSFlexGrid control,
- employ an array of user-defined types.

Chapter Objectives

The objectives of this chapter each address a different topic. We begin with multidimensional arrays, move to tables (multidimensional lists and an MSFlexGrid), and end with user-defined types. The following are the four objectives:

1. Declaring and using multidimensional arrays
2. Designing a multi-column list

3. Using the MSFlexGrid control
4. Employing an array of user-defined types

Objective 17.1 Declaring and Using Multidimensional Arrays

Besides one-dimensional arrays, Visual Basic allows for *multidimensional arrays.* Although most multidimensional arrays are limited to two or three dimensions, you can declare arrays of up to 60 dimensions with Visual Basic—provided you can visualize beyond three dimensions and have the memory to handle it. Consider the following examples of two-dimensional array declarations:

```
ReDim MyTable (9, 9) As Integer
ReDim MyTable(1 To 10, 1 To 10) As Integer
```

This array, named `MyTable`, consists of 10 rows and 10 columns and stores 100 elements. A three-dimensional array declaration is similar. It might be as follows:

```
Static MyTable (1 To 10, 1 To 10, 1 To 10) As Integer
```

This array would store 1,000 elements (10×10×10).

17.1.1 Declaring a Multidimensional Array

Like one-dimensional arrays, *static* and *dynamic* multidimensional arrays are permitted. To create a dynamic array, declare the array with a `Public` statement or a `Dim` statement at the module level, or a `Static` or `Dim` statement in a procedure, as in the following example:

```
Static MyTable() As Integer
```

Use `ReDim()` in a procedure to allocate the number of array elements:

```
ReDim MyTable (1 to 10, 1 to 10) As Integer
```

This instruction allocates memory for a 10 by 10 array.

17.1.2 Writing a Table Look-Up Procedure

Multidimensional arrays are especially useful whenever two or more array values are required to determine a value. Table look-up provides an excellent example of how this works. A value checked using the first column of a two-dimensional array leads to a value stored in the second column of the array.

Exercise 17.1 A Table Look-Up to Compute Taxes

A favorite table look-up exercise involves determining a person's income and, based on that income, looking up the percentage of tax the person must pay. After the percentage is known, the person's tax can be computed. Figure 17.1 shows the simple interface required for this exercise.

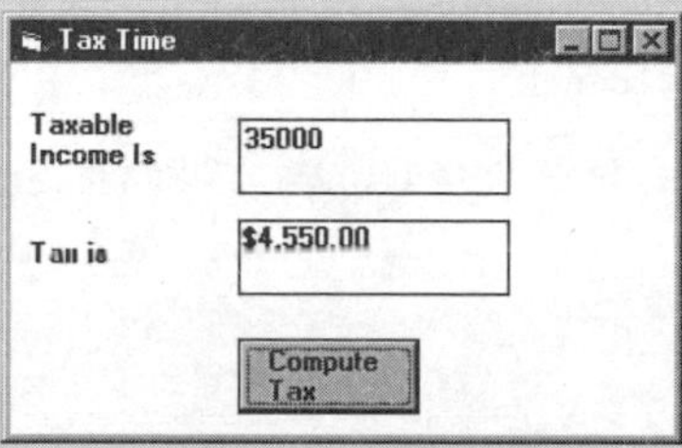

Figure 17.1

Enter your taxable income, and click Compute Tax to determine your income tax.

1 Create the interface as shown in Figure 17.1, adding one text box and three labels (the one to display the tax is named **`lblTax`**). Add the command button (**`cmdCompute`**), and change the form caption.

2 Write the `cmdCompute_Click()` procedure as follows:

```
Private Sub cmdCompute_Click()
   Static taxrate(4, 1) As Single
   Dim income As Currency
   Dim tax As Currency
   Dim flag As Boolean
   Dim i As Integer

   taxrate(0, 0) = 10000
   taxrate(1, 0) = 20000
   taxrate(2, 0) = 30000
   taxrate(3, 0) = 40000
   taxrate(4, 0) = 50000

   taxrate(0, 1) = .09
   taxrate(1, 1) = .12
   taxrate(2, 1) = .13
   taxrate(3, 1) = .14
   taxrate(4, 1) = .16
```

continues

```
    income = Val(txtIncome.Text)
    flag = False
    For i = 4 To 0 Step -1
      If income >= taxrate(i, 0) Then
        tax = taxrate(i, 1) * income
        lblTax.Caption = Format(Str$(tax), "Currency")
        flag = True
        Exit For
      End If
    Next i
    If flag = False Then
       If income >= 0 Then
         lblTax.Caption = Format("0", "Currency")
       Else
         lblTax.Caption = "Income must be positive"
       End If
    End If
End Sub
```

Do you understand from this procedure how table look up works? First a person's income level is checked to determine how many thousands of dollars are earned. After this is known, the tax percentage can be found. The tax percentage multiplied by the income determines the total amount of tax the person owes.

This program must include steps for low income or negative income. With negative income, the message "Income must be positive" is displayed. You might want to change this to $0.00.

Exercise 17.2 A Table Look-Up to Compute Taxes Based on Marital Status

This exercise adds further complexity to the problem of looking up the tax a person is required to pay. Figure 17.2 illustrates that a person's marital status affects tax liability (compare Figure 17.1 to Figure 17.2).

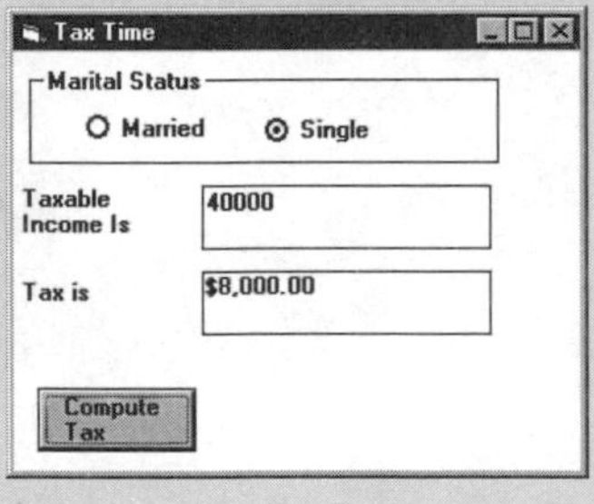

Figure 17.2

A single person's income tax is higher than a married person's income tax.

1 Change the interface by adding a marital status frame and two option buttons, one for **Married** (**`optMarried`**) and one for **Single** (**`optSingle`**).

2 Declare the array `taxrate` at the general level. This will make it global to all procedures. This statement should read

```
Dim taxrate(4, 1) As Single
```

3 Revise the `cmdCompute_Click()` procedure to call the function `taxrate1` if a person is married, and `taxrate2` if a person is single. The code in this procedure is limited to the following statements:

```
Private Sub cmdCompute_Click()
   Dim married As Integer
   If (optMarried.Value = True) Then
      Call taxrate1
   Else
      Call taxrate2
   End If
End Sub
```

4 Use a tax look-up procedure in `taxrate1()` to determine a married person's tax. This procedure reads as follows:

```
Private Sub taxrate1()
   Dim income As Currency
   Dim tax As Currency
   Dim flag As Boolean
   Dim i as Integer

   taxrate(0, 0) = 10000
   taxrate(1, 0) = 20000
   taxrate(2, 0) = 30000
   taxrate(3, 0) = 40000
   taxrate(4, 0) = 50000

   taxrate(0, 1) = .09
   taxrate(1, 1) = .12
   taxrate(2, 1) = .13
   taxrate(3, 1) = .14
   taxrate(4, 1) = .16

   income = Val(txtIncome.Text)

   flag = False
   For i = 4 To 0 Step -1
     If income >= taxrate(i, 0) Then
       tax = taxrate(i, 1) * income
       lblTax.Caption = Format(Str$(tax), "Currency")
       flag = True
       Exit For
     End If
```

continues

```
    Next i
    If flag = False Then
      If income >= 0 Then
        lblTax.Caption = Format("0", "Currency")
      Else
        lblTax.Caption = "Income must be positive"
      End If
    End If
End Sub
```

5 Use a tax look-up procedure in `taxrate2()` to determine a single person's tax. The tax rates are as follows:

```
taxrate(0, 0) = 10000
taxrate(1, 0) = 20000
taxrate(2, 0) = 30000
taxrate(3, 0) = 40000
taxrate(4, 0) = 50000

taxrate(0, 1) = .11
taxrate(1, 1) = .14
taxrate(2, 1) = .17
taxrate(3, 1) = .2
taxrate(4, 1) = .22
```

Complete the rest of this procedure on your own.

6 Run and test your program. Whose taxes are higher: taxes for single people or taxes for married people?

Objective 17.2 Designing a Multi-Column List

Besides multidimensional arrays, *multi-column lists* can be designed to look like tables, even though they function like simple lists. To create a multi-column list, change two properties: `Listbox.Columns` and `Listbox.MultiSelect`. Change the number of columns from one to the number desired, and change the `MultiSelect` property to allow for more than one item to be selected.

Exercise 17.3 Creating a Multi-Column List for Making Multiple Selections

This short exercise asks you to create a multi-column list that allows for multiple selections. Figure 17.3 illustrates the two list box designs.

Figure 17.3

A multi-column list enables the user to select from a table of items.

1 Create the design shown in Figure 17.3, adding two list boxes (**`lstSale`** and **`lstBuyMe`**), the command button (**`cmdIWant`**), two labels, and all captions.

2 For each list box, change the number of columns to 3 and `MultiSelect` to 1.

3 Write the `Form_Load()` procedure to add the items to `lstSale`:

```
Private Sub Form_Load()
   lstSale.AddItem "Kittens"
   lstSale.AddItem "Dogs"
   lstSale.AddItem "Snakes"
   lstSale.AddItem "Bunnies"
   lstSale.AddItem "Mice"
   lstSale.AddItem "Hamsters"
   lstSale.AddItem "Fish"
   lstSale.AddItem "Birds"
   lstSale.AddItem "Rats"
   lstSale.AddItem "Turtles"
   lstSale.AddItem "Worms"
   lstSale.AddItem "Bugs"
End Sub
```

4 Write the `cmdIWant_Click()` procedure:

```
Private Sub cmdIWant_Click()
   'n is the number in the list
   Dim n As Integer

   lstBuyMe.Clear
   For n = 0 To (lstSale.ListCount - 1)
      If lstSale.Selected(n) = True Then
         lstBuyMe.AddItem lstSale.List(n)
      End If
   Next n
End Sub
```

continues

This procedure initially clears the Buy Me list. It then loops through the For Sale list to determine which items are selected. The `AddItem` method is used to display the selected items on the Buy Me list. Although the list box has multiple columns, the items in the list merely wrap from one column to the next. To specify that certain tabular data be displayed in certain columns or rows, it's much easier to use a control other than the built-in list box. These include either controls bound to an underlying database or the MSFlexGrid control.

We will turn to databases and data-bound controls in Chapter 24, "Accessing and Manipulating Databases with Visual Basic," and Chapter 25, "Using Data Access Objects." We next turn to the MSFlexGrid control.

Objective 17.3 Using the MSFlexGrid Control

A third way to work with tables in Visual Basic is to utilize the MSFLXGRD.OCX component. This component displays an MSFlexGrid on the screen and lets you display information in rows and columns. The control itself looks like a table or spreadsheet. If the MSFlexGrid control is not already added to your toolbox, you can add it by choosing Project, Components, and selecting the Controls tab (see Figure 17.4). Like other controls, select the check box to add the MSFlexGrid to your toolbox.

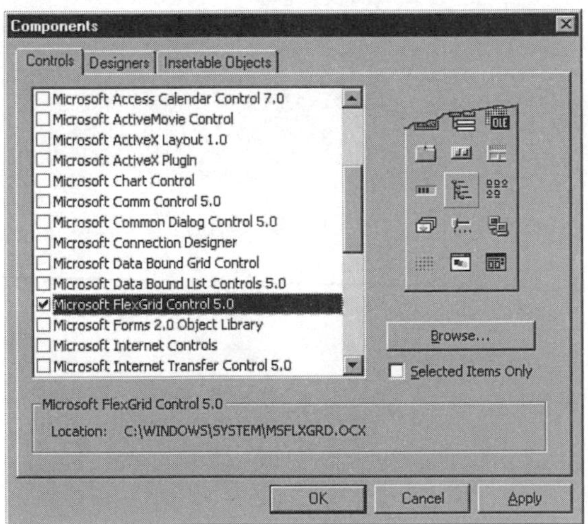

Figure 17.4

The Components Controls tab enables you to add .OCX components to your toolbox.

Table 17.1 shows four properties of the MSFlexGrid control that are especially important.

Table 17.1 MSFlexGrid Control Properties

Property	*Description*
`Rows`	Sets the total number of fixed and nonfixed rows in the MSFlexGrid.
`Cols`	Sets the total number of fixed and nonfixed columns in the MSFlexGrid.
`FixedRows`	Sets the number of fixed rows in the MSFlexGrid.
`FixedCols`	Sets the number of fixed columns in the MSFlexGrid.

What is the difference between fixed and nonfixed rows and columns? A fixed row or column is used for adding row and column descriptions to the MSFlexGrid. These rows and columns do not scroll when scroll bars are added to the MSFlexGrid and are shaded gray on the MSFlexGrid. A nonfixed row or column is used to store MSFlexGrid values. These rows and columns scroll when necessary. They are not shaded and are shown in white.

Each intersection of a row and column forms an MSFlexGrid location called a *cell.* An individual cell location can be set by using an assignment statement to the `Row` and `Col` (singular) properties. For example, the following sets the cell nonfixed location at `Grid1(1,1)`:

```
Grid1.Row = 1
Grid1.Col = 1
```

After the cell has been identified, the individual cell properties can be changed. Table 17.2 shows the cell properties.

Table 17.2 Cell Properties

Property	*Description*
`Text`	Returns or sets the text contents of a cell.
`CellPicture`	Returns or sets an image to be displayed in the current cell. These images are often icons.
`CellFont`	This group of properties sets the font characteristics of cells containing text: `CellFontName`, `CellFontSize`, `CellFontBold`, `CellFontItalic`, and so on.
`ColSel`, `RowSel`	Returns or sets the starting or ending row or column for a range of cells. These properties enable you to select a region of the MSFlexGrid control and change that region's appearance.

continues

Table 17.2 continued

Property	*Description*
`BackColor`, `BackColorBkg`, `BackColorFixed`, `BackColorSel`	Returns or sets the background color of various elements of the MSFlexGrid. By using the `ColSel` and `RowSel` properties, a region can be created.
`ColWidth`, `RowHeight`	Returns or sets the width of the specified column or row in twips.

The MSFlexGrid has many more properties than those listed here. Consult the Visual Basic online Help for full documentation. The following exercises will give you a sense of the more prominent properties.

Text can be added to an individual cell of the MSFlexGrid by identifying the cell with the `Row` and `Col` properties and then assigning a value to the control's `Text` property. Text can also be added one row at a time with the `AddItem` method. A string expression that is the argument of the `AddItem` method must then be delimited between cells by the tab character (ASCII value 9). The next exercise demonstrates these two methods of adding string data to cells.

Exercise 17.4 Entering Numbers into the MSFlexGrid

Perhaps the best way to learn how to use the MSFlexGrid control is through an example. Figure 17.5 shows the table of numbers you are asked to create.

The MSFlexGrid Control

Table						
of						
Value	1	2	3	4	5	
	2	4	6	8	10	
	3	6	9	12	15	
	4	8	12	16	20	
	5	10	15	20	25	
	6	12	18	24	30	
	7	14	21	28	35	
	8	16	24	32	40	
	9	18	27	36	45	
	10	20	30	40	50	

Figure 17.5

The MSFlexGrid control creates a two-dimensional table for you to work with.

1 Double-click the MSFlexGrid control to place the grid on the screen. Add the form caption.

2 Open the MSFlexGrid control's property box, and set the following:

```
Cols = 5
Rows = 3
```

3 Write the following `Form_Load()` procedure:

```
Private Sub Form_Load()
   Dim Row As Integer
   Dim Col As Integer
   Dim TK As String
   Dim temp As String

   Grid1.Cols = 7
   TK = Chr$(9) ' Tab
   For Row = 0 To 9          'Fill each row one at a time
      temp = "" & TK
      For Col = 0 To 4
         temp = temp & ((Row + 1) * (Col + 1)) & TK
      Next Col
      Grid1.AddItem temp     'Add entire row
   Next Row

   Grid1.FixedRows = 2    'Grid property
   Grid1.FixedCols = 2    'Grid property

   Grid1.Col = 0
   Grid1.Row = 0
   Grid1.Text = "Table"
   Grid1.Col = 0
   Grid1.Row = 1
   Grid1.Text = " of "
   Grid1.Col = 0
   Grid1.Row = 2
   Grid1.Text = "Value"
End Sub
```

In this code, the first two statements set the number of rows and columns. Following this, text is added to selected cells. The `AddItem` method adds rows, so in design time you set the number of rows to 3 and add them dynamically in a `For...Next` loop.

4 Test your program. Change the number of `FixedRows` and `FixedCols` to 1 to observe some interesting differences.

Exercise 17.5 Entering Items from Text Boxes into an MSFlexGrid

Besides adding numbers to an MSFlexGrid using a `Form_Load()` procedure, it is possible to add values to an MSFlexGrid, row by row or column by column. Figure 17.6 shows the next grid interface that you are to complete. This interface works as follows: After entering the name of the student and the student's scores, click Add to add the names to the grid; click Next to clear the text boxes to allow the next student's name and scores to be added to the grid.

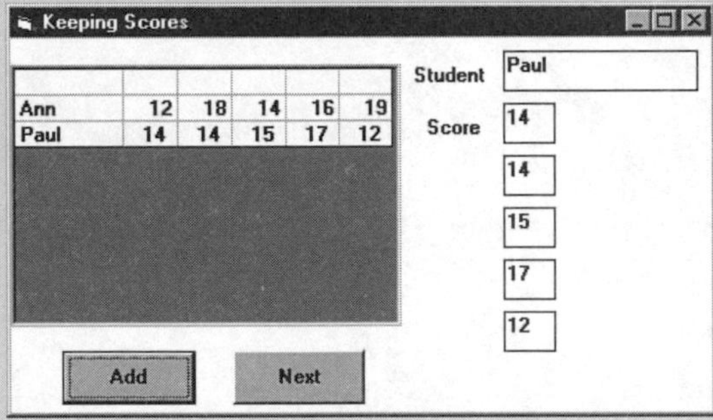

Figure 17.6

The MSFlexGrid enables you to create a spreadsheet of values.

1 Create the form shown in Figure 17.6, adding the MSFlexGrid, the two command buttons (**`cmdAdd`** and **`cmdNext`**), the name text box (**`txtName`**), and a control array of five text boxes (**`txtScore`**) to hold student scores. Add the labels and the form caption.

2 Set the `Column` property of the grid (`Cols`) to `6`. Leave the `Row` property set to `1`.

3 Write the `cmdAdd_Click()` procedure as follows:

```
Private Sub cmdAdd_Click()
   Dim temp1 As String
   Dim temp2 As String
   Dim score As Integer
   Static row As Integer

   row = row + 1
   grid1.Rows = row
   temp1 = txtName.Text & Chr$(9)

   For score = 0 To 4
      temp2 = temp2 + txtScore(score) & Chr$(9)
   Next score
   grid1.AddItem temp1 & temp2
End Sub
```

In this procedure, two temporary strings are declared: one to store the student's name and one to store the student's scores. The variable `row` is declared as type `Static`. This allows it to be incremented by 1 as each new student is added to the grid.

Within the procedure, `temp1` stores the student's name, and `temp2` stores the student's scores. A Tab key (`Chr$(9)`) is required to place a score in a different cell on the grid.

4 Write the `cmdNext_Click()` procedure as follows:

```
Private Sub cmdNext_Click()
   Dim i As Integer

   txtName = " "
   For i = 0 To 4
      txtScore(i) = " "
   Next
   txtName.SetFocus
End Sub
```

This procedure first clears the student's name and then continues to clear all student scores. The `SetFocus` instruction places the cursor in the `txtName` text box.

Objective 17.4 Employing an Array of User-Defined Types

You were briefly exposed to user-defined types in Chapter 10, "Defining the Scope of Variables and Procedures." User-defined types provide a way to add a table to a design without employing a multidimensional array, which requires that all elements and all dimensions of the array be of the same data type. *User-defined types,* on the other hand, combine variables of several different data types to form a single variable.

When an array of these variables is created, a table of mixed data types can be created. This type of array provides a program designer with an enormously useful tool in a wide variety of programming tasks. It allows related data to be held together without defining an entire class.

17.4.1 Declaration of a User-Defined Type

You will recall that a user-defined type must be defined in a standard code module with the `Type` statement:

```
[Private | Public] Type UserType
  elementname [(subscripts)] As typename
  [elementname [(subscripts)] As typename
End Type
```

This new data type can then be used in variable declarations, such as the following:

```
Dim MyType As UserType
```

Consider the following example. Type `Asset` contains three members: one member of type `String`, a second of type `Integer`, and a third of type `Currency`.

```
Type Asset
   Name As String * 20
   Location As Integer
   Value As Currency
End Type
```

The declaration using this type then might be as follows:

```
Dim Inventory As Asset
```

This declares the variable `Inventory` as being of type `Asset`.

17.4.2 A User-Defined Variable Array

After a user-defined type has been defined, it can be declared as a single variable or as an array. The following declares the variable `Part` as type `Asset`:

```
Dim Part As Asset
```

The following declares `Inventory` to be an array of 100 items, all of which are type `Asset`:

```
Dim Inventory(1 To 100) As Asset
```

This type of array is called an `array of structures` because each element in the array contains a complex data type or structure.

Exercise 17.6 Extending Options Using an Array of User-Defined Type

In the European vacation program created in Exercise 16.8, "Creating an Application with Variable Arrays," in Chapter 16, you may have noticed that the user can enter multiple choices for a single option. For instance, during selection, you can choose to have five Louvre guides and it will be added to the vacation add-ons list box. When the options are summed up, the add-ons list box on the right is cleared, and only individually selected items in the options list on the left are totaled.

A better design would toggle the add-on prices: If the user deselected an item in the left-hand vacation options list box, the price displayed in the add-ons list box would simultaneously be subtracted.

Employing a user-defined type is one means to gain such functionality. You could create a type called `VacationOption` that would hold data about the vacation add-on amount and its current selection status. You could place the following declaration in a standard module:

```
Public Type VacationOption
  Amount as Currency
  SelStatus as Boolean
End Type
```

This data structure will be used to help track the selection status of vacation add-on options.

1 Open Exercise 16.8 if you have completed it, and save it and its files under a new name, such as **p17-6.vbp**. (Otherwise, complete Exercise 16.8 first.)

2 Add a standard module to the program, and make the preceding declaration for the `VacationOption` user-defined type.

3 Make the following declaration at the form module level:

```
Dim EuroChoice(0 To 4) As VacationOption
```

4 Change the code in the `lstOptions_Click()` event to read as follows:

```
Dim amt As Currency
Dim n As Integer

If EuroChoice(lstOptions.ListIndex).SelStatus = True Then
' Remove the item
   For n = lstPrice.ListCount - 1 To 0 Step -1
      If Val(lstPrice.List(n)) =
      ➥Val(EuroChoice(lstOptions.ListIndex).Amount)
      Then
         lstPrice.RemoveItem n
         Exit For
      End If
   Next n
' change the selection status

   EuroChoice(lstOptions.ListIndex).SelStatus = False
   Exit Sub
End If

Select Case lstOptions.ListIndex
   Case 0
      amt = 455.55
      lstPrice.AddItem Str$(amt)
   Case 1
      amt = 630.75
      lstPrice.AddItem Str$(amt)
   Case 2
      amt = 120.25
      lstPrice.AddItem Str$(amt)
   Case 3
      amt = 330.33
      lstPrice.AddItem Str$(amt)
   Case 4
      amt = 765.45
      lstPrice.AddItem Str$(amt)
```

continues

```
End Select
' assign the amount to the array element
EuroChoice(lstOptions.ListIndex).Amount = amt
' toggle the selection status of the array element
EuroChoice(lstOptions.ListIndex).SelStatus = True
```

This code might be a bit difficult to follow. Look at the last two instructions. Assignments are made to the array `EuroChoice` declared at the form module level. Notice that the list box index value is used to identify the array element. Because you are using list box `ListIndex` properties to identify your array elements, you must use a zero-based array.

At the beginning of the procedure, a test is made of the array element's selection status. If the item has been selected, a loop is entered that tests whether values held in the array equal the value of the prices listed in the list box. If so, the item is removed, the selection status is set to `False`, the loop is exited, and the procedure is exited. If the item has not been previously selected, the item is added to the `lstPrice` list box, and the array element in `EuroChoice` is assigned the appropriate values.

Chapter Summary

An array with more than one dimension is called a *multidimensional array.* Declare a multidimensional array by placing a comma between the dimensions, as follows:

```
ReDim MyTable (1 to 10,1 to 10) As Integer
```

A *multi-column* list resembles a two-dimensional array; however, it functions like a simple list. To create this type of list, change the number of columns to the number desired.

The *MSFlexGrid control* places a two-dimensional array on the display screen and looks like a table or a spreadsheet. When using the MSFlexGrid control, fixed rows and columns are used for adding grid headings; nonfixed rows and columns are used to store grid data. String data can be added to cells in two ways:

- Identified individual cells with the `Cell` and `Row` properties; then, the `Text` property can be assigned a new value or read.
- Add tab-delimited strings to the MSFlexGrid control with the `AddItem` method. The `ColSel` and `RowSel` properties allow a range of cells to be selected.

User-defined types can be placed in arrays with declarations, such as the following:

```
Dim Inventory(1 To 100) As Asset
```

This allows a table of mixed data types to be created, providing the programmer a very useful tool for maintaining related data together without creating a class.

Skill-Building Exercises

1. Use a nested loop and a two-dimensional array to display the following pattern on the screen after a `Form_click()` procedure:

```
0 1 0 0 0
0 1 1 0 0
0 1 1 1 0
0 1 1 1 1
0 1 1 1 1
```

Name the array **`num()`**. Use row and column array values in your solution.

2. Use a two-dimensional array named **`scores()`** to store the bowling team scores for four players. Assign the following scores to each player:

```
                    Player
Round     1         2           3          4
No. 1     140       150         220        160
No. 2     160       220         230        120
No. 3     165       200         230        140
Avg.      155       190         226        140
```

Use a nested loop to produce the table, including the average score for each player. Use a `Form_click()` procedure to display the results. Do not store the average in the array.

3. Modify the preceding exercise by using an MSFlexGrid to record player scores. In your design, add the words **Round 1**, **Round 2**, **Round 3**, and **Average** to a set of fixed rows; add **Player 1**, **Player 2**, **Player 3**, and **Player 4** to a set of fixed columns. Write a `Grid1_MouseDown()` event that will enable you to use the right mouse button to enter player scores. When the user right-clicks, open an input box asking the user to enter a score. After all scores have been entered, compute the average score for each player.

4. This exercise and the next ask you to design a race car speedway. This initial exercise asks you to race two cars by setting the octane mixtures for each car by using two scroll bars. A start button begins the race. The first car runs the length of the track, and the time of the run is displayed. The second car then runs the length of the track, and its time is displayed. A checkered flag appears for the car with the winning time.

In your design, use a picture box for the track, a Line control to separate the two lanes on the track, pictures of cars, and a checkered flag embedded in Image controls. Use `DoEvents` to ensure that each car completes a race before the second car starts.

5. Attach an MSFlexGrid to the race car speedway of the preceding exercise to illustrate the time of each car, whether the time for the car represents the driver's best, the course record, and the number of races won by a car. If a car beats the course record, display this fact with a message box.

18

Numeric Functions

Besides event and general procedures—namely those that begin with such statements as `Private Sub` and end with `End Sub`—it is also possible to write function procedures and to use built-in functions supplied by Visual Basic. Both function and built-in function procedures return a single value. They differ from general procedures that do not return a value.

Function procedures are written by the Visual Basic designer, much like general procedures. Numeric built-in functions deal with ways of manipulating and converting numbers. For example, the built-in `Abs()` function is provided to return the *absolute value* of a number.

This chapter shows you how to write a Visual Basic function procedure, and introduces a number of built-in numeric functions. (The next chapter examines string functions.) Functions often simplify the writing of a Visual Basic program. Instead of writing a procedure, the Visual Basic programmer is able to use a numeric function to compute a value using one or more mathematical formulas and return the value of that computation. When you complete this chapter, you will be able to

- call a numeric function,
- write a numeric function procedure,
- write a recursive function procedure,
- use the built-in numeric conversion functions,
- use the built-in mathematical functions,
- use the built-in financial functions.

Chapter Objectives

The objectives of this chapter address how numeric functions work. To begin, you learn how to create a function, and you discover how a function differs from a procedure. You then are introduced to a special type of function called a *recursive function.* This is a function that calls itself. Finally, you examine the use of built-in numeric functions. These are provided by the Visual Basic language. The four chapter objectives are as follows:

1. Understanding how numeric functions work
2. Writing numeric function procedures
3. Writing a recursive function
4. Using built-in numeric functions

Objective 18.1 Understanding How Numeric Functions Work

The syntax for a Visual Basic function is similar to the syntax for a Visual Basic procedure. The following is the simplified syntax:

```
[Public | Private] [Static] Function functionname (arguments) [As type]
   function statements
End Function
```

There are differences between a function and a general procedure:

- The `Function` keyword identifies the beginning of the function; the `End Function` keywords mark the end of the function.
- Arguments are typically required by a function. These argument values are most often passed from a procedure to a function.
- The `[As type]` part of the declaration syntax specifies the type of data to be returned by a function. If a function returns a value of type `Integer`, this type should be indicated.

As an example, consider the event procedure:

```
Private Sub Form_Click()
   Dim Length As Integer
   Dim Width As Integer
   Dim Height As Integer

   Length = 4
   Width = 6
   Height = 5
   Label1.Caption = cubeIt(Length, Width, Height)
End Sub
```

The `cubeIt(Length, Width, Height)` statement represents a function call. It calls a function with the same name of `cubeIt()`. The function call passes the function three values: a value for the length, a value for the width, and a value for the height. Each argument is separated by a comma.

The function `cubeIt()` is written as follows:

```
Public Function cubeIt(Length, Width, Height) As Integer
   cubeIt = Length * Width * Height
End Function
```

The function computes the cube as 120 and returns this integer value to the general procedure. The returned value is assigned to `Label1.Caption`.

Why write a function as simple as `cubeIt()` and add it to a program? A function such as this will simplify a program if the function `cubeIt()` is called at different times in several different places in a program, or if the function simplifies the writing of the procedure.

18.1.1 Common Mistakes

A common mistake is to write the preceding procedure as follows:

```
Private Sub Form_Click()
   Dim Length As Integer
   Dim Width As Integer
   Dim Height As Integer

   Length = 4
   Width = 6
   Height = 5

   cubeIt(Length, Width, Height)
   Label1.Caption = cubeIt
End Sub
```

This leads to a syntax error. The function call, because it returns a value, must assign that value to a variable in the program, as in the following code:

```
Label1.Caption = cubeIt(Length, Width, Height)
```

Another common mistake is to write the calling statement as follows:

```
Label1.Caption = cubeIt(Length * Width * Height)
```

This will lead to an argument-count-mismatch error because the addition of the multiplication operators (instead of the commas) shows one value being passed to the function instead of three.

A third mistake is to write a function call to a procedure rather than a function:

```
Public Sub cubeIt(Length, Width, Height)
   cubeIt = Length * Width * Height
End Sub
```

This procedure will lead to a compiler error because the function call statement specifies a call to a function, not a procedure:

```
Label1.Caption = cubeIt(Length, Width, Height)
```

18.1.2 The Expanded *Function* Syntax

The initial `Function` syntax described at the beginning of this objective is actually an abbreviated one. The expanded syntax is as follows:

```
[Public | Private] [Static] Function functionname (argument list) [As type]
   [statement block]
   [functionname = expression]
   [Exit Function]
   [statement block]
   [functionname = expression]
End Function
```

The argument list can be expanded to read as follows:

```
[ByVal]variable[()][As type] [,[ByVal]variable[()] [As type]]
```

Table 18.1 explains the new terms in this syntax.

Table 18.1 The Expanded *Function* Syntax

Term	*Description*
`Static`	Signifies that the function's local values are retained between function calls.
`Private`	Indicates that the `Function` procedure is accessible only to other procedures in the module in which it exists.
`Public`	Indicates that the `Function` procedure is accessible to procedures in other modules.
`Exit Function`	Causes an immediate exit from the procedure.
`ByVal`	Indicates that the argument is passed by value rather than by reference. *Passed by value* means that a copy of the value is passed rather than the address (in memory) of the value. Use of the `ByVal` keyword is reviewed in Chapter 22, "Modular Design."

Even though `As type` in this syntax is optional, a good programming practice is to specify the type, as in the following example:

```
Public Function cubeIt(Length As Integer, Width As Integer, Height As Integer)
➥As Integer
```

This revised declaration indicates that the function will receive three values, all type `Integer`, and return a single value of type `Integer`.

Objective 18.2 Writing Numeric Function Procedures

Visual Basic contains a number of built-in functions, and libraries of functions also can be purchased. Even so, there will be times when function procedures are written by a programmer.

A common use of function procedures is to conduct an analysis of a set of numbers to return a computed value or to determine a characteristic of the set. For example, in Exercise 18.1, you are asked to use a function to return a runner's pace in minutes per mile. Following this, you are asked to write functions to determine the highest and lowest values in a set of numbers.

Exercise 18.1 Writing a Simple Function Procedure

Design a program called **Convert Race Results**, as shown in Figure 18.1. The program requires you to check either a 10 kilometer run/walk or an 8 kilometer run/walk. You are then asked to enter the time required to complete the run/walk in minutes and seconds. Given this information, the program computes your pace in *minutes per mile.*

Figure 18.1

This design asks the user to click the type of race and enter the time in minutes.

1 Add the option buttons (place in a frame), the labels, the text box, and the command button on a form.

2 Add a `txtTime_KeyPress()` procedure. Get the time from the text box and call a function, named `mph()`, to compute and return the pace of the walker/runner. Display this pace. This code is as follows:

```
Private Sub txtTime_KeyPress (KeyAscii As Integer)
   Dim result As Single
   Dim newTime As Single
   If KeyAscii = vbKeyReturn Then
      newTime = CSng(txtTime.Text)
      result = mph(newTime)
      lblPace.Caption = Str$(result)
   Else
      Exit Sub
   End If

End Sub
```

With this code, the Return key on the keyboard must be pressed to compute the pace of the runner (walker). The function `mph()` is called, and the computed value is assigned to the variable result.

3 Write the function `mph()`. This function receives the time required to walk or run the distance, and checks to determine whether the run/walk was 8 kilometers or 10 kilometers and greater than 0.

continues

```
Function mph (newTime as Single) as Single
   If (opt10K = True And newTime > 0) Then
      mph = CSng(newTime / 6.2)
   ElseIf (opt8K = True And newTime > 0) Then
      mph = CSng(newTime / 5.1)
   End If
   txtTime.SetFocus
End Function
```

This function determines if the choice is 10 kilometers or 8 kilometers and if the time is greater than 0. If it is, then the miles per hour is computed and returned to the calling procedure.

4 Write a `cmdClear()` procedure to allow you to try again before deciding to end.

```
Private Sub cmdClear_Click()
   lblPace.Caption = ""
   txtTime.Text = ""
   txtTime.SetFocus
End Sub
```

5 Test this procedure. Select the type of race you want to run, and indicate your time in minutes to complete the race.

Exercise 18.2 Returning the Highest and Lowest Values

This exercise asks you to create an array and to search the array for the highest and lowest values. Two functions are utilized in performing this search. One function searches the array for the highest value. After it is found, this value is returned. A second function searches the array for the lowest value. After it is found, this value is returned. Figure 18.2 shows the form to create.

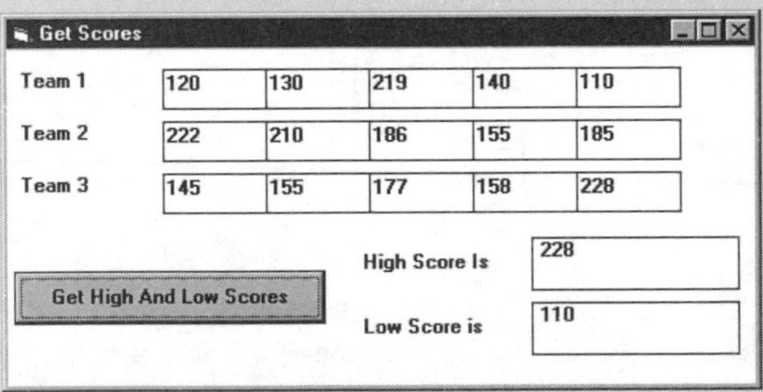

Figure 18.2

Enter all team scores in the array before pressing the Get High And Low Scores button.

1 Write a program that allows you to enter 15 scores for three teams. Design the interface, using an array of text boxes. Press the Tab key to move from one cell to the next in the table. This will allow you to enter scores one row at a time.

2 Write a `cmdScores()` procedure. This procedure (which follows) contains two function calls: one to a function named `High_Score()`, which computes and returns the highest of all scores, and another to a function named `Low_Score()`, which computes and returns the lowest of all scores.

```
Private Sub cmdScores_Click()
   Static score(0 To 14) As Integer
   Dim num As Integer
   Dim High As Integer
   Dim Low As Integer
   For num = 0 To 14
      score(num) = Val(txtScore(num).Text)
   Next num
   High = High_Score(score(), num - 1)
   lblHigh.Caption = Str$(High)
   Low = Low_Score(score(), num - 1)
   lblLow.Caption = Str$(Low)
End Sub
```

3 Write a `High_Score()` function. This function searches the array of scores to determine and return the highest value:

```
Private Function High_Score (score() As Integer, num As Integer)
   Dim max As Integer
   Dim High As Integer
   High = 0
   For max = 0 To num
      If (score(max) > High) Then
         High = score(max)
      End If
   Next max
   High_Score = High
End Function
```

4 Write a `Low_Score()` function. This function is similar to the `High_Score()` function, but returns the low score:

```
Private Function Low_Score (score() As Integer, num As Integer)
   Dim min As Integer
   Dim Low As Integer

   Low = 1000
   For min = 0 To num
      If (score(min) < Low) Then
         Low = score(min)
      End If
   Next min
   Low_Score = Low
End Function
```

continues

5 Run and test your design. As one test, have duplicate high and low scores. Does this lead to problems?

Several features of this exercise merit special emphasis. First, examine the function declarations. Function arguments that expect to receive an array must be declared with a set of parentheses. Procedures that pass an array to the function must also pass them with a set of parentheses.

Second, observe that along with the array, the array's upper bound is also passed. This permits the `Low_Score()` and `High_Score()` functions to know how many iterations of the `For` loop to execute. In this case, you use the counter variable `num`. Notice that the upper bound of the array is passed as `num-1`. Why? Because the last iteration of the `For...Next` loop in `cmdScores_Click()` increments the `num` counter variable so that it equals `15` at the end of the loop execution. By decrementing it by `1`, you can pass along the upper bound of the array.

Alternatively, you could have passed the score array to the `High` function with the following assignment expression:

```
High = High_Score(score(), UBound(score)).
```

Notice also how the variables `High` and `Low` are initialized in the `High_Score()` and `Low_Score()` functions. They are each assigned some absurd value. `High` is assigned the value `0`, and `Low` is assigned the value `1000`. These values are replaced in the first assignment in the execution of the function `For` loops. At the end of the loop's execution, the high and low values in the array are respectively determined.

Objective 18.3 Writing a Recursive Function

In the preceding exercises, you have seen procedures call functions. Functions, themselves, can call other functions. Functions can even call themselves. This is known as *recursion.* Recursive functions provide elegant solutions to several kinds of computing problems, including some complex sorting algorithms.

Some computer languages, such as Prolog, are designed to use recursion extensively. Recursive functions can, however, be very difficult to debug. We demonstrate a recursive function to determine the factorial of a number. You are asked to write this example in the first exercise in the "Skill-Building Exercises" section at the end of this chapter.

In the following procedure, the function call statement is `Label1.Caption = fact(n)`.

```
Private Sub Form_Click()
   'Program to determine the factorial of a number using
   'recursion
   Dim n As Integer
   n = 4
   Label1.Caption = fact(n)
End Sub
```

The corresponding `Function` procedure is passed the value of `n`, and is written as follows:

```
Public Function fact(n As Integer) As Integer
   If (n <= 1) Then
      fact = 1
   Else
      fact = n * fact(n - 1)      ' fact() calls itself
   End If
End Function
```

The calling statement passes control to the function `fact()`. The function uses a test condition to determine whether `n` is equal to `1`. This condition will eventually end the recursion. In the preceding example, because `n` is equal to `4`, the `else` portion of the statement executes with the following result:

```
fact = 4 * fact(3)
```

To evaluate this statement, `fact()` is called again, leading to the following statement:

```
fact = 3 * fact(2)
```

This is followed by:

```
fact = 2 * fact(1)
```

At this point, `n` is equal to `1`, and the recursive calls stop. However, the calls must then be resolved. The following are the unresolved calls:

```
(a)  fact = 4 * fact(3)
(b)  fact = 3 * fact(2)
(c)  fact = 2 * fact(1)
```

Knowing that `fact(1)` is equal to `1` and remembering that a function has static scope, this stack of unresolved calls can be resolved as follows:

```
fact(1) = 1
fact(2) = 2
fact(3) = 6
```

This leads to the answer `24`, which is determined by the expanded equations:

```
(a) fact = 2 * 1 =  2
(b) fact = 3 * 2 =  6
(c) fact = 4 * 6 = 24
```

This is the same as 4! (four factorial).

Objective 18.4 Using Built-in Numeric Functions

Visual Basic contains a number of built-in numeric and string functions (the latter are addressed in Chapter 19, "String Functions"). The three types of numeric functions discussed in the remainder of this chapter are

Numerical conversion

Mathematical

Financial

18.4.1 Numeric Conversion Functions

Numerical conversion functions convert one numeric data type to another (see Table 18.2).

Table 18.2 Numeric Conversion Functions

Function	*Description*
`Int()`	Returns the integer value of the number entered as the argument. With a negative number, it returns the first integer less than, or equal to, the number.
`Fix()`	Returns the integer value of the number entered as the argument. With a negative number, it returns the first integer greater than, or equal to, the number.
`CInt()`	Returns the rounded integer closest to the value of the argument.
`CLng()`	Returns the rounded long integer closest to the value of the argument.
`CSng()`	Converts the argument to a single-precision number.
`CDbl()`	Converts the argument to a double-precision number.
`CCur()`	Converts the argument to a number of type `Currency`.
`CStr()`	Converts the argument to a string.
`CVar()`	Converts the argument to a value of type `Variant`.

As an example of the first two numerical conversion functions, if the argument is `-8.6`, `Int()` will return `-9`, while `Fix()` will return `-8`.

Of the seven remaining types, `CCur()` is a most useful function when a value computed as type `Double` or `Single` needs to be expressed as type `Currency`. The following statements explicitly state that a value of type `Currency` is to be assigned to the variable `SalesTax`:

```
TaxRate = 0.082
SalesTax = CCur(Cus_Sale * TaxRate)
```

At times, this process of transforming data explicitly is known as *casting the data to a type.* The `CCur()` function in the preceding example casts the product of `Cus_Sale` and `TaxRate` to type `Currency`.

18.4.2 Mathematical Functions

Other, perhaps more common, mathematical functions provided with Visual Basic are shown in Table 18.3. These differ from the previously described numeric functions, which are designed to handle rounding of numbers or casting to an explicit data type.

Table 18.3 Mathematical Functions

Function	*Description*
`Exp()`	Returns the base of the natural logarithm (e) of the argument raised to specified power.
`Log()`	Returns the natural logarithm of the argument.
`Sqr()`	Returns the square root of the argument.
`Randomize()`	Seeds all random numbers by using the number entered as the argument.
`Rnd()`	Returns a random number determined by the computer.
`Abs()`	Returns the absolute value of the argument.
`Sgn()`	Returns the sign of the value of the argument.
`Atn()`	Returns the arctangent of the argument in radians.
`Cos()`	Returns the cosine of the argument in radians.
`Sin()`	Returns the sine of the argument in radians.
`Tan()`	Returns the tangent of the argument in radians.

Exercise 18.3 Taking the Square of a Number and Casting It to Type *Integer*

This brief exercise asks you to enter a number and take the square root of the number, then cast the number as an integer before it is displayed. Figure 18.3 shows the form to create.

1 Add the labels and the text box as shown.

continues

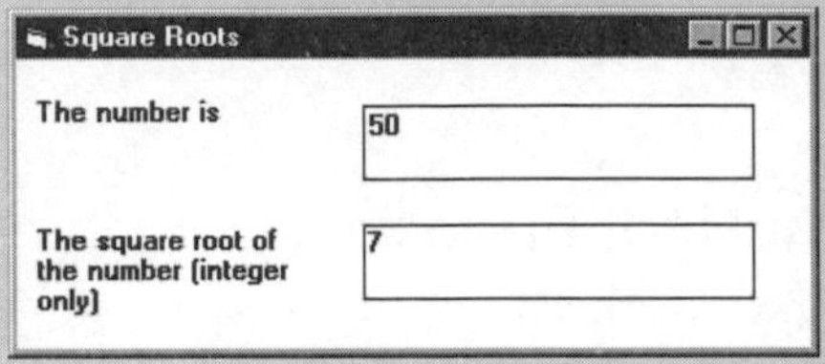

Figure 18.3

The square root of a number makes use of the `Sqr()` *function.*

2 Write a `txtNumber_Change()` procedure. Compute the square of a number entered in `txtNumber.Text`. Following this, use `CLng()` to cast the number to type `long`. Display the number in `lblSquare`.

```
Private Sub txtNumber_Change()
   Dim number As Long
   'After taking the square root, cast the number as type long

   number = CLng(Sqr(Val(txtNumber.Text)))
   lblSquare.Caption = number
End Sub
```

18.4.3 Generating Random Numbers

Although most of the mathematical functions are easy to understand, two—`Rnd()` and `Randomize()`—are somewhat unusual and require additional clarification. In many situations, these two functions are used in pairs. The syntax for `Rnd()` is as follows:

```
Num = Rnd [(number)]
```

`Rnd()` returns a number of type `Single` between `0` and `1`, and the value for `number` determines how `Rnd()` generates the random number. The relationship between `number` and the value returned is as follows:

Value of `number`	*Value Returned*
`< 0`	The same value every time, as determined by `number`
`> 0`	The next random number value in the sequence
`= 0`	The value most recently generated

To have the program generate a different random number every time a procedure is run, use the `Randomize` statement without an argument. Simply write the following:

```
Randomize      'Place this before Rnd()
```

This will initialize the random number generator before `Rnd()` is called.

Because the random number generator returns a value between `0` and `1`, it is often necessary to convert this number to a larger or smaller number. To convert the number to a range of integers, such as 1 to 6, write the following statement:

```
Int((Rnd * 6) + 1)
```

This converts the random number to one of six integers (1, 2, 3, 4, 5, or 6).

Exercise 18.4 Using the *Randomize* Statement to Roll a Die

This exercise asks you to design a program to roll a die and show the result using a picture of a die. Figure 18.4 shows the form with which you are asked to work. With this picture box and seven circles, you can display the die markers 1, 2, 3, 4, 5, and 6.

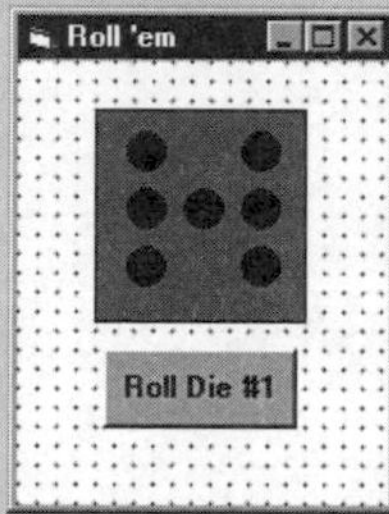

Figure 18.4

A single die can be used to display the numbers 1 to 6.

1 Design the form as shown in Figure 18.4. Use a picture box and create an array of circle shapes, numbered from 0 to 6.

2 Initialize the die to make all shapes invisible to begin. This code is as follows:

```
Private Sub initialize()
   'Make the die invisible
   pic1.Visible = False
   'Make the die markings invisible
   shape1(0).Visible = False
   shape1(1).Visible = False
   shape1(2).Visible = False
   shape1(3).Visible = False
   shape1(4).Visible = False
   shape1(5).Visible = False
   shape1(6).Visible = False
End Sub
```

continues

3 Write the `cmdRoll_Click()` procedure.

```
Sub cmdRoll_Click()
   Dim die1 As Integer

   'Set seed
   Randomize
   die1 = Int((Rnd * 6) + 1)
   Print die1

   Call initialize
   pic1.Visible = True
   Select Case die1
      Case 1
         shape1(3).Visible = True
      Case 2
         shape1(0).Visible = True
         shape1(6).Visible = True
      Case 3
         shape1(0).Visible = True
         shape1(3).Visible = True
         shape1(6).Visible = True
      Case 4
         shape1(0).Visible = True
         shape1(2).Visible = True
         shape1(4).Visible = True
         shape1(6).Visible = True
      Case 5
         shape1(0).Visible = True
         shape1(2).Visible = True
         shape1(3).Visible = True
         shape1(4).Visible = True
         shape1(6).Visible = True
      Case 6
         shape1(0).Visible = True
         shape1(1).Visible = True
         shape1(2).Visible = True
         shape1(4).Visible = True
         shape1(5).Visible = True
         shape1(6).Visible = True
   End Select
End Sub
```

4 Test this design. Run the exercise several times to determine whether you display the same or a different set of random numbers.

Exercise 18.5 Generating a Set of Random Numbers and Displaying Them on a Grid

This exercise asks you to use the MSFlexGrid control to create a table of rows and columns and to fill the rows and columns with random numbers in the range of 1 to 50. Create the interface shown in Figure 18.5.

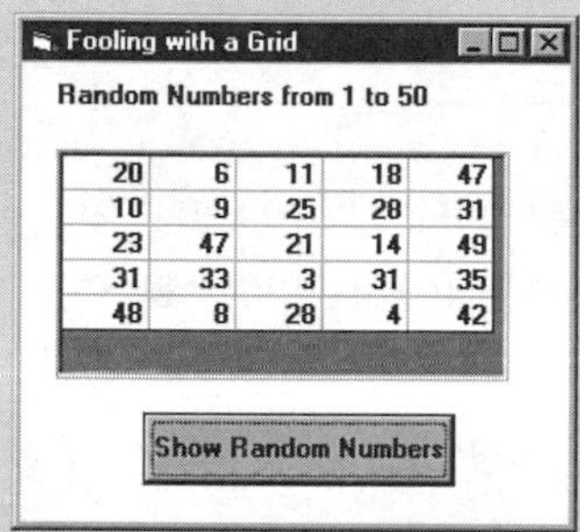

Figure 18.5

This grid will display a different set of random numbers each time the program is run.

1 For the Grid property dialog box, set the number of rows and columns at 5, turn off scroll bars, and set the number of fixed rows and fixed columns to zero. Name the grid **Grid1**.

2 Write the `cmdShow_Click()` procedure, and use a nested `For...Next` loop. Within the inner loop, get the first row value; then, get a random number and assign it to the first row and column intersection, and assign it to `Result`. Use the following code:

```
Private Sub cmdShow_Click()
   Static Result(0 To 4) As Integer
   Dim Row As Integer
   Dim Column As Integer

   For Row = 0 To 4             'Start at top row
      Grid1.Row = Row
      For Column = 0 To 4
         Randomize
         Result(Column) = Int(50 * Rnd + 1)
         Grid1.Col = Column
         Grid1.Text = Result(Column)
      Next Column
   Next Row
End Sub
```

3 Run and test this design. As one test, generate a different set of random numbers, such as from 1 to 40.

18.4.4 Financial Functions

Besides common numerical functions, Visual Basic provides a set of specialized financial functions. These are most useful when working with interest and its effect on the value of money. Table 18.4 shows financial functions. The first six are used when working with annuities, and the last four deal with cash flow.

Table 18.4 Financial Functions

Function	*Description*
`FV()`	Returns the future value of an annuity based on periodic, constant payments and a constant interest rate over a specified period.
`IPmt()`	Returns the interest payment for a given period of an annuity based on periodic, constant payments and a constant interest rate over a specified period.
`NPer()`	Returns the number of periods for an annuity based on periodic, constant payments and a constant interest rate.
`Pmt()`	Returns the payment for an annuity based on periodic, constant payments and a constant interest rate over a specified period.
`PPmt()`	Returns the principle payment for an annuity based on periodic, constant payments and a constant interest rate over a specified period.
`Rate()`	Returns the interest rate per period for an annuity.
`NPV()`	Returns the net present value of an investment based on periodic and variable cash flows (payments and receipts) and a discount rate. This function requires an array of cash flows.
`PV()`	Returns the present value of an investment based on a periodic and constant cash flow (payments and receipts) and a discount rate.
`IRR()`	Returns the internal rate of return for a series of periodic cash flow (payments and receipts).
`MIRR()`	Returns the modified internal rate of return for a series of periodic cash flow (payments and receipts).

The financial functions all require several arguments. The arguments and their meanings are found by using the Visual Basic online Help under the heading "Financial Functions." In computing the payment for a house, for example, the payment function is needed. The syntax for this function is as follows:

```
Pmt (rate, nper, pv, fv, type)
```

Table 18.5 shows the components of the syntax.

Table 18.5 The *Pmt()* Function Syntax

Component	*Description*
`rate`	The interest rate per period. If an annual rate is specified, such as 6 percent, the percentage must be divided by 100 (to make it .06) and by 12 (to reduce it to .005). This reflects the monthly payment percent.

Component	*Description*
nper	The number of payments. If years are specified, this number must be multiplied by 12 to determine the number of months.
pv	The present value of the loan. If you want to borrow $50,000 today, the present value is `-50000` (a negative number).
fv	The future value of the loan. Because you want to pay off your home (eventually), the `fv` is 0.
type	When the payment is due. Use `1` for the beginning of a month and `0` for the end of the month. End of month payments are assumed.

Exercise 18.6 Using Financial Functions to Determine Your New House Payment

Suppose you want a simple design to show you how large your house payments will be. Create the design shown in Figure 18.6.

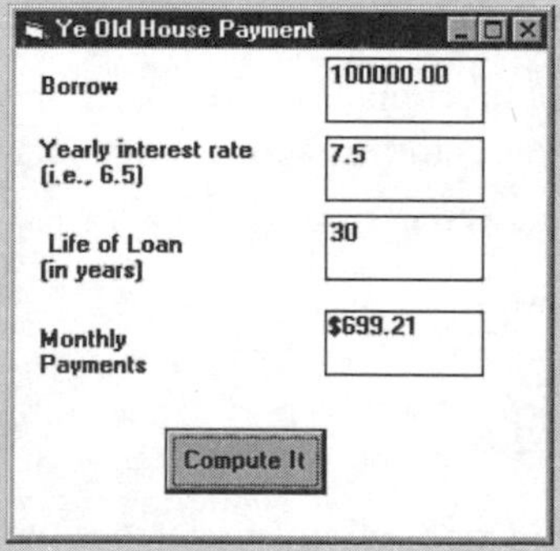

Figure 18.6

Add the amount of the loan, the interest rate, and the life of the loan to compute the monthly payment.

1 Add the three text boxes and labels (including one for the computed result), as shown in Figure 18.6. Name the text boxes, **`txtBorrow`**, **`txtRate`**, and **`txtYears`**. Name the results label **`lblPayment`**, and add a border to it. Add the command button, **`cmdCompute`**.

continues

2 Write the `cmdCompute()` procedure as follows:

```
Private Sub cmdCompute_Click()
   Dim yearIRate As Single
   Dim nperiods As Integer
   Dim borrow As Currency
   Dim payment As Currency

   yearIRate = CSng(txtRate.Text) / (12 * 100)
   nperiods = CInt(txtYears.Text) * 12
   borrow = CCur(txtBorrow.Text)

   payment = Pmt(yearIRate, nperiods, -borrow, 0, 0)

   lblPayment.Caption = Format(payment, "Currency")
End Sub
```

3 Run and test your program. Get the amount you want to borrow, the yearly interest rate, and the number of years of the loan. Compute the monthly payment.

Chapter Summary

This chapter describes the use of functions. Unlike procedures, a function must return one and only one value. The following is the expanded function syntax:

```
[Public | Private] [Static] Function functionname (argument list) [As type]
   [statement block]
   [functionname = expression]
   [Exit Function]
   [statement block]
   [functionname = expression]
End Function
```

When writing a function, the function call must reference the function name and contain the same number of arguments as the argument list of the function. The following code is a function call, containing three arguments:

```
Label1.Caption = cubeIt(Length, Width, Height)
```

The function would be written as follows:

```
Function cubeIt(Length, Width, Height) As Integer
```

As shown, the argument list is the same; the function returns a value of type `Integer`. The return instruction is placed within the body of the function, as in the following code:

```
cubeIt = Length * Width * Length
```

This value is assigned to `cubeIt()`, which in turn assigns the value to the caption in the calling statement.

Certain categories of functions can call themselves. These are known as *recursive functions.* They call themselves repeatedly until a condition is met; then, the multiple calls to the function are resolved.

Visual Basic contains a number of built-in functions. These include numeric conversion functions, mathematical functions, and financial functions. Numeric conversion functions enable the transforming of data explicitly. This is known as *casting to a type.* For example, `CCur()` will cast a value to type `Currency`.

Mathematical functions include `Rnd()` and `Randomize()`, the two functions needed in creating a unique set of random numbers. To create such a set, place the `Randomize` statement in front of the `Rnd()` function.

Financial functions provide specialized ways of analyzing money. Six financial functions are available when working with annuity problems, and four are available for use with cash-flow problems.

Skill-Building Exercises

1. Write a program to test the `cubeIt()` and `fact()` functions discussed in Objective 18.1, "Understanding How Numeric Functions Work," and Objective 18.3, "Writing a Recursive Function," respectively.
2. Write a program to call a function to compute a set of 10 random numbers between 1 and 100. After these are displayed, call the function a second time to compute a second *and different* set of 10 random numbers from 1 to 100.
3. Create a design to allow the user to enter a number from 2 to 8 and to click one of the following to raise the number to a power: square, cube, fourth, fifth, sixth, seventh, eighth. Display the result using a label. As an example, the user should be able to display "2 raised to the 8th power."
4. This exercise asks you to use a built-in numeric function to determine what your savings will be if you decide to deposit a fixed amount of money each month. You must enter the amount you want to save each month, the expected rate of interest, and the number of years you plan to save. Display the result using a label.
5. You have just won the lottery! Enter the amount you will be paid each year, the number of years you are to be paid, the yearly rate of inflation. Then, determine what that amount would be worth if you were paid a lump-sum amount today. Use a built-in numeric function. Display your result using a label.

19

String Functions

String functions deal with ways of manipulating and converting strings. For example, the `Val()` function is used to translate a string into a number, while the `Str$()` function converts a number into a string. The `Format$()` function is used to format numbers, dates, and times.

In this chapter, we consider several other types of string functions besides `Val()`, `Str$()`, and `Format$()`. Because strings are used so often in Visual Basic programming, it is important to know how they can be inspected, converted, manipulated, and compared. Built-in string functions are used for this purpose.

You learn many tasks in this chapter, after which you will be able to

- determine the length of a string;
- convert a string from upper- to lowercase and lower- to uppercase;
- convert a string to its numeric code, or an ANSI code (formerly ASCII code) to a character;
- convert a string to a specified data type;
- work with functions that return part of a string;
- use statements to modify strings;
- return a substring from a larger string;
- compare strings with the `Like` operator;
- use the string compare function;
- sort strings in ascending order;
- use the `String()` and the `Space()` functions.

Chapter Objectives

The objectives for this chapter are based on the different categories used to inspect, convert, manipulate, and compare strings, as follows:

1. Determining the length of a string
2. Understanding string conversion categories
3. Manipulating strings
4. Modifying strings
5. Comparing and sorting strings
6. Constructing strings

Objective 19.1 Determining the Length of a String

It is often necessary to inspect a string to determine its length. The function for doing this is the `Len()` function. The syntax for `Len()` is simply the following:

```
Len (string)
```

The following instructions display a value of `5`:

```
First$ = "Smith"
Print Len(First$)
```

The `Len()` function is often used to test whether string variables or object properties have been assigned values. Testing for the length of a string, for instance, is one way to determine whether the user has entered data into a text box. Consider the following procedure:

```
Private Sub txtInput_KeyPress(KeyAscii As Integer)
   Dim answer As String

   If KeyAscii = vbKeyReturn Then
      If Len(txtInput.Text) <> 0 Then
         answer = txtInput.Text
         Call testanswer(answer) ' procedure to test user input
      Else
         Print "Please input data."
      End If
   End If
End Sub
```

By using a simple test of string length, this code helps ensure that required data input is received before processing continues. The `Len()` function, combined with other string functions, can help make your application less subject to user input errors.

Objective 19.2 Understanding String Conversion Categories

String conversion categories serve to convert a string from one form to another. There are several categories of string conversion, including the following:

- Converting the case of a string from upper- to lowercase and lower- to uppercase. This is known as converting the *case* of a string.
- Converting a string to its numeric code or converting a numeric code to a string. This is often known as converting a character to an ANSI number or an ANSI code to a character.
- Converting a string to a specified data type.

In Visual Basic, there are often two versions of functions that return strings: One version returns a `Variant` of type string; the other adds a dollar sign (`$`) suffix to the function and explicitly returns a string. We usually employ functions that append the dollar sign. Using this version is more resource efficient, and it makes certain that data string conversions are taking place.

A list of functions that return strings and can carry a dollar sign follows. Some you have encountered before; many you will learn about in this chapter. These functions have the same usage and syntax as their variant versions without the dollar sign.

`Chr$()`	`CurDir$()`	`Date$()`	`Dir$()`
`Error$()`	`Format$()`	`Hex$()`	`Input$()`
`LCase$()`	`Left$()`	`LTrim$()`	`Mid$()`
`Oct$()`	`Right$()`	`RTrim$()`	`Space$()`
`Str$()`	`String$()`	`Time$()`	`Trim$()`
`UCase$()`			

19.2.1 Converting the Case of a String

Three functions are used in converting the case of a string: `LCase()`, `UCase()`, `StrConv()`. The following is the syntax for these functions:

```
LCase$ (strexpr)

UCase$ (strexpr)

StrConv (strexpr, conversion)
```

The component `strexpr` is the string expression to be converted, and `conversion` is the value specifying the type of conversion to perform. Table 19.1 shows the three `conversion` argument settings for the `StrConv()` function.

Table 19.1 The *conversion* Argument Settings for *StrConv()*

Constant	*Value*	*Description*
`vbUpperCase`	1	Converts the string to uppercase characters
`vbLowerCase`	2	Converts the string to lowercase characters
`vbProperCase`	3	Converts the first letter of every word in the string to uppercase

The instructions

```
Name$ = "Sally sue"
Print Name$
Print UCase$(Name$)
Print LCase$(Name$)
Print StrConv(Name$, vbProperCase)
```

lead to the following output:

```
Sally sue
SALLY SUE
sally sue
Sally Sue
```

19.2.2 Converting Strings to ANSI Code

Every string character is represented by an *ANSI* (American National Standards Institute) code that ranges from 65 to 90 for uppercase letters and 97 to 122 for lowercase letters. The entire ANSI code range is from 0 to 255. (See Visual Basic Help for the entire table of values, or see Appendix A, "The ANSI Character Set.") The two ANSI conversion functions are `Asc()` and `Chr$()`. `Asc()` returns the ANSI code for a string expression, while `Chr$()` returns an ANSI character. The syntax for these two functions is as follows:

```
Asc (strexpr)

Chr$ (ANSIcode)
```

As an example of the `Asc()` function, consider the following instructions:

```
Print Asc("A")
Print Asc("a")
```

These would display

```
65
97
```

As an example of the `Chr$()` function, consider the following instructions:

```
Print Chr$(65)
Print Chr$(97)
```

These would display

```
A
a
```

How is this latter function used in a Visual Basic program? A common need is to create a tab character for tab-delimited text or a variable that combines a linefeed with a carriage return. Consider the following instruction:

```
NewLine = Chr$(13) & Chr$(10)
```

`NewLine` starts on a new line because the ANSI code for `13` represents a carriage return, and the code for `10` is a linefeed. Visual Basic also provides constants for these characters (for example, `vbTab`, `vbCr`, `vbLf`, `vbCrLf`, and `vbNewLine`).

Consider the use of the `vbNewLine` constant in an instruction, such as the following:

```
FName$ = "Roger" & vbNewLine & vbNewLine
LName$ = "Rabbit"
Name$ = FName$ & LName$
MsgBox Name$
```

In the message box, the following output would appear:

```
Roger
'space'
'space'
Rabbit
```

19.2.3 Converting Strings to Numbers and Numbers to Strings

Several functions convert strings to numbers and numbers to strings. Table 19.2 lists the 10 types and their meanings.

Table 19.2 Functions for String/Number Conversion

Function	*Description*
`Val()`	Converts a string to a number
`Str$()`	Converts a number to a string
`Hex$()`	Converts a number to a hexadecimal string
`Oct$()`	Converts a number to an octal string
`CDate()`	Converts a string to type `Date`
`CInt()`	Converts a string to type `Integer`
`CLng()`	Converts a string to type `Long`
`CSng()`	Converts a string to type `Single`
`CDbl()`	Converts a string to type `Double`
`CCur()`	Converts a string to type `Currency`

Of these 10 functions, types `Val()` and `Str$()` are of the most importance because text boxes and combo boxes are used so often in getting and displaying values.

The middle two functions, `Hex$()` and `Oct$()` serve a special purpose. These are often used when manipulating memory addresses and byte values.

The last six conversion functions—`CDate()`, `CInt()`, `CLng()`, `CSng()`, `CDbl()`, and `CCur()`—are the same as those described in Chapter 18, "Numeric Functions." These *cast* functions are more explicit than `Val()`. If `Val()` cannot return a numeric value, it returns a 0. The following instructions would display `0`:

```
Number$ = "Ten"
Print Val(Number$)
```

The following instructions would generate the error `Type mismatch` because `"Ten"` is not a number:

```
Number$ = "Ten"
Print CCur(Number$)
```

Objective 19.3 Manipulating Strings

The more difficult string functions to work with are the six functions that return part of a string. Table 19.3 lists these six functions and their meanings.

Table 19.3 Functions for Returning Part of a String

Function	*Description*
`Left$()`	Returns a substring from far left
`Right$()`	Returns a substring from far right
`Mid$()`	Returns a substring
`LTrim$()`	Strips the blanks from the far left
`RTrim$()`	Strips the blanks from the far right
`TRim$()`	Strips the blanks from both ends of a string

Each of these functions contains arguments, which explains why these functions are complex.

19.3.1 The *Left$()*, *Right$()*, and *Mid$()* Functions

The syntax for the `Left$()` is as follows:

```
Left[$](strexpr, n)
```

The `strexpr` component is the string from which the substring is returned, and `n` is the number of characters to return.

For example, the following function call returns the first four characters, `What`:

```
Left$("What is your name?", 4)
```

The syntax for `Right$()` is the same as `Left$()`, with a change in the keyword:

```
Right[$](strexpr, n)
```

The following function call returns the last five characters, `name?`:

```
Right$("What is your name?", 5)
```

The syntax for `Mid$()` contains three arguments (see Table 19.4):

```
Mid[$](strexpr, start [,n])
```

Table 19.4 The Arguments for *Mid$()*

Component	*Description*
`strexpr`	The string from which the substring is returned
`start`	The character position where the substring begins
`n`	The length of the substring

The following function call returns the sixth and seventh characters (2 in total), beginning at the sixth character and displaying the word `is`:

```
Mid$("What is your name?", 6, 2)
```

If the length of the string (`n`) to return is omitted or if there are fewer than `n` characters in the text (including the character at start), all characters from the start position to the end of the string are returned. For example, consider the following statement:

```
Mid$("What is your name?", 6)
```

It will return the following:

```
is your name?
```

The beginning is the sixth character, and the last character is the last character in the string.

19.3.2 The *LTrim$()*, *RTrim$()*, and *Trim$()* Functions

The three trim functions, `LTrim$()`, `RTrim$()`, and `Trim$()`, are easy to understand. These remove the blanks from a string. The syntax for these functions are as follows:

```
LTrim[$](strexpr)

RTrim[$](strexpr)

Trim[$](strexpr)
```

As an example, consider the following code fragment:

```
message = "                This is a test."
text = Ltrim$(message)
Print text
```

It displays the following:

```
This is a test.
```

The `LTrim$()`, `RTrim$()`, and `Trim$()` functions strip blank space that is input by users. For instance, you could extend the earlier example (from Objective 19.1, "Determining the Length of a String") to test user input:

```
Private Sub txtInput_KeyPress(KeyAscii As Integer)
   Dim answer As String
   If KeyAscii = vbKeyReturn Then
      If Len(Trim$(txtInput.Text)) <> 0 Then
         answer = txtInput.Text
         Call testanswer(answer) ' procedure to test user input
      Else
         Print "Please input data."
      End If
   End If
End Sub
```

Exercise 19.1 Entering Strings and Printing Initials

This exercise asks you to enter a person's name and display the person's initials. Figure 19.1 shows how the completed form should look.

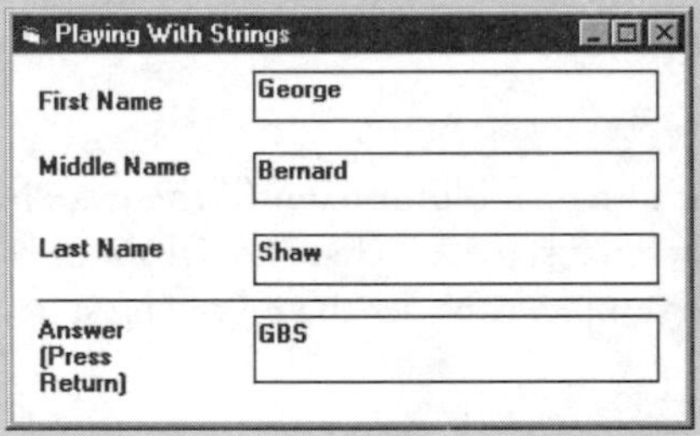

Figure 19.1

This exercise uses the `Left$()` function to display the first letter in a string.

1 Open a new project and change the `Form1` caption to **Playing with Strings**; add the four text boxes and the four labels; and use the Line tool to place a line on the form. Name the controls **`txtFirst`**, **`txtMiddle`**, **`txtLast`**, and **`txtInitials`**.

2 Write a `KeyPress()` procedure to extract the first letter from each name and add the three letters to the fourth text box.

```
Private Sub txtInitials_KeyPress (Key Ascii As Integer)
   Dim first As String
   Dim middle As String
   Dim last As String

   first = Left$(txtFirst.text, 1)
   middle = Left$(txtMiddle.text, 1)
   last = Left$(txtLast.text, 1)
   txtInitials.text = first & middle & last
End Sub
```

As indicated, `Left$()` is used to extract the first letter from each string.

3 Test your design and save this project.

Exercise 19.2 Testing Input and Displaying It in Uppercase

This exercise asks you to employ several of the newly introduced string functions. User input is tested for length and converted to uppercase for display. Figure 19.2 shows the display that you are to create.

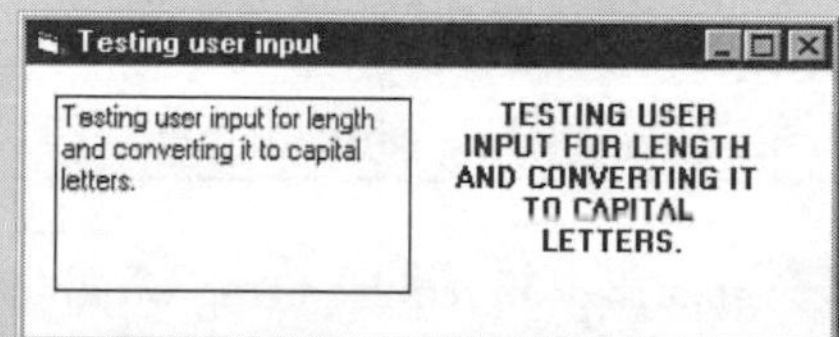

Figure 19.2

All letters typed will appear in uppercase within the label placed on the right-hand side.

1 Open a new project and change the `Form1` caption to **Testing user input**. Add a text box (**`txtInput`**) and a label (**`lblDisplay`**). Change the `MultiLine` property for `txtInput` to `True`, and the `WordWrap` property of `lblDisplay` to `True`.

2 Write a `KeyPress()` procedure for `txtInput`:

```
Private Sub txtInput_KeyPress(KeyAscii As Integer)
   Dim s1 As String
   If KeyAscii = vbKeyReturn Then
      s1 = txtInput.Text
      If Trim(Len(s1)) <> 0 Then
         s1 = UCase(Trim(s1))
         lblDisplay.Caption = s1
      Else
         lblDisplay.Caption = "Please input data"
         Exit Sub
      End If
   End If
End Sub
```

3 Test your program. What happens if you remove `Trim()` from your code and type in blank spaces before and after the typed text?

Objective 19.4 Modifying Strings

Three statements (not functions) are available to modify strings, as shown in Table 19.5.

Table 19.5 Statements for Modifying Strings

Name	*Description*
`LSet`	Used to left justify a string
`RSet`	Used to right justify a string
`Mid`	Used to merge one string with another

These statements are similar in that they all require string variables or buffers to be allocated before they can be used. They are used to replace part of one string with another. These functions have purposes other than just alignment. The `LSet` statement, for instance, can be used to prepare a string before it is saved to disk in a fixed length or random access record. This type of input/output operation is addressed in the chapters dealing with files: Chapter 20, "File-Processing Controls and Sequential File Processing," and Chapter 21, "Random-Access and Binary File Processing." These functions are also used in advanced Visual Basic to work with string addresses in calls to Windows API functions.

19.4.1 The *LSet* and *RSet* Statements

To left justify and right justify text inside a string variable, you can use `LSet` and `RSet`, respectively. The statement syntax is as follows:

```
LSet stringvar = stringexpr
RSet stringvar = stringexpr
```

The component `stringvar` is the name of a string variable, and `stringexpr` is the string expression.

As mentioned, the variable must have space allocated to handle the string expression. For instance, the following code fragment would *not* display the word `"Test"` with three leading spaces:

```
Dim s1 as String
LSet s1 = "   Test"
Print s1
```

Although declared, no space has been allocated. The following code examples would display the value of `s1`:

```
Dim s1 as String
s1 = "0123456789"
LSet s1 = "   Test"
Print s1

Dim s1 as String * 10
LSet s1 = "   Test"
Print s1
```

Observe that after using the `LSet` or `RSet` statement, text occupying the variable is replaced entirely by the newly aligned string. The leading spaces of `s1` could be removed from the display by using the `Trim()` or `LTrim()` functions.

19.4.2 The *Mid* Statement

The `Mid` statement differs from the `Mid$()` function. The `Mid$()` function, which is described in section 19.3.1, "The `Left$()`, `Right$()`, and `Mid$()` Functions," returns part of a string specified in the function arguments. The `Mid` statement replaces part of a string with another string. Like the `LSet` and `RSet` statements, space must first be allocated. The syntax for the `Mid` statement is similar to that of the `Mid$()` function:

```
Mid[$](stringvar, start [,n]) = stringexpr
```

Table 19.6 explains the syntax.

Table 19.6 Syntax of the *Mid* Statement

Component	*Description*
`stringvar`	Name of the string variable to modify.
`start`	Character position in `stringvar` where the replacement of text begins.
`n`	Number of characters to replace. If omitted, all of the string is used.
`stringexpr`	The string expression that replaces part of `stringvar`.

Consider the following code fragment:

```
message = "Rolling down the river     "
Mid$(message, 18) = "highway."
Print message
```

This displays

```
Rolling down the highway.
```

Notice that without the extra space allocated, the code above would be displayed as follows:

```
Rolling down the highw
```

Exercise 19.3 Illustrating the *LSet* Statement

In this exercise, the `LSet` statement is used to load text into a variable string buffer. The ANSI values of the loaded string are then displayed to illustrate the use of the `LSet` statement. (See Appendix A, "The ANSI Character Set," for the ANSI code.) Create the design shown in Figure 19.3.

continues

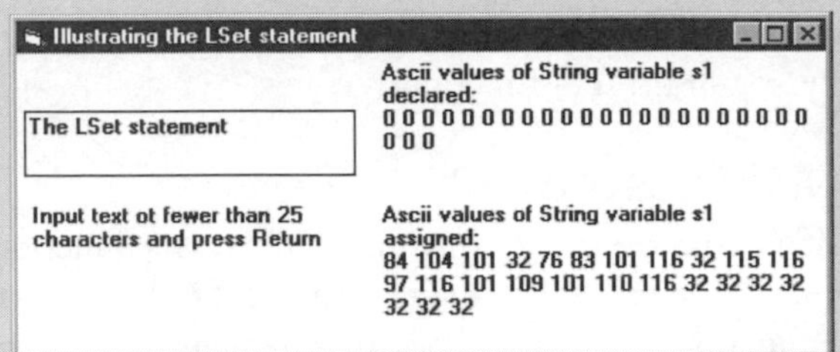

Figure 19.3

The `LSet` *statement places blank spaces (ANSI 32) into the array when the text is less than the array length.*

1 Change the caption to **Illustrating the LSet Statement**. Add a text box (**`txtInput`**) and two labels (**`lblDeclared`** and **`lblAssigned`**). Add a label with its caption set to **Input text of fewer than 25 characters and press Return.**

2 Write a `KeyPress()` procedure for `txtInput`:

```
Private Sub txtInput_KeyPress(KeyAscii As Integer)
   Dim s1 As String * 25
   Dim s2 As String
   Dim n As Integer
   If KeyAscii = vbKeyReturn Then
      For n = 1 To Len(s1)
         s2 = s2 & Str$(Asc(Mid(s1, n, 1)))
      Next
      lblDeclared.Caption = "Ascii values of String variable s1 declared: "
      ➥& vbCrLf & s2
      s2 = ""
      LSet s1 = Trim(txtInput.Text)
      For n = 1 To Len(s1)
         s2 = s2 & Str$(Asc(Mid(s1, n, 1)))
      Next
      lblAssigned.Caption = "Ascii values of String variable s1 assigned: "
      ➥& vbCrLf & s2
   End If
End Sub
```

3 Run and test the program.

In this exercise, two strings are declared: `s1`, a fixed length string of 25 characters, and `s2`, a variable length string that holds the ANSI values for characters in `s1`. Two `For...Next` loops are used to iterate through `s2`, which has been assigned string conversions of the ANSI values of the individual characters in `s1`.

In the first `For...Next` loop, all characters are `0` because `s1` has only been declared. In the second `For...Next` loop, you see the results of the `LSet` statement. All characters in `s1` have been replaced with either the inputted text from `txtInput` or with a space character (ANSI value of 32).

Exercise 19.4 Displaying Names in the Proper Case

This exercise combines many of the string functions that have been introduced to enable you to display names in the proper case, while accommodating those of Celtic ancestry. It is a bit longer than others and somewhat complex, but illustrates how string functions can be used to check and correct user input. Figure 19.4 shows the interface.

Figure 19.4

This application displays the name `McDonald` *spelled with a capital* `D` *(instead of* `Mcdonald`*).*

1 Create the interface shown in Figure 19.4. It is the same as the one in Exercise 19.1, except for the addition of a frame with the option buttons. The option buttons are part of a three-member control array named `optStatus()`.

2 Add the following `GotFocus()` procedure for the bottom text box (`txtFormat`):

```
Private Sub txtFormat_GotFocus()
   Dim first As String * 20
   Dim middle As String * 2
   Dim last As String * 25
   Dim display As String * 50
   Dim i As Integer

   first = txtFirst.Text
   first = TestString(first)
   middle = Trim(txtMiddle.Text)
   middle = StrConv(Left(middle, 1) & ".", vbProperCase)
   last = txtLast.Text
   last = TestString(last)
   For i = 0 To 2
      If optStatus(i).Value = True Then
         RSet display = optStatus(i).Caption & ":  " & Trim(first) & "
         ➥" & middle &  " " & Trim(last)
         Exit For
      End If
   Next
   txtFormat.Text = display
End Sub
```

continues

This procedure executes when the user inputs a name and then tabs to the `txtFocus` text box, which generates a `GotFocus()` event. In this event, the user's first and last names are passed to the `TestString()` function and displayed, along with his or her status, near the middle of `txtFormat` using the `RSet` function.

3 Add the `TestString()` function at the form module level:

```
Public Function TestString(pass As String) As String
   Dim i As Integer
   Dim CapCount As Integer
   pass = Trim(pass)

   If Len(pass) <> 0 Then      ' test for empty inputs
      If Left(pass, 2) = "Mc" Or Left(pass, 3) = "Mac" Then     ' test for
      ➥' Celts
        For i = 1 To Len(pass)    ' count number of capital letters
           If Asc(Mid$(pass, i, 1)) >= 65 And Asc(Mid$(pass, i, 1)) <= 90
           ➥Then
             CapCount= CapCount + 1
        End If
        Next
        If CapCount <= 2 Then   ' permit no more than 2 capital letters
           TestString = pass
        Else
            TestString = Trim(StrConv(pass, vbProperCase))
        End If
      Else
          TestString = Trim(StrConv(pass, vbProperCase))
      End If
   Else
      TestString = ""
   End If
End Function
```

The `TestString()` function permits those with last names starting with *Mc* or *Mac* to have a second capital letter placed in the name. Those who, by mistake, enter more than two capital letters have their input placed in proper case with the `StrConv()` function.

19.4.3 The *InStr()* Function

A unique string function is `InStr()`, which returns the numerical starting position of a substring within a larger string. The syntax has two forms:

```
InStr([start,], strexpr1, strexpr2)
```

```
InStr(start, strexpr1, strexpr2 [,compare])
```

Table 19.7 explains the components of the syntax.

Table 19.7 The Syntax of the *InStr()* Function

Component	*Description*
start	Indicates the position at which the search should begin.
strexpr1	The string expression to be searched.
strexpr2	The string expression to be searched for.
compare	Specifies the string Compare method for which the two compare arguments are 0 and 1. With 0, the comparison is case sensitive. With 1, the comparison is not case sensitive.

Consider the following code fragment:

```
message = "Rolling down the river"
newmess = InStr(message, "river")
Print newmess
```

This would display 18.

The number indicates the position in the string where the word *river* begins. The following code fragment would not find the substring "river":

```
message = "Rolling down the liver"
newmess = InStr(message, "river")
Print newmess
```

In this case, the function returns 0. Likewise, the following code will return 0 because the compare argument makes the comparison case-sensitive:

```
message = "Rolling down the River, 0"
newmess = InStr(message, "river")
Print newmess
```

Objective 19.5 Comparing and Sorting Strings

There are two ways to compare strings: using relational operations, including the Like operator, and using the string compare function.

19.5.1 The *Like* Operator

The Like operator returns a Boolean value of True or False. The following is the syntax for an instruction that contains the Like operator:

```
strexpr Like pattern
```

The component pattern can include the wild cards ?, *, #, [charlist], and [!charlist], as explained in Table 19.8.

Table 19.8 Wild Cards for *pattern*

Character	*Description*
`?`	Any one character
`*`	Any group of characters
`#`	Any digit
`[charlist]`	Any one character in a list of characters
`[!charlist]`	Any one character not in a list of characters

As examples, the following comparisons would all return `True`:

```
"Cat" Like "?at"

"Cat" Like "C*"

"12 Rounders" Like "1# Rounders"

"12 Rounders" Like "12 [BHR]ounders"

"12 Rounders" Like "12 [!ABCDEFG]ounders"
```

19.5.2 The *StrComp()* Function

The string compare function, `StrComp()`, is designed to compare one string with another. The syntax for this function is as follows:

```
StrComp(strexpr1, strexpr2 [,compare])
```

This is very similar to the second form of the `InStr()` function. As with `InStr()`, compare options are `0` or `1`, with `0` indicating the string is not case-sensitive and `1` indicating it is case-sensitive. Table 19.9 lists the return values of this function and their comparisons.

Table 19.9 Return Values of the *InStr()* Function

Return	*Comparison*
-1	strexpr1 < strexpr2
0	strexpr1 = strexpr2
1	strexpr1 > strexpr2
NULL	strexpr1 = NULL or strexpr2 = NULL

Consider the following code fragment:

```
message1 = "Rolling down the river"
message2 = "Rolling down the highway"

num = StrComp(message1, message2)
Print num
```

This code would display the value `1`. Why `1`? Because the *r* in the word *river* is greater than the *h* in the word *highway*.

19.5.3 The *Option Compare* Statement

The `Option Compare` statement is an alternative way of setting case sensitivity. The syntax is as follows:

```
Option Compare { Binary | Text }
```

Placing the following statement in the declarations section of a form indicates that string (text) comparisons are no longer case-sensitive:

```
Option Compare Text
```

19.5.4 Sorting Strings

The easiest way to sort a list of strings is to use a list box or combo box and set the `Sorted` property to `True`. This will place added items in alphabetical order. Many third-party products provide libraries (DLLs) to which you can pass an array, even very large arrays, and the libraries' functions will sort them for you. For small arrays, however, do not hesitate to sort them yourself.

Exercise 19.5 Sorting Strings

One of the easiest sort algorithms is a *bubble sort.* It is not the fastest sort algorithm, but for small arrays or arrays that are nearly sorted, it is quite suitable. It is also one of the easiest to code. The best way to learn about a bubble sort is to use it. Create the interface shown in Figure 19.5.

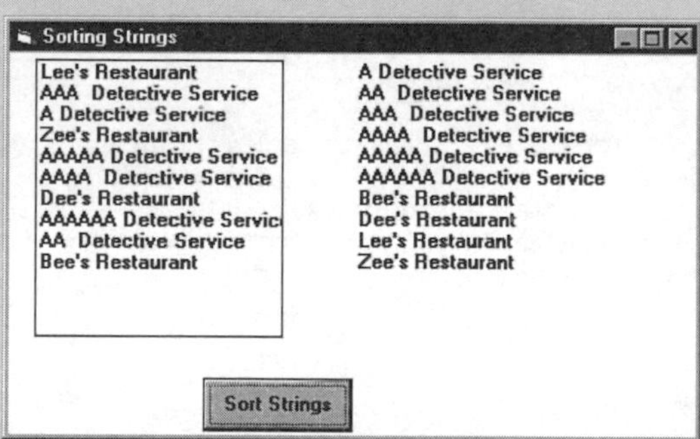

Figure 19.5

This display shows the list before and after the bubble sort procedure.

1 Add a list box (**`lstElements`**) to the left-hand side and to the right, and a large label with its `WordWrap` property set to `True`.

continues

2 In the `Form_Load()` event, add the following code:

```
Private Sub Form_Load()
   lstElements.AddItem "Lee's Restaurant"
   lstElements.AddItem "AAA  Detective Service"
   lstElements.AddItem "Zee's Restaurant"
   lstElements.AddItem "AAAAA Detective Service"
   lstElements.AddItem "AAAA  Detective Service"
   lstElements.AddItem "Dee's Restaurant"
   lstElements.AddItem "AAAAAA Detective Service"
   lstElements.AddItem "AA  Detective Service"
   lstElements.AddItem "Bee's Restaurant"
   lstElements.AddItem "A Detective Service"
End Sub
```

3 In the `cmdSort_Click()` event, enter the following code:

```
Private Sub cmdSort_Click()
   Dim i As Integer
   Dim topindex As Integer
   Dim test As Integer
   Dim Temp As String
   Dim NextItem As Integer
   Dim loopcount As Integer

   ReDim sortarray(lstElements.ListCount - 1) As String

' assign array elements
   For i = 0 To lstElements.ListCount - 1
      sortarray(i) = lstElements.List(i)
   Next

' perform bubble sort
   NextItem = 0
   Do While NextItem < UBound(sortarray)
      topindex = UBound(sortarray)
      Do While topindex > NextItem
         test = StrComp(sortarray(topindex), sortarray(topindex - 1), 1)
         If test = -1 Then
            Temp = sortarray(topindex)
            sortarray(topindex) = sortarray(topindex - 1)
            sortarray(topindex - 1) = Temp
         End If
         topindex = topindex - 1
      Loop
      NextItem = NextItem + 1
   Loop

' display results
   lblDisplay.Caption = ""
   Temp = ""
   For i = LBound(sortarray) To UBound(sortarray)
      Temp = Temp & sortarray(i) & Chr(13) & Chr(10)
   Next
   lblDisplay.Caption = Temp
End Sub
```

4 Run and test the program. Check the sort algorithm by setting the list box `Sorted` property to `True`.

As you study this code, its operation should become clear to you. Almost all sort algorithms are based on nested loops. Data is successively looped through, tested, and then placed.

In the bubble sort, two nested loops are used. A counter variable (`NextItem`) is used so the outer loop iteration is equal to the remaining number of elements in the array (`Sortarray`). In the inner loop, the variable `topindex` is at first equal to the top element of the array. It is then compared (using the `StrComp` function) to the element immediately below it.

If `topindex` encounters an array element with a value less than its own (closer to the front of the alphabet), it swaps places with it. This new element is compared to array elements; if it also encounters a value lower than its own, they swap places. By the completion of the first iteration of the outer loop, the lowest array element (closest to the front of the alphabet) has been identified and placed at the beginning of the array.

The second outer loop iteration only processes array elements from the second position from the end of the array. By completion of the second outer loop iteration, the second lowest array element has been identified. This process continues until the last loop is only comparing the last and next-to-last element of the array. These successively smaller loops "bubble" highest values to the top of the array.

Objective 19.6 Constructing Strings

Finally, two functions are designed to construct strings: the `String$()` function and the `Space$()` function. The syntax for the `String$()` function consists of two formats:

```
String[$](number, charcode)
```

```
String[$](number, string)
```

Table 19.10 explains the syntax.

Table 19.10 The Syntax of the *String$()* Function

Component	*Description*
`number`	Indicates the length of the string
`charcode`	Indicates the ANSI code
`string`	Indicates the first character to place in the return string

The `Space$()` function, in contrast to the `String$()` function, returns a string of blank spaces. The syntax is as follows:

```
Space[$](number)
```

The `number` component is the number of blank spaces to return.

As an example of these final two functions, consider the following code fragment:

```
Print String$(10, "*") + Space$(10) + String$(10, "*")
```

This will display the following:

```
**********          **********
```

The `String$()` function has special use in advanced Visual Basic. It is frequently employed to create a string buffer and address in memory so that values from Windows API functions can be returned.

Chapter Summary

String functions deal with ways of manipulating and converting strings. Determining the length of a string is accomplished using the `Len(string)` function.

A more complex set of string functions deals with converting a string from one form to another. Converting the case of a string is accomplished using `LCase$()`, `UCase$()`, and `StrConv()`. Converting a character to its ANSI value is accomplished using `Asc()`. Converting an ANSI value to a character is accomplished using `Chr$()`. Ten functions convert strings to numbers and numbers to strings: `Val()`, `Str$()`, `Hex$()`, `Oct$()`, `CDate()`, `CInt()`, `CLng()`, `CSng()`, `CDbl()`, and `CCur()`.

Ways of manipulating strings also feature a number of different functions. `Left$()`, `Right$()`, and `Mid$()` return part of a string. `LTrim$()`, `RTrim$()`, and `Trim$()` remove blank spaces from a string.

Ways to justify the display of a string are to use two statements (not functions). These are `LSet` and `RSet`. The `Mid` statement is used to merge one string with another.

A unique string function is `InStr()`. This function returns the position of a substring within a larger string.

The `Like` operator and the `StrComp()` function are two ways to compare strings. The `StrComp()` function is especially useful for sorting sets of strings. The return values from this function are as follows:

Return	*Comparison*
-1	strexpr1 < strexpr2
0	strexpr1 = strexpr2
1	strexpr1 > strexpr2
NULL	strexpr1 = NULL or strexpr2 = NULL

If a sort requires ignoring case sensitivity, place `Option Compare {Binary|Text}` in your code.

Finally, `String$()` and `Space$()` represent two ways of constructing strings. `String$()` displays a string of characters; `Space$()` displays a string of blank spaces.

Skill-Building Exercises

1. For this exercise, use the same form as the one used for Exercise 19.1. The purpose of this exercise is to determine if a name is too long. Check the length of each part of the person's name:

   - If the first name is greater than 10 characters, print "The first name is too long."
   - If the middle name is greater than two characters, print "The middle initial is too long."
   - If the last name is greater than 15 characters, print "The last name is too long."

   Use an `If...Then` or `Select Case` structure in your procedure. Do not use a `GoTo` statement to exit the structure. Instead, set a flag to `1`, and change it to `0` if a name is found to be too long.

2. The `App.Path` property is a string property that can be read to determine where your application resides. Write a procedure that examines the passed string of the `App.Path` and then prints out "Drive: (drive letter)" on one line and "Path: (application path minus drive)" on the line below. In your procedure, use the `Len()`, `Instr()`, `Left$()`, and `Mid$()` functions. For a challenge, extend this exercise to accommodate network drives that have double-slash path identifiers (for example, M:\\).

3. Create a string of 20 characters using the `String$()` function. Then, using the `LSet` statement, place the string expression `"Testing"` within it. Finally, use the `Instr()` function to display the position where you first encounter an empty space. *Hint:* the ANSI value for a space is 32.

4. Write a single function that places the following string expressions in the order of first name, middle initial, and last name, and in the correct case.

   ```
   Jones, Amos Borden

   McDonald, Angus Stewart

   Miller, Lewis Allen
   ```

5. Write a routine that examines a string and sorts the letters into ascending order by their ANSI values.

File-Processing Controls and Sequential File Processing

If the data needed by a program is to be saved, it must be stored on a disk. With sequential file processing, the computer is required to store two types of files on a disk drive:

- **Program files**—These files contain Visual Basic or application-specific instructions. These include .VBP, .VBW, .FRM, .CLS, and .BAS files.
- **Data files**—These files contain the data to read to and from program files. These files will include text files (.TXT) and data files (.DAT).

This chapter guides you through many tasks having to do with file processing. It begins with an introduction to four file-processing controls:

- Drive List Box control
- Directory List Box control
- File List Box control
- Common Dialog control

Next, it addresses how to write to, and read from, a sequential file.

When you complete this chapter, you will be able to

- add drive, directory, and file list boxes to a design;
- use the Common Dialog control;
- set the path for the drive and directory;
- use the `App` object and its properties;
- incorporate file error-handling instructions in a design;
- understand the difference between the Common Dialog control's `Filter`, `FilterIndex`, and `Action` properties;

- write to and read from a file of sequential records;
- know how to open and close a file;
- understand the difference between `Input #` and `Line Input #`;
- design and use a sequential record;
- write the instructions to kill (delete) a file.

Chapter Objectives

The preceding lengthy list of tasks is distilled into the following five objectives. The first introduces you to the three file controls as well as the Common Dialog control. The second explains the use of a built-in object known as the `App` object (short for *application*). The final three objectives explore sequential file processing, including the tasks of error trapping, creating a sequential file, and removing a sequential file when it is no longer needed.

1. Using built-in Visual Basic file-processing controls
2. Understanding the purpose of the `App` object
3. Adding error-handling procedures to file processing
4. Understanding design concepts of sequential file records
5. Writing, reading, or appending to a sequential file

Objective 20.1 Using Built-in Visual Basic File-Processing Controls

Many Visual Basic applications require information about the current disk drive (to determine which drive is open), the directories stored on the disk placed in the current drive, and the files stored within a directory or placed on the disk. This drive-directory file system helps to explain why Visual Basic places four controls on the toolbox to deal with file processing. These are the Drive List Box, the Directory List Box, the File List Box, and the Common Dialog controls.

20.1.1 The Three Individual File System Controls

Three file-system controls stored in the Visual Basic toolbox are designed to provide specific file-system information. Each tool, shown as a separate icon on the Visual Basic toolbox, serves a special purpose:

- **Drive List Box control**—Displays available disk drives on a computer. Click the down arrow to reveal the available drives.
- **Directory List Box control**—Displays the directories created for a disk placed in a disk drive. Click a folder to reveal the contents of a particular directory.
- **File List Box control**—Displays the files stored within the directories, including the root directory. The scroll bar appears when the number of stored files is large.

These tools tell you what files (the file list box) are stored on a specific drive (the drive list box) within the named directory (the directory list box). As shown in Figure 20.1, the folder named VB is on drive C. It contains a number of files, beginning with ADDSCCUS.DLL.

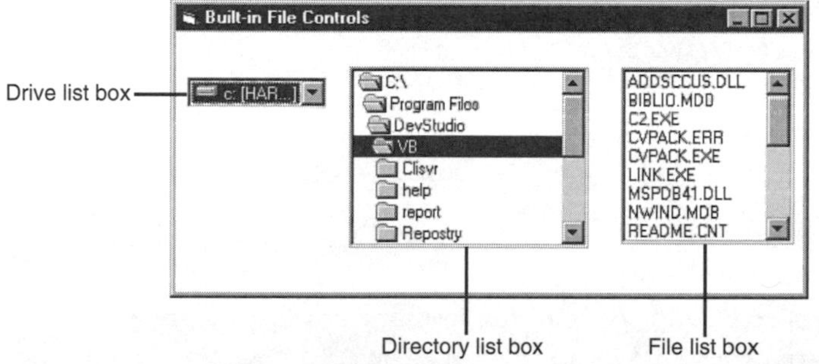

Figure 20.1

The three file controls enable you to examine the contents of a drive, the directories on the drive, and the files within a directory.

Exercise 20.1 Using Drive, Directory, and File List Boxes

This exercise asks you to design your own unique file-processing system by using the drive, directory, and file list boxes. Figure 20.2 shows the form that you are to create. With this design, you can change the disk drive, open directories on the drive, and view files stored in a directory.

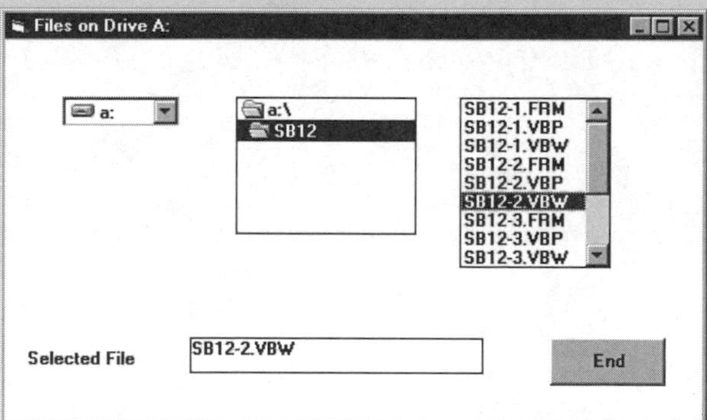

Figure 20.2

This interface shows the files stored on drive A and the file that is selected.

1 Create the interface as shown in Figure 20.2. Add a drive list box, a directory list box, and a file list box. Add a label with a border to store the file selected (`lblFile`), and the single command button (`cmdEnd`). Write the `End` procedure, add the captions, and add the descriptive label.

continues

2 Write the following `Form_Load()` procedure:

```
Private Sub Form_Load()
  'Return the current path for the current drive
  'Set the path of the drive
  Drive1.Drive = CurDir$
  'Set the path of the directory
  Dir1.Path = CurDir$
  'Set the path of the file
  File1.Path = CurDir$
  'Set the file filter pattern
  File1.Pattern = "*.*"
End Sub
```

This procedure is somewhat difficult to follow. The keyword `CurDir$[("drive")]` returns the path for the specified drive. If no drive is specified, `CurDir$` returns a string for the current drive. In the example, the directory and file path are set for the current directory. Observe that `Drive1`, `Dir1`, and `File1` are controls placed on the design. The `*.*` wild card indicates that all files with this pattern are to be selected.

3 Write the `Drive1_Change()` procedure to control a change in a disk drive.

```
Private Sub Drive1_Change()
  Dim Msg As String
  'Need message if incorrect directory results
  On Error Resume Next
  Dir1.Path = Drive1.Drive
  Msg = "Error in setting directory path"
  If Err Then
    MsgBox Msg, 48, "Drive/Directory Error"
    Drive1.Drive = Dir1.Path
  End If
On Error GoTo 0
End Sub
```

In this code, the following instruction sets the path of the directory to the current drive:

```
Dir1.Path = Drive1.Drive
```

The remaining instructions deal with the handling of errors. For example, an error will result if drive A is selected and there is no diskette in the drive. We will turn to the writing of error-handling instructions in Objective 20.3, "Adding Error-Handling Procedures to File Processing." For now, do not worry if you do not understand how error-handling works.

4 Write the `Dir1_Change()` procedure to control a change in the directory.

```
Private Sub Dir1_Change()
  'Code to determine the files in a directory
  File1.Path = Dir1.Path
  If File1.FileName <> "" Then
    lblFile.Caption = File1.FileName
```

```
    Else
      lblFile.Caption = File1.Pattern
    End If
  End Sub
```

In this code, the following instruction sets the path of the file to the path of the current directory:

```
File1.Path = Dir1.Path
```

If there is no file name, the text displayed is either a specific file name or a file pattern.

5 Write the `File1_Click()` procedure to display in the text box a file selected in the file list box. This procedure is limited to a single instruction:

```
Private Sub File1_Click()
  'Click to select and display the file
  lblFile.Caption = File1.FileName
End Sub
```

This instruction tells the user to click the file name to display it in the label caption.

6 Run and test your design. Study what you have done. The three key instructions are as follows:

```
Dir1.Path = Drive1.Drive   'Set the path of the directory from the drive
File1.Path = Dir1.Path     'Set the path of the file from the directory
lblFile.Caption = File1.FileName  ' Select the file
```

20.1.2 The Common Dialog Control

A fourth control when working with files is the Common Dialog control. As shown in Figure 20.3, this control opens the standard Windows dialog box that integrates the drive, directory, and file list boxes. It permits the designer to control the types of files (in the Files of Type list) to display. In the example in Figure 20.3, the file box filter is set to list files of type icons (.ICO).

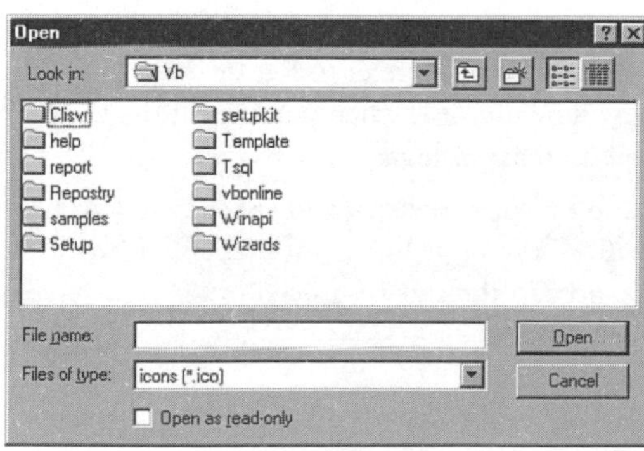

Figure 20.3

The Common Dialog control combines the functions of the three built-in file controls.

Exercise 20.2 Using the Common Dialog Control

Design an application that allows you to open a file, inspect icons stored in the c:\vb\...\samples directory, and load an icon on your display using an image box. (There are also several icon files stored on your disk.) Figure 20.4 illustrates how the dialog box is represented on the form. Figure 20.5 shows the use of the completed design.

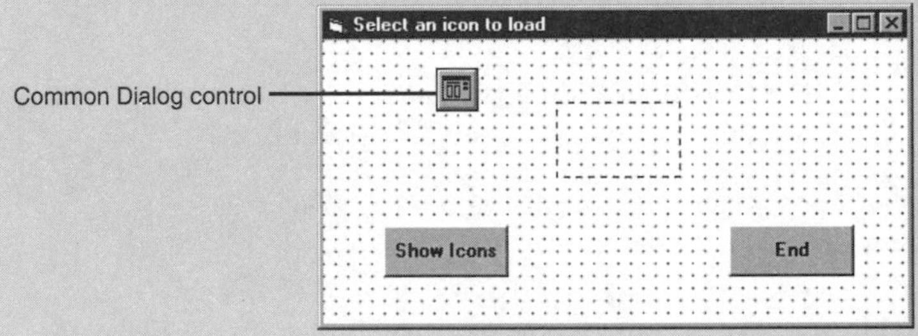

Figure 20.4

The Common Dialog control is invisible at runtime and can be placed at any position on a form.

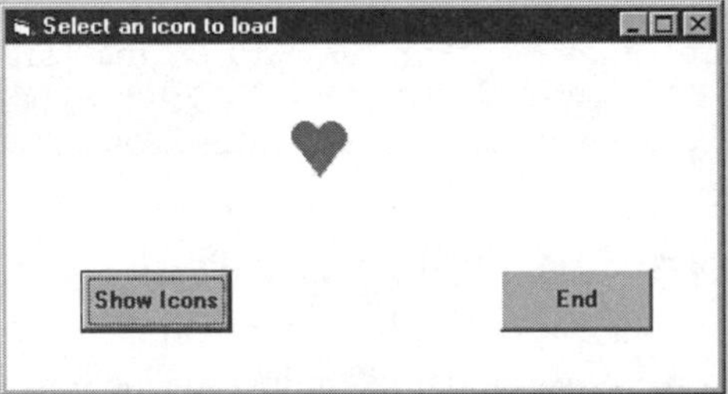

Figure 20.5

The interface opens the Common Dialog control to allow an icon to be displayed on the image box.

1 Click the Common Dialog control (refer to Figure 20.4). Like the Timer control, this control will be invisible when you run the application. Change the `DialogTitle` to **Select an icon to load**.

2 Add the two command buttons (**`cmdShow`** and **`cmdEnd`**) and an image box (**`Image1`**) to the form. Figure 20.4 shows this form (with labels added). Change the caption of the form, and add the `cmdEnd` procedure.

3 Write the `Form_Load()` procedure as follows:

```
Private Sub Form_Load()
  'This statement changes the drive to the path
  'of the application.  App is an object, with
  'properties, but no methods
  ChDrive App.Path
  'Changes the directory to the path of the application.
  ChDir App.Path
End Sub
```

The meaning of this code is explained in Objective 20.2, "Understanding the Purpose of the `App` Object."

4 Write the Show Icons (`cmdShow_Click()`) procedure:

```
Private Sub cmdShow_Click()
  On Error GoTo ErrHandler
  cmDialog1.Filter = _
  "All Files (*.*)|*.* |Text Files (*.txt)|*.txt|icons (*.ico)|*.ico"
  cmDialog1.FilterIndex = 3
  cmDialog1.Action = 1
  If cmDialog1.Filename <> "" Then
     Image1.Picture = LoadPicture(cmDialog1.Filename)
  End If
ErrHandler:
  Exit Sub
End Sub
```

5 Run your design. When you click Show Icons, the Common Dialog control should appear, asking you to indicate the icon file that you want to inspect. Click the file and then click OK to remove the dialog box and to display the icon. Figure 20.5 illustrates the display of an icon in the image box.

Objective 20.2 Understanding the Purpose of the *App* Object

The `App` object (short for *application*) is a global object available at runtime. It is used to determine or specify information about the title of the application, its path, and the name of its executable and Help files. When working with files, the `App` object is often paired with the `ChDir` and `ChDrive` statements.

`ChDir` is used to change the default directory on a specific drive. The syntax is limited to the following:

```
ChDir path
```

The `path` component is a string expression written to identify which directory becomes the default directory.

`ChDrive` is used to specify a default (disk) drive. The syntax is as follows:

```
ChDrive drive
```

The `drive` component is a string expression to identify the default drive.

Consider how `ChDrive` and `ChDir` are used with the `App` object. The following instructions set the default of the drive and the directory:

```
ChDrive App.Path
ChDir App.Path
```

The first instruction sets the default drive to the same drive that stores the application. The second instruction sets the default directory to the same directory that stores the application.

`ChDrive` and `ChDir` can also be used to set a specific path, such as in the following code:

```
ChDrive "b:\"
ChDir "b:\ch22\"
```

Objective 20.3 Adding Error-Handling Procedures to File Processing

The Common Dialog procedure contains a number of new properties and concepts. These include error-handling instructions, `Filter` and `FilterIndex` properties, and the Common Dialog control `Action` property. We begin with the most difficult subject: error handling.

Error handling is added to file processing code to deal with errors if and when they occur. For example, a procedure might require that data be written to drive A on the computer, which presupposes that a diskette has been placed in the drive and is ready to receive data. However, what happens if a diskette has not been inserted? Without error-handling instructions, the program will crash!

20.3.1 Error Handling with the *On Error* Statement

Error handling makes use of a variety of statements, beginning with the `On Error` statement. This statement enables an error-handling routine or section of the code where error-handling is treated. The three forms of this statement are as follows:

```
On Error GoTo line
On Error Resume Next
On Error GoTo 0
```

The *On Error GoTo* Line Syntax

The `On Error GoTo` line statement means that when an error is discovered by Visual Basic, the program will go to a specified line number or to a line label. We will always refer to a line label.

The following statement tells you that when an error is discovered, the `On Error` statement will enable an error-handling routine and *go to* (or jump to) the statement that begins with the label `ErrHandler`:

```
On Error GoTo ErrHandler
```

The `ErrHandler` code, in turn, might read as follows:

```
ErrHandler:     'Colon is required
  Select Case Err
     Case 53:  Msg = "Error 53:  The file does not exist."
     Case 68:  Msg = "Error 68:  Drive is not available."
     Case 71:  Msg = "Error 71:  The disk is not ready."
     Case Else:  Msg = "Error" & Err & " occurred."
  End Select
```

Where do these error numbers come from? The list of error numbers and their meanings is found using Visual Basic Help under the heading "Trappable Errors." They include such messages as "File not found" (53) and "Disk not ready" (71). In writing code, select the error message to report which type of error will most likely occur, with the explanation given through use of the message box.

The *On Error Resume Next* Syntax

The `On Error Resume Next` statement indicates that when a runtime error occurs, control should be transferred to the statement that follows the error:

```
On Error Resume Next
```

In other words, the program is allowed to continue even though it contains a runtime error.

The *On Error GoTo 0* Syntax

In contrast to `On Error Resume Next`, the `On Error GoTo 0` statement (zero, not capital *O*) is easier to understand:

```
On Error GoTo 0      'zero, not capital O
```

This statement disables any enabled error handler. Use it to turn off error trapping.

The *Resume* Statement

The `Resume` statement does not require the `On Error` statement; however, it must appear within the error-handling routine (or an error will occur). The syntax for this statement contains three forms:

```
Resume [0]
Resume Next
Resume line
```

Table 20.1 explains the syntax.

Table 20.1 The *Resume* Statement Syntax

Component	*Description*
`0`	Indicates that program execution is to resume with the statement that caused the error

continues

Table 20.1 continued

Component	*Description*
`Next`	Indicates that program execution is to resume with the statement immediately following the statement that caused the error
`line`	Indicates that program execution is to resume at a designated line label or line number

20.3.2 Common Dialog Control Properties

Besides error-handling routines written for the Common Dialog control, object properties can also be set. These include the `Filter` property, the `FilterIndex` property, and the `Action` property.

The *Filter* Property

The `Filter` property for the Common Dialog control specifies the types of files that are to be displayed in the dialog box's file list box. In Exercise 20.2, the instruction was given to display all files, only text files, or only icon files. The following syntax was used:

```
[form.]CMDialog.Filter [ =  description1|filter1|description2|filter2...]
```

Both a description and filter arguments are required, with a pipe (|) separating the two, as follows:

Description	*Separator Filter*	*Arguments*
Text Files (.TXT)	|	.TXT

When writing this instruction, *don't include spaces before or after the pipe.* If you do, these spaces will be included in the description and filter values.

The *FilterIndex* Property

The `FilterIndex` property specifies the default filter. When writing the `FilterIndex` property, the index for the first defined filter is 1, the second is 2, and so forth. To set the default to the third filter, use the following code:

```
[form.]CmDialog.FilterIndex = 3
```

This indicates that `icon(*.ico)|*.ico` files should be displayed as the default.

Common Dialog Control Methods

The Common Dialog control can display the dialog boxes listed in Table 20.2, using the specified method.

Table 20.2 Common Dialog Control Dialog Boxes

Method	*Description*
`ShowOpen`	Shows the Open dialog box
`ShowSave`	Shows the Save As dialog box
`ShowColor`	Shows the Color dialog box
`ShowFont`	Shows the Font dialog box
`ShowPrinter`	Shows the Print or Print Options dialog box
`ShowHelp`	Invokes the Windows Help engine

The specific setting is determined at runtime rather than at design time. For example, the following statement opens the Color dialog box:

```
cmDialog1.ShowColor
```

To open the Save As dialog box, use the following statement:

```
cmDialog1.ShowSave
```

Objective 20.4 Understanding Design Concepts of Sequential File Records

Although Visual Basic provides four controls for providing information about disk drives, directories, and files, it does not provide a specific control for storing data on files. Program instructions are required to do all of this. In Visual Basic, file access can be done in three ways: random, sequential, and binary, with random being the default.

All three types feature the processing of records. A *record* is the unit of data to be processed by the computer at one time. Perhaps the best way of understanding the makeup of a record is by examining a user-defined type. Consider the structure of a patient record in a hospital.

```
Type Patient
   PatientID     String * 9
   LastName      String * 15
   FirstName     String * 15
   Ward          Integer
   DoctorID      Integer
End Type
```

In this example, each element of the record type is called a field. Thus, `PatientID` is a field. The collection of data for all fields for one person represents a single record. The collection of data for all fields for all persons then represents a data file.

20.4.1 Sequential File Access

We begin our discussion of file access by examining *sequential file access,* which is especially useful for text files in which each character in the file represents either a character (including blank spaces) or a text formatting character (such as a new line).

Text processing is limited to reading and writing strings of text, rather than sets of records. With sequential access, two files are typically required: one from which to read data and one to write data to. It is not possible to read from, and write to, the same sequential file at the same time. If, for example, you want to insert a record into a sequential file, it is necessary to read from one file and write to a second file until the desired file location is reached, at which time a new record is inserted and written to the second file (not the first). Once this is done, all remaining records must be read from the first file and written to the second.

Another method is to load the file completely into memory and then insert a record at the correct position or add records to the end of the file, and write the entire file to disk.

Although designed for text files, sequential access can also be used in processing records. In this instance, the fields of the record must be defined and placed in read and write instructions.

Functions and Statements

The functions and statements shown in Table 20.3 and Table 20.4, respectively, are used in sequential file processing.

Table 20.3 The Functions for Sequential File Processing

Function	*Description*
`Dir`, `Dir$`	Returns the name of the file or directory that matches a specific pattern
`EOF`	Returns a value to indicate whether the end-of-file (`EOF`) has been reached
`FileCopy`	Copies a file
`FileDateTime`	Returns a string that indicates when the file was created or last modified
`FileLen`	Returns the length of a file in bytes
`GetAttr`	Returns an integer that indicates the attributes of a file, directory, or volume label
`Input`	Returns characters read from a sequential file
`Loc`	Returns the current position within an open file
`LOF`	Returns the size of an open file in bytes
`Seek`	Sets the position in a file for the next read or write
`SetAttr`	Sets the attribute information for a file

Table 20.4 The Statements for Sequential File Processing

Statement	*Description*
`Open`	Enables input or output to a file
`Close`	Closes input or output to a file
`Reset`	Closes all files opened with the open statement and writes the contents of all file buffers to disk
`Input #`	Reads characters from a sequential file and assigns data to variables
`Line Input #`	Reads a line from a sequential file into a String or `Variant` variable
`Print #`	Writes display-formatted data to a sequential file
`Write #`	Writes raw data to a sequential file

Many of the functions and statements can be understood by reading their descriptions in Table 20.3 and Table 20.4, respectively. Others require some clarification. The following sections help you become familiar with many of the functions and statements used in sequential file processing.

The *Open* Statement

Before a file can be used, it must be opened. The syntax for the Open statement is as follows:

```
Open file For[ mode] [Access access][lock] As [#]filenumber[Len = reclength]
```

Table 20.5 explains the components of the syntax.

Table 20.5 Syntax for the *Open* Statement

Component	*Description*
`file`	The file name or path plus the file name.
`mode`	One of the following: `Append`, `Binary`, `Input`, `Output`, or `Random`.
`Access`	One of the following: `Read`, `Write`, or `Read Write`.
`lock`	Specifies which file operations are permitted: `Shared`, `Lock Read`, `Lock Write`, or `Lock Read Write`.
`filenumber`	An integer from 1 to 511, inclusive. A file number is required.
`reclength`	The record length, which for sequential files is the number of characters buffered.

The purpose of the `Open` statement is twofold: to create a new file and to open a file. Once opened, one of the following operations can be performed:

- Input (read) characters from a file.
- Output (write) characters to a file.
- Append (write) characters to an existing file.

Remember: you must open a file before any I/O operations can be performed on it.

20.4.2 Writing Text to a File

Before writing data to a file, the data must be in a form ready for output, and a file must be opened for outputting or appending. Only then can the `Print #` or `Write #` statements be used.

The question arises as to which approach is best. With `Write #`, Visual Basic separates each value (number and string) with a comma and puts quotation marks around string expressions. This statement is used when writing a sequential record to a file. The `Print #` statement does not do this. It is used to write one string per line. For this reason, use `Write #` when writing sequential records to a file, which must be read with the `Input #` statement later on.

The *Write #* Statement

The syntax of the `Write #` statement is as follows:

```
Write # filenumber [,expression list]
```

The component `filenumber` is the number used in the `Open` statement, and `expression list` lists the fields of the record to be written to a file.

For example, a `Write #` instruction must be written as follows:

```
Write #1, FirstName$, LastName$, JobRank%, Start%
```

In this instance, each record consists of four fields.

The *Print #* Statement

The syntax of the `Print #` statement is as follows:

```
Print # filenumber, [[{Spc(n) | Tab(n)}][expression][{;|,}]
```

The component `filenumber` is the number used with the `Open` statement, and `expression` is the numeric or string expressions to be written to a file.

If writing to a file using text placed in a text box, a `Print #` instruction might be written as follows:

```
Print #1, Text1.Text
```

20.4.3 Reading Data from a File

Data placed on a sequential file can be read using the `Input #` or `Line Input #` statements. The `Input #` statement reads a list of numeric or string expressions from a file. This statement is used when the exact type of data is known in advance. This is in contrast to the `Line Input #` statement, which reads a text file one line at a time, reading all characters up to a carriage return and line feed combination. In reading the file, it does not retain the carriage return and line feed.

The *Input #* Statement

The syntax for the `Input #` statement is as follows:

```
Input # filenumber, variable list
```

The component `filenumber` is the number used in the `Open` statement, and `variable list` is a comma-delimited list of the assigned variable values read from the file.

As an example, an `Input #` statement must be written as follows:

```
Input #1, FirstName$, LastName$, JobRank%, Start%
```

The *Line Input #* Statement

The syntax for the `Line Input #` statement is as follows:

```
Line Input # filenumber, variablename
```

The component `filenumber` is the number used in the `Open` statement, and `variablename` is the name of the variable used to receive a line of text from the file.

As an example, a `Line Input #` instruction might read as follows:

```
Line Input #1, TextLine
```

`TextLine` represents one line of text.

The *Close* Statement

After all records have been read from, or written to, a file, the file should be closed. The close statement is simply:

```
Close [[#] filenumber] [,[#] filenumber]...
```

If no arguments are provided, the statement closes all open files. To close a single file, such as file #1, write the following:

```
Close #1
```

To close all open files, type the following instruction:

```
Close
```

Exercise 20.3 Using *Input #* and *Write #* to Design a File Note Pad

Design a file note pad that enables you to write notes and save them on a disk. For this exercise, you use the `Input #` and `Write #` statements and the enable and disable commands to disallow reading from a file until a note has been created and written to the file.

1 Add the text box and the three command buttons, as illustrated (see Figure 20.6). Add all captions and write the `End` procedure. If the files are to be placed on a specific drive, set `ChDrive` and `ChDir` in the `Form_Load` procedure.

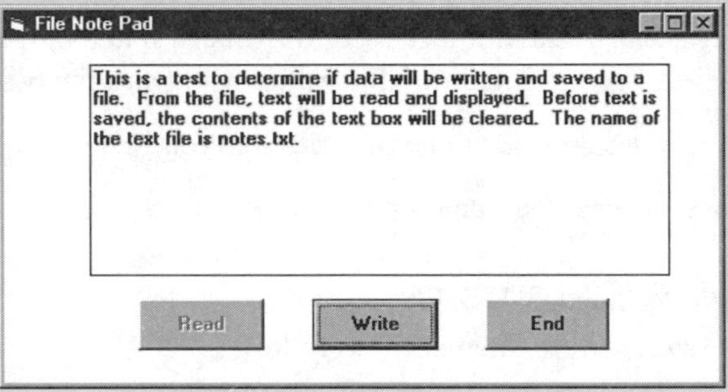

Figure 20.6

This design enables text to be entered, written to a disk, and read back from the disk.

2 Allow the text box to accept multiple lines.

3 Disable the Read command button.

4 Write the `cmdWrite_Click()` procedure (changing the file path as necessary) as follows:

```
Private Sub cmdWrite_Click()
   'These instructions create a program titled notes.txt
   'Error-handling instructions
   Dim Msg As String
   On Error GoTo ErrorWrite
   Open "notes.txt" For Output As #1
   Write #1, txtNotes.Text
   Close #1
   txtNotes.Text = ""
   cmdRead.Enabled = True
   cmdWrite.Enabled = False
   Exit Sub
ErrorWrite:
   Select Case Err
     Case 53:  Msg = "Error 53:  The file does not exist"
     Case 68:  Msg = "Error 68:  Drive is not available"
```

```
      Case 71:  Msg = "Error 71:  The disk is not ready"
      Case Else:  Msg = "Error" & Err & " occurred."
    End Select
    MsgBox Msg
    Exit Sub
End Sub
```

When writing this code, do not forget to add the `Exit Sub` instruction prior to the error-handling code. If you do not add this statement, a runtime error will be indicated because the label (in our case, `ErrorWrite:`) is reached, but not called.

5 Write the `cmdRead_Click()` procedure (changing the file path as necessary). The instructions for this program are as follows.

```
Private Sub cmdRead_Click()
  Dim LineText As String
  'Error-handling instructions
  Dim Msg As String
  On Error GoTo ErrorRead

  Open "notes.txt" For Input As #2
  Do Until EOF(2)
    Input #2, LineText
    txtNotes.Text = txtNotes.Text & LineText
  Loop
  Close #2
  cmdWrite.Enabled = True
  cmdRead.Enabled = False
  Exit Sub
ErrorRead:
  Select Case Err
    Case 53:  Msg = "Error 53:  The file does not exist"
    Case 68:  Msg = "Error 68:  Drive is not available"
    Case 71:  Msg = "Error 71:  The disk is not ready"
    Case Else:  Msg = "Error" & Err & " occurred."
  End Select
  MsgBox Msg
  Exit Sub
End Sub
```

In this procedure, the end-of-file (`EOF`) function requires the `filenumber` as the argument. It continues to read from the file until the end-of-file marker is returned.

6 Run and test the program. Add the text shown in Figure 20.6, or any text you want.

Exercise 20.4 Using the *Line Input #* Statement

Modify the note pad program to enable it to accept line input. Use line input when you want to read individual lines from a note pad, include blank lines in your text, or write in outline form. Figure 20.7 shows the user interface for this design (and the preceding sentence rendered as separate lines).

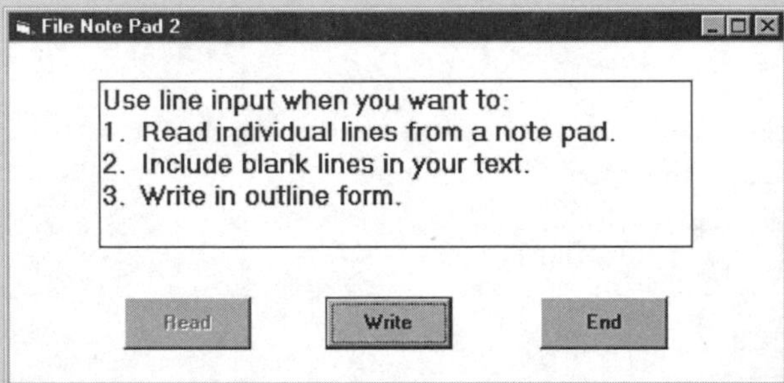

Figure 20.7

The `Line Input #` *statement reads lines of text as they are written to disk.*

1 Modify the `cmdRead_Click()` procedure written for Exercise 20.3 as follows:

```
Private Sub cmdRead_Click()
  Dim LineText As String
  'Error-handling instructions
  Dim Msg As String
  On Error GoTo ErrorRead
  Open "notes2.txt" For Input As #2
  Do Until EOF(2)
    Line Input #2, LineText
    txtNotes.Text = txtNotes.Text & LineText & Chr(13) & Chr(10)
  Loop
  Close #2
  cmdWrite.Enabled = True
  cmdRead.Enabled = False
  Exit Sub
ErrorRead:
  Select Case Err
    Case 53:  Msg = "Error 53:  The file does not exist"
    Case 68:  Msg = "Error 68:  Drive is not available"
    Case 71:  Msg = "Error 71:  The disk is not ready"
    Case Else:  Msg = "Error" & Err & " occurred."
  End Select
  MsgBox Msg
  Exit Sub
End Sub
```

2 Modify the `cmdWrite_Click()` procedure to create the notes2.txt file:

```
Private Sub cmdWrite_Click()
   'These instructions create a program entitled notes2.txt
   'Error-handling instructions
   Dim Msg As String
   On Error GoTo ErrorWrite

   Open "notes2.txt" For Output As #1
   Print #1, txtNotes.Text
   Close #1
   txtNotes.Text = ""
   cmdRead.Enabled = True
   cmdWrite.Enabled = False
   Exit Sub
ErrorWrite:
      Select Case Err
         Case 53:  Msg = "Error 53:  The file does not exist"
         Case 68:  Msg = "Error 68:  Drive is not available"
         Case 71:  Msg = "Error 71:  The disk is not ready"
         Case Else:  Msg = "Error" & Err & " occurred."
      End Select
      MsgBox Msg
      Exit Sub
End Sub
```

3 Run and test this design. Add the text shown in Figure 20.7, or add your own lines of text.

Objective 20.5 Writing, Reading, or Appending to a Sequential File

Although random and binary are the preferred methods of working with records, sequential files can and should be used for such tasks as creating a written record of processing. In design, the fields important to a record must be identified. This is followed by the design of a data-input interface that allows for the entry of data for a record, and the design of a report that lists the data entered.

To delete a file from a disk, use the `Kill` statement (terrible name, I agree). The following is the syntax for this statement:

```
Kill filespec
```

The component `filespec` is the name of the file to delete.

Observe the following rule: *`Kill` will not work with an open file. A file must be closed before the `Kill` statement can be applied.*

The `Kill` statement is used in Exercise 20.6.

Exercise 20.5 Writing and Reading Records Using a Sequential File

Design an input screen for entering inventory information and for writing inventory records to a sequential file. After records are placed on a file, they can be read and displayed.

Figure 20.8 shows the data-entry screen used for capturing records. Each inventory record contains four fields: product name, product price, product cost, and product inventory. After a record has been entered, it is written to a file. Figure 20.9 shows the report produced after all records have been entered into processing (four records in the example).

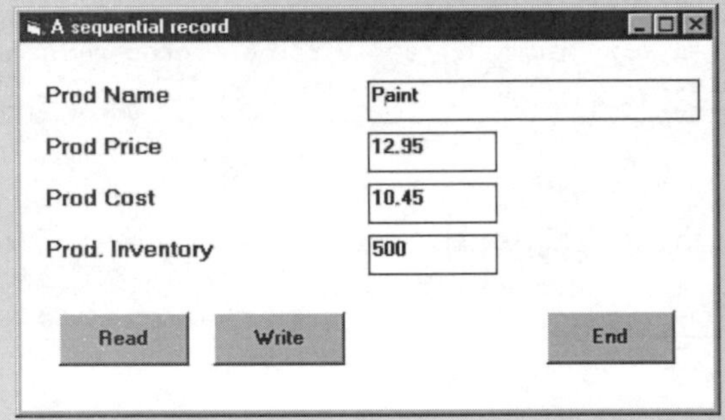

A sequential record

Prod Name Paint

Prod Price 12.95

Prod Cost 10.45

Prod. Inventory 500

Read Write End

Figure 20.8

For this application, add the name of the product and its price, cost, and quantity in inventory.

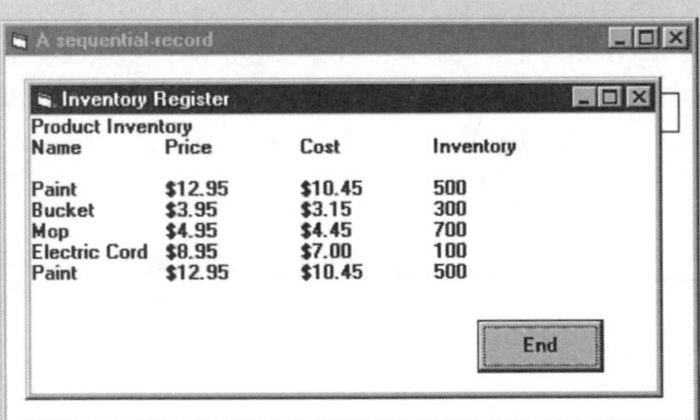

Figure 20.9

This Inventory Register is produced from the records stored on the sequential file.

1 Create the interface shown in Figure 20.8. Add four text boxes, four labels, and four command buttons. Add the captions for all. Write the `End` procedure, but not the `Kill` procedure (which is described in Exercise 20.6). Add a `Close` statement to the `End` procedure. Name the form **`frmDataEntry`**.

2 Write the `cmdWrite_Click()` procedure as follows:

```
Private Sub cmdWrite_Click()
  'These instructions create a program titled prod.tex
  Static count As Integer
  Dim Msg As String
  On Error GoTo ErrorWrite
  txtProductName.SetFocus
  If txtProductName.Text = "" Then
    Close #1
    Exit Sub
  End If
  count = count + 1
  If count <= 1 Then
    Open "prod.txt" For Append As #1
  End If
  Write #1, txtProductName.Text, Val(txtProductPrice.Text), _
  Val(txtProductCost.Text), Val(txtProductInv.Text)
  txtProductName.Text = ""
  txtProductPrice.Text = ""
  txtProductCost.Text = ""
  txtProductInv.Text = ""
  Exit Sub
ErrorWrite:
  Select Case Err
    Case 53:  Msg = "Error 53:  The file does not exist"
    Case 68:  Msg = "Error 68:  Drive is not available"
    Case 71:  Msg = "Error 71:  The disk is not ready"
    Case Else:  Msg = "Error" & Err & " occurred."
  End Select
  MsgBox Msg
  Exit Sub
End Sub
```

In this procedure, the file is opened only once and is not closed. This allows more than one record to be appended to the file. The file is opened in Append mode to allow for one record after another to be added to the file. After a record has been added to the file, all text fields are cleared.

3 Add a second form, `frmReport` (see Figure 20.9), and write the `End` procedure for this form. Return to the first form.

4 Write the `cmdRead_Click()` procedure:

```
Private Sub cmdRead_Click()
  Dim pname As String
  Dim price As Currency
  Dim cost As Currency
  Dim pinventory As Long
  Dim Msg As String
```

continues

```
    On Error GoTo ErrorRead
    frmReport.Show
    Close
    frmReport.Print "Product Inventory"
    frmReport.Print "Name", "Price", "Cost", "Inventory"
    frmReport.Print

    Open "prod.txt" For Input As #2
    Do Until EOF(2)
      Input #2, pname, price, cost, pinventory
      frmReport.Print pname, Format(price, "$#,###.00"), _
      Format(cost, "$#,###.00"), Format(pinventory, "#######")
    Loop
    Close
    Exit Sub
  ErrorRead:
    Select Case Err
      Case 53:  Msg = "Error 53:  The file does not exist"
      Case 68:  Msg = "Error 68:  Drive is not available"
      Case 71:  Msg = "Error 71:  The disk is not ready"
      Case Else:  Msg = "Error" & Err & " occurred."
    End Select
    MsgBox Msg
    Exit Sub
  End Sub
```

5 Write the following `Form_Load()` procedure to set the focus on the data-entry form:

```
Private Sub Form_Load()
  ChDrive App.Path
  ChDir App.Path
  txtProductName.SetFocus
End Sub
```

6 Run and test your design. The next exercise asks you to use the same design, but to add the `Kill` procedure code.

Exercise 20.6 Adding a *Kill* File Procedure

This exercise illustrates how to write a `Kill` procedure to delete a file from disk. Use the design interface shown in Figure 20.8, but add a new button between the Write button and the End button named **Kill** (**`cmdKill`**).

1 Add a `cmdKill_Click()` procedure to your previous design:

```
Private Sub cmdKill_Click()
  Dim Msg As String
  Dim FileName As String
  Dim Ans As Integer
  On Error GoTo Errhandler
```

```
    FileName = UCase(InputBox("Enter file name to delete"))
    If Len(FileName) Then
      Ans = MsgBox("Sure you want to delete " & FileName & "?", 4)
      If Ans = 6 Then
        Msg = "Deleting " & FileName & " from disk"
        Kill FileName
      Else
        Msg = FileName & "not deleted"
      End If
    Else
      Msg = "File name not known"
    End If
    MsgBox Msg
    Exit Sub
  Errhandler:
    If Err = 53 Then
      Msg = "Sorry, file could not be found"
    Else
      Msg = "Sorry, unable to delete file"
    End If
    Resume Next
  End Sub
```

2 Run and test your design. To avoid deleting prod.txt, the file created for the last exercise, consider creating a new one, such as prod2.txt.

Chapter Summary

This initial chapter on file processing was written to introduce you to the drive-directory-file sequence used by Visual Basic, as well as the four toolbox controls designed to enable you to add file selection functionality to your applications. The first three tools are individual file controls, and the fourth integrates the three file controls:

- Drive List Box control
- Directory List Box control
- File List Box control
- Common Dialog control

The `App` object is a global object available at runtime. This object is used, in part, to set the default drive and directory. The following instruction sets the default drive to be the same as the path of the application:

```
ChDrive App.Path
```

Error handling is added to file processing code to trap errors, such as when a drive does not respond to a request, or when a file does not exist. Visual Basic contains a large number of trappable errors.

The instruction

```
On Error GoTo ErrorSection
```

must be matched with the label

```
ErrorSection:
```

found later in the code.

Common Dialog control properties include the `Filter` property (which determines how files are displayed), the `FilterIndex` property (which determines the default files), and the `Action` property (which displays one of the six types of dialog boxes of the Common Dialog control).

Sequential access files are especially useful for reading and writing text files. Before a file can be used, it must be opened. Once opened, it is possible to read or write to a file. To write to a sequential file, use the `Write #` or the `Print #` statements; to read from a sequential file, use the `Input #` or `Line Input #` statements (the pound sign (#) is followed by the number of the file).

After using a file, close it with the `Close` statement.

Besides text, records can be written to, or read from, a file. Writing records to a file features the use of the Append mode. When reading from a file, individual fields must be read.

Deleting a file from a disk requires the `Kill` statement. `Kill` will not work unless a file is closed.

Skill-Building Exercises

1. Place a Common Dialog control on a form and set the `Action` property to display the Color dialog box after a command button `Click()` event. When the dialog box appears, select a color to change the background color of the form. Use the following statement:

```
Form1.Backcolor = cmDialog1.Color
```

2. Place a Common Dialog control, a text box, and three command buttons—`cmdRead`, `cmdWrite`, and `cmdEnd`—on a form. Before starting this exercise, make a copy of the original file so you can return and restart; then, perform the following steps:

    1. When the user clicks `cmdRead`, call a procedure named `OpenMe()` that displays text boxes to open as the `FilterIndex` default. Open notes.txt, the file created in this chapter, display the stored data in the text box, and close the file. Disable the `Read` and enable the `Write`.
    2. Change the text in the text box.
    3. When the user clicks `cmdWrite`, call a second procedure named `CloseMe()`. Open a second file, write the revised file to Notes1.txt, and close the file. Enable the `Read` and disable the `Write`.
    4. Test your design. Open Notes1.txt to determine if your revision was changed.

3. Create a program to read the file layout.txt stored on your disk. Use a Common Dialog control to open this file. Read the contents of the file, line by line, to display in a list box the fields of the file. Data that corresponds to this layout is found in cities1.txt. (You will use this second file in the final exercise.)

    Allow the user to display the contents of this layout, but not to change the contents of the file. *Warning:* Make copies of the layout.txt file in case it is destroyed.

4. Modify Exercise 20.5 to allow you to display the records stored in prod.txt and to append new records to the file. (Remember to make a backup copy of this file.) Use a Common Dialog control to open a file and to display the records currently stored on it. Add the following record to the file:

```
Light       $6.50            $4.50                300
```

    Write this new record to the file. As a test, display the contents of the revised file.

5. Design a program that displays three types of information stored in the cities1.txt file. (Remember to make a backup copy of this file.) These fields are the area name (name of the city, for example, Portland), the name of the state, and the 7/1/94 Resident All-Age Population Estimate. The layout of this file is shown in the file layout.txt.

To complete this exercise, you will have to read the file line by line, followed by the use of the `Instr()` and `Mid$()` functions. These functions will enable you to parse the file to extract the desired information. Place your output on a grid, containing three columns and 215 rows. Use column widths of 2,000, 1,500, and 1,200.

As a test, the following should be the first line of output displayed on the grid:

```
New York city          NY                 7333253
```

21

Random-Access and Binary File Processing

Besides sequential file processing, which is addressed in Chapter 20, "File-Processing Controls and Sequential File Processing," the two more common types of file processing in Visual Basic are *random-access file processing* (the default) and *binary file processing.*

With random-access file processing, fixed-length records are required and all records must correspond to one type. Fixed-length records necessitate the use of fixed-length strings and the calculation of the record length. Binary file processing allows for more flexibility than random-access processing. With binary access, the bytes in the file can represent anything. Binary files also prevent wasted space: They store only the exact number of bytes required in processing. The disadvantage of binary access is that it requires a greater amount of code to handle input/output (I/O) operations, as you will soon discover.

You learn several tasks that that are related to creating and using random-access and binary files. When you complete this chapter, you will be able to

- use the `Get` statement,
- use the `Put` statement,
- design and use a random-access and a binary file,
- add records to and display records stored in a random-access or binary file,
- delete a random-access or binary file using the `Kill` statement,
- determine the length and the number of records stored in a random-access file,
- determine the size of each field of a record stored in a binary file,
- find a single record stored in a binary file.

Chapter Objectives

This chapter is divided into two parts. The first part examines random-access file processing and illustrates how to create and use this type of file. The second part examines binary file processing, using the same examples. In this way, it becomes possible to easily compare the two different methods of file processing. The following are the two corresponding objectives:

1. Understanding the concept of the random-access file
2. Understanding the concept of the binary file

In working through these two objectives, you will become familiar with the important differences between these two types of direct-access files.

Objective 21.1 Understanding the Concept of the Random-Access File

Random-access files store identical records, with each record containing the same fields and each field requiring the same number of bytes. Because random-access processing requires data of one type, a user-defined data type is required of all records with more than one field. This type definition is placed in a .BAS file.

21.1.1 Functions and Statements

Like sequential files, random-access processing features functions and statements. The following functions and statements are the same as those used in sequential file processing:

`Dir()`	`FileDateTime()`	`LOF()`
`Dir$()`	`FileLen()`	`Seek`
`EOF()`	`GetAttr()`	`SetAttr`
`FileCopy`	`Loc()`	

To this list, add the `FreeFile()` function, a function that returns the next valid file number.

The functions of particular importance to random-access processing are the `Seek` statement and `Loc()` function. The following instruction, for example, moves the file pointer to the tenth record stored on the second file:

```
Seek #2, 10
```

This will be the next record to read or write.

The following instruction returns the current location in a file:

```
Loc(#2)
```

In the example, `Loc()` would return `9` because this is the current record, with `10` being the next record to process.

Table 21.1 lists and explains the statements that are important to random-access processing.

Table 21.1 Statements for Random-Access Processing

Statement	*Description*
`Open`	Enables input and output to a file
`Close`	Closes input and output to a file
`Get`	Reads from a file into a variable
`Put`	Writes from a variable to a disk file
`Type...End Type`	Defines a user-defined type containing one or more elements

Of these, `Open`, `Close`, and `Type...End Type` were encountered in Chapter 20. With `Open`, the syntax becomes the following:

```
Open file For Random As filenumber Len = recordLen
```

`Len` is used to specify the fixed size of a record. With `Close` and `Type...End Type`, no changes in the syntax are required.

21.1.2 The *Get* and *Put* Statements

Two new statements are the `Get` and `Put` statements. `Get` means to read (to *get*) from the file, while `Put` means to write (to *put*) to the file. Let's consider the `Get` statement first.

The *Get* Statement

The following is the syntax for the `Get` statement:

```
Get [#]filenumber, [recnumber], varname
```

Table 21.2 explains the components of the syntax.

Table 21.2 Syntax of the *Get* Statement

Component	*Description*
`filenumber`	The number used in the `Open` statement to open the file.
`recnumber`	The number of the record to be read. If `recnumber` is omitted, the next record or byte following the last `Get` or `Put` statement (or pointed to by the last `Seek` function) is read. Even so, you must include delimiting commas, as in `Get #2,,Employee`.
`varname`	The name of the variable used to receive the input from the file.

As an example, if you know that an Employee is a structure of type `Person`, the following statement instructs the program to get the tenth record stored on file #2 and to assign the stored values to a variable `Employee` of type `Person`:

```
Get #2, 10, Employee
```

After a record is read, values can be obtained from individual record elements using the familiar *dot* (.) notation. If type `Person` is defined as

```
Type Person
  FirstName As String * 15
  MiddleName As String * 1
  LastName As String * 15
  Dept_No As Integer
End Type
```

and declared as

```
Dim Employee As Person
```

then the instructions

```
Get #2, 10, Employee
Print Employee.FirstName
```

will display the employee's first name, provided the record can be read from the file.

The *Put* Statement

The `Write` statement equivalent for random-access processing is `Put`. The `Put` syntax features the same terms as the `Get` statement (refer to Table 21.2):

```
Put [#]filenumber, [recnumber], varname
```

Besides a similar syntax, `Get` and `Put` complement one another in other ways. With `Put`, if the record number is omitted, the default record is the last one written (or put), plus one.

Exercise 21.1 Creating and Displaying the Contents of a Random-Access File

Using `Get` and `Put`, create a small file of random-access records for a sports organization. Besides entering personal information, you are asked to add the person's favorite sport. Figure 21.1 shows the data-entry interface. Figure 21.2 shows the second form listing the three records you are asked to place in the file.

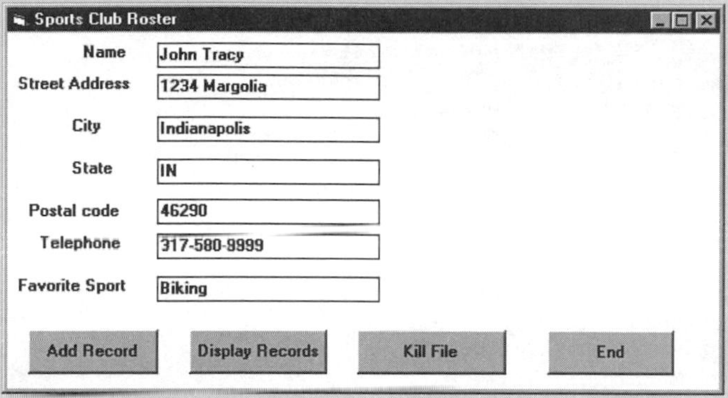

Figure 21.1

This form display is used to enter an entire record for the Sports Club Roster.

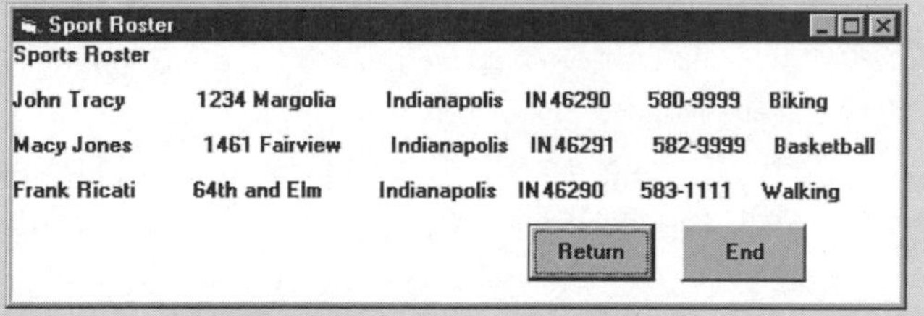

Figure 21.2

This second form display shows the Sports Club membership.

1 Create the user interfaces shown in Figure 21.1 and Figure 21.2. On `Form1` (`frmEnter`), add the seven text boxes and seven labels. The names of the text boxes are **`txtPname`**, **`txtStreet`**, **`txtCity`**, **`txtState`**, **`txtZip`**, **`txtPhone`**, **`txtSport`**.

   Add the four command buttons: **`cmdAdd`**, **`cmdDisplay`**, **`cmdKill`**, and **`cmdEnd`**. Add captions to all controls, including the form. Write the `End` procedure. On `Form2` (`frmShow`), add the two command buttons, add the captions, and write the `End` procedure.

2 Add a .BAS module, and define the following user-defined type within it:

```
Type roster
  pname As String * 20
  street As String * 20
  city As String * 15
  state As String * 2
  zip As String * 10
  phone As String * 12
  sport As String * 20
End
```

continues

3 Write the following `Add Record` procedure:

```
Private Sub cmdAdd_Click()
  Dim member As roster
  Dim RecordLen As Integer
  Static NumRec As Integer

  'Keep a count of the number of records added
  Static count As Integer
  count = count + 1

  'Assign values for a record
  member.pname = txtPname.Text
  member.street = txtStreet.Text
  member.city = txtCity.Text
  member.state = txtState.Text
  member.zip = txtZip.Text
  member.phone = txtPhone.Text
  member.sport = txtSport.Text

  If member.pname = "" Then
     txtPname.SetFocus
     Exit Sub
  End If
  RecordLen = Len(member)
  If count <= 1 Then
    Open "sport1.dat" For Random As #1 Len = RecordLen
    NumRec = LOF(1) / RecordLen
  End If

  Put #1, count, member

  txtPname.Text = ""
  txtStreet.Text = ""
  txtCity.Text = ""
  txtState.Text = ""
  txtZip.Text = ""
  txtPhone.Text = ""
  txtSport.Text = ""

  txtPname.SetFocus
End Sub
```

After data is entered, the length of the record can be determined. Following this, the file is opened if `count` is less than or equal to `1`, and the number of records stored on file are determined. The new record is placed at the end of the file. In this procedure, the file is opened the first time the user attempts to add a record and remains open throughout the procedure.

4 Add the following `Display Records` procedure:

```
Private Sub cmdDisplay_Click()
  Dim member As roster
  Dim RecordLen As Integer
  Dim NumRec As Integer
  Dim count As Integer
```

```
    'Close any open files
    Close
    frmShow.Cls
    frmShow.Show
    frmShow.Print "Sports Roster"
    frmShow.Print

    'Determine the length of a record
    RecordLen = Len(member)
    'Open the file
    Open "sport1.dat" For Random As #1 Len = RecordLen
    'Determine the number of records
    NumRec = LOF(1) / RecordLen

    'Loop to get all records
    For count = 1 To NumRec
      Get #1, count, member
      frmShow.Print member.pname;
      frmShow.Print member.street;
      frmShow.Print member.city;
      frmShow.Print member.state;
      frmShow.Print member.zip;
      frmShow.Print member.phone;
      frmShow.Print member.sport
      frmShow.Print
    Next count

    Close
  End Sub
```

This procedure is similar to the `Add Record` procedure. The length of a record and the number of records stored on file are determined. The value of `NumRec` is placed in the loop to display all records on file.

5 Write the `Kill` record procedure, using the code learned in Chapter 20:

```
Private Sub cmdKill_Click()
  Dim Msg As String
  Dim FileName As String
  Dim Ans As Integer

  'This procedure contains error-handling code
  On Error GoTo Errhandler

  FileName = UCase(InputBox("Enter file name to delete"))
  If Len(FileName) Then
     Ans = MsgBox("Sure you want to delete " & FileName & "?", 4)
     If Ans = 6 Then
       Msg = "Deleting " & FileName & " from disk"
       'File is deleted here
       Kill FileName
     Else
       Msg = FileName & " not deleted"
     End If
```

continues

```
    Else
      Msg = "File name not known"
    End If
      MsgBox Msg

    txtPname.SetFocus

    Exit Sub

  Errhandler:
    If Err = 53 Then
      Msg = "Sorry, file could not be found"
    Else
      Msg = "Sorry, unable to delete file"
    End If
    Resume Next
  End Sub
```

6 Add the `Return` procedure to `frmShow` to return to `frmEnter` and give the focus to the person's name.

```
Private Sub cmdReturn_Click()
  frmEnter.Show
  frmShow.Hide
  frmEnter!txtPname.SetFocus
End Sub
```

7 Write a `Form_Load()` procedure to show `frmEnter` at loading time, give `txtPname` the focus, and set the correct file path.

```
Private Sub Form_Load()
  Load frmEnter
  frmEnter.Show
  txtPname.SetFocus

  ChDrive App.Path
  ChDir App.Path

End Sub
```

This is one possible design of the program. In writing file design programs, it is important to specify which control has the focus at all times. In addition, decisions must be made regarding when to open and close files. If a file is already open, it cannot be opened a second time.

Exercise 21.2 Printing a Special Report

Add a third form to the design that you created in Exercise 21.1 in order to find and list those persons who have the same favorite sport. Figure 21.3, Figure 21.4, and Figure 21.5 illustrate the new interfaces. The Next Record and Previous Record buttons are disabled. (You will enable them when you complete the second exercise in the "Skill-Building Exercises" section at the end of this chapter.) The file used in this exercise is sport2.dat, which is a copy of the sport1.dat.

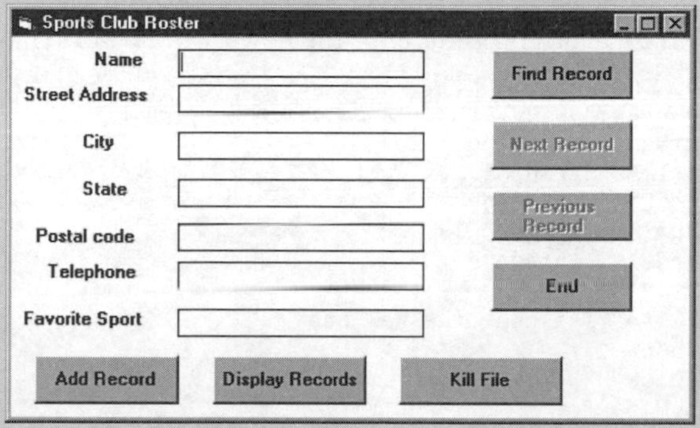

Figure 21.3

The expanded data-entry display contains Find Record, Next Record, and Previous Record buttons (in addition to the buttons of the original display shown in Figure 21.1), with the Find Record button enabled.

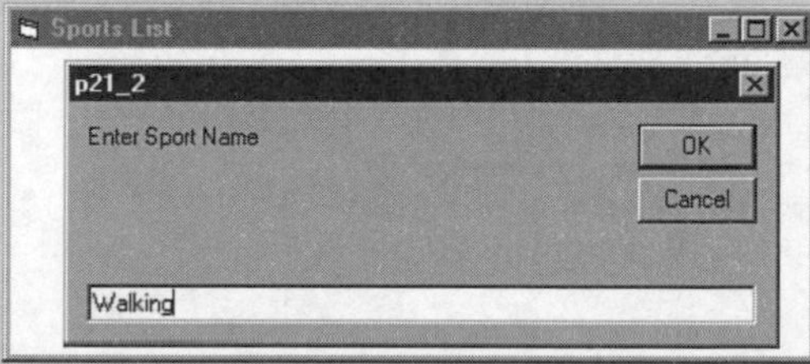

Figure 21.4

The Enter Sport Name form provides a text box for the user to enter the name of a sport.

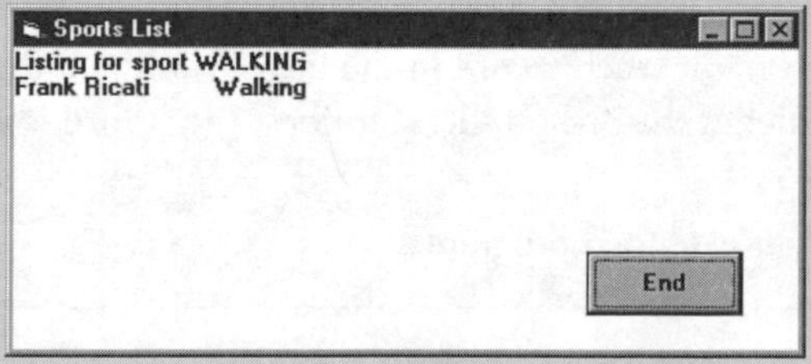

Figure 21.5

The Sports List form shows those members whose favorite sport matches the entered sport's name.

continues

1 Modify the **Sports Club Roster** interface as shown in Figure 21.3, adding the three additional command buttons: **`cmdFind`**, **`cmdNext`**, and **`cmdPrevious`**.

2 Add the **Sports Roster** form (`frmShow`) to the project (refer to Figure 21.2).

3 Write the `Find Record` procedure for the new form (`frmList`) as follows:

```
Private Sub cmdFind_Click()
  Dim member As roster
  Dim SportName As String * 20
  Dim Msport As String * 20
  Dim RecordLen As Integer
  Dim RecNum As Integer
  Dim Records As Integer

  'Hide all forms except the new one
  frmList.Show
  frmShow.Hide
  frmEnter.Hide

  'Enter the sport you want to find
  SportName = UCase(InputBox("Enter Sport Name"))

  RecordLen = Len(member)
  Open "sport2.dat" For Random As #1 Len = RecordLen
  frmList.Print "Listing for sport " & SportName

  RecNum = LOF(1) / RecordLen

  For Records = 1 To RecNum
    Get #1, Records, member
    'Determine the member's sport
    Msport = member.sport
    Msport = UCase(Msport)

    If Msport = SportName Then
      frmList.Print member.pname;
      frmList.Print member.sport
    End If
  Next Records
End Sub
```

In this procedure, an input box is used to determine the name of the sport. After the entry of the name, each record in the file is examined to determine if the sport name matches the one on file. If it does, the record is displayed on the Sports List form.

4 Run and test this extended program.

Objective 21.2 Understanding the Concept of the Binary File

Binary file processing stores identical records according to their fields, but allows for variable-length record size. If, for example, a person's name is allocated 20 bytes but only requires 10, then 10 bytes of storage would be wasted with random-access storage. On the other hand, only the characters entered are stored with binary storage. Whether a person's name requires 10 bytes or 50 bytes, for example, only the required amount of storage space is allocated.

Because binary file processing requires data of one type, a user-defined type is required of all records with more than one field. When defining a type, use variable-length strings instead of fixed-length strings.

21.2.1 Functions and Statements

Like random-access and sequential files, binary file processing features functions and statements. The following functions and statements are identical to those for random-access processing:

```
Dir()         FileDateTime()    Loc()
Dir$()        FileLen()         LOF()
EOF()         FreeFile()        Seek
FileCopy      GetAttr()         SetAttr
```

Table 21.3 lists and explains the statements that are important to binary file processing.

Table 21.3 Statements for Binary File Processing

Statement	*Description*
`Open`	Enables input and output to a file
`Close`	Closes input and output to a file
`Get`	Reads from a file into a variable
`Put`	Writes from a variable to a disk file
`Input`	Returns characters read from a file

These statements are more like random-access processing than sequential file processing. The `Input #` statement cannot be used.

With `Open`, the syntax change is as follows:

```
Open file For Binary As filenumber
```

`Len = recordLen` is not specified. If a record length is provided, it is ignored.

With `Close`, no change in syntax is required, nor is there a change in the syntax of the `Type` definition statement, with the following exception: Fixed-length fields are no longer required.

Writing Variable-Length Strings

With records of variable length, determining the exact size of string fields within a record is required before the field is written to storage. We use `StrSize` to store the length of the string, and two `Put` statements to store this length and the actual string. Thus, the following code stores the length of the string (as an integer) just prior to the storage of the member's name:

```
StrSize = Len(member.pname)
Put #1, , StrSize
Put #1, , member.pname
```

Reading Variable-Length Strings

Reading from a binary file is simplified if the string size is stored in front of each variable-length string. In processing, three statements are needed: one to get the size of the string, another to size the field based on `StrSize`, and still another to get the value of the stored string. In the following example, the second instruction allocates space for the exact size of the member's name:

```
Get #1, , StrSize
member.pname = String$(StrSize, " ")
Get #1, , member.pname
```

A simpler form is to use the `Input` statement. The following instructions also get the size of the string, but use a single statement to size and assign the value of the stored string to the member's name:

```
Get #1, , StrSize
member.pname = Input(StrSize, 1)
```

21.2.2 Determining Record Locations

When record length is variable, the toughest part of binary file processing is determining record locations. One way of determining where a record exists involves the use of a file-location or record lookup table. Consider the following code, placed in a .BAS module:

```
Type rosterTable
  mname As String
  FileLoc As Integer
End Type
```

Also consider the corresponding code placed in a procedure:

```
Const FILESIZE = 100
Static memberTable (1 to FILESIZE) As rosterTable
```

This code makes it possible to search through the `memberTable` to look up a person's name (`mname`) and find the byte location of the person's record (`FileLoc`).

A modification to the `Add Record` procedure is required to load this type of table if the table is prepared as records are added. The `Find Record` procedure must also be modified. Consider the following code for the latter case:

```
Const FILESIZE = 100
Static memberTable (1 to FILESIZE) As rosterTable
Dim member as roster

Dim count As Integer
Dim memName As String

Position = 1       'Position is a global variable

Open "sport3.dat" For Binary As #1

For count = 1 To FILESIZE
  If EOF(1) = True Then
    memberTable(count).mname = "END_OF_FILE"
    Exit For
  End If
  memberTable(count).FileLoc = Position

  Get #1, position, StrSize
  member.pname = String$(StrSize, " ")
  memberTable(count).mname = member.pname

  Position = Seek (#1)
Next count
```

In this example, the variable named `position` is a global variable that is updated each time a binary record is read. The following instruction gets and saves the current file position:

```
Position = Seek (#1)
```

With each new record, the updated position is assigned to the record's `FileLoc` element. Why is this table helpful? To look up a record, enter a person's name and find the file location position. Use this position directly in locating a record on file.

Exercise 21.3 Building a Binary File

Revise the Sports Club Roster in Exercise 21.1, using a binary file in place of a random-access file. The interfaces shown in Figure 21.1 and Figure 21.2 remain the same.

This exercise does not require a separate record lookup table. Exercise 21.4 asks you to find a single member's record placed in the file.

1 Use the same interface as Exercise 21.1.

2 Modify the user-defined type, placed in a .BAS file, as follows:

```
Type roster
  pname As String
  street As String
  city As String
  state As String
  zip As String
```

continues

```
  phone As String
  sport As String
End Type
Public position As Integer
```

The `position` component is a global variable of type `Integer`.

3 Rewrite the `Add Record` procedure. Include the following statements:

```
Private Sub cmdAdd_Click()
  Dim member As roster
  Dim StrSize As Integer

  'Keep a count of the number of records added
  Static count As Integer
  count = count + 1

  'Assign values for a record
  member.pname = txtPname.Text
  member.street = txtStreet.Text
  member.city = txtCity.Text
  member.state = txtState.Text
  member.zip = txtZip.Text
  member.phone = txtPhone.Text
  member.sport = txtSport.Text

  If member.pname = "" Then
    txtPname.SetFocus
    Exit Sub
  End If

  'Open file
  If count <= 1 Then
    Open "sport3.dat" For Binary As #1
  End If

  'Determine size of field
  StrSize = Len(member.pname)
  'Write field to disk
  Put #1, , StrSize
  Put #1, , member.pname

  StrSize = Len(member.street)
  Put #1, , StrSize
  Put #1, , member.street

  StrSize = Len(member.city)
  Put #1, , StrSize
  Put #1, , member.city

  StrSize = Len(member.state)
  Put #1, , StrSize
  Put #1, , member.state

  StrSize = Len(member.zip)
  Put #1, , StrSize
  Put #1, , member.zip
```

```
    StrSize = Len(member.phone)
    Put #1, , StrSize
    Put #1, , member.phone

    StrSize = Len(member.sport)
    Put #1, , StrSize
    Put #1, , member.sport
    txtPname.Text = ""
    txtStreet.Text = ""
    txtCity.Text = ""
    txtState.Text = ""
    txtZip.Text = ""
    txtPhone.Text = ""
    txtSport.Text = ""

    txtPname.SetFocus
  End Sub
```

4 Rewrite the `Display Records` procedure. Include the following instructions:

```
Private Sub cmdDisplay_Click()
  Dim member As roster
  Dim count As Integer
  Dim StrSize As Integer
  Const FILESIZE = 10

  'Close any open files
  Close
  frmShow.Cls
  frmShow.Show
  frmShow.Print "Sports Roster"

  position = 1
  Open "sport3.dat" For Binary As #1

  For count = 1 To FILESIZE
    If EOF(1) = False Then
      'Get size of field
      Get #1, position, StrSize
      'Display field
      member.pname = String$(StrSize, " ")
      Get #1, , member.pname

      Get #1, , StrSize
      member.street = String$(StrSize, " ")
      Get #1, , member.street

      Get #1, , StrSize
      member.city = String$(StrSize, " ")
      Get #1, , member.city

      Get #1, , StrSize
      member.state = String$(StrSize, " ")
      Get #1, , member.state

      Get #1, , StrSize
      member.zip = String$(StrSize, " ")
      Get #1, , member.zip
```

continues

```
        Get #1, , StrSize
        member.phone = String$(StrSize, " ")
        Get #1, , member.phone

        Get #1, , StrSize
        member.sport = String$(StrSize, " ")
        Get #1, , member.sport

        'Set the file position
        position = Seek(1)

        frmShow.Print
        frmShow.Print member.pname,
        frmShow.Print member.street,
        frmShow.Print member.city,
        frmShow.Print member.state,
        frmShow.Print member.zip,
        frmShow.Print member.phone,
        frmShow.Print member.sport
      Else
        Exit For
      End If
    Next count
    'Close the file after displaying all records
    Close
  End Sub
```

5 Run and test your program.

Exercise 21.4 Adding a Record Lookup Table and Procedure

Add a lookup table by which to find a person's record following the entry of the person's name. Use the same interface as designed for Exercise 21.2; however, modify `form3` (`frmList`). Figure 21.6 shows the revised form.

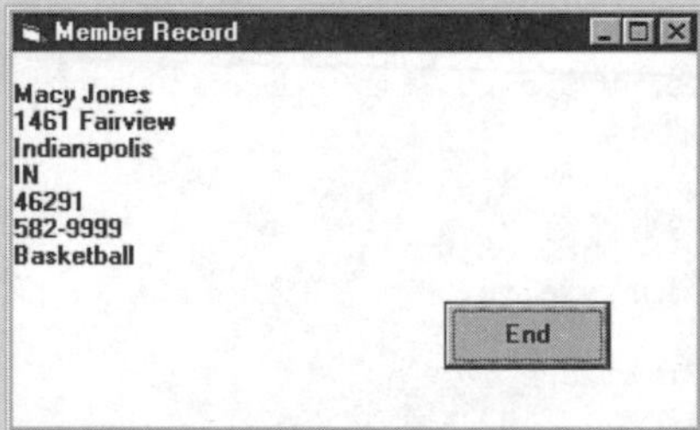

Figure 21.6

This output display shows the member information of the name retrieved from the binary file.

1 Make the `frmList` caption change.

2 Add a second user-defined type to the .BAS module. The type is defined as follows:

```
Type rosterTable
  mname As String
  fileLoc As Integer
End Type
```

This type will be used to store each member's name and file location.

3 Modify the `Find Record` procedure using the new user-defined type. The first section of the code fills the table with data; the second section uses the table to find the record location of a member.

```
Private Sub cmdFind_Click()

  Const FILESIZE = 10
  'Declare an array of records
  Static memberTable(1 To FILESIZE) As rosterTable

  Dim member As roster
  Dim count As Integer
  Dim memName As String
  Dim i As Integer
  Dim location As Integer
  Dim StrSize As Integer

  'Close all open files
  Close
  'Set the initial position
  position = 1
  Open "sport4.dat" For Binary As #1

  frmList.Show
  frmShow.Hide
  frmEnter.Hide
  For count = 1 To FILESIZE
    If EOF(1) = True Then
      memberTable(count).mname = "END_OF_FILE"
      Exit For
    End If

    memberTable(count).FileLoc = position
    Get #1, position, StrSize
    member.pname = String$(StrSize, " ")
    Get #1, , member.pname

    memberTable(count).mname = member.pname
    Get #1, , StrSize
    member.street = String$(StrSize, " ")
    Get #1, , member.street

    Get #1, , StrSize
    member.city = String$(StrSize, " ")
    Get #1, , member.city
```

continues

```
    Get #1, , StrSize
    member.state = String$(StrSize, " ")
    Get #1, , member.state

    Get #1, , StrSize
    member.zip = String$(StrSize, " ")
    Get #1, , member.zip

    Get #1, , StrSize
    member.phone = String$(StrSize, " ")
    Get #1, , member.phone

    Get #1, , StrSize
    member.sport = String$(StrSize, " ")
    Get #1, , member.sport

    'Set the file position
    position = Seek(1)
  Next count
  memName = InputBox("Enter member's name")

  'Search for name in array
  For i = 1 To count - 1
    If memName = memberTable(i).mname Then
     location = memberTable(i).FileLoc
    End If
  Next i

  'Get member information
  Get #1, location, StrSize
  member.pname = String$(StrSize, " ")
  Get #1, , member.pname

  Get #1, , StrSize
  member.street = String$(StrSize, " ")
  Get #1, , member.street

  Get #1, , StrSize
  member.city = String$(StrSize, " ")
  Get #1, , member.city

  Get #1, , StrSize
  member.state = String$(StrSize, " ")
  Get #1, , member.state

  Get #1, , StrSize
  member.zip = String$(StrSize, " ")
  Get #1, , member.zip

  Get #1, , StrSize
  member.phone = String$(StrSize, " ")
  Get #1, , member.phone

  Get #1, , StrSize
  member.sport = String$(StrSize, " ")
  Get #1, , member.sport
```

```
        frmList.Print
        frmList.Print member.pname
        frmList.Print member.street
        frmList.Print member.city
        frmList.Print member.state
        frmList.Print member.zip
        frmList.Print member.phone
        frmList.Print member.sport

        Close
    End Sub
```

4 Run and test your program.

In testing your program, observe that the name must be typed exactly as shown in the roster, as there is no built-in error checking. Developing a "bulletproof" program is left as an assignment (the fourth exercise in the "Skill-Building Exercises" section at the end of this chapter).

Chapter Summary

This chapter considers two alternatives to sequential file processing: random-access and binary file processing. In Visual Basic, random-access file processing is the default. Random-access files store identical records containing the same fields requiring the same number of bytes. This leads to a fixed-length record. Compared to a sequential file, a random-access file allows direct access to a single record without reading or writing all records stored prior to the desired record.

In contrast, a binary file does not contain fixed-length records, but rather records sized to fit the number of bytes needed. A binary file usually requires less storage space than a random-access file, but requires more programming instructions.

The fields of a random-access record are defined with a user-defined type. The record length can then be computed using the `Len()` function. The `LOF()` function (the size of the file in bytes) divided by the record length determines the number of stored records.

`Seek #` and `Loc()` are of particular importance to random-access and binary file processing. `Seek` moves the file pointer to a record specified in a file, while `Loc()` returns the current location in a file.

`Get` statements are used to read a record from a random-access file; `Put` statements are used to write a record to a random-access file. Once a record is read, individual fields of the record can be displayed using the *dot* notation, as in the following example:

```
Print Employee.FirstName    'Where Employee is a record and FirstName is a field
```

Binary file processing stores identical records in terms of their fields, but allows for variable-length record size. Like random-access processing, `Get` is used to read and `Put` to write records to a file; however, in writing, two `Put` instructions are used: one to store the size of the field and another to store the actual field. In reading a record, `Get` is used to first read the string size. Following this, a second statement reads the stored field value.

The toughest part of binary file processing is determining record locations. One way of doing this is to create a record lookup table to store the record identifier and its file location. Prior to using `Seek`, the table is searched to locate the file location position. Following this, it is possible to directly retrieve a desired record.

Skill-Building Exercises

1. A main advantage of random-access processing is that a single record can be identified, read from storage, modified, and written back to storage. Modify the program used in Exercise 21.1 and Exercise 21.2 to allow a record to be displayed, modified, and written back to storage. As a test, change Macy Jones's ZIP code to 46292.

2. Modify the program used in Exercise 21.1 and Exercise 21.2 to allow for the next record in the file to be displayed as well as the previous record in the file.

3. Modify the program used in Exercise 21.1 and Exercise 21.2 to allow for a record to be deleted from the file and the file to be rewritten. In your design, copy all records following the one you want to delete, "down" one record position. This will overwrite the record you want to delete. Some of code for this procedure is as follows:

   ```
   Dim member As roster
   Dim count As Integer
   For Count = Position To NumRec -1
     Get #1, Count + 1, member
     Put #1, Count, member
   Next Count
   ```

   This procedure will leave a duplicate record at the end of the file. To remove the duplicate record, copy all valid records from the original to a new file, Delete (`Kill`) the old file, and use the `Name` statement to rename the file with the name of the original file. The following is the syntax for this statement:

   ```
   Name oldfilespec As newfilespec
   ```

4. Make the binary file program "bulletproof." If the user enters an incorrect user name, indicate that this name is not on file and ask the user to try again. Data entry should handle both uppercase and lowercase entries.

5. Make the random-access file program in Exercise 21.2 more modular. Call procedures to add records to and read records from a file, passing arguments (such as the file name) as needed.

Part III Applications

Modular Design

When developing Visual Basic applications, modular design is encouraged through the use of a hierarchy of forms and careful segmentation of application functionality into code and class modules. Applications often are constructed in *layers.* The different Visual Basic module types are then used to build these layered Visual Basic applications.

This chapter includes exercises that review the use of multiple forms in a larger application. It also clarifies the passing of values between procedures and between procedures and functions. Specifically addressed are the differences between *call by reference* (by address) and *call by value.* A "traditional" approach to a sample application is compared and examined with a new, more *object-oriented* approach.

You learn many tasks in this chapter. After you work through the exercises, you will be able to

- create a login window;
- add password control to a design;
- design a visual chart of a Visual Basic application;
- show, hide, load, and unload forms;
- use call by reference;
- pass an array to a procedure;
- use call by value;
- work with arrays of user-defined types;
- work with an object collection;
- use the MSFlexGrid control to display the results of processing;
- segment an application using standard code modules;
- segment an application using class modules.

Chapter Objectives

This chapter condenses the preceding lengthy list of tasks into four objectives that address the use of modules and layers in program design:

1. Understanding the significance of modular design
2. Understanding the difference between call by reference and call by value
3. Using code modules in the processing/business-rule and data layers
4. Using class modules in the processing/business-rule and data layers

Objective 22.1 Understanding the Significance of Modular Design

Modular design is the breaking down of a computer program into recognizable, functional components, as opposed to creating one long application. The complexity of building applications for graphic-oriented operating systems, such as Windows, places a premium on employing reusable software modules. Ideally, these components can be used to rapidly deploy software that is reliable and readily understandable to those who must maintain it. Needless to say, software development often falls short of this ideal. But adhering to modular design principles makes the chances of meeting this ideal much greater. The benefits of modular design include making it easier to

- define the solution to the design problem,
- communicate to others the internal construction (or organization) of the design,
- specify the functionality of a coded procedure by designing each procedure (or function) to accomplish a specific task,
- assign programming tasks to a group of people,
- write computer instructions (leading to more rapid development of computer applications),
- isolate problems better during program design and testing,
- maintain computer programs when errors occur.

Perhaps the most important benefit is the third: specifying the functionality of a coded procedure. When designing a Visual Basic program, use a form to convey a specific function, write procedures or functions to perform one and only one task, and use standard and class modules to clarify the importance of application specifics and where they are defined. With a modular design, it is possible to clearly specify the functions and organization of an application.

22.1.1 The Layers of an Application

One frequently used approach to software development is to think about an application as being composed of several layers, as follows:

- **User interface layer**—The *user interface layer* refers to the forms and visual components by which the user navigates through the program and executes the program's functions. This layer ideally is easily learned, logical, and obvious in the execution of the program's core functionality. It retrieves user input and displays the results of processing.
- **Processing or business-rule layer**—This layer comprises the modules that retrieve and write persistent data on a disk, maintain program security, and perform calculations on data. It is called the *business-rule layer* because it is here that the policies of the enterprise are often implemented in code. One rule might specify the difference between the markups for wholesale and retail customers. Another rule might state that prompt payment of invoices entitles the customer to a discount. These rules are often encapsulated in this middle layer.
- **Data layer**—Consisting of the application's persistent (stored) data, the *data layer* often is represented by a relational database. (Databases are covered in Chapter 24, "Accessing and Manipulating Databases with Visual Basic," and Chapter 25, "Using Data Access Objects.")

There is no hard and fast rule for employing these layers. It is better to think of them as a logical architecture, not a preordained model that must be followed. Business rules, for instance, are frequently handled at the data layer as stored procedures in the database. Speed requirements might necessitate bypassing a business-rule module to directly access data.

This chapter is organized around the concept of layers to help you become familiar with this idea. Although a *layered* approach is often explicitly used in developing client/server applications, this design approach is not limited to client/server database programs. Multimedia, game, and even stand-alone applications can employ similar modular design.

22.1.2 Using Form Modules to Build the Interface Layer

Because a Visual Basic application can be designed using numerous forms, the organization of the forms collection needs to be understood by others. A forms-based structure chart provides such a picture. As shown in Figure 22.1, one way of depicting the interface layer of a design is by tracing the design hierarchy. In the example, only two symbols are used: one to depict forms and one to depict code segments. As illustrated, this application begins with a Welcome window (which is a login window) and proceeds to the Main Menu window. From the Main Menu window, it is possible to search by name, category, or keyword, and to obtain help. Selecting any of these four choices leads to another form. (The link to Help is indicated by the encircled 1s in the diagram.)

Figure 22.1

A forms-based structure chart shows how the various forms of a design are related.

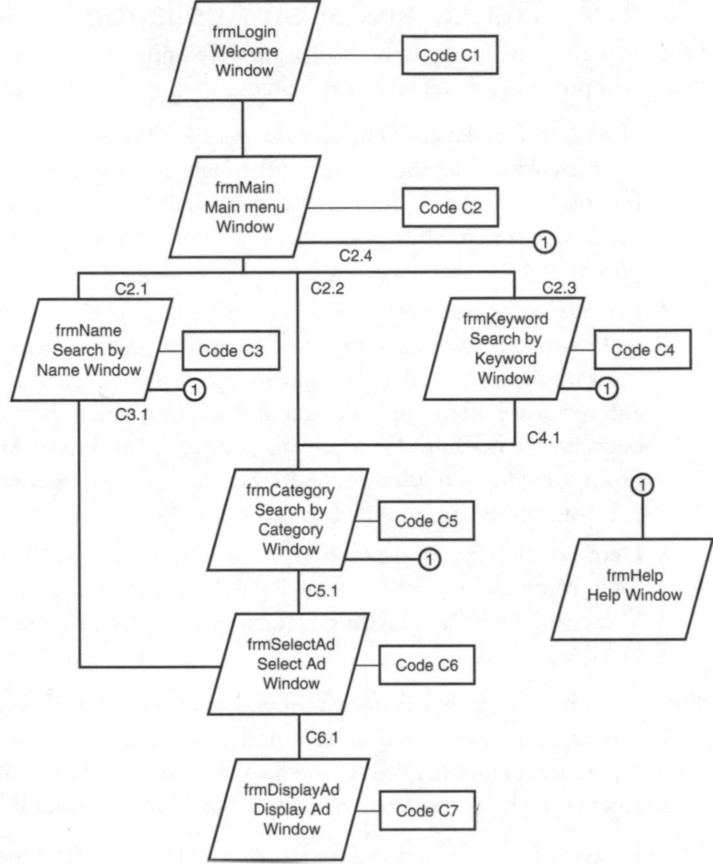

22.1.3 A Login Window and a Main Window

Most multiple-form designs feature a login window and a main window. The login window is added to provide security of access to a design. Unless the user knows the correct login name and password, access to a design is denied. The main window serves as the "home base" in a design. This window illustrates the various ways to navigate through a design. With many designs, a return-to-main-window option is provided to always allow the user to return to this "home base."

Exercise 22.1 Creating a Login Window

Create a login window to screen users from access to an application. The login name is assigned within the code; however, a search of a file of login names is more typical in practice. Figure 22.2 and Figure 22.3 show the login window and the Main Menu window, respectively.

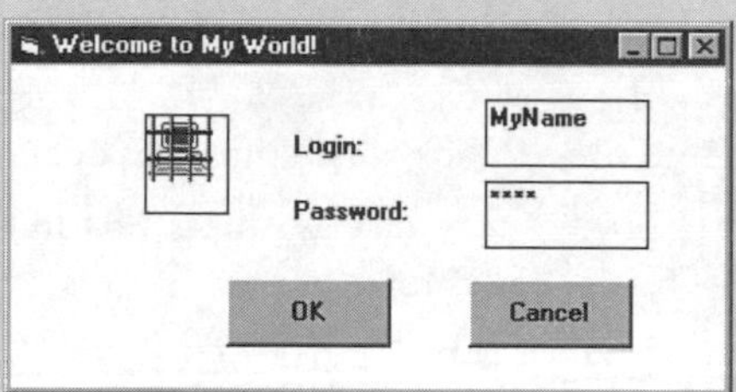

Figure 22.2

The login window requires the user to enter a user name and coded password.

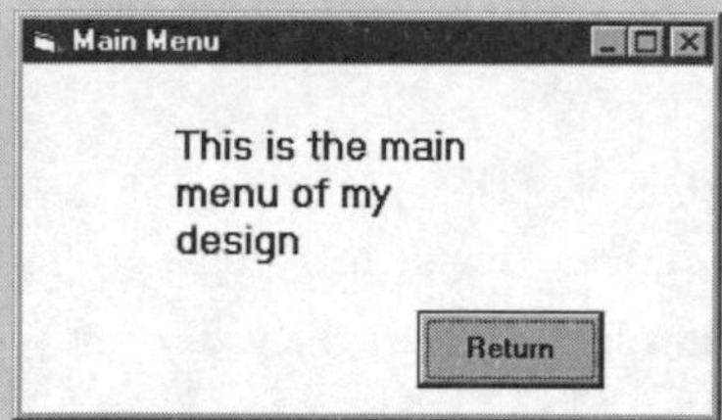

Figure 22.3

A correct password opens the Main Menu window.

1 Create the forms **frmLogin** and **frmMain**, as shown.

2 Write the following code for the OK command button.

```
Private Sub cmdOK_Click()
   Dim message1 As String
   Dim message2 As String
   Dim message3 As String
   Dim options As Integer

  'Password is Open
   message1$ = "Please enter your password"
   message2$ = "Login name not found, please try again"
   message3$ = "Login error"

   If txtPassWord.Text = "" Then
     Beep
     MsgBox message1, 1, message3
      txtPassWord.SetFocus
   ElseIf (txtPassWord.Text = "Open") Or (txtPassWord.Text = "open") Then
     frmLogin.Hide
     frmMain.Show
   Else
     MsgBox message2, 1, message3
   End If
End Sub
```

continues

In this procedure, an error allows the user to try again; however, the user must enter the password—the word *Open* or *open*—to gain access to `frmMain`. After an error, focus is given to the text box where the error occurred.

3 Write a `Form_Load()` procedure to display empty text in the text boxes when `frmLogin` appears. This code is as follows:

```
Private Sub Form_Load()
   txtLogin.Text = ""
   txtPassWord.Text = ""
End Sub
```

4 Write the following `cmdReturn_Click()` procedure for `frmMain`.

```
Private Sub cmdReturn_Click()
   frmMain.Hide
   frmLogin.Show
End Sub
```

Once again, the visible form is hidden before the new form appears. Making the password invisible requires a change to a single property: `PasswordChar`.

5 Go to the password text box. Change the `PasswordChar` property on `frmLogin` to an asterisk (*).

6 Test your modified design. Use a different `PasswordChar`.

22.1.4 Navigating and Documenting a Multiple-Form Application

Modular design requires that the user be able to navigate through multiple forms. You have already used most of the available methods and statements to control forms and their displays. As a review, Table 22.1 lists these methods and statements and their meanings.

Table 22.1 Methods and Statements for Controlling Forms

Method/Statement	*Description*
`Hide` method	Hides a form, but does not unload it
`Show` method	Shows a hidden form
`Load` statement	Loads a form (or control) into memory
`Unload` statement	Unloads a form (or control) from memory

Recall that the `Load` statement adds a form to memory, while `Unload` removes a form from memory. If a form is used more than once during runtime, it is wise to hide, but not unload, the form following its use. This improves program performance. If a form is used infrequently, it is best to unload it from memory because this saves memory.

Exercise 22.2 Working with Several Forms

This exercise asks you to work with five forms named `frm1` through `frm5`. It demonstrates the use of the `Unload` statement. Figure 22.4 shows the relationship among the various forms. As you load and unload forms, the count of the number of forms stored in memory is displayed.

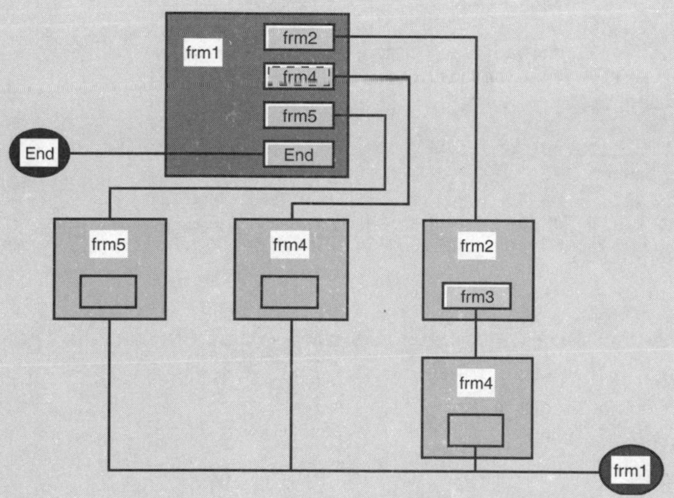

Figure 22.4

From form 1 you can move to form 2, form 4, and form 5, but not to form 3.

1 Add four buttons to `frm1` (**`cmdForm2`**, **`cmdForm4`**, and **`cmdForm5`**), add the `End` procedure and an output label (**`lblOutput`**), and write the following code:

```
Private Sub cmdForm2_Click()
   frm1.Hide
   frm2.Show
End Sub

Private Sub cmdForm4_Click()
   frm1.Hide
   frm4.Show
   frm4.txt1.SetFocus
End Sub

Private Sub cmdForm5_Click()
   frm1.Hide
   Load frm5
   frm5.Show
End Sub
```

continues

2 Add a button (**cmdShowForm3**) to `frm2`, and write the following procedure:

```
Private Sub cmdShowForm3_Click()
   frm2.Hide
   frm3.Show
End Sub
```

3 Add an image of your choice (**Image1**) to `frm3`, and write the following procedure:

```
Private Sub Image1_Click()
   Dim i As Integer
   frm3.Hide
   'This will unload the form
   'It will not appear in the count
   Unload frm3

   frm1.lblOutput.Caption = ""
   frm1.Show

   For i = 0 To Forms.Count - 1
      frm1.lblOutput.Caption = frm1.lblOutput.Caption & Forms(i).Caption
   Next i
End Sub
```

4 Add another image (of your choice) to `frm4`, and write a similar procedure, leaving out the `Unload` statement this time:

```
Private Sub Image1_Click()
   Dim i As Integer

   frm4.Hide
   frm1.Show
   frm1.cmdForm2.SetFocus
   frm1.lblOutput.Caption = ""

   For i = 0 To Forms.Count - 1
     frm1.lblOutput.Caption = frm1.lblOutput.Caption & Forms(i).Caption
   Next i
End Sub
```

5 Add a final image (of your choice) to `frm5`, and write a similar procedure, adding the `Unload` statement:

```
Private Sub Image1_Click()
   Dim i As Integer
   frm5.Hide
   'This will unload Form 5
   'It will not appear in the count

    Unload frm5
    frm1.Show
    frm1.cmdForm4.SetFocus
    frm1.lblOutput.Caption = ""
    For i = 0 To Forms.Count - 1
      frm1.lblOutput.Caption = frm1.lblOutput.Caption & Forms(i).Caption
    Next i
End Sub
```

6 Run and test your program.

> If you load `frm3` or `frm5`, the forms index displayed will not reflect this change. Unloading removes a form from the index, indicating that it is no longer part of the runtime program.

Objective 22.2 Understanding the Difference Between Call by Reference and Call by Value

Central to modular design is the built in capability to create procedures and functions, which are, in turn, called by a procedure or function. In this way, the modules that make up an application are able to interact and pass data among themselves. The `Call` statement serves to call a procedure. With either procedures or functions, arguments can be passed from the calling statement. For procedures, the syntax for the `Call` statement is

```
[Call] name [(argumentlist)]
```

The component `argumentlist` is optional.

The argument list is received by the called procedure and follows another syntax. The complete syntax for a procedure is as follows:

```
 [Private | Public] [Static] Sub name [(arglist)]
      [statements]
    [Exit Sub]
     [statements]
End Sub
```

The argument list (`arglist`) must be used when the call to the procedure contains arguments. This `arglist` must contain the same number of arguments as those found in the calling statement.

The complete syntax for a `Function` is only slightly more complex, with the `As type` clause added to the syntax:

```
 [Public | Private] [Static] Function name [(arglist)] [As type]
     [statements]
     [name = expression]
     [Exit Function]
     [statements]
     [name = expression]
End Function
```

As with procedures, if the call to the function contains arguments, the called function must also contain the same number of arguments in the `arglist` as those found in the calling statement. The `As type` clause specifies the type of data returned by the function. Why is this `As type` clause used? With a procedure, no values are returned following processing. With a function, one and only one value is returned. The `As type` clause indicates the type of data being returned.

When passing arguments to a procedure or a function, you can pass arguments *by reference* or *by value.*

22.2.1 Call by Reference

Call by reference, the default, means that a copy of the address of the variable (not its value) is passed to the receiving procedure or function. Argument values passed by address can be directly altered by the procedure or function.

Exercise 22.3 Passing Values by Address

This exercise asks you to pass three values to a function named `calc_max`; the purpose of the function is to find the highest number and return it to the calling procedure. Figure 22.5 shows the data-entry screen for entering the numbers, while Figure 22.6 shows the highest value entered.

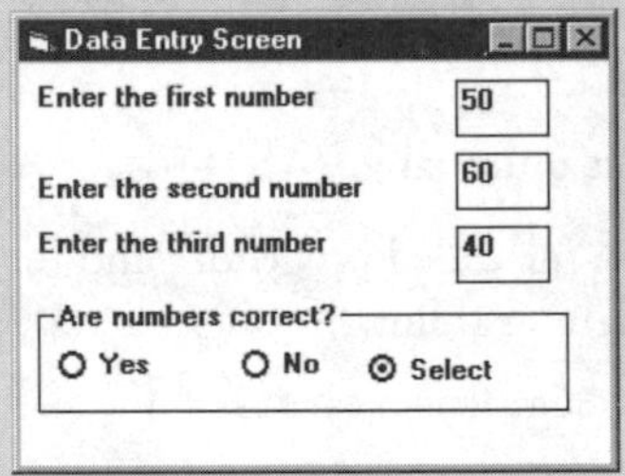

Figure 22.5

Enter three numbers, and when you are satisfied, click Yes in the section labeled Are numbers correct?

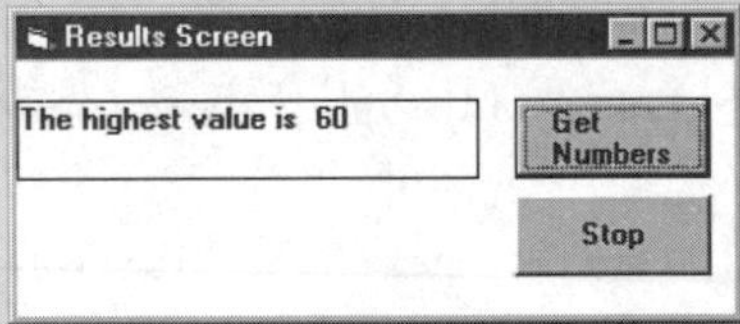

Figure 22.6

The function examines all three numbers and returns the highest value.

1 Create the forms **`frmResults`** and **`frmDataEntry`**. The text boxes on `frmDataEntry` represent a control array named `txtNum(0 to 3)`. Place the option buttons (**`optYes`**, **`optNo`**, and **`optChoice`**) in a frame. Write the `End` procedure (shown as Stop) for `frmResults`.

2 Write the following `cmdGetNum_Click()` procedure in the `frmResults` module:

```
Private Sub cmdGetNum_Click()
   frmResults.Hide
   frmDataEntry.Show
   frmDataEntry!optChoice.SetFocus
End Sub
```

3 Write the `optYes_Click()` and `optNo_Click()` procedures in the `frmDataEntry` module:

```
Private Sub optYes_Click()
   Dim num1, num2, num3 As Integer
   Dim max As Integer

   num1 = CInt(txtNum(0).text)
   num2 = CInt(txtNum(1).text)
   num3 = CInt(txtNum(2).text)

   max = calc_max(num1, num2, num3)

   frmDataEntry.Hide
   frmResults.Show
   frmResults.lblResult.Caption = "The highest value is " & Str$(max)
End Sub
```

This first procedure reveals a function call with three values passed to a function named `calc_max()`. When the maximum value is returned, it is displayed on `frmResults` (not `frmDataEntry`).

```
Private Sub optNo_Click()
   Dim i As Integer
   For i = 0 To 2
      txtNum(i).text = ""
   Next i

   txtNum(0).SetFocus
   optChoice.SetFocus
End Sub
```

This second procedure clears all text boxes, and allows the user to start again.

4 Write the `optChoice_Click()` procedure:

```
Private Sub optChoice_Click()
   txtNum(0).SetFocus
   txtNum(0).text = 0
   txtNum(1).text = 0
   txtNum(2).text = 0
End Sub
```

This code sets all values in the text box to zero, and gives the topmost text box the focus.

5 Write the `calc_max()` function in the `frmDataEntry` module as follows:

```
Private Function calc_max (num1, num2, num3)
   If num1 > num2 And num1 > num3 Then
      calc_max = num1
```

continues

```
    ElseIf num2 > num3 Then
        calc_max = num2
    Else
        calc_max = num3
    End If
End Function
```

In this function, the three numbers are compared to one another to determine which number is the largest. This number is returned to the calling procedure.

Exercise 22.4 Passing an Array to a Procedure

Arrays can be passed to a procedure, much like a single variable, using call by reference. With arrays, two style changes must be remembered:

- The array name in the calling statement must be followed by empty parentheses.
- The called procedure must specify the array name.

In this exercise, you will pass an array to a procedure where it is filled with letters. Following this, you will pass the filled array to a second procedure where the stored values are printed (see Figure 22.7).

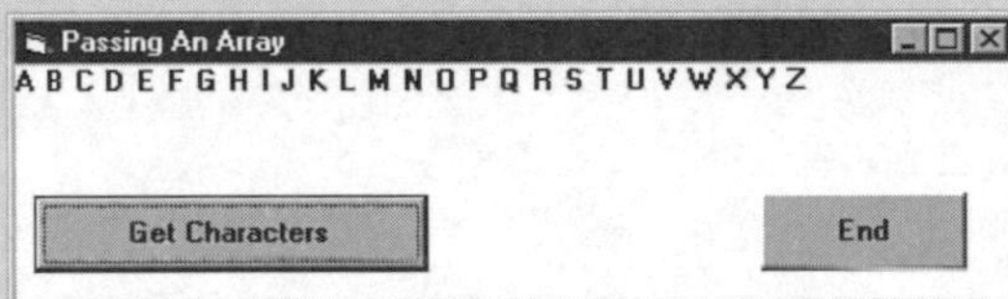

Figure 22.7

The Get Characters `Click()` *event calls two procedures: one to fill an array with letters and a second to display the array of letters on the form.*

1 Add the two command buttons, **`cmdGetChar`** and **`cmdEnd`**, and write the `End` procedure.

2 Write the following `cmdGetChar` procedure:

```
Private Sub cmdGetChar_Click()
   Static letters(0 To 25) As String * 1
   Call fun1(letters())
   Call fun2(letters())
End Sub
```

3 Write the following `fun1()` and `fun2()` procedures:

```
Private Sub fun1 (letters() As String * 1)
   'This procedure fills the array with letters
   Dim i As Integer
   For i = 0 To 25
     letters(i) = Chr$(i + 65)
   Next i
End Sub

Private Sub fun2 (letters() As String * 1)
   'This procedure prints the filled array of letters
   Dim i As Integer
   For i = 0 To 25
     Print letters(i) & " ";
   Next i
End Sub
```

The `fun1` procedure fills the array with letters and returns control to the calling function, `cmdGetChar`. The `fun2` procedure displays the array of letters and returns control to the calling function.

22.2.2 Call by Value

Call by value means that a copy of the variable (not its address) is passed to the receiving procedure or function. Arguments passed by value cannot be altered by the procedure or function. Instead, only the copy can be altered.

To pass an argument by value, two special rules are required:

- In the calling statement, insert parentheses around the argument in the call statement, as in the following example:

  ```
  Call Reverse ((my_age), my_name)
  ```

  In this example, the variable `my_age` (placed within parentheses) is passed by value; the variable `my_name` is passed by address.

- The second special rule is found in the called function. With call by value, the `ByVal` keyword is placed in the procedure's parameter list, as in the following example:

  ```
  Private Sub Reverse (ByVal my_age, my_name)
  ```

  This clearly indicates that `my_age` is an argument passed by value. As before, `my_name` is an argument passed by reference.

Exercise 22.5 Passing Arguments by Value

Make a call to a procedure, named `Reverse`, that changes your entered name, but not your entered age. Figure 22.8 shows the interface to create.

continues

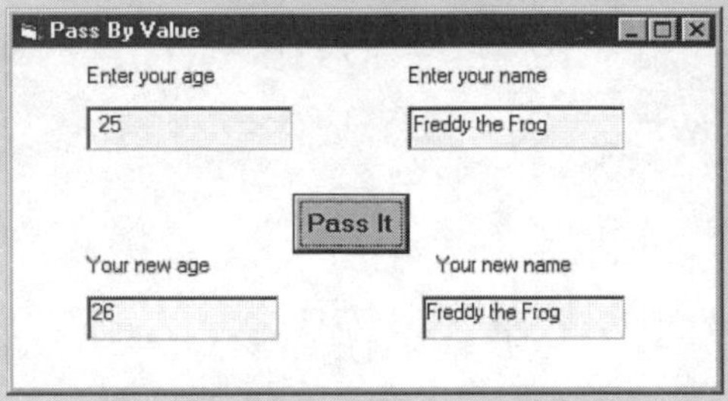

Figure 22.8

With pass by value, your age is copied before it is passed, which leads to multiple values.

1 Create the interface shown in Figure 22.8. The top row contains two text boxes (**`txtAge1`** and **`txtName1`**); the bottom row contains two labels (**`lblAge2`** and **`lblName2`**).

2 Write the following `cmdPassIt_Click()` procedure:

```
Private Sub cmdPassIt_Click()
   Dim my_age As Integer
   Dim my_name As String
   'Assign the initial values of age and name
   my_age = Val(txtAge1.Text)
   my_name = txtName1.Text

   'Pass my_age by value
   'Pass my_name by address
   Call Reverse((my_age), my_name)
   txtAge1.Text = Str$(my_age)
   txtName1.Text = my_name
End Sub
```

3 Write the following `Reverse()` procedure:

```
Private Sub Reverse (ByVal my_age, my_name)
   'Add ByVal in the parameter list to indicate
   'passing by value
   'Passing by address requires no special reference
   'My_age will have an initial and an after value
   my_age = my_age + 1
   'Sorry!  You're name will always be Freddy the frog
   my_name = "Freddy the frog"

   lblAge2.Caption = my_age
   lblName2.Caption = my_name
End Sub
```

4 Run the program. Observe that after the procedure is called, the new values of `txtAge1` and `txtName1` are displayed, using the labels. Only one of these will retain the original value passed to the function named `Reverse`. Test this program by reversing the argument passed by value and the argument passed by address.

22.2.3 Choosing the Best Method

The decision to use call by value or call by reference is sometimes determined in advance for you. For example, when making Windows API (application program interface) calls, the `ByVal` designation is typically required. As another example, you typically avoid passing large arrays by value in order to conserve memory. Instead, large arrays are usually passed by reference.

Besides saving memory, passing by reference is faster because variables need not be copied. For these reasons, passing by reference is the most common calling convention, and it is the Visual Basic default.

Objective 22.3 Using Code Modules in the Processing/Business-Rule and Data Layers

You have worked with forms and employed statements and methods to navigate through a multiple-form application. This section presents the use of a standard code module to perform the business-rule services typically found in the processing or business-rule layer of an application. This layer of an application performs the following tasks:

- Retrieves information from the user
- Makes calculations
- Reads and writes persistent data to and from a database or file
- Returns the results of processing to the user interface layer

22.3.1 Placement of Code in Standard Code Modules

With single forms, references to objects do not require the full reference, which includes the form name. This differs from a multiple-module application. In that environment, references to objects in other modules—such as controls placed on a form module—require a full reference to the object. You might, for instance, want to pass a control as a parameter to a procedure in a standard code module. Or, you might simply want to assign a new value to one of the control's properties. Any reference to the control requires that the module in which it resides be referenced.

Changing the caption property of a label named `lblDisplay` on a form module named `frmInvoice`, for example, would require the full reference when the assignment statement came from another module:

```
frmInvoice.lblDisplay.Caption = "New order!"
```

Let's assume the label is to be passed to a public procedure in another module:

```
Public Sub ChangeCaption(DisplayLabel as Label, NewCaption as String)
    DisplayLabel.Caption = NewCaption
End Sub
```

The call to the procedure would require a full reference:

```
Call ChangeCaption(frmInvoice.lblDisplay, "New Order!")
```

22.3.2 Segmenting an Application with Standard Code Modules

Until the recent advent of object-oriented software development, *modular design* meant short procedures with precise functions that could be readily documented and maintained. All data validation procedures might be placed together in one group of procedures, data access procedures in another group, and data presentation procedures in yet another. In Visual Basic, the use of a set of standard modules is still seen in a majority of applications.

Building your own class modules within Visual Basic is a relatively new feature. Class modules existed in Visual Basic 4.0, but the advanced class design capabilities of Visual Basic 5.0 have moved the language much closer to a genuine object-oriented language.

Exercise 22.6 Using Forms and a Standard Module to Create a Ski Rental Shop Application

Using standard code modules to segment an application into a logical, layered design remains the heart of most Visual Basic programs. This exercise demonstrates how standard code modules can be used for this purpose.

The ski rental application is composed of two forms and one standard module. One form is as an order form for ski rental equipment, and the other form, the invoice form, displays the ordered rental items and a total. The invoice form uses the MSFlexGrid control. Because of space limitations, the data layer in this application is represented by a global array of user-defined types rather than a random-access file. The data layer in most applications is represented by the stored tables of a relational database.

In this somewhat lengthy exercise you use a standard code module as the primary means to build the processing/business-rule layer of an application. The user may select ski items to rent, and if desired, provide the customer a standard 15 percent discount. Figure 22.9 and Figure 22.10 show the two forms for this application.

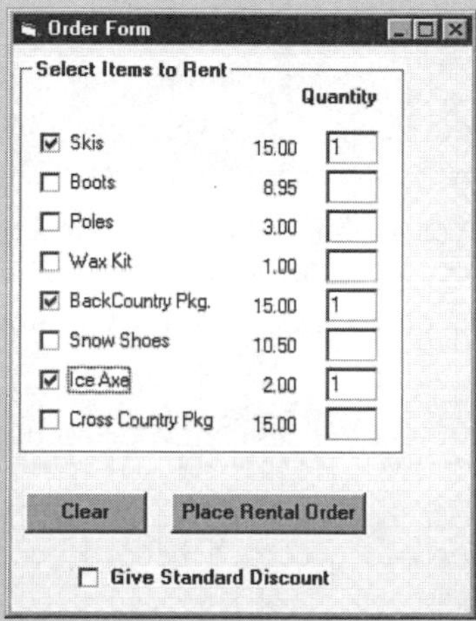

Figure 22.9

The ski rental order form provides a selection of rental items.

Invoice Form

Item	Quantity	Each	Total
Skis	1	$15.00	$15.00
BackCountry Pkg.	1	$15.00	$15.00
Ice Axe	1	$2.00	$2.00
Total:			**$32.00**

Figure 22.10

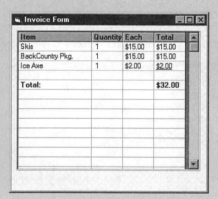

The ski rental invoice shows the items that were selected and their total cost.

1 Add the two forms (`frmOrder` and `frmInvoice`) to a project. Save these as **f 22-6A.frm** and **f 22-6B.frm**, respectively.

2 Place a frame on `frmOrder`, and change the caption to **Select Items to Rent**.

3 On `frmOrder`, place three control arrays inside the frame: an array of check boxes named **`chkItem()`**, an array of labels without borders named **`lblPrice()`**, and an array of text boxes named **`txtQuantity()`**. All three control arrays have a subscript range of `0` to `7` running from top to bottom.

continues

4 Add the two command buttons, **cmdClear** and **cmdOrder**. Change the captions as shown.

5 Add a final check box to the form below the two command buttons. Name the check box **chkDiscount**, and change the caption as shown.

6 Place an MSFlexGrid control on frmInvoice, and name the grid control InvGrid. (Remember that to add the MSFleaxGrid control to your toolbox, select Project, Components and make the selection from the Custom Controls tab.)

7 Add a standard code module to your project, and save it to disk as **m22-6.BAS**.

8 In the declaration section of the standard code module, make the following declarations:

```
Type orderitem
   description As String
   quantity As Integer
   fee As Currency
   Linetotal As Currency
End Type

Public skirental() As orderitem
Private Const StdDiscount = 0.85
```

Observe that skirental() is declared as an array of type orderitem with global scope. The constant represents the multiplier for the standard 15 percent discount.

9 In the frmOrder.cmdOrder_Click() event, enter the following code:

```
Private Sub cmdOrder_Click()
   Dim i As Integer
   'Variable for checking array redeclaration
   Dim arraynum As Integer
   'Variable for tracking assignment to global array
   Dim assigncheck As Integer
   arraynum = 0
   For i = chkItem.LBound To chkItem.UBound
      If chkItem(i).Value = 1 Then arraynum = arraynum + 1
   Next i
   If arraynum = 0 Then
      Exit Sub
   Else
      'This sets the size of the array and the type
      ReDim skirental(1 To arraynum) As orderitem
   End If

   assigncheck = 1
   For i = chkItem.LBound To chkItem.UBound
      If chkItem(i).Value = 1 Then
         skirental(assigncheck).description = chkItem(i).Caption
         skirental(assigncheck).fee = Val(lblPrice(i).Caption)
         skirental(assigncheck).quantity = CInt(txtQuantity(i).Text)
         skirental(assigncheck).Linetotal = skirental(assigncheck).fee _
         * skirental(assigncheck).quantity
         assigncheck = assigncheck + 1
      End If
```

```
    Next i
    If chkDiscount.Value = 1 Then
       Call CalculateRental(frmInvoice.InvGrid, True)
    Else
       Call CalculateRental(frmInvoice.InvGrid, False)
    End If
  End Sub
```

This code examines how many items the user has checked, and it redimensions `skirental`, the array declared as a user-defined type. Each element of `skirental` is populated with four items of information: the rental item, the rental price, the number of items ordered, and the line total.

The `CalculateRental` procedure is then called, passing two arguments: the grid on which the results of processing are to be displayed and a Boolean argument indicating whether a standard discount should apply to the order total.

10 Add the `CalculateRental()` procedure to the standard code module:

```
Public Sub CalculateRental(grid As MSFlexGrid, Discount As Boolean)
   Dim i As Integer
   Dim RunTotal As Currency
   Dim addstring As String
   Dim TabChar As String * 1
   Dim rowcounter As Integer

   TabChar = Chr(9)

   'Unload all previously loaded items from grid
   Call ClearGrid(grid)
   rowcounter = 1
   For i = LBound(skirental) To UBound(skirental)
     addstring = ""
     addstring = skirental(i).description & TabChar
     addstring = addstring & Str$(skirental(i).quantity) & TabChar
     addstring = addstring & Format(skirental(i).fee, "Currency") & TabChar
     addstring = addstring & Format(skirental(i).Linetotal, "Currency")

     grid.AddItem addstring, rowcounter
     rowcounter = rowcounter + 1
   Next

  'calculate total
  RunTotal = 0
  For i = LBound(skirental) To UBound(skirental)
     RunTotal = RunTotal + skirental(i).Linetotal
  Next i

  'add underline
  grid.Row = rowcounter - 1
  grid.Col = 3
  grid.CellFontUnderline = True

  'display total on grid
```

continues

```
    If Discount Then
      grid.Row = rowcounter + 1
      grid.Col = 0
      grid.CellFontBold = True
      grid.Text = "Total: "
      grid.Row = rowcounter + 1
      grid.Col = 3
      grid.CellFontBold = True
      grid.Text = Format(RunTotal, "Currency")

      grid.Row = rowcounter + 3
      grid.Col = 0
      grid.CellFontBold = True
      grid.Text = "Total w/Discount: "
      grid.Row = rowcounter + 3
      grid.Col = 3
      grid.CellFontBold = True
      grid.Text = Format(RunTotal * StdDiscount, "Currency")

    Else
      grid.Row = rowcounter + 1
      grid.Col = 0
      grid.CellFontBold = True
      grid.Text = "Total: "
      grid.Row = rowcounter + 1
      grid.Col = 3
      grid.CellFontBold = True
      grid.Text = Format(RunTotal, "Currency")
    End If
    'erase the global array

    ReDim skirental(0)    ' clear the array !!
  End Sub
```

Using the properties of the grid control, this code loops through the global array, building individual strings to add to the grid control. The total is calculated, and, depending upon the value of the discount argument, the order total is displayed or the order total and discount total are displayed. The array is then cleared.

The procedure also contains code to format individual grid cells, such as underlining the last order item and placing the total and discount amounts in bold.

11 Every time the `CalculateRental()` procedure is called, the grid is cleared with the `ClearGrid()` procedure. Add it to the standard code module:

```
Public Sub ClearGrid(grid As MSFlexGrid)
    grid.Clear
    Call Addheaders(grid)
End Sub
```

12 The MSFlexGrid control's `Clear` method clears all text, including text on fixed rows. The fixed row is used for displaying column headers in this application. The `Addheaders()` procedure resets the values for the fixed row. Add it to the standard code module:

```
Public Sub Addheaders(grid As MSFlexGrid)
  ' insert column headers
```

```
    grid.Row = 0
    grid.Col = 0
    grid.CellFontBold = True
    grid.Text = "Item"
    grid.Row = 0
    grid.Col = 1
    grid.CellFontBold = True
    grid.Text = "Quantity"
    grid.Row = 0
    grid.Col = 2
    grid.CellFontBold = True
    grid.Text = "Each"
    grid.Row = 0
    grid.Col = 3
    grid.CellFontBold = True
    grid.Text = "Total"
  End Sub
```

13 When the application is first launched, code is needed to set the grid control's properties. Add the following code to the standard code module:

```
Public Sub PrepGrid(grid As MSFlexGrid)
  ' set columns
  grid.FixedCols = 0
  grid.FixedRows = 1
  grid.Cols = 4
  grid.Rows = 100
  ' set width

  grid.Width = 4600
  ' set column widths

  grid.ColWidth(0) = 1800
  grid.ColWidth(1) = 800
  grid.ColWidth(2) = 800
  grid.ColWidth(3) = 800

  ' insert column headers
  Call Addheaders(grid)
End Sub
```

14 Return to `frmOrder`, and add the `form_Load()` event to set up the application:

```
Private Sub Form_Load()
  frmInvoice.Show
  Call PrepGrid(frmInvoice.InvGrid)
End Sub
```

15 Add code to the `chkItem_Click()` event to add a default quantity of `1` when an item is selected and clear the quantity text box when the item is deselected:

```
Private Sub chkItem_Click(Index As Integer)
   If chkItem(Index).Value = 1 Then
     txtQuantity(Index).Text = "1"
   Else
     txtQuantity(Index).Text = ""
   End If
End Sub
```

continues

16 Add code to allow the user to clear choices in the `cmdClear_Click()` event

```
Private Sub cmdClear_Click()
   Dim i As Integer
   For i = chkItem.LBound To chkItem.UBound
     chkItem(i).Value = 0
     txtQuantity(i).Text = ""
   Next i
   Call ClearGrid(frmInvoice.InvGrid)
End Sub
```

17 Add code to the `txtQuantity_LostFocus()` event to test whether the user has entered a numeric value in the `txtQuantity` text boxes:

```
Private Sub txtQuantity_LostFocus(Index As Integer)
   Dim test As String
   test = txtQuantity(Index).Text
   If Not IsNumeric(test) And Len(test) <> 0 Then
      MsgBox "Please enter an integer value.", 0, "Ski rental order"
      txtQuantity(Index).Text = ""
      txtQuantity(Index).SetFocus
   End If
End Sub
```

As you work with this code, observe that the `CalculateRental()` procedure performs the main business-rule functions. It receives data, calculates the totals checking for discount pricing, and returns the results of processing to the user.

The code is modularized by having frequently called functions placed in a standard code module, such as the procedures formatting the grid control. As more forms are added to this application, these forms would have these procedures available to them.

As lengthy as this exercise is, it does not contain code for dealing with a persistent data layer. Suppose, for instance, that data on the products offered for rent were maintained in a database. Code to populate the order form would need to be written as would code for modifying the product list.

A real world application would also need to have customer information, including form of payment information. Code to write and read this kind of information would normally be placed in the business-rule layer of an application. Segmenting these procedures in standard code modules makes the code more modular.

Objective 22.4 Using Class Modules in the Processing/Business-Rule and Data Layers

Using modules in applications takes a more complex turn when employing class modules. This requires an analysis of the object classes that should make up the application. Volumes are written on object-oriented design. A wide variety of notation systems exist for modeling object classes and the relationships among classes. These topics go well beyond the scope of this book. You should be aware, however, that the kind of object-oriented analysis that formerly went into large projects coded in C++ is now being used for large scale Visual Basic projects.

You can use class modules to segment an application. Recall from Chapter 13, "Class Modules," that you created an `OrderItem` class and then placed newly instantiated objects within that class. That objects collection provided the beginnings of an `Order` class. In Exercise 22.7, you will modify the `OrderItem` class created in Chapter 13 and build an `Order` class for the ski rental shop using Visual Basic's intrinsic `Collection` object.

Think of the different attributes and functions of the `Order` class as compared to the `OrderItem` class. The `Order` class should contain all line item information and display the order when requested. It should be able to total the order and apply a discount if requested. It needs to know on which grid to display the information and be able to manipulate the grid as needed. In contrast, the `OrderItem` class needs to hold information about the ski items: the unit price, the quantity ordered, and the line item total. It is the responsibility of the `OrderItem` class to calculate line items by multiplying the quantity by their unit prices.

With these concepts in mind, let's revamp the ski rental shop application using class modules.

Exercise 22.7 Building Classes to Enhance the Ski Rental Shop Application

In this lengthy exercise, you will build two classes: an `OrderItem` class and an `Order` class. It uses the same forms as those in Exercise 22.6.

1 Save the two forms, `frmOrder` and `frmInvoice`, as **f22-7A.frm** and **f22-7B.frm**, respectively.

2 Add a new standard code module to the project, change its `Name` property to **m22_7**, and save it to disk as **m22-7.BAS**. (You might want to add the code module from Exercise 22.6 to the project during design time in order to copy and modify its code. Much of the code from m22-6.BAS will be placed inside the `Order` class.) The new standard module, m22-7.BAS, contains only a single global declaration:

```
Public RentalOrder As Order
```

3 In the `Load()` and `Unload()` events for `frmOrder`, add code that creates a single instance of the `Order` class and removes it from memory when the form is unloaded:

```
Private Sub Form_Load()
  frmInvoice.Show
  Set RentalOrder = New Order
End Sub
```

continues

```
Private Sub Form_Unload(Cancel As Integer)
  Set RentalOrder = Nothing
End Sub
```

4 Add a class module to the project, name it `OrderItem`, and save it to disk as **cls22-A.cls**. (You might want to add the `OrderItem` class module built in Chapter 13 for Exercise 13.4, "Creating and Using an `OrderItem` Class," and modify its code.) The entire code for the class module is listed below. It comprises only accessor and mutator methods (`Property Get` and `Property Let`, respectively) and class initialization code:

```
Private m_ProductName As String
Private m_Quantity As Double
Private m_UnitPrice As Currency
Private m_ItemTotal As Currency

Public Property Get ProductName() As String
   ProductName = m_ProductName
End Property

Public Property Let ProductName(ByVal vNewValue As String)
   m_ProductName = vNewValue
End Property

Public Property Get Quantity() As Double
   Quantity = m_Quantity
End Property

Public Property Let Quantity(ByVal vNewValue As Double)
   m_Quantity = vNewValue
End Property

Public Property Get UnitPrice() As Currency
   UnitPrice = m_UnitPrice
End Property

Public Property Let UnitPrice(ByVal vNewValue As Currency)
   m_UnitPrice = vNewValue
End Property

Public Property Get ItemTotal() As Currency
   m_ItemTotal = m_UnitPrice * m_Quantity
   ItemTotal = m_ItemTotal
End Property

Private Sub Class_Initialize()
   m_ProductName = "Product"
   m_UnitPrice = 0
   m_Quantity = 0
   m_ItemTotal = 0
End Sub
```

Observe that the `ItemTotal` property is a read-only property because it is derived from the `Quantity` and `UnitPrice` properties.

5 In the `frmOrder.cmdOrder_Click()` event, add code that instantiates new instances of the `OrderItem` class and adds values to its key properties:

```
Private Sub cmdOrder_Click()
   Dim i As Integer
   Dim numitems As Integer
   Dim Lineitem As orderitem
   numitems = 0

   For i = chkItem.LBound To chkItem.UBound
      If chkItem(i).Value = 1 Then numitems = numitems + 1
   Next i

   If numitems = 0 Then ' check to see that data has been entered
      Exit Sub
   Else
      For i = chkItem.LBound To chkItem.UBound
        If chkItem(i).Value = 1 Then
           Set Lineitem = New orderitem
           Lineitem.ProductName = chkItem(i).Caption
           Lineitem.UnitPrice = Val(lblPrice(i).Caption)
           Lineitem.Quantity = CInt(txtQuantity(i).Text)
           RentalOrder.AddOrderItem Lineitem
        End If
      Next i
   End If

   Call RentalOrder.DisplayOrder
End Sub
```

Observe that the last line invokes the `AddOrderItem` and `DisplayOrder` methods of the single instance of the `Order` class, `RentalOrder`. Let's begin building the order class.

6 Add a new class module to the project, name it **`Order`**, and save it to disk as **cls22-7B.cls**.

7 The `Order` class will have as a private data member an intrinsic Visual Basic `Collection` object. It also needs to manipulate a grid control in order to display the results of processing. In addition, the class must retain information about the amount of the standard discount, and whether a discount should be applied to an order. Make the following declarations in the General declaration section of the class:

```
Private colOrder As New Collection
Private m_Count As Long
Private m_Discount As Boolean
Private m_grid As MSFlexGrid
Private Const StdDiscount = 0.85
```

8 Initialize these data members, and prepare the grid when the `Order` class is initialized by adding the following code to the `Initialize()` event:

continues

```
Private Sub Class_Initialize()
  Set m_grid = frmInvoice.InvGrid
  m_Count = 0
  m_Discount = False
  Call PrepGrid
End Sub
```

Observe how the `Set` statement is used with the MSFlexGrid object. This allows you to reference the grid within the class without passing it to the grid manipulation procedures.

9 The `Order` class essentially "wraps" a class around the intrinsic Visual Basic `Collection` object. Because the collection is private to the `Order` class, you need to replicate some of the collection's methods and properties. Add a read-only `Count` property and an `AddOrderItem()` method to replicate the collection's `Add` method:

```
Public Property Get Count() As Long
   m_Count = colOrder.Count
   Count = m_Count
End Property

Public Sub AddOrderItem(newone As orderitem)
  colOrder.Add newone
End Sub
```

10 Add accessor and mutator methods for the `Discount` property:

```
Public Property Get Discount() As Boolean
   Discount = m_Discount
End Property

Public Property Let Discount(ByVal vNewValue As Boolean)
   m_Discount = vNewValue
End Property
```

11 Add a `ClearOrder()` method that allows the collection to be cleared of objects and also clears the grid:

```
Public Sub ClearOrder()
  Dim i As Integer
  For i = colOrder.Count To 1 Step -1
     colOrder.Remove i
  Next
  Call ClearGrid
End Sub
```

12 Add the following grid manipulation procedures. (You might want to copy and modify this code from Exercise 22.6.) Observe that all of the grid manipulation procedures are `Private`.

```
Private Sub ClearGrid()
   m_grid.Clear
   Call Addheaders
End Sub
```

```
Private Sub Addheaders()
  ' insert column headers
  m_grid.Row = 0
  m_grid.Col = 0
  m_grid.CellFontBold = True
  m_grid.Text = "Item"
  m_grid.Row = 0
  m_grid.Col = 1
  m_grid.CellFontBold = True
  m_grid.Text = "Quantity"
  m_grid.Row = 0
  m_grid.Col = 2
  m_grid.CellFontBold = True
  m_grid.Text = "Each"
  m_grid.Row = 0
  m_grid.Col = 3
  m_grid.CellFontBold = True
  m_grid.Text = "Total"
End Sub

Private Sub PrepGrid()
  ' set columns
  m_grid.FixedCols = 0
  m_grid.FixedRows = 1
  m_grid.Cols = 4
  m_grid.Rows = 100
  ' set width

  m_grid.Width = 4600
  ' set column widths

  m_grid.ColWidth(0) = 1800
  m_grid.ColWidth(1) = 800
  m_grid.ColWidth(2) = 800
  m_grid.ColWidth(3) = 800

  ' insert column headers
  Call Addheaders

End Sub
```

13 As the final piece of the `Order` class, add the `DisplayOrder()` method. Again, you might want to copy and then modify code from Exercise 22.6, but note the differences, especially the use of the `For Each...Next` syntax to iterate through the collection and read `Object` property values:

```
Public Sub DisplayOrder()
   Dim i As orderitem
   Dim n As Integer
   Dim RunTotal As Currency
   Dim addstring As String
   Dim TabChar As String * 1
   Dim rowcounter As Integer

   TabChar = Chr(9)
```

continues

```
'Unload all previously loaded items from grid
Call ClearGrid
rowcounter = 1
For Each i In colOrder
  addstring = ""
  addstring = i.ProductName & TabChar
  addstring = addstring & i.Quantity & TabChar
  addstring = addstring & Format(i.UnitPrice, "Currency") & TabChar
  addstring = addstring & Format(i.ItemTotal, "Currency")
  m_grid.AddItem addstring, rowcounter
  rowcounter = rowcounter + 1
Next

'calculate total
RunTotal = 0
For Each i In colOrder
   RunTotal = RunTotal + i.ItemTotal
Next i

'add underline
m_grid.Row = rowcounter - 1
m_grid.Col = 3
m_grid.CellFontUnderline = True

'display total on grid

If m_Discount Then
  m_grid.Row = rowcounter + 1
  m_grid.Col = 0
  m_grid.CellFontBold = True
  m_grid.Text = "Total: "
  m_grid.Row = rowcounter + 1
  m_grid.Col = 3
  m_grid.CellFontBold = True
  m_grid.Text = Format(RunTotal, "Currency")

  m_grid.Row = rowcounter + 3
  m_grid.Col = 0
  m_grid.CellFontBold = True
  m_grid.Text = "Total w/Discount: "
  m_grid.Row = rowcounter + 3
  m_grid.Col = 3
  m_grid.CellFontBold = True
  m_grid.Text = Format(RunTotal * StdDiscount, "Currency")

Else
  m_grid.Row = rowcounter + 1
  m_grid.Col = 0
  m_grid.CellFontBold = True
  m_grid.Text = "Total: "
  m_grid.Row = rowcounter + 1
  m_grid.Col = 3
  m_grid.CellFontBold = True
  m_grid.Text = Format(RunTotal, "Currency")
End If
```

```
        ' remove items from collection

      For n = colOrder.Count To 1 Step -1
        colOrder.Remove n
      Next

   End Sub
```

14 Return to `frmOrder` and add the following code to the `cmdClear_Click()` and `chkDiscount_Click()` events:

```
Private Sub cmdClear_Click()
   Dim i As Integer
   For i = chkItem.LBound To chkItem.UBound
      chkItem(i).Value = 0
      txtQuantity(i).Text = ""
   Next i
   Call RentalOrder.ClearOrder
End Sub

Private Sub chkDiscount_Click()
   If chkDiscount.Value = 1 Then
      RentalOrder.Discount = True
   Else
      RentalOrder.Discount = False
   End If
End Sub
```

15 Run and test the program. Be sure to remove any extra modules added from Exercise 13.4 and Exercise 22.6.

Although lengthy, this exercise shows how segmenting your application into classes makes the application easier to extend and maintain. As your application expands, adding functionality to the right class makes the application conceptually manageable.

As you work more with classes and include them in larger applications, you will come to appreciate that correct design of classes, correct designation of class functionality, and clear designation of class relationships takes on added importance. In fact, it is often the key to software success.

Chapter Summary

Modular design is the breaking down of a computer program into several discrete units, rather than writing one long application. It also involves using several forms, rather than one. Program modules consist of procedures and functions that are either bound to forms, stored in standard code modules, or encapsulated in objects defined by class modules.

Perhaps the most important reason for modular design is to specify the functionality of a coded procedure. Use a form to convey a specific function, and write procedures or functions to perform one and only one task. Often, you might want to think of an application as having a user-interface layer, a processing/business-rule layer, and a data layer. This logical construct can assist you in building modular, readily understood applications.

In multiple-module applications, an initial form is often the login window. This window is used to provide security to a design. Use the `PasswordChar` property to make a password invisible to the person typing (or to the person trying to steal the password).

Modular design requires an understanding of how to load, show, hide, and unload forms. If a form is hidden, it remains in memory; if unloaded, it is removed from memory.

Call by reference and call by value are essential to modular design. Both allow the Visual Basic programmer to pass values from one procedure to another. Call by reference, the default, passes a copy of the address of the variable to a function or procedure. This allows the value of the argument to be directly altered by the function or procedure. Call by value passes a copy of the variable to the receiving function or procedure. The original value cannot be altered if passed by value, only the copy. To call by value, place the variable to be passed in parentheses, as in the following example:

```
Call Reverse ((my_age), my_name)
```

In this statement, `my_age` is passed by value; `my_name` is passed by reference.

Call by value features the `ByVal` keyword in the called function or procedure:

```
Private Sub Reverse(ByVal my_age, my_name)
```

Both standard modules and class modules can be used to help segment an application, especially in constructing the processing/business-rule and data layers of an application.

Skill-Building Exercises

1. Design a login window that allows a user to access `frmSystem` if the password ButterBun is typed, but only `frmMain` if the password RobRun is typed. The system form gives the user access to the entire design; the main form provides only limited access.
2. Modify Exercise 22.5 to allow you to change your name and address from old to new, showing both on the same display. `Name` should not be changed, only the address.
3. Modify Exercise 22.4. Use .BAS files to store the functions `fun1()` and `fun2()`. As before, display all letters.
4. Pass an array of numbers to a procedure. Within the procedure, sort the numbers from lowest to highest. Return to the calling procedure, and display the sorted list of numbers.
5. Modify Exercise 22.7 so that the user can select from a range of discounts between 10 and 40 percent. Present the choices in increments of five percent. Also allow the user to have no discount.

23

Creating an ActiveX DLL

In Chapter 22, "Modular Design," you used class modules to revamp the ski rental program (see Exercise 22.7, "Building Classes to Enhance the Ski Rental Shop Application"). The class modules defined the properties and methods of the objects created from those classes. In this manner, you made your programs more modular.

Among the many benefits of modular design is the promotion of code reuse. In Visual Basic, it is possible to employ a module in more than one application. An existing form module, standard code module, or class module can be added to a project in design time. Even so, the added module must usually be modified to become fully integrated into the new project. Form module names, for instance, must often be changed to make them distinct. Class modules usually require some modification to fit into the new application's design.

Visual Basic also makes the code and capabilities of an entire project available to other applications. The multiple modules of a project can be used to create a custom software component. You have already used components created for you by others. Every time you use a Visual Basic control, you use a component. When you used the MSFlexGrid control, for example, you added this custom control to the Visual Basic toolbox from the Custom Control tab of the Components dialog box. In advanced Visual Basic, you can build your own ActiveX controls. An ActiveX control is a control you design and add to your toolbox. This type of control can be used by you as well as by other Visual Basic developers.

Software components provide the principal means by which Visual Basic promotes code reuse. Creating your own ActiveX control in Visual Basic is a major undertaking, but ActiveX DLLs and ActiveX EXE components can be built fairly easily. Because of a recent development, Visual Basic can now also create ActiveX documents for World Wide Web applications for distribution of computing applications on corporate intranets (see Chapter 28, "Building Internet Applications with ActiveX Documents").

ActiveX DLLs are often referred to as *code components*. They frequently contain no visual interface at all. Instead, they perform vital processing for any application that chooses to use the services the DLL provides. These functions might involve numerically intense calculation routines, sorting routines, string parsing, or standard database query management.

The exercises in this chapter teach you several tasks. After you work through the exercises, you will be able to

- use the References dialog box to examine type libraries registered on your computer;
- add a reference of a type library to a Visual Basic project;
- examine objects exposed by the type library with the Object Browser;
- learn the difference between in-process and out-of-process servers;
- create an ActiveX DLL server application that performs various string functions;
- write a client application that uses an ActiveX DLL;
- write a function that parses a delimited string and places it inside an array;
- create a Visual Basic group so that you can work on multiple projects simultaneously;
- write a sort algorithm to return a sorted array from an ActiveX DLL;
- use Visual Basic's application Setup Wizard to create a project dependency file, listing the files that need to be distributed with your application.

Chapter Objectives

This chapter focuses on the creation of ActiveX DLLs. This type of component is easier to develop than other types of software components. ActiveX DLLs also provide some decided speed advantages over most ActiveX EXE projects. (We will look at using ActiveX EXE components in Chapter 26, "Working with Objects from Microsoft Office," especially the ActiveX EXE components exposed by the Microsoft Office suite.) This chapter is organized around four objectives:

1. Understanding ActiveX components and their type libraries
2. Creating and using an ActiveX DLL
3. Creating and using a Visual Basic group
4. Determining an application's file dependencies

These objectives make a preliminary introduction to the topics that are central to the architecture of ActiveX components and the understanding of how ActiveX components can be used by many applications.

Objective 23.1 Understanding ActiveX Components and Their Type Libraries

How is it that multiple applications can use the same software component? How does an application know the properties, methods, and events of an object if the object isn't first loaded into memory with the application? Isn't it wasteful to have multiple instances of the same component occupying memory?

Answering these questions has occupied developers of systems and application software for the last decade and longer. Think of the differences between an old DOS program and an operating system, such as Windows. In the DOS world, developers frequently use libraries of reusable code in their applications. There are a handful of sort algorithms that nearly every program uses at one time or another. Why code them again? The same goes for many standard string parsing functions.

In a single-tasking environment, such as older versions of DOS, the linking functions of compilers bring libraries of code into an executable file. They might make the executable file a little larger, but that, after all, is just space on the hard disk.

23.1.1 A Brief Sketch of DLLs

In a shared memory environment, such as Windows 3.0 and 3.1 (or DOS with shared memory manager programs), or a multitasking environment, such as a 32-bit Windows operating systems, the situation becomes more complicated. With more than one program running, you do not want them occupying precious memory with the same code. Operating system functions, functions to use the built-in Windows dialog boxes, or common mathematical functions need to be shared by all running applications. This is where DLLs come in.

The term DLL stands for *dynamic-link library*. When programs use a DLL, the functions in the library are not linked into the application when it is compiled. Instead, functions in the dynamic-link library are identified and called *when needed* by the running application. Multiple running applications can call the same library, but only one instance of the library is loaded into memory. In the Windows operating system itself, the User32.DLL in the Windows\System subdirectory services every running application with core operating system functionality.

It is important to remember that DLLs are a type of executable. Visual Basic controls have an .OCX extension. This stands for *OLE control extension,* the predecessor of ActiveX. However, .OCX files are a type of DLL, and sometimes .EXE applications resemble DLLs as well. They are executables that service other applications. In the Windows operating system itself, the GDI.EXE application, for instance, performs drawing functions on the screen and printer.

When you add a custom control to your Visual Basic toolbox, you essentially identify the library that will act as the engine for that control when it is used by your application. Another running application might be using the same control and the same engine. In this example, the .OCX control (a type of DLL) acts as a *server* to a *client* application.

This is the heart of component-based development. Software components act as servers to client applications. Visual Basic allows you to develop these types of servers.

23.1.2 ActiveX DLLs and Type Libraries

ActiveX DLLs are DLLs of a special type. They conform to what is known as the *Component Object Model* (COM), the technology upon which ActiveX is based. The COM defines a set of protocols (usually referred to as *interfaces*) by which objects communicate with one another. This communication includes the manner in which objects inquire about each other's properties methods and events. The COM also defines how communication will occur *within a process* and *outside a process.*

Every time an application is launched in 32-bit Window operating systems, it is launched within a *process.* The full details of a process are quite complex, but a process includes the executing program, a set of memory addresses, and a least one *thread:* a part of a process that runs program instructions. A process can have more than one thread, which allows a single process to have different parts of a program running concurrently. All the programs in this text are single-threaded.

ActiveX DLLs are *in-process* servers. That means they operate within the same process as your application. This usually means dramatically faster program performance than *out-of-process* communication, the type of communication that occurs when using an ActiveX EXE, a topic covered in Chapter 26.

Because ActiveX DLLs are software components, they can be used by several applications. However, for this to occur, they must make themselves known. Like other software components, ActiveX DLLs make themselves known to other Windows applications by being *registered* in the Windows Registry.

You have probably already encountered the Windows Registry in working with other Windows programs or when installing new programs on your computer. The Registry is essentially a large database that contains information about the hardware on the computer, the configuration of different peripherals, the different users on the machine, and environmental settings for software on the computer. It also holds information about the various software components on the computer.

Each of the software components registered in the Windows Registry contains a unique identification. The components usually receive their identification when they are compiled, using what is known as the GUID standard or *globally unique identifier.* The GUID standard is a long alphanumeric string that identifies a software component, and is created by an algorithm made universally available to software component developers. When you compile an ActiveX DLL, a GUID for that component is automatically created for you. This then allows the software component to be installed on other computers and registered.

By being registered in the Registry, ActiveX software components define object data types that can be declared and bound to object variables. In Chapter 22, "Modular Design," you defined two object data types in class modules: an `OrderItem` class and an `Order` class. You made object variable declarations of these objects and bound them with such an instruction as

```
Dim Lineitem as OrderItem
Set Lineitem = New OrdeItem
```

or

```
Dim skirental as New Order
```

Your program knew what these datatypes were because you defined them within class modules inside your project.

When you create an ActiveX DLL, other applications know the datatypes and the interface the ActiveX DLL exposes because the DLL also defines a *type library*. There are different type libraries for ActiveX components. These include type libraries for ActiveX controls, ActiveX DLLs, ActiveX EXEs, and explicit libraries that have .OLB or .TLB extensions.

The type libraries available to you on your computer are listed in the References dialog box, which you access by choosing Project, References. Once a reference is selected, you can examine the objects described by the type library by using the Object Browser. We turn to using these two important features next.

23.1.3 Working with the Reference Dialog Box and Object Browser

You have already used the Object Browser to examine classes that you built in Chapter 13, "Class Modules," and perhaps to examine the classes you built in Chapter 22, "Modular Design." Because those class modules were defined in the projects themselves, references to the classes were already available to the Object Browser. Generally, however, using the Object Browser to examine objects defined in type libraries is a two-step process:

1. Make a *reference* to the type library by using the References dialog box.
2. Examine the properties, methods, and events of the object in the Object Browser.

If the type library you are examining is an ActiveX control, you need only select the control from the Components dialog box. A reference to the control's type library is then automatically added to the References dialog box.

Exercise 23.1 Using the Reference Dialog Box and Object Browser

In this exercise, you will use the Reference dialog box to select a type library. You will then examine some of the objects described in the type library with the Object Browser.

1 Start a new project. Do not save it to disk because you will only use the utilities associated with type libraries.

2 Choose Project, References. You will see the References dialog box, as shown in Figure 23.1.

continues

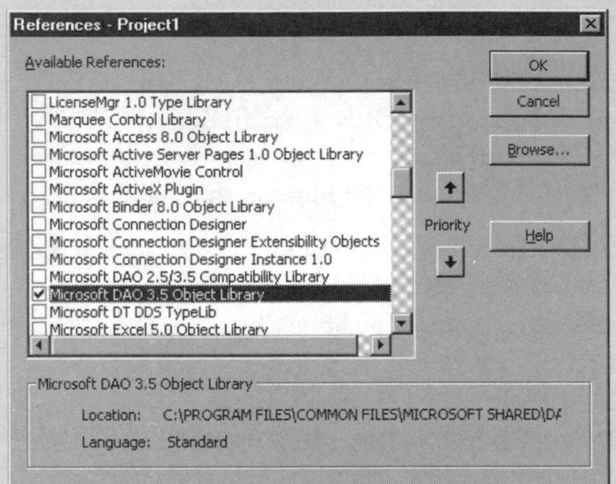

Figure 23.1

Use the References dialog box to select another application's objects that you want available in your code by setting a reference to that application's object library.

3 Assuming you have either the Professional or Enterprise editions of Visual Basic, select one of the Microsoft DAO libraries (for example, DAO 3.0 or DAO 3.5). Either of these is the type library for Data Access Objects, a topic addressed in Chapter 25, "Using Data Access Objects." It exposes most of the functionality of the Access database, but without the visual interface of Access. This is often referred to as the *Access database engine.* Select the DAO library and press OK. You now have a reference to this type library in the project.

4 Press F2 and open the Object Browser. Alternatively choose View, Object Browser from the Visual Basic menu.

5 Select the DAO library and examine the Database class; then, examine the `OpenRecordset` member of this class (see Figure 23.2).

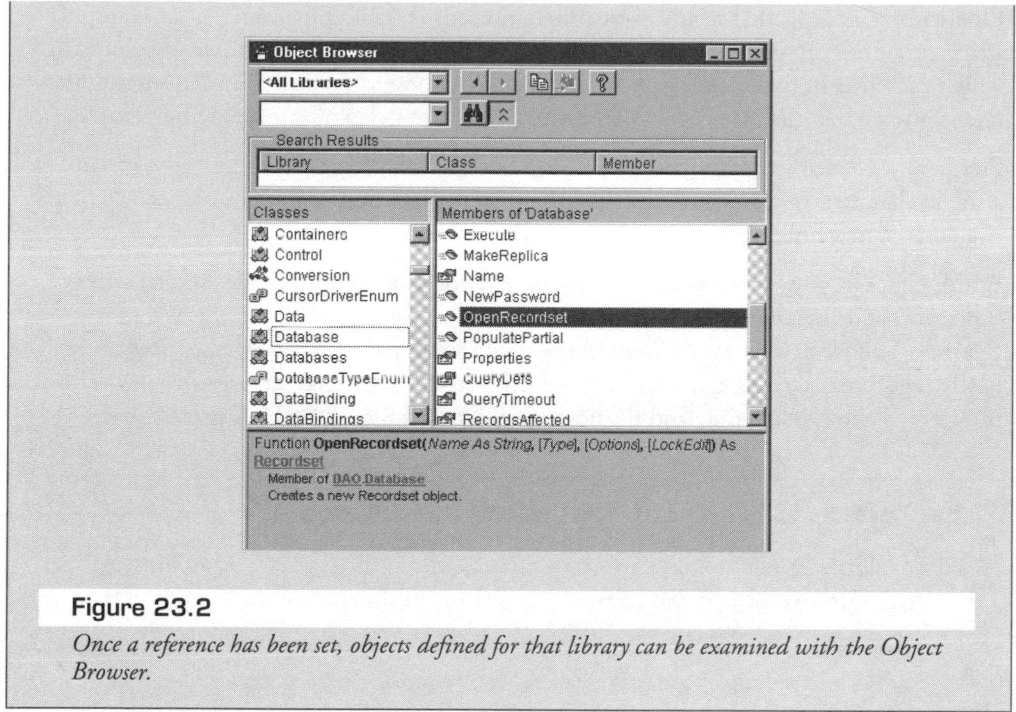

Figure 23.2

Once a reference has been set, objects defined for that library can be examined with the Object Browser.

Objective 23.2 Creating and Using an ActiveX DLL

When you create an ActiveX DLL project, you also create a type library that other applications can use when that project is compiled. These DLLs are referred to as *code components.* They are often used to encapsulate numerically intensive operations or to provide pre-coded functions to applications.

23.2.1 The Uses of ActiveX DLLs

The database capabilities of Access are encapsulated in the DAO library. Database functions are so often needed by programs that these libraries are among the most frequently used by Visual Basic programmers. Nearly every program needs to store data to disk. If the program serves a typical enterprise, the database will need to be queried often for data on a regular basis. The power of a relational database to retrieve data with ease makes the DLLs that encapsulate the Access database engine into *Data Access Objects* (DAO) a premier feature of the Visual Basic language.

One set of functions that nearly every program needs is string functions. As an example, string functions are used to parse delimited files. Delimited files are sequential records in which each line in the file has its fields delimited by some character. The delimiter in the line is often a comma, a semicolon, or a tab. Path strings also often need to be parsed.

Frequently needed string functions include a function to provide a count of the instances of a single character in a string, a function to return the position of the *n*th occurrence of a character, and a function to place the contents of a delimited string into an array.

In this chapter, you will construct a DLL server application that provides several string functions, and then compile it and observe how it registers itself within the Windows Registry. Following this, you will build the client application and use the services of the newly registered ActiveX DLL. In building the ActiveX DLL, pay close attention to the `Name` property of the class module and the `Name` property of the ActiveX DLL project itself.

Exercise 23.2 Building an ActiveX DLL

This project asks you to build an ActiveX DLL, and is the first in a series of two exercises. After you build and register the ActiveX DLL, you will use it in Exercise 23.3.

1 Start a new Project inside Visual Basic. When the New Project dialog box appears, select the ActiveX DLL icon (see Figure 23.3).

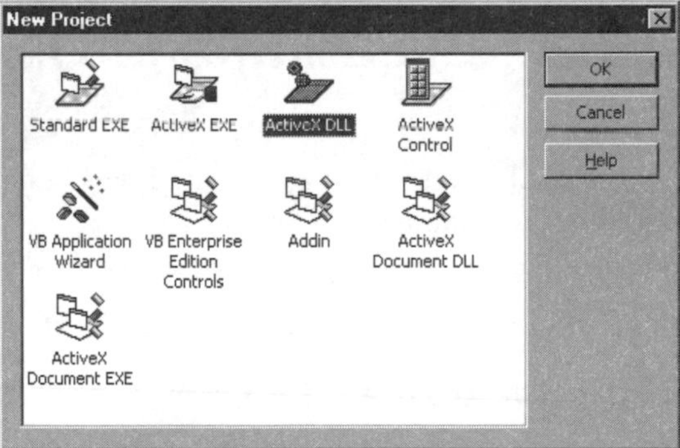

Figure 23.3

The ActiveX DLL icon indicates that you plan to design and register a DLL on your computer.

2 The new project will include a general section and a single class. Press the F4 key to examine the properties of the Class module. Make no changes at this time. Leave the `Name` property (`Class1`) and the `Instancing` property (5—`MultiUse`).

3 Perform the following two-step procedure:

   1 Change the name of the Class module to **`SFunctions`**.

   2 Save it to disk as **clsString.cls**.

   The `Name` property of the Class module will become the name of the set of objects that you instantiate from this class.

4 Again, follow a two-step procedure:

   1 Go to the Project Properties dialog box. (Right-click the project in the Project Explorer and select Properties, or choose Project, Properties from the Visual Basic menu.)

   2 Change the `Name` property of the project to **`Str_Functions`**. *Important:* The `Name` property of the project will become the name of the type library when the DLL is compiled. You need not specify a Startup object for the DLL. The client applications start the DLL when it is needed.

5 Save the project to disk as **Str_Functions.vbp**.

6 Add two string functions to the `SFunctions` class module. One function, `GetCharCount`, returns the number of occurrences of a single character within a string. The other function, `GetCharPos`, returns the *n*th position of a single character within a string.

```
Public Function GetCharPos(s1 As String, testchar As String, pos As
➥Integer) As Integer
  ' this function tests for the nth occurrence of a character
  Dim i As Integer
  Dim slength As Integer
  Dim testpos As Integer
  If Len(testchar) > 1 Then
    MsgBox "Search item must be a single character.", 0, "String Functions"
    Exit Function
  End If
  testpos = 0
  slength = Len(s1)
  For i = 1 To slength
    If Mid$(s1, i, 1) = testchar Then
      testpos = testpos + 1
        If testpos = pos Then
          GetCharPos = i
          Exit Function
        End If
    End If
  Next
  GetCharPos = 0 ' character not found
End Function
```

continues

```
Public Function GetCharCount(s1 As String, examinechar As String) As
➥Integer
  Dim i As Integer
  Dim slength As Integer
  Dim charcount As Integer
  If Len(examinechar) > 1 Then
    MsgBox "Search item must be a single character.", 0, "String Functions"
    Exit Function
  End If
  charcount = 0
  slength = Len(s1)
  For i = 1 To slength
    If Mid$(s1, i, 1) = examinechar Then
      charcount = charcount + 1
    End If
  Next
  GetCharCount = charcount
End Function
```

7 Compile the DLL by selecting File, Make Str_Functions.dll from the Visual Basic menu. Click OK to accept the name Str_Functions.dll.

8 As Visual Basic compiles the application, it provides the DLL a new GUID. This unique naming of the class ID occurs every time the project is recompiled. Don't worry about whether your class module might somehow create a duplicate ID: The algorithm that generates this ID could be run every few seconds for thousands of years before generating a duplicate ID.

9 To observe that you have created a registered class, you can use the Regedit.exe Windows Registry utility provided with Windows (under the Windows directory). Examine the HKey_Classes_Root tree with this utility. You should see the `Str_Functions` type library registered. *Important:* Do not change any settings in the Registry. An errant assignment in the Registry can make your computer, or programs within it, inoperable.

Exercise 23.3 Building the Client Application

To test the newly created DLL, you need to build a client application that uses it. In this exercise, you build a Standard EXE project, set a reference to the type library created by the `Str_Functions` DLL, and then use the string functions the DLL provides.

1 Create the interface you see in Figure 23.4. It consists of a text box (**`txtDisplay`**) with its `MultiLine` property set to `True`, a combo box (**`cboSelect`**), a command button (**`cmdOK`**), and two labels (**`lblCommaCount`** in the upper-right position of the form and **`lblCommaPos`** immediately below `lblCommaCount`) with their `BorderStyle` properties set to `Fixed Single`. The other labels offering prompts or

descriptions retain their default names. Save the form to disk as **f23-3.frm** and the project as **p23-3.vbp**.

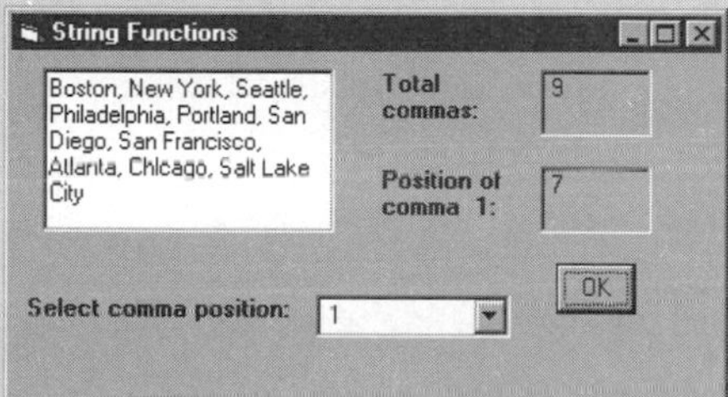

Figure 23.4

The interface for testing the `Str_Functions` ActiveX DLL is designed to count the total commas and the position of the selected comma.

2 Choose Project, References to make a reference to the new `Str_Functions` type library by selecting the type library in the References dialog box (see Figure 23.5).

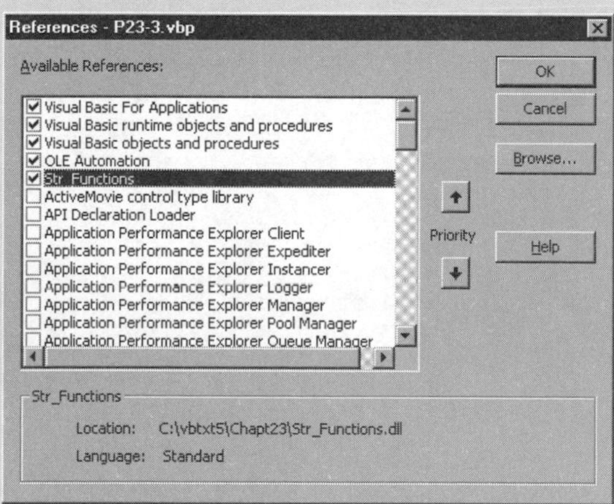

Figure 23.5

Find the `Str_Function` reference and click it to add it to the library for the project.

3 Once a reference has been made, use the Object Browser to examine the `SFunctions` member of the `Str_Functions` type library. You should see listed the two methods you wrote for the DLL: `GetCharPos` and `GetCharCount`. When you select these functions inside the Object Browser, the parameters of these functions should be displayed.

continues

4 At the module level of `Form1`, make the following declarations:

```
Dim strExamine As String
Dim strDLL As Str_Functions.Sfunctions
```

5 In the Load event of `Form1`, add the following code to populate the text box with cities, to populate the combo box with numbers, and to create an instance of the `SFunctions` class:

```
Private Sub Form_Load()
  Dim i As Integer
  strExamine = "Boston, New York, Seattle, "
  strExamine = strExamine & "Philadelphia, Portland, San Diego, San
  ➥Francisco, "
  strExamine = strExamine & "Atlanta, Chicago, Salt Lake City"
  txtDisplay.Text = strExamine
  For i = 1 To 8
    cboSelect.AddItem Str$(i)
  Next
  cboSelect.ListIndex = 0
  Set strDLL = New SFunctions
End Sub
```

6 In the `Form_UnLoad()` event add code to remove the instance of the DLL:

```
Private Sub Form_Unload(Cancel As Integer)
  Set strDLL = Nothing
End Sub
```

7 In `cmdOK_Click()`, add code to call the functions in the DLL:

```
Private Sub cmdOK_Click()
  Dim pos1 As Integer, pos2 As Integer
  Dim passpos As Integer
  Dim msg As String
  strExamine = txtDisplay.Text
  passpos = CInt(cboSelect.Text)
  pos1 = strDLL.GetCharCount(strExamine, ",")
  pos2 = strDLL.GetCharPos(strExamine, ",", passpos)
  lblCommaCount.Caption = Str$(pos1)
  lblPrompt.Caption = "Position of comma " & Str$(passpos) & ":"
  lblCommaPosition.Caption = Str$(pos2)
End Sub
```

8 Test the program by adding cities with comma delimiters to the text box.

If you run into errors, you might need to go back to the Str_Functions.vbp project and correct them in the DLL. Remember to recompile the DLL before testing it again in the client application. You will learn how to work with two projects simultaneously in Objective 23.3, "Creating and Using a Visual Basic Group."

23.2.2 Fully Qualified Declarations

Observe the manner in which the module level declaration of `SFunctions` was made. It was declared as

```
Dim strDLL As Str_Functions.SFunctions
```

This is referred to as a *fully qualified declaration.* It specifies both the type library for the object as well as the object name. Suppose two type libraries had the same name for an object and were referenced within a project. It is possible that an object variable could be bound to the wrong object. By specifying the type library in which the class is defined, you avoid these errors.

23.2.3 The *Instancing* Property of Class Modules in ActiveX DLLs

When you add class modules to Standard EXE projects, no `Instancing` property appears. A Standard EXE project has only a private instancing capability: Other applications are not given access to type information about the Standard EXE and cannot create new instances of the program. With ActiveX DLLs and ActiveX EXE projects, the way the components are created must be specified.

For most ActiveX DLLs, the instancing should be set to `MultiUse`. This allows other applications to access the same DLL: One instance of your component can provide services to another application. Table 23.1 explains the different instancing choices.

Table 23.1 Instancing Options for ActiveX DLLs

Property	*Setting*	*Description*
`Private`	1	The default. Other applications aren't allowed access to type library information about the class, and cannot create instances of it. Private objects are only for use within your component.
`PublicNotCreatable`	2	Other applications can use objects of this class only if your component creates the objects first. Other applications cannot use the `CreateObject` function or the `New` operator to create objects from the class.
`SingleUse`	3	Allows other applications to create objects from the class, but every object of this class that a client creates starts a new instance of your component. Not allowed in ActiveX DLL projects.
`GlobalSingleUse`	4	Similar to `SingleUse`, except that properties and methods of the class can be invoked as if they were simply global functions. Not allowed in ActiveX DLL projects.
`MultiUse`	5	Allows other applications to create objects from the class. A component can provide any number of objects created in this fashion.

continues

Table 23.1 continued

Property	*Setting*	*Description*
`GlobalMultiUse`	6	Similar to `MultiUse`, with one addition: properties and methods of the class can be invoked as if they were simply global functions. It isn't necessary to explicitly create an instance of the class first because one will automatically be created.

It is important to remember that ActiveX DLL projects must include at least one class that is publicly creatable.

Objective 23.3 Creating and Using a Visual Basic Group

Creating a separate ActiveX server application and a separate client application might have seemed a little unwieldy to you. This is especially true if the ActiveX DLL project contains errors. Visual Basic eases this situation by allowing you to gather different projects together into a *Visual Basic group*. A group allows you to easily switch between projects and to test a DLL as it is developed.

23.3.1 The Startup Project

Visual Basic groups have a crucial property: You must specify which project is the startup project. Usually this is the client application. Pay attention to this important property as you work through the next exercise.

Exercise 23.4 Using a Visual Basic Group to Expand the ActiveX DLL

In this project, you will create a Visual Basic group to expand and test the capabilities of the `Str_Functions` ActiveX DLL.

1. If p23_3.vbp is not open, reopen it.
2. Save f23-3.frm as **f23-4.frm**. Save p23-3.vbp as **p23-4.vbp**. Change the project name (select Project Properties) from p23_3 to **p23_4**.
3. Choose File, Add Project from the Visual Basic menu. When the Add Project dialog box appears, choose the Existing tab and add the Str_Functions.vbp project.

   You have now created a Visual Basic group!

4 Choose File, Save Project Group As. Save the project group as StringTest.vbg. (*.vbg* stands for Visual Basic group.) The Project Explorer should now appear as it does in Figure 23.6.

Observe the Visual Basic title menu as you set focus on the two projects in the Project Explorer. The title will toggle between the two projects in the group.

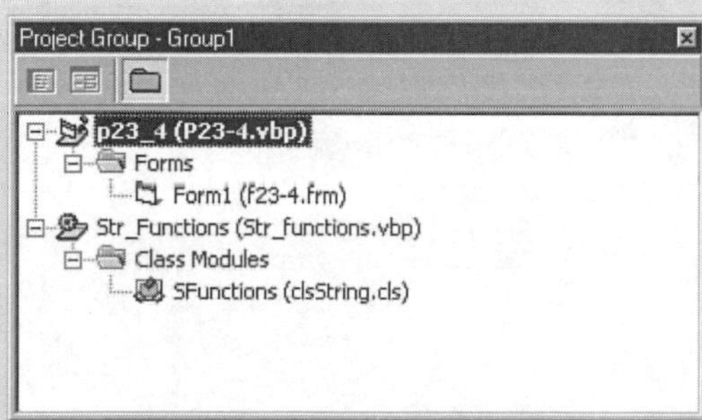

Figure 23.6

With Visual Basic groups, you can work on several projects simultaneously.

5 Select project p23_4 in the Project Explorer, and click the right mouse button. Click the top menu item: Set as Start Up. This specifies the P23_4.vbp project as the startup project in the group. The project should now be in bold inside the Project Explorer.

6 Select `Form1` inside the p23_4 project, and add a list box to the form (see Figure 23.7). Name the list box **`lstCities`**.

7 Change the `cmdOK_Click()` event on `Form1` to the following code:

```
Private Sub cmdOK_Click()
  Dim pos1 As Integer, pos2 As Integer
  Dim passpos As Integer
  Dim msg As String
  Dim retArray As Variant
  Dim i As Integer
  strExamine = txtDisplay.Text
  passpos = CInt(cboSelect.Text)
  pos1 = strDLL.getCharCount(strExamine, ",")
  pos2 = strDLL.GetCharPos(strExamine, ",", passpos)
  lblCommaCount.Caption = Str$(pos1)
  lblPrompt.Caption = "Position of comma " & Str$(passpos) & ":"
  lblCommaPosition.Caption = Str$(pos2)
  retArray = strDLL.ParseString(strExamine, ",")
```

continues

```
    lstCities.Clear
    For i = LBound(retArray) To UBound(retArray)
      lstCities.AddItem Trim(retArray(i)), i - 1
    Next
  End Sub
```

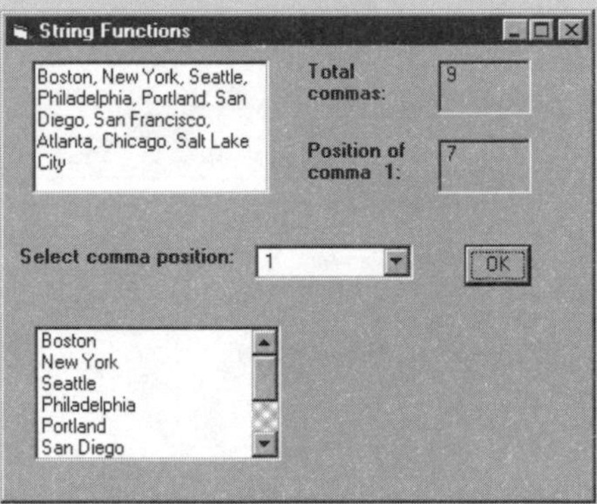

Figure 23.7

This interface allows you to test various comma positions and parse a delimited string.

This code calls the ParseStrings method of the SFunctions class, which parses a delimited string and loads it into an array. It has two arguments: the delimited string being parsed and the delimiting character. After the array is returned from the DLL, it is processed and displayed in the lstCities list box.

8 Switch the focus in the Project Explorer to the `Str_Functions` project. The title bar of Visual Basic changes with the changeover. Add a routine that parses a delimited string to the `SFunctions` class module:

```
Public Function ParseString(s1 As String, delimiter As String)
  Dim charnums As Integer
  Dim i As Integer
  Dim slength As Integer
  Dim firstchar As Integer, secondchar As Integer
  Dim OrdPosChar As Integer
  Dim strhold As String
  OrdPosChar = 0  ' This variable handles the array element tracking.
                  'Initialize to zero to get the number of delimiting
                  'characters.
  charnums = GetCharCount(s1, delimiter)
  slength = Len(s1)

  ' dimension a one-based array to handle the parsed elements
  ReDim returnarray(1 To charnums + 1) As String
  ' this for loop handles the first delimiting character,
  ' the last delimiting character, and all intervening characters
```

```
    For i = 1 To slength
      If Mid$(s1, i, 1) = delimiter Then
        OrdPosChar = OrdPosChar + 1 ' we have an item, increment the array
                                    ' tracker
        Select Case OrdPosChar
          Case 1    ' first delimiter encountered
            firstchar = i
            returnarray(1) = Left(s1, i - 1)
            If charnums = 1 Then ' check for case of only two items
               strhold = Trim(Right$(s1, Len(s1) - i))
            End If
            If Len(strhold) > 1 Then
               returnarray(OrdPosChar + 1) = Trim(Right(strhold, Len(strhold)
               ➥- 1))
            End If

          Case charnums ' handle the last two items in the string
            secondchar = i
            strhold = Mid$(s1, firstchar + 1, secondchar - (firstchar + 1))
            If Len(strhold) <> 0 Then returnarray(OrdPosChar) = Trim(strhold)
            ' test for trailing characters
            strhold = Right$(s1, (Len(s1) - i) + 1)
            If Len(strhold) > 1 Then
               returnarray(OrdPosChar + 1) = Trim(Right(strhold, Len(strhold)
               ➥- 1))
            End If

          Case Else
            secondchar = i
            ' extract the string portion
            strhold = Mid$(s1, firstchar + 1, secondchar - (firstchar + 1))
            If Len(strhold) > 0 Then returnarray(OrdPosChar) = Trim(strhold)
            firstchar = secondchar  ' get ready for next encounter with
                                    ' delimiting character
        End Select
      End If
    Next
    ParseString = returnarray()    ' return the array
End Function
```

This parsing routine first counts the number of delimiting characters in the string passed to it by the `GetCharCount` function. It then dimensions a one-based array based on this calculation. Following this, it loops through the string, examining it one character at a time. If the character is the delimiting character, it checks to see if it is the first occurrence of the character. If so, it assigns the value of the `For...Next` loop counter variable to the `firstchar` variable, and populates the first element of the array. If only two items are in the string, the second element of the array is populated.

The next time a delimiting character is encountered, the loop counter variable is assigned to the `secondchar` variable. The string portion between `firstchar` and `secondchar` is retrieved and placed in the array. The `firstchar` variable is then

continues

assigned the value of the `secondchar` variable to prepare for the next encounter with a delimiting character.

Processing continues until the last delimiting character is encountered. When this occurs, the last two elements of the array are populated. The routine checks the length of the string in order to handle an instance where the string ends with a delimiting character.

9 Test the new program. Changes in the DLL can be tested without the DLL being compiled when the project is part of a Visual Basic group. This greatly speeds application development. Once you are satisfied with the performance of the DLL, it can be recompiled.

Examine the p23-4.vbp file in a text editor, such as Notepad. You should see a reference to the GUID for the `Str_Functions` type library near the top of the file. In the application compiled on the author's machine, the first lines of the P23-4.vbp file read as follows:

```
Type=Exe
Reference=*\G{00020430-0000-0000-C000-
➥000000000046}#2.0#0#D:\WINNT40\System32\STDOLE2.TLB#OLE
Automation Reference=*\G{55D0EE96-FC3C-11D0-A4E0-
➥00A0C90D3A89}#4.0#0#Str_Functions.dll#Str_Functions
Form=f23-4.frm
Startup="Form1"
Type=Exe
```

Examine the GUID for the `Str_Functions` type library compiled on your computer.

23.3.2 Adding a Sort Algorithm to the ActiveX DLL

DLLs are frequently used for numerically intensive tasks, which makes them ideal for sort routines. With these routines, you often need to return sorted strings or numbers from functions. If you are placing strings inside a list box, the task is easy: The list box can sort the strings for you. But if sorted strings need to be placed inside another control, such as a grid, or to be printed, you need to sort them yourself.

Exercise 23.5 Adding a Bubble Sort Routine to the ActiveX DLL

This exercise employs a bubble sort algorithm that you used in Chapter 19, "String Functions." When the user chooses, a parsed string is returned in sorted order.

1 Start an entirely new Standard EXE project.

2 Select the default project created (`Project1`) in the Project Explorer, and remove it by selecting File, Remove Project from the Visual Basic menu. No project should now be displayed in the Project Explorer.

3 Add project 23-4.vbp to the project by selecting File, Add Project. Save f23-4.frm to disk as **f23-5.frm**. Save p23-4.vbp to disk as **p23-5.vbp**. Rename the project **P23_5**.

4 Add the `Str_Functions` project and create a Visual Basic group. Save the Visual Basic group to disk as String&Sort.vbg.

5 Make sure that P23_5 is the startup project. Shift the focus to `Form1` in P23_5. Remove the list box (`lstCities`) from the form. Add a Line control to the form and keep the default name of `Line1`. Also add a check box (**`chkSort`**). As shown by Figure 23.8, make the form a bit taller than it was before.

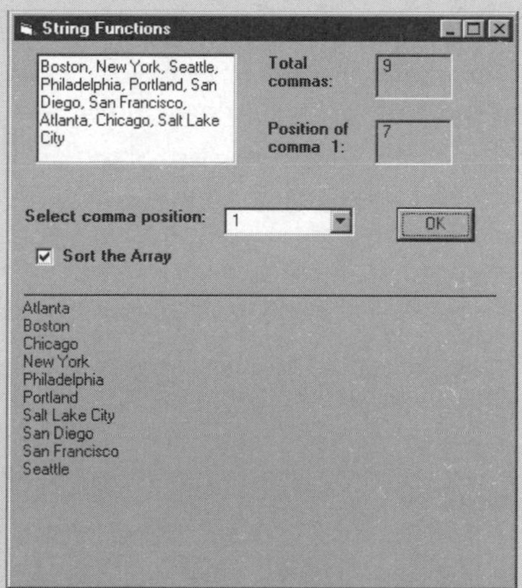

Figure 23.8

This display shows the sorted list following the use of the bubble sort algorithm placed within the DLL.

6 Change the `cmdOK_click()` procedure to read as follows:

```
Private Sub cmdOK_Click()
  Dim pos1 As Integer, pos2 As Integer
  Dim passpos As Integer
  Dim msg As String
  Dim retArray As Variant
  Dim i As Integer
  strExamine = txtDisplay.Text
```

continues

```
    ' change the sort property for the DLL
    If chkSort.Value = 1 Then
      strDLL.Sort = True
    Else
      strDLL.Sort = False
    End If
    passpos = CInt(cboSelect.Text)
    pos1 = strDLL.GetCharCount(strExamine, ",")
    pos2 = strDLL.GetCharPos(strExamine, ",", passpos)
    lblCommaCount.Caption = Str$(pos1)
    lblPrompt.Caption = "Position of comma " & Str$(passpos) & ":"
    lblCommaPosition.Caption = Str$(pos2)
    retArray = strDLL.ParseString(strExamine, ",")
    Cls
    CurrentY = Line1.Y1 + 50
    For i = LBound(retArray) To UBound(retArray)
      CurrentX = 100
      Print Trim(retArray(i))
    Next
  End Sub
```

This code now adds a value to a `Sort` property to the `strDLL` object. It specifies whether you want to have the parsed string returned sorted or unsorted.

7 Add the new features to the `SFunctions` class. Shift the focus in the Project Explorer to the `Str_Functions` project. In the `SFunctions` class module, declare a `Private` data member at the module level, and add a `Property Let` and `Property Get` procedure as well as initialization code:

```
Private m_Sort As Boolean ' declare at the module level

Public Property Get Sort() As Boolean
  Sort = m_Sort
End Property

Public Property Let Sort(ByVal vNewValue As Boolean)
  m_Sort = vNewValue
End Property

Private Sub Class_Initialize()
  m_Sort = False
End Sub
```

This code should look familiar. The `Sort` property holds a Boolean value indicating whether or not the client application wants to have a sorted or unsorted array returned. The data member is declared privately and accessed through `Property Let` and `Property Get` procedures; the data member is initialized in the class module's `Initialize()` event.

8 Replace the last line of code in the `ParseString` procedure with code that tests whether the array should be returned sorted:

```
If m_Sort = True Then
  ParseString = ReturnBubbleSort(returnarray())
Else
  ParseString = returnarray()
End If
```

9 Add the `ReturnBubbleSort` routine to the `SFunctions` module:

```
Private Function ReturnBubbleSort(strArray As Variant) As Variant
  Dim NextItem As Integer
  Dim topindex As Integer
  Dim i As Integer, j As Integer
  Dim temp As Variant
  Dim tindex As Integer, bindex As Integer
  Dim test As Integer
  tindex = UBound(strArray)
  bindex = LBound(strArray)
  NextItem = bindex

  Do While NextItem < tindex
    topindex = tindex
      Do While topindex > NextItem
        test = StrComp(strArray(topindex), strArray(topindex - 1), 1)
        If test = -1 Then
           temp = strArray(topindex)
           strArray(topindex) = strArray(topindex - 1)
           strArray(topindex - 1) = temp
        End If
        topindex = topindex - 1
      Loop
      NextItem = NextItem + 1
  Loop
  ReturnBubbleSort = strArray
End Function
```

The function is declared `Private` and is invoked only if the value of the object's `Sort` property has been set to `True` and cannot be called by the client application.

10 Run and test the program. Add comma delimited city names to the text box. Depending upon whether the user checks the check box, the parsed string can be returned sorted or unsorted.

23.3.3 Design and Documentation Considerations in an ActiveX DLL

You have expanded the `SFunctions` class that is defined in the `Str_Functions` type library over the course of this chapter. To make this expansion even more attractive, you can readily add other string functions to this ActiveX DLL. It can then be shared with other Visual Basic programmers.

Classes are somewhat *self-documenting* by being exposed by the Object Browser. ActiveX DLLs, however, require documentation as they become larger and more complex. This is especially true for DLLs made up of multiple class modules. Other programmers will want to know the meaning and use of different function parameters. Often there is an interdependency between functions and object properties. This means that certain properties must be set before a function can be effectively invoked. These interdependencies need to be documented.

As you build larger DLLs, you will confront a fundamental design question: Should many arguments be passed to a DLL function or should values for class properties first be set and then functions operate on these class data members? As with most things, there is a tradeoff.

Keeping your DLL easy to use is one consideration. Being able to set properties first and then invoke functions usually makes the DLL easier to understand and work with. But if you think your ActiveX DLL might, at a later time, be recompiled into an ActiveX EXE, other considerations come into play.

As you will learn in Chapter 26, "Working with Objects from Microsoft Office," ActiveX EXE components provide enormous flexibility. They suffer in performance compared to DLLs, however, because they work as *out-of-process* servers, while DLLs are *in-process* servers. If there is a chance your DLL will need to be recompiled to become an ActiveX EXE, passing many arguments to a function can dramatically improve performance.

Objective 23.4 Determining an Application's File Dependencies

This final objective deals with files you need to distribute with your application. When you create an ActiveX DLL and use it in a program, you need to be sure that it is distributed and registered on the end user's computer. This is just one of the files to be distributed. Depending on the type libraries that are referenced in your application, the number of files to be distributed can vary dramatically.

Let's be frank: *Visual Basic installations can be difficult.* Visual Basic provides a Setup Wizard for creating distribution disks, and it usually works well. But you can encounter problems if there are name conflicts in the libraries, software components fail to register, or the installation program generated by the application Setup Wizard encounters an unhandled error.

There are several third-party tools on the market that allow robust installation scripts to be written, and these can more than pay for themselves in a short period. You might have written the best program in the world, but if the user can't install it, all you have done is made the end user angry.

This objective is not intended to provide a thorough review of Visual Basic's application Setup Wizard. It is intended to show how you can use the Setup Wizard to generate a dependency file, so you can examine your program and its dependent files before moving to the creation of installation disks. Visual Basic dependency files have a .DEP file extension.

Exercise 23.6 Generating a Dependency File with the Application Setup Wizard

This exercise asks you to use the application Setup Wizard to create a dependency file. This file lists the supporting files needed by your application when it is distributed to end-user machines and which should be included on your program's installation disks.

1 Go back to Start Programs, Microsoft Visual Basic 5.0 program group on your computer, and Start the Setup Wizard from the Visual Basic 5.0 startup directory (not the Visual Basic 5.0 program). This wizard is separate from the startup of Visual Basic itself. After an introductory screen, you should see the dialog box displayed in Figure 23.9.

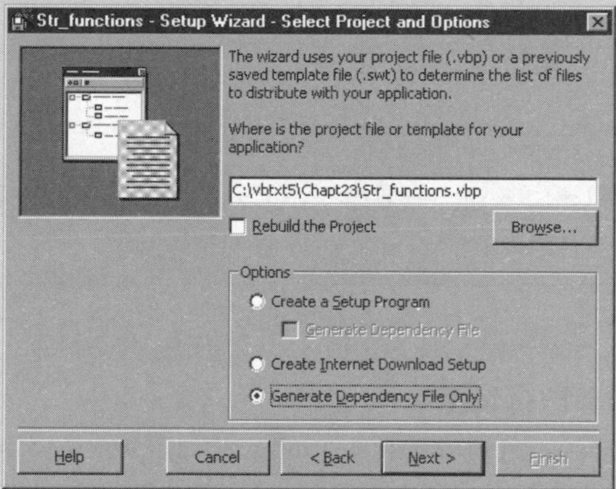

Figure 23.9

This dialog box allows you to generate a dependency file.

2 Select the Browse button and choose the Str_Functions.vbp project (from where you placed it), and click Open.

3 Select the Generate Dependency File Only option button shown in Figure 23.9.

4 Continue clicking through the application Setup Wizard until you see the generated list of files in the dialog box displayed in Figure 23.10

5 Continue clicking through the Setup Wizard until you click the Finish button. A Str_Functions.DEP file should now be in the project directory. This text file lists information about all of the files your ActiveX DLL project depends upon in order to run properly.

continues

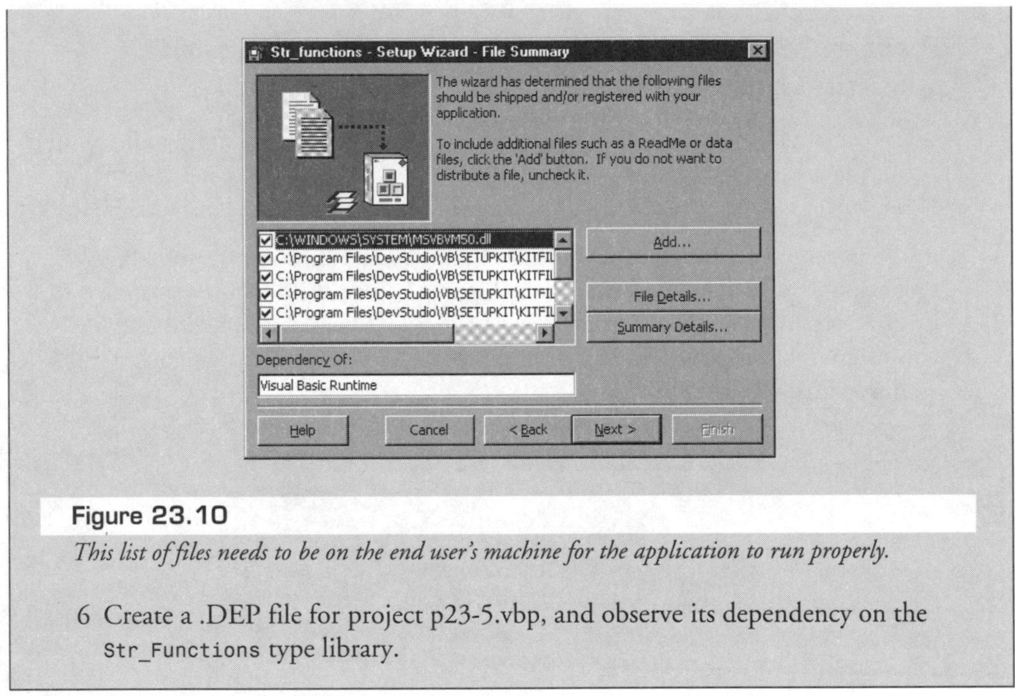

Figure 23.10

This list of files needs to be on the end user's machine for the application to run properly.

6 Create a .DEP file for project p23-5.vbp, and observe its dependency on the `Str_Functions` type library.

Chapter Summary

In this chapter, you developed your first true software component. You defined a component that could be created by other applications and then provided services to those other applications. This component acted as a *server* to *client* applications.

Software components are defined in type libraries. ActiveX type libraries can have .OCX, .DLL, and .EXE extensions. Some ActiveX type libraries have .OLB and .TLB extensions. These type libraries define classes of objects and are registered within the Windows Registry. Once registered, type libraries are available for inspection and use by other applications.

You make type libraries available to your project by creating a reference to that type library in the References dialog box. Objects defined in that type library can then be bound to object variables in your application.

ActiveX DLLs created in Visual Basic mainly use class modules to create code components. These components usually provide processor or computationally intensive services to client applications. In an ActiveX DLL project, the `Name` property of publicly instanced modules becomes the class names of objects that can be created from that DLL. The `Name` property of the ActiveX DLL project specifies the name of the type library for those objects when the ActiveX DLL is compiled. The `Instancing` property specifies how the ActiveX DLL can be created. For ActiveX DLLs, the most frequently used setting for the `Instancing` property is `MultiUse`.

Multiple projects in Visual Basic can be placed in Visual Basic groups. These groups, when saved to disk, have a .vbg extension. Visual Basic groups aid the development of ActiveX DLL projects by permitting both the server application (the ActiveX DLL project) and the client application (usually a Standard EXE project) to be worked on simultaneously in the Visual Basic design environment.

ActiveX DLL applications are *in-process servers.* This enhances their performance because they work within the same operating system process as the client application.

Visual Basic's application Setup Wizard can be used to create a dependency file. This file lists the supporting files needed by your application when it is distributed to end-user machines and which should be included on your program's installation disks. Dependency files have a .DEP extension.

Skill-Building Exercises

1. Create a client application that uses the `Str_Functions` type library, but uses a semicolon rather than a comma to parse a string of the names of the ten people you most admire.
2. Add a function to the `SFunctions` class module that returns the initials of the ten people you most admire. *Hint:* The space character has an ASCII value of 32.
3. Create an ActiveX DLL that includes a function that returns the average of an array of numbers passed to the function. Create a client application that passes an array of 50 randomly generated numbers between 1 and 10,000 to the ActiveX DLL function, and displays the numbers and their average.
4. Add a function to the ActiveX DLL developed in the preceding exercise that returns an array of numbers passed to the function sorted in ascending value. Write a client application that displays 50 unsorted, randomly generated numbers on one MSFlexGrid control and the sorted numbers on another grid. Also display the average of these numbers.
5. Generate a dependency file for the project in the preceding exercise.

24

Accessing and Manipulating Databases with Visual Basic

Visual Basic includes a special control for accessing external databases, named, appropriately, the Data control. It allows a user to access an existing database and to navigate through the stored data with little or no programming. Visual Basic also contains Visual Data Manager: an application that allows a user to create and repair a database from within the Visual Basic design environment.

When you complete this chapter, you will have a good understanding of what a database is and how it works, and be able to

- use the Data control,
- open a database,
- add to a database,
- edit a record stored in a database,
- use `Recordset` methods,
- write a query to retrieve a set of prescribed facts stored in a database,
- use Visual Data Manager to build a database,
- query a database built using Visual Data Manager.

Chapter Objectives

In this chapter, we initially explore the use of the Data control to open existing databases. Following this, we examine building a small database from within Visual Basic using the Visual Data Manager. The objectives of this chapter focus on database concepts and the use of Visual Basic's database features:

1. Understanding relational database terminology
2. Using the Data control

3. Querying data stored in a database
4. Building a database using Visual Data Manager

Objective 24.1 Understanding Relational Database Terminology

Before learning how to use the Data control and the Visual Data Manager, you need to understand some fundamental terminology and concepts regarding the structure of a database. To Visual Basic, *database structure* means a relational view or structure. A relational view of a database depicts a database as nothing more or less than a collection of tables, in which the rows of each table represent database *records,* and the columns of each table refer to the *fields* or *attributes* of a record.

Consider the relational database tables shown in Figure 24.1. These tables contain customer information and sales representative information. Information for one customer is placed in a single row. The information shown for a customer includes the customer number, the customer name, the customer telephone number, and the number of the sales representative assigned to the customer. Information for one sales representative is also placed in one row. The information shown for a sales representative includes the sales representative's number, name, and sales.

Figure 24.1

The structure of a relational database shows fields as columns and records as rows.

CUSTOMER : Table

Cust_no	Cust_name	Cust_phone	Sales_rep
1123	J.J. Hardy	503-444-3333	1400
1124	D.D. Smiley	203-345-3333	1100
1125	W.W. Wear	507-343-9999	1200
1126	R.R. Tumbo	503-465-7777	1400
1127	A.A. Arvo	503-222-3333	1400
1128	K.K. Klind	507-333-2222	1200
1129	E.E. Rams	203-456-2222	1100

Record: 8 of 8

SALESPERSON : Table

Sales_rep	Rep_name	Rep_Sales
1100	F.M. Flowers	345000
1200	P.J. Mongomery	345222
1400	D.D. Happy	556999

Record: 4 of 4

24.1.1 Important Terms

To begin, let's review the terms that are important to database tables. A database table contains data about a specific type of entity or object, such as a customer or salesperson. Figure 24.1 shows two database tables designed for use with Microsoft Access, a database software package. The following are essential database terms:

- **Table**—A table is a logical structure used to group sets of related information. In the example shown in Figure 24.1, customer information is kept separate from salesperson data. The two types of information are not mixed and placed in a single table.

- **Record**—A record is a grouping of a set of attributes describing each person, place, or item in a database. In Figure 24.1, a record is a collection of data describing each customer. For each customer, there is one and only one record. Likewise, there is one and only one record for each sales representative.
- **Field**—A field is an attribute of a record. For example, an attribute of a customer is the customer name; another attribute is the number of the sales representative. For each field, there is one and only one entry.
- **Primary key**—Each record in a relational database must be unique. This is accomplished by assigning a unique ID or number to each record, which is called the primary key. Records are stored based on the primary key. For example, in storing a customer's record, a unique customer number is assigned (`Cust_no`). A unique key separates two customers with the same name. Likewise, each sales representative is assigned a unique primary key (`Sales_rep`). No two members of the sales force can have the same `Sales_rep` number.
- **Foreign key**—Foreign keys are fields in common between tables. In our example, the `Sales_rep` number in the sales representative table is a primary key, while the `Sales_rep` number in the CUSTOMER table is a foreign key. By knowing the `Sales_rep` number in the CUSTOMER table, it becomes possible to link to the SALESPERSON table and to retrieve stored information for that number.
- **Indexes**—Access to a database record is often made faster through the use of an index assigned to fields other than the primary key field.

  Suppose you want to retrieve a customer record by entering the name of the customer, rather than the unique `Cust_no`. To accomplish this, you can create an index based on customer names, keeping this index separate from the database table. Before retrieving a record from the table, the index is searched for the record containing the specified customer name. If the search is successful, the record location of the customer is identified. This allows for direct access of the customer's record.
- **Queries**—Queries are questions asked of a database. In Visual Basic, an English-like language named *Structured Query Language* (SQL) is used to retrieve data from a database. The following is an example of SQL:

  ```
  SELECT Cust_Name, Cust_phone
  FROM Customer
  WHERE Customer.Cust_Name = 'D.D. Smiley'
  ```

  This SQL query can be read as follows: Display the customer name and the customer telephone number from the customer table for the customer named D.D. Smiley.

24.1.2 Setting Relationships

Besides defining tables, records, fields, primary and foreign keys, and indexes, constructing a relational database requires the setting of relationships between tables, using primary and foreign keys. The relationships permitted include one to one (1:1), one to many (1:N), and many to many (M:N). The way of depicting a one-to-many relationship is shown in Figure 24.2. This diagram depicts what is termed the *database schema.*

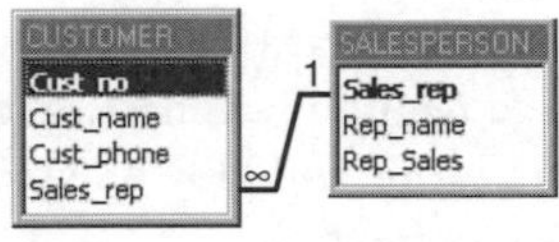

Figure 24.2

In this one-to-many relationship, one sales representative is responsible for many customers.

In this instance, each sales representative number can be found in many customer records. Stated as a rule, *each sales representative can be responsible for several customers.*

However, for each customer there is one and only one sales representative, as indicated by the number 1. Stated as a rule, *for each customer, there is one and only one sales representative.*

Objective 24.2 Using the Data Control

The relational database native to Visual Basic is the Jet database engine. This underlying database is the same database engine used by Microsoft Access. With Visual Basic, it is possible to build your own database. Such a database will have an .MDB extension just as an Access database has an .MDB extension. It is possible from Visual Basic to open and manipulate a database developed in Access, and the reverse is also true.

The underlying Jet engine can also be used in combination with specially designed software *drivers* to open and manipulate other databases, including Btrieve, Paradox, dBASE, and FoxPro. Learning to exploit the Jet database engine will take up most of Chapter 25, "Using Data Access Objects," and Chapter 26, "Working with Objects from Microsoft Office."

24.2.1 Data Control Properties

The easiest way to connect to Visual Basic's built-in database capabilities is to use the Data control. By simply setting Data control properties, it is possible to connect to a database without writing a single line of code. Several Data control properties are key to the control's functionality. As you work through these properties, you might want to place a Data control on a form and manipulate it as you read along.

The first property to observe is `DatabaseName`. Its complete syntax is the following:

```
[form.]datacontrol.DatabaseName [= pathname]
```

The component `pathname` is a string expression describing the location of the database file. For a database named Design1.MDB, you could assign it to a Data control, named `Data1`, with such an expression as the following:

```
Data1.DatabaseName = "C:\data\design1.mdb"
```

An alternative form is to write

```
Data1.DatabaseName = App.Path & "\design1.mdb"
```

With this form, it is assumed that the database is stored with the Visual Basic application.

In design time, the default database extension in the `databaseName` property dialog box is determined by the `Connect` property. Examine this property and you will see it's possible to connect to Paradox and dBASE databases, Excel and Lotus spreadsheets, and ASCII text databases. It is also possible to connect to ODBC (Open Database Connectivity) databases, such as SQL Server and Oracle. The default is Microsoft Access. With this database, no `Connect` property need be set.

After setting the path and the `Connect` properties, a third Data control property, the `RecordSource` property, must be set. The `RecordSource` must correspond to a type of record supported by the database. For example, in a database consisting of several tables, select the table you want to draw records from. The following is the complete syntax for this property:

```
[form.]datactl.RecordSource [= {tablename | sqlstatement | queryname} ]
```

The `RecordSource` can be a specific database table (`tablename`), a structured query language query (`sqlstatement`), or the name of a stored query (`queryname`). In the Jet engine, this last option is known as a `QueryDef` object, which is explored in Chapter 25.

Closely associated with the `RecordSource` property is the `RecordSetType` property. The complete syntax for this property is as follows:

```
datactrl.RecordsetType [= value ]
```

Table 24.1 shows the settings for `value`.

Table 24.1 Settings for *value* in the *RecordSetType* Property

Constant	*Value*	*Description*
`vbRSTypeTable`	0	A table-type `Recordset`.
`vbRSTypeDynaset`	1	The default. A dynaset-type `Recordset`.
`vbRSTypeSnapshot`	2	A snapshot-type `Recordset`.

As specified by the Jet database engine, a `Recordset` can be one of the following types:

- **Table-type `Recordset`**—A data set that is part of the underlying database and can be updated
- **Dynaset-type `Recordset`**—A data set that is returned from a query and can be updated
- **Snapshot-type `Recordset`**—A data set that can be from either a table or a query, but is read-only

The `Recordset` object is explored in detail in section 24.2.4, "Working with `Recordset` Objects," later in this chapter.

24.2.2 Adding Bound Controls

Once the properties have been established for the Data control, *bound controls* can be added to display the contents of database records. A bound control is a *data-aware* control. When a control is bound, Visual Basic provides field values from the database to that control. With a bound control, moving to a different record in the database leads to its display on a form.

Bound controls are not limited to text boxes. Visual Basic allows check boxes, image boxes, labels, picture boxes, text boxes, and other controls to be bound to the Data control.

Two important properties when adding bound controls to the database are the `DataSource` and the `DataField`. The `DataSource` property refers to the source of the database records to be returned and displayed. For example, if the data control is to return customer records and place one field in a text box, the database for that text box will be set or bound to a customer. The `DataField` property refers to the specific field within the record to be bound to a control. It specifies the field to be displayed by a control, such as a text box. The syntax for a DataField is as follows:

```
[form.] control.DataField [=fieldname]
```

Thus, if the `txtCustomer` text box is to display the field `customer name`, the property should be set as

```
txtCustomer.DataField = "customer name"
```

Exercise 24.1 A Beginning Database

Design a form to open a database and display the returned records using bound controls. The name of the database is Design1.MDB and is supplied on the disk with this book. Figure 24.3 shows the form that you are to create.

The schema for this database is shown in Figure 24.4. As indicated, a customer places an order for a single product. When placing an order, the order number, customer number, product number, date required, and quantity ordered must be recorded.

Figure 24.3

The Visual Basic form contains a single Data control named `Customer`.

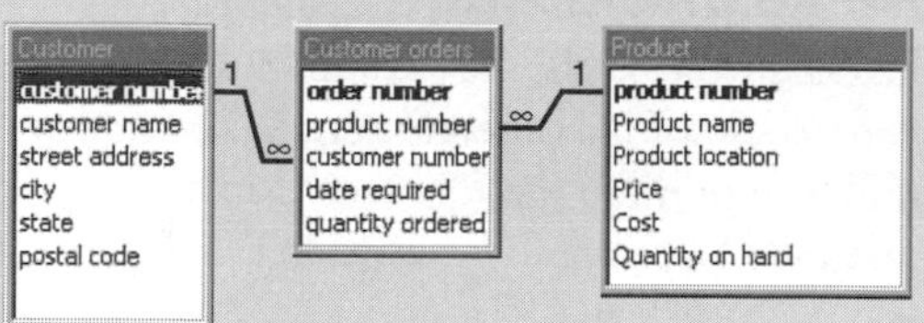

Figure 24.4

The Design1.MDB schema contains three tables: a customer table, a product table, and an orders table.

1 Place a Data control on the form (the default name is Data1), the six text boxes, and the six labels. Add the captions to the labels. The text box names are **`txtID`**, **`txtName`**, **`txtStreet`**, **`txtCity`**, **`txtState`**, and **`txtZip`**.

2 Make sure a copy of the Design1.MDB file is in the same directory as your Visual Basic project (for example, Ch24\Exercises\). In the `DatabaseName` property of the Data control, click the ellipses button, and select the Design1.MDB database in the dialog box that appears.

```
<drive>:\[<pathname>\]DESIGN1.MDB
```

Alternatively, you can type in the full path for the database. For example, if the database is stored on the C drive, you might type C:\VBTXT5\DATA\ DESIGN1.MDB.

3 Change the `RecordSource` property for the Data control to `Customer` by selecting the table from the drop-down list.

4 Change the `RecordSetType` to `0 ' table`. This indicates that you want to examine the Customer table in the database saved as DESIGN1.MDB.

5 Bind all text boxes to the Data control. This can be done in two ways. Click the form and drag the cursor to enclose all text boxes. Then, set the `DataSource` property to `Data1`. The alternative is to set the `DataSource` to `Data1` for each control, one at a time.

6 Once controls are bound, set each data field. For the `txtID`, set the `DataField` property to customer number; for the `txtName`, set the `DataField` to customer name; and so on. Click the `DataField` box with the mouse to move to the next setting or make the selection from the drop-down list box.

7 Run and test your design. You should be able to display the six customer records stored in this database by navigating through the database with the Data control.

24.2.3 The *Refresh* Method

When working with the Data control, the `Refresh` method is needed when opening a database or reopening a database with a different set of Data control properties. Also use `Refresh` whenever you want to move to the first record in a `Recordset`. With this method, the Data control automatically loads the first record and makes it current. The syntax for this method is as follows:

```
[form.] datacontrol.Refresh
```

The component `datacontrol` specifies the database name and record source.

If the `Refresh` method fails to open the database, a trappable error will result. The following code can be used to determine the cause of failure should `Refresh` fail to open a database:

```
Data1.Refresh
If Data1.Database Is Nothing Then
   MsgBox "Database opening failed"
End If
If Data1.Recordset Is Nothing Then
   MsgBox "Database recordset failed"
End If
```

24.2.4 Working with *Recordset* Objects

When working with a database, Visual Basic treats the rows of data returned to the Data control as a `Recordset` object. This object, like most others, features a number of methods, which are explained in Table 24.2.

Table 24.2 Methods of the *Recordset* Object

Method	*Description*
`AddNew`	Clears the buffer in preparation for creating a new record in a table or dynaset
`Delete`	Deletes the current record in a specified table or dynaset
`Edit`	Opens the current record in a specified `Recordset` for editing
`Update`	Saves the contents of the copy buffer to a specified table or dynaset
`FindFirst`	Locates the first record that satisfies specified criteria and makes it the current record
`FindLast`	Locates the last record that satisfies specified criteria and makes it the current record
`FindNext`	Locates the next record that satisfies specified criteria and makes it the current record
`FindPrevious`	Locates the previous record that satisfies specified criteria and makes it the current record
`MoveFirst`	Moves to the first record in a specified `Recordset` and makes it the current record

Method	*Description*
`MoveLast`	Moves to the last record in a specified `Recordset` and makes it the current record
`MoveNext`	Moves to the next record in a specified `Recordset` and makes it the current record
`MovePrevious`	Moves to the previous record in a specified `Recordset` and makes it the current record
`Seek`	In a table-type `Recordset`, moves to the indexed record specified by method arguments

Besides methods that are applied to the `Recordset` object, `Recordset` object properties can also be referenced or changed. Consider the following instruction:

```
data1.Recordset("order number") = CInt(txtOrdNum.Text)
```

This instruction assigns the integer value of the string placed in the text box to the `order number` field of a specific record in a designated `Recordset`. `Recordset` objects also have `BOF` (beginning of file) and `EOF` (end of file) properties. The following instruction would help prevent navigating past the end of a database:

```
Data1.Recordset.MoveNext
If Data1.Recordset.EOF then
   MsgBox "End of File"
   Data1.Recordset.MovePrevious
Else
   '   code to update status bar and/or add a new record
End if
```

As long as you are using the Data control to move through records, you needn't worry about this. The Data control prevents these kinds of errors automatically.

There are three other principal sources of errors that you must remain mindful of when modifying data in database tables, as follows:

- **Data validation errors**—These errors occur when you seek to assign data to a field that the field is not designed to handle. Incorrect data types can generate these types of errors. Assigning a string to a field that expects a numeric value, for instance, will usually generate an error of this type.

  Other data validation errors occur when data violates rules put in place when the database table was designed. Suppose, for instance, you have a `Ship_Date` field for the Order table. You might enter a data validation rule that says the `Ship_Date` value cannot precede the date placed in the `Date_Ordered` field. Microsoft Access allows you to place these type of range limitations on numeric data fields.
- **Entity integrity errors**—These errors occur when you attempt to add a second record with the same primary key as a record currently stored in the database.
- **Referential integrity errors**—These complex errors occur when you seek to modify data in one table, but it violates the data set in another table. Take a look again at the schema in Figure 24.2. If you seek to remove a sales representative who is assigned to a

customer, such a deletion will not be allowed. Before the sales representative can be deleted, all references to that person in the Customer table must be changed.

In the schema shown in Figure 24.4, referential integrity errors also occur if you attempt to add a customer number to the orders table when that number does not represent a stored customer's primary key.

Exercise 24.2 Adding and Deleting a Record in a Database

Besides displaying records stored in a database, Visual Basic can be used directly in modifying the contents of stored records. This exercise asks you to open the customer orders table in the database, add a record, and delete a record. A copy of the database, named DESIGN2.MDB, has been placed on file on the student disk. This database has the same schema as Design1.mdb. Figure 24.5 shows the modified form for this exercise.

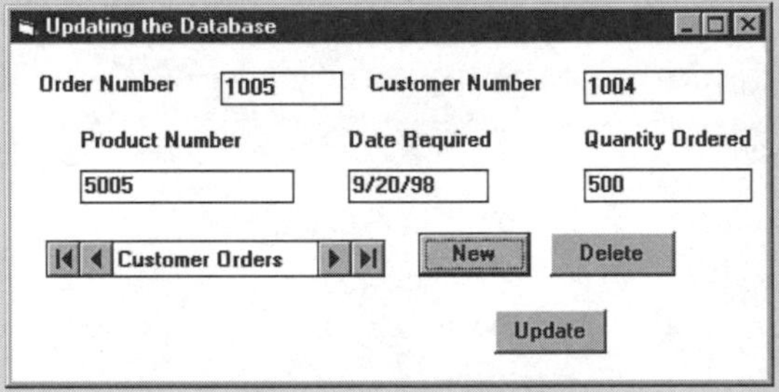

Updating the Database

Order Number 1005 Customer Number 1004

Product Number Date Required Quantity Ordered

5005 9/20/98 500

Customer Orders New Delete

Update

Figure 24.5

The data form layout contains three new buttons: New, Update, and Delete.

1 Design the form as shown in Figure 24.5. For the Data control, make the following settings:

```
DatabaseName = "C:\design2.mdb    'or set your own path"
RecordSource = "Customer orders"
RecordSetType =  0 ' Table
```

Name the new commands **cmdNew**, **cmdUpdate**, and **cmdDelete**. Name the text boxes **txtOrdNum**, **txtCustNum**, **txtProdNum**, **txtDate**, and **txtQuantity**. Set the DataFields and DataSource for each text box. Add all labels as shown.

2 Write the cmdNew_Click() procedure as follows:

```
Private Sub cmdNew_Click()
   cmdUpdate.Enabled = False
   data1.RecordSource = "Customer orders"
   data1.Refresh
```

```
    txtOrdNum.SetFocus

    data1.Recordset.AddNew
    cmdNew.Enabled = False
    cmdUpdate.Enabled = True
End Sub
```

The keywords important to this code include the following:

- **RecordSource**—Sets the table from which to draw records.
- **Refresh**—Updates the data structure of a Data control and goes to the beginning of the `Recordset`.
- **RecordSet.AddNew**—Adds a buffer in memory for a new record in the specified `RecordSource` that clears all bound controls.

3 Write the `cmdUpdate_Click()` procedure as follows:

```
Private Sub cmdUpdate_Click()
  On Error GoTo cmdUpdateErr:

  Data1.Recordset("order number") = CInt(txtOrdNum.Text)
  Data1.Recordset("customer number") = CInt(txtCustNum.Text)
  Data1.Recordset("product number") = CInt(txtProdNum.Text)
  Data1.Recordset("date required") = CDate(txtDate.Text)
  Data1.Recordset("quantity ordered") = CInt(txtQuantity.Text)
  Data1.Recordset.Update
  cmdNew.Enabled = True

  Exit Sub
cmdUpdateErr:
  MsgBox Str$(Err) & "  " & Err.Description, 0, "Error in Update"
End Sub
```

In this case, the `data1.Recordset.Update` instruction complements the `data1.Recordset.AddNew` instruction. It assigns values to the record in memory and writes the data to the database.

4 Write the `cmdDelete_Click()` procedure as follows:

```
Private Sub cmdDelete_Click()
   data1.RecordSource = "Customer orders"
   data1.Recordset.Delete
   data1.Refresh
End Sub
```

The key instruction is `data1.Recordset.Delete`. The `Delete` method removes the current record from the database.

5 Add a record to the database. To avoid violating integrity rules, the following are the values that can be added for product and customer numbers:

continues

Product Number	*Customer Number*
5000	1000
5001	1001
5002	1002
5003	1003
5004	1004
5005	

Add the following data to the data entry screen. If order number 10059 has been entered already, delete this order and fill in the new information for 10059.

Order number = **10059**

Date required = **7/15/98**

Quantity ordered = **500**

6 After adding the data, click Update.

7 Find record 10059, and click Delete to remove the record from the database.

8 Place a product number or customer number that does not exist in the Product or Customer tables, respectively, and observe the error that occurs.

Exercise 24.3 Adding Editing to a Design

Add one control to the preceding design: a command that allows you to edit a record, as shown in Figure 24.6. With editing, the `Edit` method is used, which locks the record to prevent changes by others. Once changes are made, the `Update` method is required. With either `Edit` or `AddNew`, the "page" around a record is locked until the `Update` method has completed writing to the database.

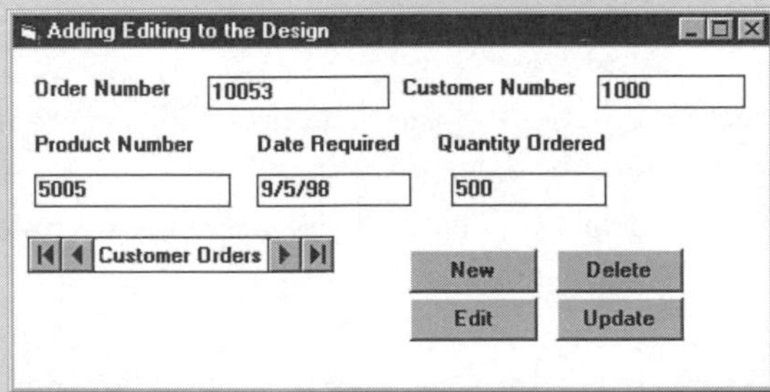

Figure 24.6

The Edit button is used to lock a record in place while it is being revised.

1 Add the command button as shown to the design, and change the name to `cmdEdit`. This exercise uses the Design3.mdb, which is found on the student disk. Make sure all of the appropriate text boxes are bound to the correct fields of the Customer Order table.

2 Write the following `cmdEdit_Click()` procedure:

```
Private Sub cmdEdit_Click()
   cmdNew.Enabled = False
   cmdUpdate.Enabled = False

   data1.RecordSource = "Customer orders"
   data1.Recordset.Edit

   cmdUpdate.Enabled = True
   cmdEdit.Enabled = False
End Sub
```

3 Write the following `cmdUpdate_Click()` procedure:

```
Private Sub cmdUpdate_Click()
  On Error GoTo cmdUpdateErr:
  Data1.Recordset("order number") = CInt(txtOrdNum.Text)
  Data1.Recordset("customer number") = CInt(txtCustNum.Text)
  Data1.Recordset("product number") = CInt(txtProdNum.Text)
  Data1.Recordset("date required") = CDate(Trim(txtDate.Text))
  Data1.Recordset("quantity ordered") = CInt(txtQuantity.Text)

  Data1.Recordset.Update
  cmdNew.Enabled = True
  cmdEdit.Enabled = True

Exit Sub
cmdUpdateErr:
  MsgBox Str$(Err) & "  " & Err.Description, 0, "Error encountered in
  ➥Update"
End Sub
```

In this example, the instruction `Recordset.Edit` allows a record stored in the database to be revised. However, the `Edit` method alone is not sufficient for updating. Once changes are made to the record, the `Update` method must be employed to record the changes.

Objective 24.3 Querying Data Stored in a Database

Visual Basic supports a query language known as *Structured Query Language* (SQL). Although a complete discussion of this language falls outside the scope of this text, we can provide an overview of some important query concepts.

SQL is a scripting language that can be used to create tables, define the relationship among tables, and, most importantly for Visual Basic developers, manipulate the data held in tables.

The most frequent use of SQL is to extract data from a database and place the data in a result set. The snapshot (read-only) and dynaset (updatable) types of `Recordsets` are often result sets from this type of query. Read-only queries begin with a `SELECT` statement, which is written to allow you to select a set of records that match a prescribed set of criteria.

When using SQL, use SQL statements that conform to the following form (using the `SELECT` statement as an example):

```
SELECT the following items
FROM this table
[WHERE(match this or these criteria)]
```

Observe that the `WHERE` clause is optional. For example, the following code features the wild card (*):

```
SELECT *
FROM product;
```

The query can be read as follows: *Display all values for all products stored in the product table.*

Likewise, consider the following statements:

```
SELECT order_number
FROM customer_orders
WHERE quantity_ordered >= 1000;
```

They can be read as follows: *Display all order numbers contained in the customer orders table where the quantity ordered is greater than, or equal to, 1,000.*

Exercise 24.4 Using a Query to Retrieve a Set of Records

Use a SQL statement to retrieve a set of records from the product table. Specifically, you are asked in this exercise to examine a subset of customer orders. Figure 24.7 and Figure 24.8 show the designs of the two forms for this exercise. Use DESIGN4.MDB for this exercise, which is found on the student disk.

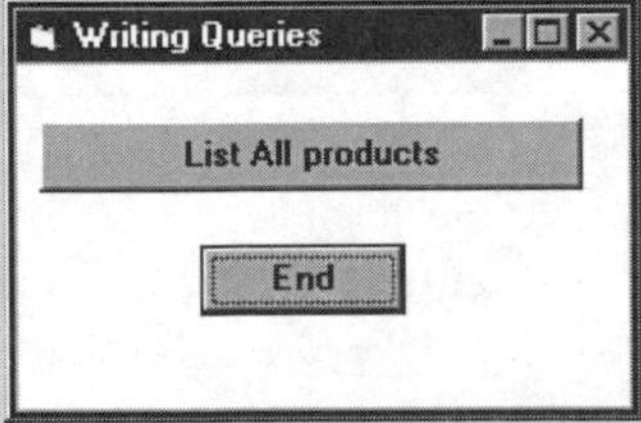

Figure 24.7

The Writing Queries form contains a button showing the name of the query to execute.

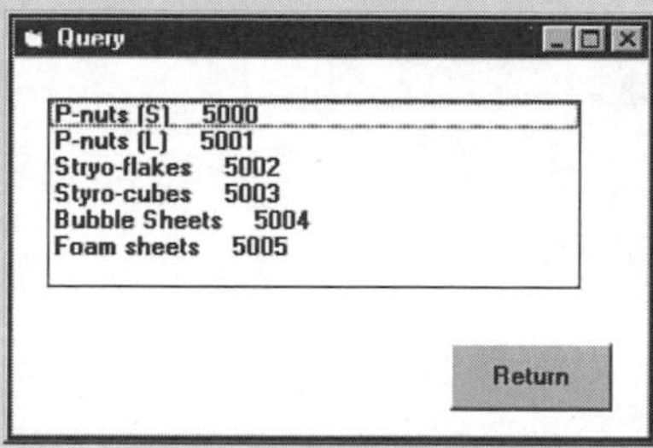

Figure 24.8

The Query form shows the results of the executed query.

1 Design **`frmOpen`** and **`frmQuery`** as shown, adding the `End` procedure to `frmOpen` and the `Return` procedure to `frmQuery`. Name the **List All Products** button **`cmdQuery`**. The list box on `frmQuery` is named **`lstResults`**. The command button on `frmQuery` is named **`cmdReturn`**. Save `frmOpen` to disk as **f24-4a.frm** and `frmQuery` to disk as **f24-4b.frm**.

2 Set the Data control properties on `frmQuery` as follows:

```
RecordsetType = 2 ' Snapshot
```

3 Write the `cmdQuery_Click()` procedure as follows:

```
Private Sub cmdQuery_Click()
   frmOpen.Hide
   frmQuery.Show

   frmQuery!Data1.DatabaseName = App.Path & "\design4.mdb"

   'Execute query
   frmQuery!Data1.RecordSource = "SELECT [Product name], [Product number]
   ➥FROM Product"
   frmQuery!Data1.Refresh

  ' Clear the List box
  frmQuery.lstResults.Clear

   'Retrieve data from database
   Do While Not frmQuery!Data1.Recordset.EOF
      If Not IsNull(frmQuery!Data1.Recordset(0)) Then
         frmQuery!lstResults.AddItem frmQuery!Data1.Recordset("Product
         ➥Name") & "    " & frmQuery!Data1.Recordset("Product Number")
      End If
      frmQuery!Data1.Recordset.MoveNext
   Loop
   frmQuery!Data1.Refresh
End Sub
```

There are several new concepts introduced by this procedure:

continues

- The record source is represented by the following query:

```
"SELECT [Product name], [Product number] FROM Product"
```

  With this syntax, field names with spaces in them must have brackets around them.
- The `data1.Recordset.MoveNext` method moves from one record to the next in the `Recordset`.
- The following statement retrieves the first field from the record:

```
data1.Recordset(0)
```

  If the second field is to be displayed, use the following:

```
data1.Recordset(1)
```

- The loop continues until the `Recordset` reaches the end-of-file marker (`EOF`). At that point, processing stops.
- The `AddItem` method displays records that match the specifications of the query.

4 On the second form, write the `cmdReturn` procedure as follows:

```
Private Sub cmdReturn_Click()
   frmQuery.Hide
   frmOpen.Show
End Sub
```

5 For the Data control, set the `Visible` property to `False`.

6 Test your query design.

Care must be taken not to enter product numbers or customer numbers that do not already exist in the Product or Customer tables when modifying records in the Customer Order table. Otherwise, errors will result. For example, it is not possible to process an order for a customer who is not found in the database, or for a product that does not exist. One way to prevent these kinds of errors is to present data in related tables to the user in combo boxes.

Exercise 24.5 Using Queries to Populate Combo Boxes

Modify Exercise 24.3 to permit a user to navigate through records with the Data control, but *not* enter new values for product number or customer number while navigating. Valid choices are presented from which to make selections when a user enters `AddNew` or Edit mode.

To accomplish these twin functions, replace the text boxes for inputting customer number and product number with labels. On top of these labels, place combo boxes with their `Visible` properties set to `False`. When the user enters `AddNew` or Edit mode, these combo boxes will be made visible. Populate the combo boxes by adding another

Data control to the form and setting its `Visible` property to `False`. Use DESIGN5.MDB for this exercise. Figure 24.9 shows the interface during runtime in `AddNew` mode.

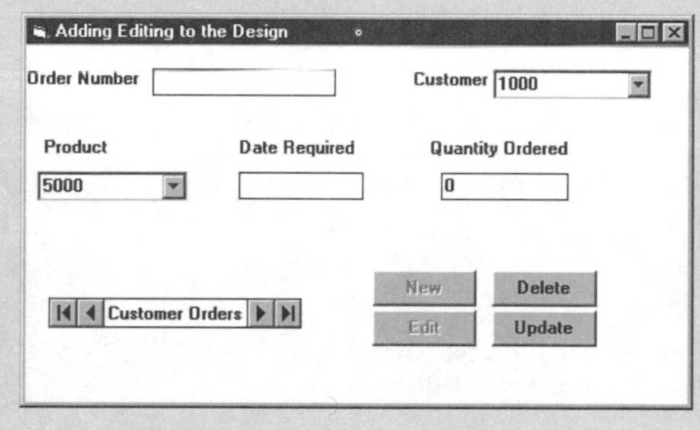

Figure 24.9

The revised form now contains two combo boxes: one for the product table and one for the customer table.

1 Save the form and project from Exercise 24-3 under a new name (for example, save f24-3.frm as **f24-5.frm** and p24-3.vbp as **p24-5.vbp**). Change the `Name` property of the project to **24_5**.

2 Remove the two text boxes `txtCusNum` and `txtProdNum` from the form, and add two labels in their places: **`lblCustomer`** and **`lblProduct`**. Remove the caption from these labels, and set the `BorderStyle` property to *1* (`Fixed Single`). Set the `DataSource` property for these labels to `Data1` and the `DataField` property to `customer number` and `product number`, respectively.

3 Add two combo boxes to the form, and size them so they are directly over the labels. Name these combo boxes **`cboCustomer`** and **`cboProduct`**. Set the `Style` property for the two boxes to `0` (`Dropdown List`), and remove any value from the `Text` property.

4 Add a second Data control to the form, and set its `Visible` property to `False`.

5 In the `Form_Load()`, event add the following code:

```
Private Sub Form_Load()

  Data2.DatabaseName = App.Path & "\design5.mdb"
  ' populate the product combo box
  Data2.RecordSource = "Product"
  Data2.RecordsetType = 2 ' snapshot
  Data2.Refresh
```

continues

```
  Do While Not Data2.Recordset.EOF
    cboProduct.AddItem Data2.Recordset("Product number")
    Data2.Recordset.MoveNext
  Loop
  cboProduct.ListIndex = 0

 'populate the customer combo box

  Data2.RecordSource = "Customer"
  Data2.RecordsetType = 2 ' snapshot
  Data2.Refresh
  Do While Not Data2.Recordset.EOF
    cboCustomer.AddItem Data2.Recordset("Customer number")
    Data2.Recordset.MoveNext
  Loop
  cboCustomer.ListIndex = 0

  Data1.RecordSource = "Customer orders"
  Data1.RecordsetType = 0 ' table
  Data1.Refresh

End Sub
```

6 The `cmdEdit_Click()` event should now read as follows:

```
Private Sub cmdEdit_Click()
  cmdNew.Enabled = False
  cmdUpdate.Enabled = False

  Data1.RecordSource = "Customer orders"
  Data1.Recordset.Edit

  cmdUpdate.Enabled = True
  cmdEdit.Enabled = False
  cboCustomer.Visible = True
  cboProduct.Visible = True
  lblCustomer.Visible = False
  lblProduct.Visible = False
End Sub
```

7 The `cmdNew_Click()` event should now read as follows:

```
Private Sub cmdNew_Click()
  cmdEdit.Enabled = False
  Data1.RecordSource = "Customer orders"

  txtOrdNum.SetFocus


  Data1.Recordset.AddNew

  cmdNew.Enabled = False
  cmdUpdate.Enabled = True
  cboCustomer.Visible = True
  cboProduct.Visible = True
  lblCustomer.Visible = False
  lblProduct.Visible = False
End Sub
```

8 The `cmdUpdate_Click()` event should now read as follows:

```
Private Sub cmdUpdate_Click()
  On Error GoTo cmdUpdateErr:
  Data1.Recordset("order number") = CInt(txtOrdNum.Text)
  Data1.Recordset("customer number") = CInt(cboCustomer.Text)
  Data1.Recordset("product number") = CInt(cboProduct.Text)
  Data1.Recordset("date required") = CDate(Trim(txtDate.Text))
  Data1.Recordset("quantity ordered") = CInt(txtQuantity.Text)

  Data1.Recordset.Update
  cmdNew.Enabled = True
  cmdEdit.Enabled = True
  cboCustomer.Visible = False
  cboProduct.Visible = False
  lblCustomer.Visible = True
  lblProduct.Visible = True
Exit Sub

cmdUpdateErr:
  MsgBox Str$(Err) & "  " & Err.Description, 0, "Error encountered in
  ➥Update"
End Sub
```

9 Run and test the program. When you click New, the combo box drop-down arrows appear. Click New and add a new order. Click Edit, followed by Update. Delete the order you added.

Visual Basic also provides special data-bound custom controls to make the type of interface in Exercise 24.5 easier to create. Examine the online Help system of Visual Basic, and explore the possibilities with the data-bound Combo Box, List Box, and Grid controls.

Objective 24.4 Building a Database Using Visual Data Manager

Besides opening and changing a database created by an external database application, you can create a database from within Visual Basic. This is accomplished by choosing Add-Ins, Visual Data. VisData, an abbreviation for Visual DataTools, is supplied as an executable with Visual Basic 5 and provides a graphical user interface that allows you to view and edit objects inside ODBC databases. If, for some reason, VisData does not appear, use Windows Explorer to find the file, and double-click to start the program.

24.4.1 Creating a Database

When the VisData window appears, choose File, New to open a new database. A dialog box will appear, asking you to choose one of several databases. Select Access and the version supported by your computer. A second dialog box will then ask you to supply the name of the database. We called our database CLASSES.MDB. A copy of this database is placed on the student disk. You can examine this copy using Access before you get started. If your version of Access is newer than the copy, you must convert it. Perform the following steps:

1. Close the copy of CLASSES.
2. Choose Tools, Database Utilities, and select Convert Database.
3. Name the database **CLASSES1**.

The procedure that we used to create this database follows. To replicate this procedure, give your database a new name.

After naming the database, tables can be added. To add a table, click the database window and click the right mouse button to open the dialog box shown in Figure 24.10. We typed ClassList for our table name.

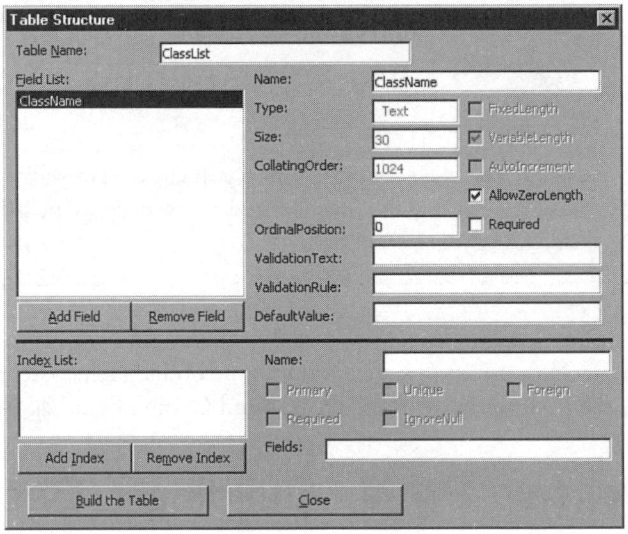

Figure 24.10

The Table Structure dialog box appears by clicking the right mouse button and selecting New_table.

Fields for a table can be defined once the table is created. To create a field, click Add Field. As shown in Figure 24.11, the field name, field type, and field size are used in defining a field. The field name is a field of your choice. We selected `ClassNo` for a field in the ClassList table.

The field types include Boolean, Byte, Integer, Long, Currency, Single, Double, Date/Time, Text (not string), Binary, and Memo. To add a fixed-length string to a table, select Text. To add an image, select Binary. To add a variable-length string, select Memo. The field size is used with variable length strings. For example, with `ClassNo`, a field of type Integer can be selected.

Notice that you can set field validation rules, designate whether data is required for the field, and specify whether the field can contain zero-length values.

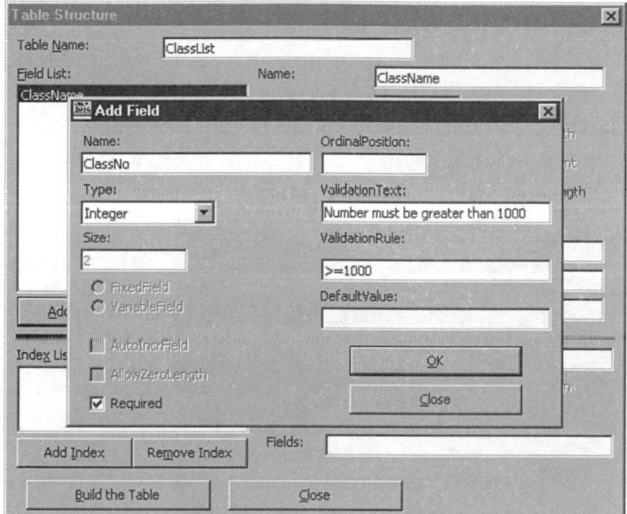

Figure 24.11

The Add Field dialog box appears when you click the Add Field button shown in the Table Structure dialog box.

Once the fields for the relational table have been established, individual fields can further be edited by making changes directly on the Table Structure dialog box. To do this, highlight the field to change, make the change, and press the Tab key.

After fields and data types have been established, a primary index and secondary indices can be set. Figure 24.12 shows this dialog box, as well as the additional fields we have added to the table. To reach it, click Add Index from the Table Structure dialog box. In Figure 24.12, you see the primary index being set. To set a secondary index, make similar settings—except do not click Primary index because there can only be one primary index for a table.

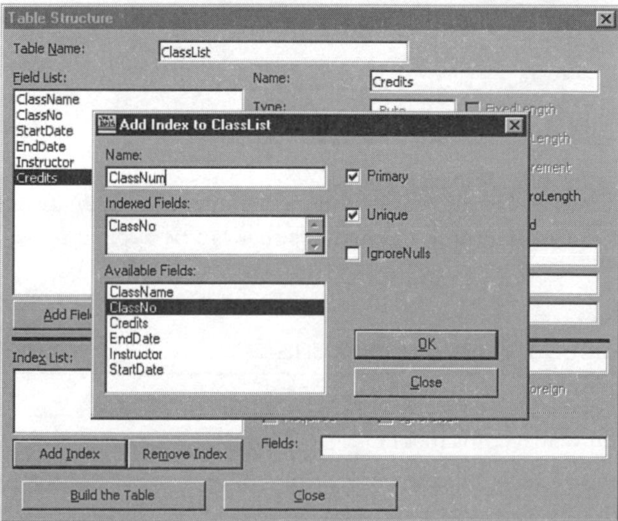

Figure 24.12

The Add Index to ClassList dialog box appears when you click the Add Index button in the Table Structure dialog box.

24.4.2 Entering Data into the Database

Data must be added to each table in the database following table, field, and index definitions. To add data to a table, click the Build the Table button in the Table Structure dialog box, and double-click the ClassList icon that appears in the Database window. This opens the ClassList data entry window (see Figure 24.13).

Figure 24.13

Double-click the ClassList table icon to open the table to the ClassList data entry window.

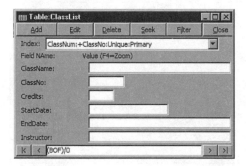

With the data entry window facing you, click Add to open the Update dialog box, fill in the data for a complete record, and click the Update button (see Figure 24.14). At this point, the dialog box shown in Figure 24.13 reappears. To add a second record, click Add, fill in the data, and click Update.

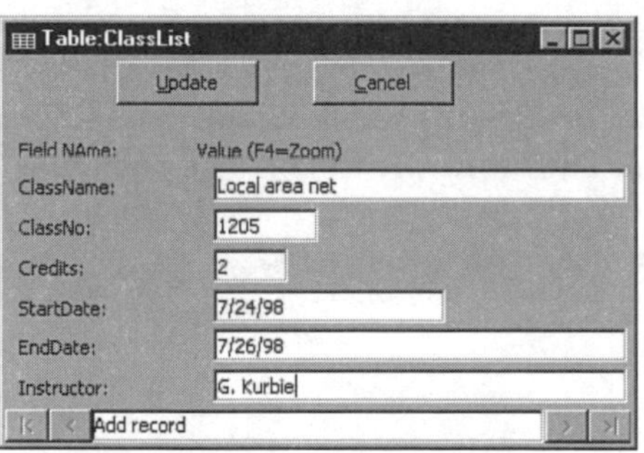

Figure 24.14

Click Add to move to the Update dialog box, add the data, and click Update.

The Table Access window shown in Figure 24.13 contains six buttons, which are explained in Table 24.3. (By now you realize that these are methods that can be applied to a `Recordset`.)

Table 24.3 Table Access Window Buttons

Button	*Description*
Add	Used to add new records to a table
Edit	Used to revise a record in the table
Delete	Used to delete a record from the table

Button	*Description*
Seek	Invokes a `Seek` method using the primary index
Filter	Places a filter on the current table-type `Recordset` and loads the newly filtered `Recordset` into a dynaset/snapshot form
Close	Closes the table

Exercise 24.6 Using VisData to Create a New Database

This exercise asks you to explore the use of VisData. The database for this exercise is named Classes.MDB.

1. Choose Add-Ins, Visual Data Manager.
2. Open the database Classes.MDB stored on disk (an Access database).
3. Scroll through the records stored in this database.
4. Add a record to the database, using the data shown in Figure 24.14.
5. Scroll through the database again.
6. Delete the record you added.

24.4.3 Building Queries with VisData

In addition to building tables, setting field properties, and defining table relationships, VisData also permits you to test and save queries. A separate SQL Statement window allows you to write, test, and save queries. These saved queries are `QueryDef` objects, which are explored in greater depth in Chapter 25, "Using Data Access Objects." Saved queries can be used to create `Recordsets`. Once saved, they will appear in the drop-down list box in the `RecordSource` property of a Data control.

Exercise 24.7 Using the VisData SQL Statement Window

1. Open the Classes.MDB database from VisData.
2. Without opening the ClassList table, enter the SQL statement you see in Figure 24.15.
3. Click Execute to test the query. When the dialog box asks if this is a SQL Pass Through query, click No.
4. Save this query. When asked for a name, call it **`Query1`**.
5. Return to the Database window to test the saved query. Double-click `Query1` and click Yes when asked, "Use DynaSet?" Experiment with other types of queries.

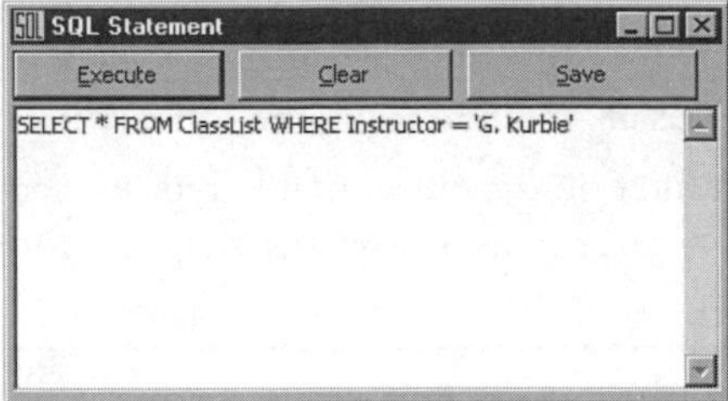

Figure 24.15

This is one possible SQL query you can write to determine the type of information stored in the ClassList table.

Exercise 24.8 Combining Visual Basic with the Created Database

A database created by Visual Basic can be used like a database created externally. This exercise asks you to list the records stored in the database Classes.MDB. Figure 24.16 shows the interface you are asked to use.

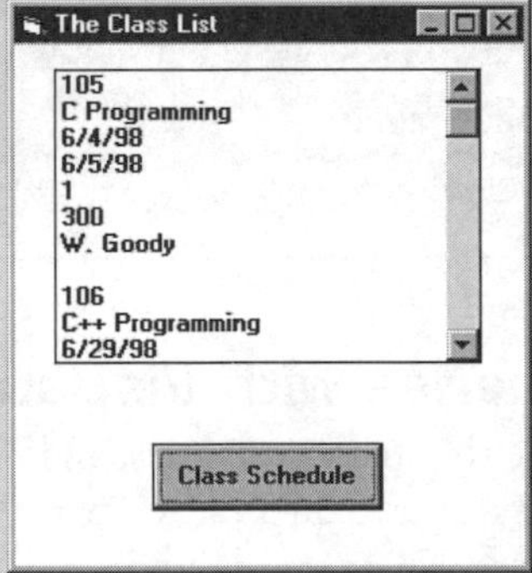

Figure 24.16

The Class List is displayed by using a query to process the data in the ClassList table.

1 Create the interface as shown, adding a Data control.

2 Set the following Data control properties:

```
DatabaseName = "C:\Classes.MDB  ' place correct path to your database
RecordSource = "ClassList"
RecordSourceType = 2 ' snapshot
Visible = False
```

3 Write the following `cmdSchedule_Click()` procedure:

```
Private Sub cmdSchedule_Click()
   data1.RecordSource = "Select * from classlist"
   data1.Refresh
   Do While Not data1.Recordset.EOF
      If Not IsNull(data1.Recordset(0)) Then
```

```
            lstResult.AddItem data1.Recordset("ClassNo")
            lstResult.AddItem data1.Recordset("ClassName")
            lstResult.AddItem data1.Recordset("StartDate")
            lstResult.AddItem data1.Recordset("EndDate")
            lstResult.AddItem data1.Recordset("Credits")
            lstResult.AddItem data1.Recordset("Cost")
            lstResult.AddItem data1.Recordset("Instructor")
            lstResult.AddItem ""
        End If
        data1.Recordset.MoveNext
    Loop
End Sub
```

In this example, a loop reads all records on file. When a null value is found, processing stops. After the contents of a record have been displayed, the `MoveNext` method is required to advance to the next record in the `Recordset`. This method explains how Visual Basic moves from record to record in a relational file.

Chapter Summary

Although this chapter presents many topics, they all center on how Visual Basic implements the relational database model through the use of the Jet engine and makes the database easy to use through the Data control. The chapter also introduces the Visual Data Manager tool for building database tables and testing and saving queries.

In the relational model, databases are constructed from a series of tables. Tables, themselves, are composed of rows and columns in which rows represent individual records and columns represent the fields of those records. Tables have a primary key index, which represents a field that has a unique value for every record in the table. Tables can also have foreign keys, which represent data shared in common between tables. In Visual Basic, queries are written in SQL to retrieve data from the database or manipulate the database itself.

The Data control allows easy access to Visual Basic and other databases. Important properties for the Data control include the `DataBaseName` property, identifying the path to the database; the `RecordSource` property, identifying whether the Data control contains data from a table or a query; and the `RecordSetType` property, identifying whether the data contained by the Data control is an updatable data set from a table (table-type `Recordset`), query (dynaset-type `Recordset`), or a read-only data set (snapshot-type `Recordset`).

Data-aware controls can be bound to a Data control by setting their `DataSource` property to a Data control in design time and their `DataField` property to an appropriate field in the Data control's `Recordset`.

Key methods for working with `Recordsets` include `AddNew`, `Edit`, `Update`, and `Delete`—the only one that can write data to the database is the `Update` method. Loops can be used in conjunction with the `Recordset Move` methods to iterate through a `Recordset` so that data can be displayed or amended.

The Visual Data Manager tool helps you build Jet engine databases. Table elements can be designed and relationships between tables set. Queries can also be tested and saved. Saved queries become `QueryDef` objects, and can be referenced for use later as `RecordSources` for the Data control.

Skill-Building Exercises

1. A database called Biblio.MDB comes with Visual Basic 5 software. Add a Data control to a form, and bind three labels to the three fields in the Authors table so the user can scroll through the table.
2. Using Biblio.MDB, write a program that lists all authors whose last names begin with the letters *A* through *L* in one list box, and those whose last names begin with the letters *M* through *Z* in another list box.
3. On your disk is a program called Cinema2.MDB. Use VisData to examine this database. Write a program that lists film titles and the years released from the Film Titles table in a list box.
4. Write a program that allows new records to be added to the Directors table in Cinema2.MDB.
5. Write a program that lists film titles from Cinema2.MDB in a list box. When the user clicks a film title, display the director's name by using the `Seek` method. *Hint:* Place the Director ID from the Film Titles table in a separate array. When the user clicks the list box, use the `ListIndex` property to retrieve the director ID from the array, and the `Seek` method in the Director's Table to retrieve the Director's name. This exercise requires two Data controls.

25

Using Data Access Objects

In Chapter 24, "Accessing and Manipulating Databases with Visual Basic," you used the Data control to open and access the Jet engine. This is the simplest method for opening databases, and the Data control must be used if you want to bind controls to data. However, large data sets bound to Data controls can appreciably degrade performance. For full flexibility and programmatic control, the Jet engine enables the developer to exploit Data Access Objects (DAOs): In large database applications, this is how the Jet engine is used.

Because the Jet engine is a full-featured database, you will perform a large number of tasks relating to DAOs in this chapter. Even with the following extensive list, we do not cover all of the DAOs. Instead, we cover the essential core of the Jet engine. With this foundation, you should be able to work with the DAOs not specifically addressed. After you complete the nine exercises of this chapter, you will be able to

- review the Jet engine object model;
- iterate through the Jet engine object collection hierarchy using `For Each...Next` structures;
- use the `OpenDatabase` method on the `Workspace` object;
- use the `OpenRecordset` method with database and other DAO objects;
- use `Recordset Move` methods;
- understand the varieties of syntax in reading and writing field values;
- use the `Seek` method on indexed table fields;
- use `Find` methods on `Recordsets`;
- use `AddNew`, `Edit`, and `Update` methods with `Recordsets`;
- open and use a `QueryDef` object;
- create a `QueryDef` object;
- open and use a `Parameter` query.

Chapter Objectives

The preceding extensive list of tasks is incorporated into the following four objectives:

1. Understanding the Jet engine and working with DAO collections
2. Opening databases and `Recordsets`
3. Working with the `Recordset` object
4. Learning to create, retrieve, and use `QueryDef` objects

Objective 25.1 Understanding the Jet Engine and Working with DAO Collections

The Jet engine features an object model based on *collections.* A collection is an ordered set of items that can be referred to as a unit. It provides a convenient way to refer to a related group of items as a single object.

Implicit in the design of an object model is the notion of a *collection hierarchy.* An object model contains a set (or collection) of objects. Each of the objects in the set (or collection) can also contain a collection. At the top of the collection hierarchy is a single object. Below that object, a series of other *nested* object collections can proliferate.

In Chapter 12, "Object Types, Variables, and Collections," you iterated through an application's `Forms` collection and then through each form's `Controls` collection. In that hierarchy, the application contained a collection of forms. Each form, in turn, contained a collection of controls. The *map* of collections in an application is thus called the *application object model.*

25.1.1 The Jet Engine Object Model

The object model for the Jet database engine, like other object models, displays a hierarchy of objects (see Figure 25.1). At the top of the object model is the `DBEngine`. All parts of the Jet engine are nested collections contained within the `DBEngine`. In Visual Basic, collections are always referenced with the plural (for example, `TableDefs`) and individual members with the singular (for example, `TableDef`).

Examine the Jet database model carefully. The `Database` object is nested within the `Workspace` object. The `Workspace` object acts as a container for open databases and provides security for the database. Users and groups are also collections nested within the `Workspace` object. These objects (which you will not be working with in this chapter) contain properties and methods for providing user passwords and permissions.

Other DAO objects should be familiar to you from Chapter 24, including the following:

- **`TableDef`**—A stored table definition.
- **`QueryDef`**—A stored query definition.

- **`Recordset`**—An object that presents a *cursored* view into a database table or a query result. Rows of data are stored in a buffer, and one record—the current record—is pointed to at a time.
- **`Index`**—A stored index associated with a `TableDef` or table-type `Recordset`.
- **`Field`**—A column of data that is contained in a `TableDef`, `QueryDef`, `RecordSet`, or `Index`.

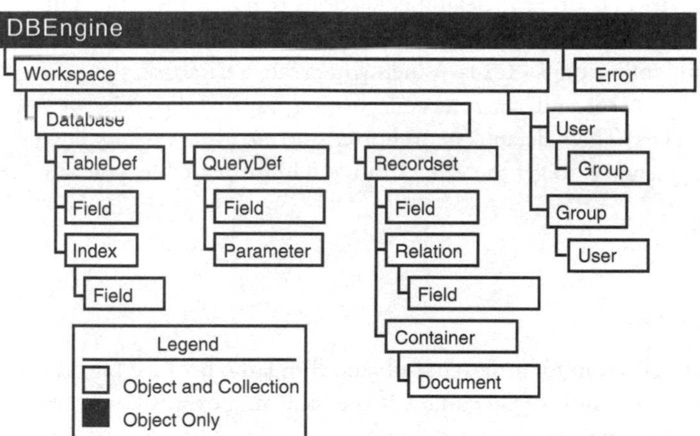

Figure 25.1
The Jet engine object model contains a collection of `Workspaces` and within each `Workspace`, a collection of `Databases`.

Appendix B, "Data Access Object Collections, Properties, and Methods," contains a list of the core DAOs. In addition to the tables you define yourself, the Jet engine maintains system tables for each .MDB database that you create. These tables are not generally accessible, and serve such Jet engine tasks as tracking object database structures for database repair and maintaining information on database security. These tables have an `MSys` prefix.

Although not illustrated, the DAO contains a `Connections` collection under the `Workspace` object. A `Connection` object permits the use of *ODBC Direct*, a library of database drivers written to bypass the Jet engine and connect directly with large, server-hosted, ODBC-compliant databases, such as SQL Server and Oracle. These new drivers provide much faster performance when accessing these types of databases than earlier versions of the Jet engine. We focus on Jet databases exclusively. Once you have learned to work with the object hierarchy of `Database` objects and `Recordset` objects, using ODBC Direct is not difficult.

25.1.2 Collection Syntax and Collection Defaults

A collection behaves somewhat like an array. Unlike the `Collection` object you have used with objects instantiated from class modules, DAO collections have a zero-based index. You can access individual members by referencing this index and by navigating through nested collections using the dot (.) navigation operator. For example, consider the following code fragment:

```
DBEngine.Workspaces(0).Databases(0).TableDefs(0).Fields("ProductName")
```

It refers to the ProductName field in the first TableDef of the TableDefs collection, the first Database of the Databases collection, and the first Workspace of the Workspaces collection of the DBEngine.

Fortunately, fields are not usually referenced this way. Instead, you typically work with Recordsets that provide an easier syntax.

Referencing individual object values through long nested collections can make code complicated to read and maintain. Most DAOs have default collections that ease this task. The default collection for a Database object, for instance, is TableDefs. For TableDef, Index, and Recordset objects, the default collection is Fields. When you create a database, these collections are stored much like Forms and Controls collections are stored. For QueryDef objects, the default is Parameters. These defaults help shorten the *navigational path* down the collection hierarchy. For instance, to get the value from a field from a table, you can use

```
tblProduct![Product Name]
```

rather than

```
tblProduct.Fields![Product Name]
```

because Fields is the default collection for a TableDef object. Similarly, because TableDefs is the default collection for Database objects, you could use the following expression to read or assign a value to the Product Name field of the current record:

```
db!tblProduct![Product Name]   ' where db is a variable of type Database
```

25.1.3 Iterating Through Database Collections with *For Each...Next* Statements

To work with collections, you can use the For Each...Next statement, as follows:

```
For Each element In group
   [statements]
     [Exit For]
   [statements]
Next [element]
```

In this syntax, element is a counter variable of an object type contained within the collection. To iterate through a database TableDefs collection, you would use code similar to the following:

```
Dim tdef as TableDef
For Each tdef in db.TableDefs
   Print tdef.Name
   ' or other code you want to execute on collection members
Next
```

This syntax and its usage will become clearer when you work through Exercise 25.1. For the exercises that follow, you use the Cinema.mdb database, a sample database that has been placed on the companion disk. This database is a simple three-table database. In one table are listed movie titles and their years of release. In another table are five movie directors and data about them. A third table consists solely of the movie title ID numbers and the directors' ID numbers. The tables, their primary and foreign keys, and their relationships are diagrammed in Figure 25.2.

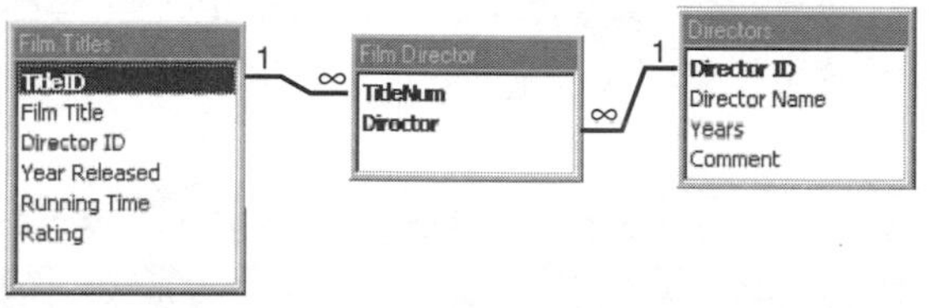

Figure 25.2

The Cinema database schema indicates that a film director is associated with one or more films, and a film is directed by one or more directors.

Exercise 25.1 Iterating Database Collections

In this lengthy exercise, you work with the main DAO collections of the Cinema database. You use several DAO methods that have not yet been introduced, but the code should be understandable. In Objective 25.2, "Opening Databases and `Recordsets`," we cover these new Visual Basic statements and methods in detail and examine their syntax.

1 Design the interface illustrated in Figure 25.3. Use three list boxes (**`lstTable`**, **`lstFields`**, and **`lstFieldData`**). Use a label to store the query (**`lblSQL`**), and name the combo box **`cboQueryDef`**.

2 Examine Project, References and click Microsoft DAO 3.5 Object Library if this reference is not checked.

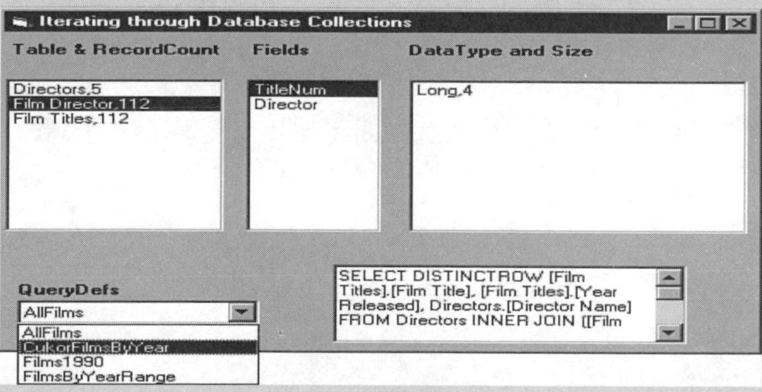

Figure 25.3

This interface is designed to show the internal makeup of the Cinema database.

continues

3 At the form module level, make the following declarations:

```
Dim db As DAO.Database
```

4 In the `Form_Load()` event, enter the following code:

```
Private Sub Form_Load()
   Dim tdef As TableDef
   Dim qdef As QueryDef
   ' set correct path if necessary
   Set db = DBEngine.Workspaces(0).OpenDatabase(App.Path & "\cinema.mdb")

   For Each tdef In db.TableDefs
      'display non-system tables in list box
      If Left(tdef.Name, 4) <> "MSys" Then lstTables.AddItem tdef.Name &
      ➥"," & tdef.RecordCount
   Next
   'display querydefs in combo box
   For Each qdef In db.QueryDefs
      cboQueryDef.AddItem qdef.Name
   Next
   cboQueryDef.ListIndex = 0
End Sub
```

5 In the `lstTables_Click()` event, enter the following code:

```
Private Sub lstTables_Click()
   Dim tablename As String
   Dim testtable As TableDef
   Dim tfield As Field
   Dim position As Integer

   lstFields.Clear           ' clear other list boxes
   lstFieldData.Clear

   ' get tablename from lstbox
   tablename = lstTables.List(lstTables.ListIndex)
   position = InStr(tablename, ",")
   tablename = Left(tablename, position - 1)

   Set testtable = db.TableDefs(tablename)

   For Each tfield In testtable.Fields
      lstFields.AddItem tfield.Name
   Next
End Sub
```

6 In the `lstFields_Click()` event, enter the following code:

```
Private Sub lstFields_Click()
   Dim tablename, fieldname As String
   Dim testtable As TableDef
   Dim tfield As Field
   Dim position As Integer
   Dim dtype As String
   lstFieldData.Clear
   ' get tablename form lstbox
   tablename = lstTables.List(lstTables.ListIndex)
```

```
    position = InStr(tablename, ",")
    tablename = Left(tablename, position - 1)
    fieldname = lstFields.List(lstFields.ListIndex)

    Set testtable = db.TableDefs(tablename)
    Set tfield = testtable.Fields(fieldname)
    dtype = GetDataName(tfield.Type)
    lstFieldData.AddItem dtype & "," & tfield.Size
End Sub
```

7 In the `cboQueryDef_Click()` event, enter the following code:

```
Private Sub cboQueryDef_Click()
   Dim qdef As QueryDef
   For Each qdef In db.QueryDefs
      If Trim(cboQueryDef.Text) = qdef.Name Then lblSQL.Caption = qdef.SQL
   Next
End Sub
```

8 The `GetDataName()` function code is as follows:

```
Public Function GetDataName(dtype As Integer) As String
   Select Case dtype
      Case 1
         GetDataName = "Boolean"
      Case 2
         GetDataName = "Byte"
      Case 3
         GetDataName = "Integer"
      Case 4
         GetDataName = "Long"
      Case 5
         GetDataName = "Currency"
      Case 6
         GetDataName = "Single"
      Case 7
         GetDataName = "Double"
      Case 8
         GetDataName = "Date/Time"
      Case 10
         GetDataName = "Text"
      Case 11
         GetDataName = "OLE Object"
      Case 12
         GetDataName = "Memo"
   End Select
End Function
```

In this exercise, a `For Each...Next` loop is used to successively iterate through several DAO collections. Let's examine the various events and functions.

continues

The *Form_Load()* Event

In the `Form_Load()` event, the database was opened using the `Workspace OpenDatabase` method. An object variable of type `TableDef` was declared. It was then set to refer to all members of the database's `TableDefs` collection in a `For Each...Next` loop. Each member's name and `RecordCount` were added to the `lstTables` list box.

Notice that the `For Each...Next` syntax permits you to reference object variables without using the `Set` statement. Remember: Object variables are references to objects, not objects themselves.

Binding DAO Variables

DAO variables, like any other object variable, must be bound to a particular object before object properties can be read or assigned or object methods invoked. Consider the following code:

```
Dim tdef as TableDef

For Each tdef In db.TableDefs
   lstTables.AddItem tdef.Name & "," & tdef.RecordCount
Next
```

This code is the equivalent of

```
Dim  as integer
Dim tdef as TableDef
For  = 0 to db.TableDefs.Count - 1
   Set tdef = TableDefs()
   lstTables.AddItem tdef.Name & "," & tdef.RecordCount
Next
```

The `For Each...Next` syntax allows you to iterate through the collection without a counter variable or the `Set` statement. Of course, you could also use a `For...Next` loop and simply reference the collection index, as follows:

```
Dim  as integer
Dim tdef as TableDef

For  = 0 to db.TableDefs.Count - 1
   lstTables.AddItem tdefs().Name & "," & tdefs().RecordCount
Next
```

The List Box *Click()* Events

The list box `Click()` events perform similar operations down the DAO hierarchy. In the `lstTables_Click()` event, you set a parent collection reference from the `lstTables` list box. A `For Each...Next` loop then lists fields from the selected table's `Fields` collection. The `lstFields_Click()` event sets two object variables from list box values. Once `Table` and `Field` object references are set, you can retrieve individual field properties, such as data type and size.

Alternatively, you could have used nested `For Each...Next` loops. To list all fields in the table selected by the user in the `lstTables` list box, you could have written the code as follows:

```
Dim tablename As String
Dim testtable As TableDef
Dim tfield As Field
Dim position As Integer

' get tablename form lstbox
tablename = lstTables.List(lstTables.ListIndex)
position = InStr(tablename, ",")
tablename = Left(tablename, position - 1)
For each testtable in db.TableDefs
   If tablename = testtable.name then
      For Each tfield In testtable.Fields
         lstFields.AddItem tfield.Name
      Next
   End if
Next
```

This exercise serves as the beginnings of a generic application that can *map* any database. You can use Visual Basic code to iterate through any database and examine core DAOs. For instance, change the single line of code in the `Form_Load()` event that references the cinema.mdb to another .MDB database file, such as Design1.MDB. This exercise should be able to list the databases' tables, fields, field data types, field sizes, and stored queries for any .MDB database stored on your computer.

Objective 25.2 Opening Databases and *Recordsets*

In Exercise 25.1, you used several new statements to open databases and `Recordsets`. These statements are used every time you employ DAO methods, so they deserve careful review.

25.2.1 Opening Databases with the *Workspace OpenDataBase*

To open a database, declare a `Database` object and set that object by employing the `OpenDataBase` method of the `Workspace` object. If only one `Workspace` is used, which is typical for a single user on a workstation, make the following kind of object variable declaration and assignment:

```
Dim db as DAO.database
Set db = DBEngine.Workspaces(0).OpenDatabase("MyDB.MDB")
```

Because the `DBEngine` is a single object, the following instruction will also work:

```
Set db = Workspaces(0).OpenDatabase("MyDB.MDB")
```

The full syntax of the method is as follows:

```
Set database = [workspace].OpenDatabase(dbname[, exclusive[, read-only
➥[,connect]]])
```

Table 25.1 explains the components of the `OpenDataBase` method.

Table 25.1 Syntax of the *OpenDataBase* Method

Component	*Description*
`database`	A variable of a `Database` object data type that represents the `Database` object that you're opening.
`workspace`	A variable of a `Workspace` object data type that represents the existing `Workspace` object that will contain the database.
`dbname`	A string expression that is the name of an existing database file or registered ODBC data source name. If the file name has an extension, it's required. If your network supports it, you can also specify a network path, such as \\MYSERVER\MYSHARE\MYDIR\MYDB.MDB.
`exclusive`	A Boolean value that is `True` if the database is to be opened for exclusive (non-shared) access, and `False` if opened for shared access. If you omit this argument, the database is opened for shared access.
`read-only`	A Boolean value that is `True` if the database is to be opened for read-only access, and `False` if opened for read/write access. If you omit this argument, the database is opened for read/write access.
`connect`	A string expression used for opening the database. This string constitutes the ODBC connect arguments. You must supply the exclusive and read-only arguments to supply a source string. (See the `connect` property (DAO) documentation using Help for the appropriate syntax.)

25.2.2 Opening *Recordsets* from a *Database* Object

After a `Database` object has been set, the `Database` object's `OpenRecordset` method is used to open a database. `Recordsets` can be of three types, which you should be familiar with from working with the Data control: table-type, dynaset-type, or snapshot-type.

The syntax for opening a `Recordset` for a Jet engine database is as follows:

```
Set variable = database.OpenRecordset (source, [type], [options], [lockedits])
```

Table 25.2 shows the components of this syntax.

Table 25.2 Syntax of the *OpenRecordset* Method when Used with a *Database* Object

Component	*Description*
`variable`	A variable that has been declared as an object of data type `Recordset`.
`database`	The name of an existing DAO object you want to use to create the new `Recordset`.
`source`	A String specifying the source of the records for the new `Recordset`. The source can be a table name, a query name, or a SQL statement that returns records. For table-type `Recordset` objects, the source can only be a table name.
`type`	If you don't specify a type, `OpenRecordset` creates a table-type `Recordset` if possible. If you specify an attached table or query, `OpenRecordset` creates a dynaset-type `Recordset`. One of the following `Integer` constants defines the data type of the new `Recordset` object:
	`dbOpenTable`—Opens a table-type `Recordset` object.
	`dbOpenDynaset`—Opens a dynaset-type `Recordset` object.
	`dbOpenSnapshot`—Opens a snapshot-type `Recordset` object.
`options`	Any combination (or none) of the following `Integer` constants specifying characteristics of the new `Recordset`, such as restrictions on other users' capabilities to edit and view it:
	`dbDenyWrite`—Other users can't modify or add records.
	`dbDenyRead`—Users can't view records (table-type `Recordset` only).
	`dbReadOnly`—Limited to viewing records; other users can modify them.
	`dbAppendOnly`—Limited to appending new records (dynaset-type `Recordset` only).
	`dbInconsistent`—Inconsistent updates are allowed (dynaset-type `Recordset` only).
	`dbConsistent`—Consistent updates are allowed (dynaset-type `Recordset` only).
	`dbForwardOnly`—The `Recordset` is a forward-only scrolling snapshot. `Recordset` objects created with this option only support the `MoveNext` method to move through the records.

continues

Table 25.2 continued

Component	*Description*
	`dbSQLPassThrough`—The Microsoft Jet database engine query processor is bypassed. The query specified in the `OpenRecordset` source argument is passed to an ODBC back-end server for processing.
	`dbSeeChanges`—Generates a runtime error if another user is changing data you are editing.
`lockedits`	You can use one of the following constant for the `lockedits` argument:
	`dbReadOnly`—Prevents users from making changes to the `Recordset` (default for `ODBCDirect Workspaces`). You can use `dbReadOnly` in either the `options` argument or the `lockedits` argument, but not both. If you use it for both arguments, a runtime error occurs.
	`dbPessimistic`—Uses pessimistic locking to determine how changes are made to the `Recordset` in a multiuser environment. The page containing the record you're editing is locked as soon as you use the `Edit` method (default for Microsoft Jet `Workspaces`).
	`dbOptimistic`—Uses optimistic locking to determine how changes are made to the `Recordset` in a multiuser environment. The page containing the record is not locked until the `Update` method is executed.

Exercise 25.2 Using the Three *OpenRecordset* Types

Create the interface shown in Figure 25.4 to open and view a table, a dynaset, and a snapshot.

1 At the form module level, make the following declarations:

```
Dim db As DAO.Database
Dim ws As DAO.Workspace
```

2 In the `Form_Load()` event, enter the following code:

```
Set ws = DBEngine.Workspaces(0)
Set db = ws.OpenDatabase(App.Path & "\cinema.mdb")
```

3 In the `cmdOpen_Click()` event, enter the following code:

```
Private Sub cmdOpen_Click()
   Dim rsTbl, rsDyna, rsSnap, rs  As Recordset
   Dim SQL1 As String
   Dim SQL2 As String
```

```
    SQL1 = "SELECT  [Film Title] FROM [Film Titles] Order by [Film Title]"
    SQL2 = "SELECT  [Film Title], [Year Released] FROM [Film Titles]
    Where [Year + Released] = 1990"

    Set rsTbl = db.OpenRecordset("Directors", dbOpenTable)
    Set rsDyna = db.OpenRecordset(SQL1, dbOpenDynaset)
    Set rsSnap = db.OpenRecordset(SQL2, dbOpenSnapshot)

    ' clear list boxes
    lstTable.Clear
    lstDynaSet.Clear
    lstSnapShot.Clear

    ' add recordset results to list boxes
    Do While Not rsTbl.EOF
       lstTable.AddItem rsTbl![Director Name]
       rsTbl.MoveNext
    Loop

    Do While Not rsDyna.EOF
       lstDynaSet.AddItem rsDyna.Fields(0).Value
       rsDyna.MoveNext
    Loop

    Do While Not rsSnap.EOF
       lstSnapShot.AddItem rsSnap![Year Released] & ": " & rsSnap!
       [Film Title].Value
       rsSnap.MoveNext
    Loop

    close recordsets
    rsTbl.Close
    rsDyna.Close
    rsSnap.Close
End Sub
```

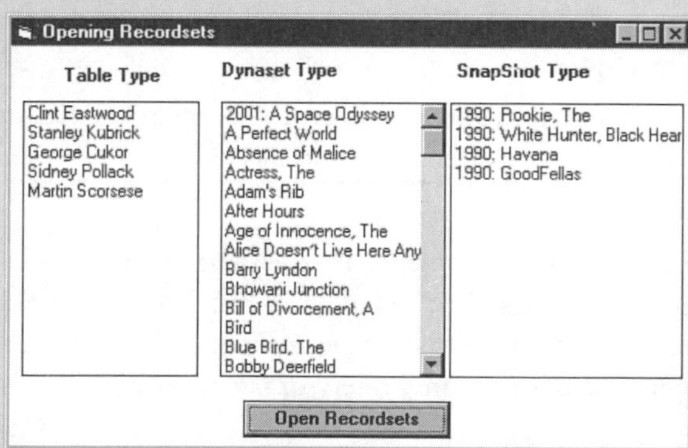

Figure 25.4

This interface allows you to compare and contrast the three `Recordset` *types.*

continues

4 Run and test the program.

When you no longer need access to a Recordset, it should be closed with the Close method. However, if you close the database, all Recordsets associated with that database are also closed. In this exercise, you could add the instruction db.Close to the Form_Unload() event to ensure that the database closes when you exit the application. Alternatively you could set the Database object variable to Nothing, as in

```
Set db = Nothing
```

25.2.3 *OpenRecordset* Method for Other DAO Objects

The OpenRecordset method can also be used on existing TableDef, QueryDef, and Recordset objects. The syntax is a little different from that used with a Database object:

```
Set variable = object.OpenRecordset([type],[options],[lockedits])
```

Table 25.3 explains the components of this syntax.

Table 25.3 Syntax of the *OpenRecordset* Method when Used with DAO Objects

Component	*Description*
variable	A variable that has been declared as an object of data type Recordset.
object	The name of an existing QueryDef, Recordset, or TableDef object that you want to use to create the new Recordset. If object refers to a dynaset or snapshot-type Recordset object, the type of the new object is the same as that of the Recordset specified by object. If object refers to a table-type Recordset object, the type of the new object is a dynaset-type Recordset.

The type, option, and lockedits arguments are the same as in the syntax for opening a Recordset for a Jet engine database (see Table 25.2), but often the type argument is omitted. Visual Basic determines the Recordset type automatically. The most frequent use of this OpenRecordset syntax is with a stored query or QueryDef object, as in the following expression

```
QueryDef.OpenRecordset
```

Exercise 25.3 Using the *OpenRecordset* Method with a *TableDef* Object

This exercise employs a small change to Exercise 25.2: It uses the `OpenRecordset` method of the `TableDef` object.

1 Change the name of the list box `lstSnapShot` to **`lstTableDef`**, and change the code in the `cmdOpen_Click()` event accordingly:

```
Private Sub cmdOpen_Click()
   Dim rsTbl, rsDyna, rs  As Recordset
   Dim tdef As TableDef
   Dim SQL1 As String

   SQL1 = "SELECT  [Film Title] FROM [Film Titles]
   Where [Year Released] Between 1959 and 1970"

   Set rsTbl = db.OpenRecordset("Film Titles", dbOpenTable)
   Set rsDyna = db.OpenRecordset(SQL1, dbOpenDynaset)
   Set tdef = db.TableDefs("Directors")
   Set rs = tdef.OpenRecordset()

   Do While Not rsTbl.EOF
      lstTable.AddItem rsTbl![Film Title]
      rsTbl.MoveNext
   Loop

   Do While Not rsDyna.EOF
      lstDynaSet.AddItem rsDyna.Fields(0).Value
      rsDyna.MoveNext
   Loop

   Do Until rs.EOF
      lstTabledef.AddItem rs![Director Name] & ";  " & rs![Years]
      rs.MoveNext
   Loop

   rsTbl.Close
   rsDyna.Close
   rs.Close
End Sub
```

2 Test this code to determine whether the `OpenRecordset` method works as you expected.

Objective 25.3 Working with the *Recordset* Object

The `Recordset` object should be familiar to you from working with the Data control. It is the object you deal with most frequently when working with the Jet engine. It is important, therefore, to review its methods thoroughly.

25.3.1 *Recordset Move* Methods

The Move methods (MoveFirst, MoveLast, MovePrevious, and MoveNext) are similar to those used with the Data control. If you use the visible Data control to navigate through a cursored Recordset, you can easily move to the first record, last record, previous record, and next record.

As you navigate through a Recordset, the BOF (beginning of file) and EOF (end of file) Recordset properties need to be tested. If the values turn to True, you must reset the cursor to the preceding record (in the cause of EOF) or the next record (in the case of BOF).

The BOF property *does not* become True when the cursor arrives at the first record in the Recordset; it becomes True only when on the first record and a MovePrevious method is invoked. The same is true for the EOF property: It becomes True when you are on the last record and a MoveNext method is invoked. Exercise 25.4 shows how the BOF and EOF properties can be used in conjunction with the Move methods.

Exercise 25.4 Using *Recordset Move* Methods

This exercise demonstrates one way of avoiding errors by navigating past the beginning or end of a Recordset. It uses BOF and EOF properties to test the position of the Recordset cursor.

1 Construct the interface shown in Figure 25.5. It consists of four buttons (**cmdFirst**, **cmdPrevious**, **cmdNext**, and **cmdLast**) and two labels (**lblTitle** and **lblStatus**).

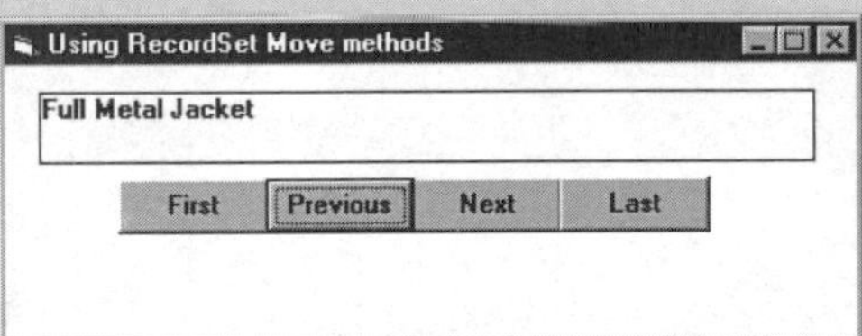

Figure 25.5

The Move methods enable you to move to the first or last record in a database as well as to the next and previous record.

2 Make the following declarations at the form module level:

```
Option Explicit
Dim db As DAO.Database
Dim ws As DAO.Workspace
Dim rsSnap As DAO.Recordset
```

3 In the `Form_Load()` event, enter the following code:

```
Private Sub Form_Load()
   Dim SQL As String
   ' open database which is declared at form module level
   Set ws = DBEngine.Workspaces(0)
   Set db = ws.OpenDatabase(App.Path & "\cinema.mdb")

   SQL = "SELECT  [Film Title] FROM [Film Titles]
   Where [Year Released] > 1980 Order By [Film Title]"

   Set rsSnap = db.OpenRecordset(SQL, dbOpenSnapshot)

   If Not IsNull(rsSnap.Fields(0).Value) Then
      lblTitle.Caption = rsSnap.Fields(0).Value
   End If
End Sub
```

4 In the first, last, next, and previous command button `Click()` events, enter the following code:

```
Private Sub cmdFirst_Click()
   rsSnap.MoveFirst
   lblTitle.Caption = rsSnap("Film Title")
   lblStatus.Caption = "Beginning of file"
End Sub
Private Sub cmdLast_Click()
   rsSnap.MoveLast
   lblTitle.Caption = rsSnap("Film Title")
   lblStatus.Caption = "End of file"
End Sub

Private Sub cmdNext_Click()
   lblStatus.Caption = ""
   rsSnap.MoveNext
   If rsSnap.EOF Then
      rsSnap.MovePrevious
      lblStatus.Caption = "End of file"
   End If
   lblTitle.Caption = rsSnap("Film Title")
End Sub

Private Sub cmdPrevious_Click()
   lblStatus.Caption = ""
   rsSnap.MovePrevious
   If rsSnap.BOF Then
      rsSnap.MoveNext
      lblStatus.Caption = "Beginning of file"
   End If
   lblTitle.Caption = rsSnap![Film Title]
End Sub
```

5 Test your code. What are the purposes of the `BOF` and `EOF` tests?

25.3.2 Varieties of Syntax in Accessing Field Values

In Exercise 25.4, you used two different means of accessing field values: In the `Form_Load()` event, you referenced the `Fields` collection index, while in the command button `Click()` events, you used the default collection for the `Recordset` object in the expression

```
lblTitle.Caption = rsSnap("Film Title")
```

The syntax varieties for accessing field values are even more numerous. For a field name with a single word, such as *Title,* the varieties include the following:

```
lblTitle.Caption = rsSnap!Title
lblTitle.Caption = rsSnap!Title.Value
lblTitle.Caption = rsSnap![Title]          ' brackets or quotes
                                             necessary for field names
lblTitle.Caption = rsSnap![Title].Value    ' with spaces
lblTitle.Caption = rsSnap(0)
lblTitle.Caption = rsSnap(0).Value
lblTitle.Caption = rsSnap("Title")
lblTitle.Caption = rsSnap("Title").Value
```

These examples do not include statements in which the default `Fields` collection is referenced, as in

```
lblTitle.Caption = rsSnap.Fields("Title")
lblTitle.Caption = rsSnap.Fields("Title").Value
```

Due to greater simplicity in the code, it is generally faster to use the exclamation point operator, as in

```
rsSnap!Title
```

than the field name in parentheses, as in

```
rsSnap("Title")
```

25.3.3 Using *MoveLast* Before Checking *RecordCount*

When a `Recordset` is opened, the `Recordset`'s `RecordCount` property cannot be ascertained unless a `MoveLast` method is invoked first. This populates the `Recordset` with data and allows the `RecordCount` to be determined. Before trying to read the `RecordCount` property, always use a `MoveLast` method, but be aware that in large data sets, a `MoveLast` method can have an impact on performance because the entire `Recordset` is populated. To determine the `RecordCount`, use code such as the following:

```
myrecordset.MoveLast
Print myrecordset.RecordCount
```

25.3.4 Using the *Seek* Method with *Recordsets*

The `Seek` method can be used to locate records in a table-type `Recordset` when the search criteria are performed upon an identified table index. It is the fastest means of locating a record. `Recordsets` also have `Find` methods (`FindFirst`, `FindLast`, `FindPrevious`, and

FindNext), which use criteria equivalent to SQL WHERE clauses. Using a SQL SELECT query or a Parameter query is often faster in these situations. When you need to locate a record and it is possible to open a table-type Recordset to retrieve the record, the Seek method should be your first choice. The Seek method's syntax is as follows:

```
table.Seek comparison, key1, key2...
```

Table 25.4 explains this syntax.

Table 25.4 Syntax of the *Seek* Method

Component	*Description*
table	The name of an existing table-type Recordset object or Table object that has a defined index as specified by the Recordset or Table object's Index property
comparison	One of the following string expressions: "<", "<=", "=", ">=", or ">"
key1, key2...	One or more values corresponding to the underlying table's Primary property setting

This syntax is a little different, especially regarding the placing of the comparison operators within quotation marks. For a Recordset object variable named tbl, a Seek method would look like the following:

```
Dim tbl as Recordset
Set tbL = db.OpenRecordset("Film Titles", dbOpenTable) tbl.Index = "Primary Key"
tbl.Seek "=", "Raging Bull"
If tbl.NoMatch = True then
   lblStatus.Caption = "Film not located."
Else
   ' write code to retrieve record field values
End if
```

Using the Seek method is a three step process:

1. Identify the Index upon which you are using the Seek method.
2. Check the Recordset's NoMatch property, a Boolean value property, to see whether your search was successful.
3. Retrieve values from the fields of the record if you are successful in locating it.

At first glance, the Seek method's syntax is a bit odd, but its speed quickly induces you to use it whenever possible.

Exercise 25.5 Using the *Seek* Method

In this exercise, film titles are selected in a list box. The user can then choose a title to receive more information about the film. The left-hand list box displays film titles; the right-hand list box displays information about the selected film (see Figure 25.6).

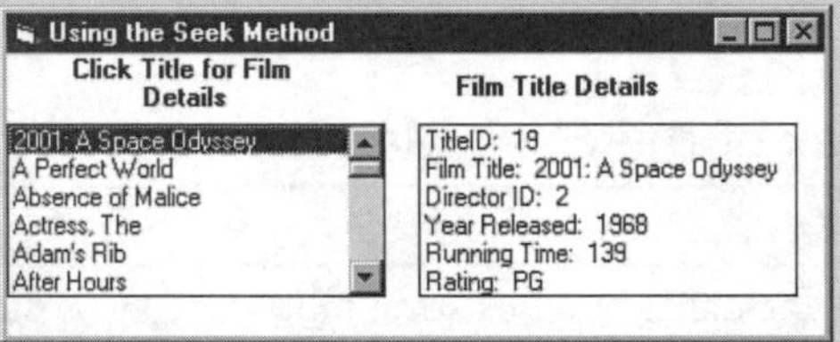

Figure 25.6

The `Seek` method is used to provide details about a selected film.

1 Create the interface illustrated in Figure 25.6. The list box on the left is named **`lstFilmTitles`** and has its `MultiSelect` property set to `0` (`None`). The list box on the right is named **`lstFilmDetails`**.

2 At the form module level, make these object variable declarations:

```
Dim db As DAO.Database
Dim ws As DAO.Workspace
Dim rs As DAO.Recordset
```

3 In the `Form_Unload()` event, enter

```
db.Close
```

4 In the `Form_Load()` event, enter code that selects all film titles and orders them alphabetically:

```
Private Sub Form_Load()
   Dim rsSQL As Recordset
   Dim SQL As String
   Dim strSearch As String
   Dim position As Integer

   Set ws = DBEngine.Workspaces(0)
   Set db = ws.OpenDatabase(App.Path & "\cinema.mdb")'change path as
                                                     'necessary
   Set rs = db.OpenRecordset("Film Titles", dbOpenTable)
   Set rsSQL = db.OpenRecordset("Select [Film Title] FROM [Film Titles]
   ➥ORDER BY [Film Title]", dbOpenSnapshot)

   Do Until rsSQL.EOF
      lstFilmTitles.AddItem rsSQL![Film Title]
      rsSQL.MoveNext
   Loop
   rsSQL.Close
End Sub
```

5 In the `lstFilmTitles_Click()` event, enter the following code:

```
Private Sub lstFilmTitles_Click()
   Dim tfield As Field
   Dim criteria As String


   lstFilmDetails.Clear
   ' set index for Seek method
   rs.Index = "Title"
   criteria = Trim(lstFilmTitles.Text)

   rs.Seek "=", criteria

   If rs.NoMatch = True Then
    lblStatus.Caption = "Film not located"
   Else
     For Each tfield In rs.Fields
        If Not IsNull(tfield.Value) Then
          lstFilmDetails.AddItem tfield.Name & ":  " & tfield.Value
        End If
     Next
   End If
End Sub
```

In the `lstFilmTitles_Click()` event, you identify the index to use for the seek. In this instance, it is an index named "Title". You then use list box values to perform the seek. Once the seek is successful, a `For Each...Next` structure iterates through every field in the `Recordset` to display values for the current record.

Notice in Exercise 25.5 that you only retrieved film information. Suppose you also wanted to retrieve information about the film's director. One way of doing this is to write a SQL query. This, however, would require a query that would "join" the Film Title table to the Director table through the Film Director table.

If you want to retrieve information from only one record, a `Seek` method is much faster. From the Film Titles table, you can retrieve the `TitleId`. This value can then be used to retrieve the director number from the Film Director table. The `DirectorID` value can then be used to retrieve director information from the Director table. The next exercise demonstrates how this can be accomplished.

Exercise 25.6 Using the *Seek* Method on Multiple Tables

This exercise uses the interface of the previous exercise (see Figure 25.6).

1 Save Exercise 25.5 under a new name, and make the following declarations at the form module level:

continues

```
Dim db As DAO.Database
Dim ws As DAO.Workspace
Dim rsFilms As DAO.Recordset
Dim rsDirectors As DAO.Recordset
Dim rsFilm_Director As DAO.Recordset
```

2 Change the `Form_Load()` event to populate the `Recordsets` with data:

```
Private Sub Form_Load()
    Dim rsSQL As Recordset
    Dim SQL As String
    Dim strSearch As String

    Set ws = DBEngine.Workspaces(0)
    Set db = ws.OpenDatabase(App.Path & "\cinema.mdb")
    Set rsFilms = db.OpenRecordset("Film Titles", dbOpenTable)
    Set rsDirectors = db.OpenRecordset("Directors", dbOpenTable)
    Set rsFilm_Director = db.OpenRecordset("Film Director", dbOpenTable)
    Set rsSQL = db.OpenRecordset("Select [Film Title] FROM [Film Titles]
    ➥ORDER BY [Film Title]", dbOpenSnapshot)

     ' list film titles in list box
    Do Until rsSQL.EOF
      lstFilmTitles.AddItem rsSQL![Film Title]
      rsSQL.MoveNext
    Loop
    rsSQL.Close
End Sub
```

3 Change the `lstFilmTitles_Click()` event to read as follows:

```
Private Sub lstFilmTitles_Click()

    Dim tfield As Field
    Dim criteria As String

    lstFilmDetails.Clear
    criteria = Trim(lstFilmTitles.Text)
    rsFilms.Index = "Title"
    rsFilms.Seek "=", criteria
    If rsFilms.NoMatch = True Then
       lblStatus.Caption = "Film not located"
    Else

    For Each tfield In rsFilms.Fields
       If Not IsNull(tfield.Value) Then
          lstFilmDetails.AddItem tfield.Name & " - " & tfield.Value
       End If
    Next
    End If

    ' retrieve film number
    rsFilm_Director.Index = "TitleNum"
    rsFilm_Dirf¨tpª.Seek "=", rsFilms!TitleID
    If rsFilm_Director.NoMatch = True Then
           Exit Sub
```

```
        Else
           ' retrieve director id...
           rsDirectors.Index = "PrimaryKey"
           ' and use it to find director name and info
           rsDirectors.Seek "=", rsFilm_Director!Director
           If rsDirectors.NoMatch = True Then
              Exit Sub
           Else
              lstFilmDetails.AddItem "Director:  " & rsDirectors![Director
              ➥Name]
              lstFilmDetails.AddItem "Director Years:  " & rsDirectors![Years]
           End If
        End If
    End Sub
```

4 Compare the new display to the old one. How has the display of information changed?

25.3.5 Using the *Find* Methods with *Recordsets*

The `Find` methods locate records as well. `Recordset` `Find` methods work only on dynaset- and snapshot-type `Recordsets` (not table-type). The search direction for the `FindFirst` and `FindNext` methods moves toward the end of the `Recordset`. Each method stops at the first record that meets the criteria. `FindFirst` starts the search at the beginning of the `Recordset`, and `FindNext` starts the search from the current record.

`FindLast` and `FindPrevious` mirror this behavior. The search direction moves toward the beginning of the database with the method starting the search from either the end of the `Recordset` or from the current record.

If you want to find more than one record using the criteria, it is often faster to use a SQL statement and create a new `Recordset`. You can then scroll through the `Recordset` with `Move` methods. The `Find` method syntax is as follows:

```
recordset.{FindFirst | FindLast | FindNext | FindPrevious} criteria
```

Table 25.5 explains this syntax.

Table 25.5 Syntax of the *Find* Method

Component	*Description*
`Recordset`	The name of an existing dynaset- or snapshot-type `Recordset`, `Dynaset`, or `Snapshot` object.
`criteria`	A string expression (like the `WHERE` clause in a SQL statement without the word `WHERE`) used to locate the record.

Using the `Find` method instead of the `Seek` method in Exercise 25.6, you could change the code in the `Form_Load()` event to open a snapshot-type, instead of a table-type, `Recordset`. You could then use the following code in the `lstFilmTitles_Click()` event:

```
Private Sub lstFilmTitles_Click()
   Dim tfield As Field
   Dim criteria As String
   Dim film As String

   lstFilmDetails.Clear

   film = Trim(lstFilmTitles.List(lstFilmTitles.ListIndex))
   criteria = "[Film Title] = " & Chr(39) & film & Chr(39)
   rsFilm.FindFirst criteria
   If rsFilm.NoMatch = True Then
      lblStatus.Caption = "Film not located"
   Else
      For Each tfield In rsFilm.Fields
         If Not IsNull(tfield.Value) Then
            lstFilmDetails.AddItem tfield.Name & ":  " & tfield.Value
         End If
      Next
   End If
End Sub
```

In SQL syntax, string literals need to be enclosed in quotation marks. In this case, we concatenated the search criteria using the `Chr()` function. For search expressions that contain apostrophes, this can be troublesome. You might, for instance, encounter an error for film titles that have a possessive adjective, as in the George Cukor film *Adam's Rib.* If possible, try to search on numeric expressions. This avoids inordinate amounts of string parsing and is generally faster and less prone to error. Better yet, use the `Seek` method.

25.3.6 Using the *AddNew*, *Edit*, *Update*, and *Delete* Methods

The DAO methods for adding, editing, updating, and deleting records are essentially the same as those with the Data control. If you use unbound controls, however, more code is required to check data entry errors, ensure data types are not mismatched, and keep controls that provide views of the `Recordset` "refreshed." If you are familiar with SQL, its syntax is often faster for modifying data than DAO methods. Use the SQL Reference section of Help to investigate the `INSERT INTO` and `UPDATE` syntax of SQL to get fast performance.

Exercise 25.7 Editing and Updating a Database Record

In Figure 25.7, you see an interface with added buttons for performing operations to edit and add data for the Film Titles table. Because the code for this application is extensive, most of it has been retained in Exercise 25.7 on the companion disk (P25_7.VBP). The key DAO methods, however, are the same as those used with the Data control.

You cannot update a `Recordset` unless the `Recordset`'s `AddNew` or `Edit` methods have been invoked. In the application, the Update button is enabled only after one of these buttons has been depressed, and at the end of the `Update` procedure, the Update button is disabled.

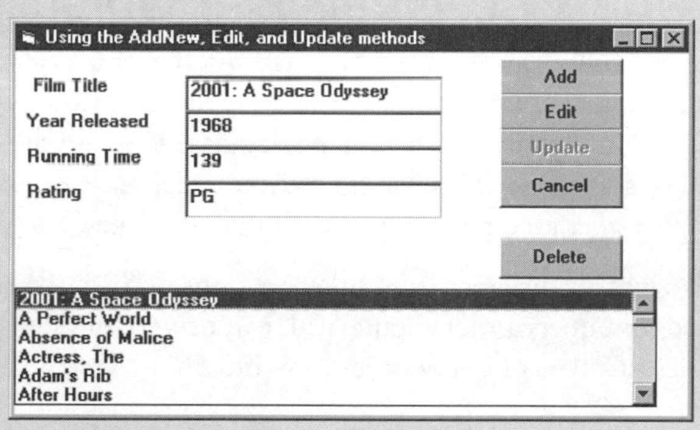

Figure 25.7

This interface allows you to add or edit a record and to update the database.

1 Load the code from the companion disk.

2 Write the `cmdUpdate_Click()` event, as follows:

```
Private Sub cmdUpdate_Click()

    On Error GoTo cmdUpdateErr:

    If Len(Trim(txtTitle.Text)) <> 0 Then rsTbl![film title] =
    ➥Trim(txtTitle.Text)
    If Len(Trim(txtYear.Text)) <> 0 Then rsTbl![Year Released] =
    ➥CInt(txtYear.Text)
    If Len(Trim(txtRunTime.Text)) <> 0 Then rsTbl![Running Time]=
    ➥CInt(txtRunTime.Text)
    If Len(Trim(txtRating.Text)) <> 0 Then rsTbl![Rating] =
    ➥Trim(txtRating.Text)
    rsTbl.Update
    cmdUpdate.Enabled = False
    cmdAdd.Enabled = True
    cmdEdit.Enabled = True
    cmdDelete.Enabled = True
    RebuildList
    ClearText
    Exit Sub
cmdUpdateErr:
    MsgBox Err & "    " & Error.Description, 0, "Error in Data Update"
    Resume Next
End Sub
```

Missing from this exercise are the numerous data access errors that you can trap. Go to the Visual Basic online Help to review the list of these errors.

Objective 25.4 Learning to Create, Retrieve, and Use *QueryDef* Objects

In the Cinema database, there are a number of stored queries or `QueryDef` objects. Using Access or Visual Data Manager, it is possible to create your own stored queries. One of the easiest ways to build a stored query in a Jet engine database is to use the query building tool included with the Access database. This feature of Access is a *query by example* (QBE) tool. With this tool, complex queries, especially joins of multiple tables, are easy to build.

The Jet engine SQL syntax is different from other versions of SQL, especially in how joins are handled. A nested join query in Jet engine SQL can be very difficult to decipher. The Access database makes this type of query easier to write.

25.4.1 Opening and Using a *QueryDef* Object

Retrieving a `Recordset` based on a `SELECT` query is a matter of iterating through the `QueryDefs` collection and using the `OpenRecordset` method with the identified `QueryDef` object.

Exercise 25.8 Using a *QueryDef* Object

This exercise opens all stored queries that do not have parameters. The user can then execute these queries and observe the SQL code upon which they are based.

1 Build the interface you see in Figure 25.8. It contains a list box for listing `QueryDef` objects (**`lstQueryDef`**) and a list box for displaying query results (**`lstResult`**). Add labels **`lblSQL`**, **`lblStatus`**, and **`lblRecordCount`**. The label `lblStatus` appears below the Previous and Next buttons to display the beginning or end of the file.

2 At the form module level, make the following declarations:

```
Dim db As DAO.Database
Dim rs As DAO.Recordset
```

3 In the `Form_Load()` event, enter code to list `QueryDef` objects that do not have parameters:

```
Private Sub Form_Load()
   Dim qdef As QueryDef
                              ' alter path as appropriate
   Set db = Workspaces(0).OpenDatabase(App.Path & "cinema.mdb")
   For Each qdef In db.QueryDefs
      If qdef.Parameters.Count = 0 Then
         lstQueryDef.AddItem qdef.Name
      End If
   Next
End Sub
```

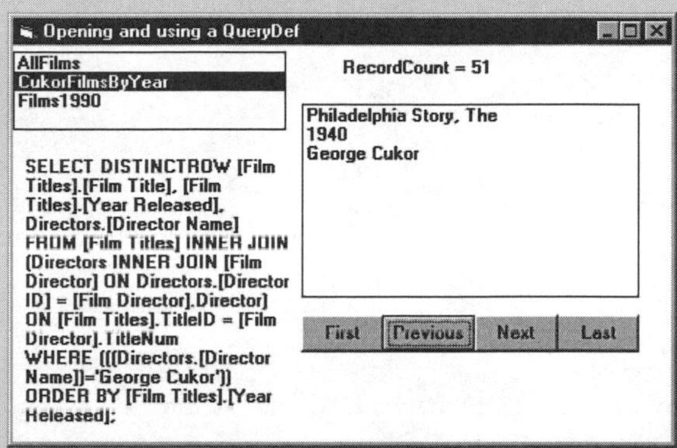

Figure 25.8

This interface displays the query used in processing and the records found through the use of the query.

4 In the `lstQueryDef_Click()` event, enter the code to identify the query, open a `Recordset` based on that query, display the query's SQL syntax, and display the first record in the `Recordset` returned by the query:

```
Private Sub lstQueryDef_Click()
   Dim qdef As QueryDef
   Dim f As Field
   Dim s As String

   s = Trim(lstQueryDef.List(lstQueryDef.ListIndex))

   For Each qdef In db.QueryDefs
      If qdef.Name = s Then
         Set rs = qdef.OpenRecordset
         lblSQL.Caption = qdef.SQL
      End If
   Next
   lstResult.Clear
   For Each f In rs.Fields
      lstResult.AddItem f.Value
   Next

   rs.MoveLast
   lblRecordCount.Caption = "RecordCount = " & rs.RecordCount
   rs.MoveFirst
End Sub
```

5 Add the appropriate code for the navigation buttons:

```
Private Sub cmdFirst_Click()
   Dim f As Field
   lstResult.Clear
   rs.MoveFirst
```

continues

```
      For Each f In rs.Fields
         lstResult.AddItem f.Value
      Next
      lblStatus.Caption = "Beginning of file"
   End Sub

   Private Sub cmdLast_Click()
      Dim f As Field
      lstResult.Clear
      rs.MoveLast
      For Each f In rs.Fields
         lstResult.AddItem f.Value
      Next
      lblStatus.Caption = "End of file"
   End Sub

   Private Sub cmdNext_Click()
      Dim f As Field
      lblStatus.Caption = ""
      lstResult.Clear
      rs.MoveNext
      If rs.EOF Then
         rs.MovePrevious
         lblStatus.Caption = "End of file"
      End If
      For Each f In rs.Fields
         lstResult.AddItem f.Value
      Next
   End Sub

   Private Sub cmdPrevious_Click()
      Dim f As Field
      lblStatus.Caption = ""
      lstResult.Clear
      rs.MovePrevious
      If rs.BOF Then
         rs.MoveNext
         lblStatus.Caption = "Beginning of file"
      End If
      For Each f In rs.Fields
         lstResult.AddItem f.Value
      Next
   End Sub
```

6 Test your design. Can you understand the logic of the SQL?

25.4.2 Opening and Using a *Parameter* Query

The power of stored queries becomes more impressive the more you work with them. Individual permissions can be granted to individual queries. In this manner, it takes a relatively short amount of time to build a *view* of a database customized for an individual or group of individuals.

One of the most flexible means of providing users with individual views of a database is to use a `Parameter` query. In a `Parameter` query, an individual value is not known until the user provides it. The following exercise shows how the `Parameters` collection of a `QueryDef` object can be used to execute a `Parameter` query.

Exercise 25.9 Using a *Parameter* Query

In this exercise, you simply modify the previous code so that all queries, including `Parameter` queries, are included. The `InputBox()` function serves to retrieve user input for the parameters (see Figure 25.9).

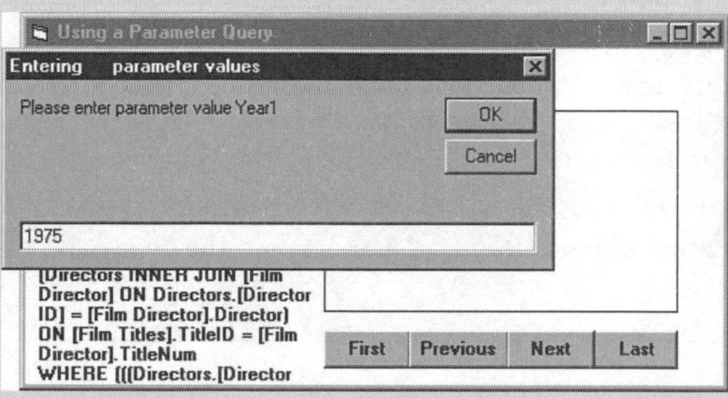

Figure 25.9

A `Parameter` query requires an input box for the user to supply a needed value.

1 Change the `Form_Load()` event so that it reads as follows:

```
Private Sub Form_Load()
   Dim qdef As DAO.QueryDef
   Set db = Workspaces(0).OpenDatabase(App.Path & "\cinema.mdb")
   For Each qdef In db.QueryDefs
      lstQueryDef.AddItem qdef.Name
   Next
End Sub
```

2 In the `lstQueryDef_Click()` event, change the code to read:

```
Private Sub lstQueryDef_Click()
   Dim qdef As QueryDef
   Dim param As Parameter
   Dim queryname As String
   Dim f as field
   Dim ret As Variant

   queryname = Trim(lstQueryDef.List(lstQueryDef.ListIndex))
   Set qdef = db.QueryDefs(queryname)
   If qdef.Parameters.Count = 0 Then
      Set rs = qdef.OpenRecordset(dbOpenSnapshot)
```

continues

```
    Else
       For Each param In qdef.Parameters
          ret = InputBox("Please enter parameter value " & param.Name,
          ➥"Entering parameter values")
          param.Value = ret
       Next
          Set rs = qdef.OpenRecordset(dbOpenSnapshot)
    End If
    lstResult.Clear
    For Each f In rs.Fields
       lstResult.AddItem f.Value
    Next
    lblSQL.Caption = qdef.SQL
    rs.MoveLast
    lblRecordCount = "Record Count: " & rs.RecordCount
    rs.MoveFirst
 End Sub
```

3 Run the program and click the `FilmsByYearRange` query in the `lstQuery` list box. Enter two dates, such as 1975 and 1992, when the `InputBox()` function requests them.

If you are familiar with SQL, you might have built *dynamic* SQL statements in which SQL strings are concatenated with user input values. As a general rule, *when working with Jet engine queries, it is better to use a* `Parameter` *query, as these types of queries tend to execute faster.*

Chapter Summary

In this chapter, you examined the properties, methods, and functionality of the core objects that make up the Jet database engine.

The Jet engine object model is based upon a hierarchy of objects and object collections. At the top of the hierarchy is the single object `DBEngine`. Below it is the collection of `Workspace` objects that provide security, user management, and data transaction functions. Among the object collections contained within the `Workspace` object is the `Database` collection. The `Database` object contains collections of objects that received our major focus: `TableDef`, `Recordset`, `QueryDef`, and `Field`. These objects and others within the Jet engine are known as Data Access Objects (DAOs).

To iterate through Jet engine objects, the `For Each...Next` structure can be used. With this structure, it is possible to map an entire database, examining individual objects and their properties as well as the objects and collections contained within them.

Databases are opened with the `OpenDatabase` method of the `Workspace` object. `Recordsets` (table-, dynaset-, or snapshot-type) can be created with the `OpenRecordset` method of the `Database` object. `TableDef` and `QueryDef` objects have a `OpenRecordset` method. All the results of `OpenRecordset` methods provide a record-by-record or cursored view of data. Whichever record has been moved to or found becomes the current record. The

OpenDatabase and OpenRecordset methods have a list of arguments that can be used to determine such behaviors, including whether or not the opened database or data set should be read-only or read/write.

To navigate through a Recordset, you can use the Move methods (MoveFirst, MoveLast, MoveNext, MovePrevious) or the Find methods (FindFirst, FindLast, FindNext, FindPrevious). The fastest way to move to a specific record is to use a table-type Recordset and the Seek method on an indexed field.

The process of adding, editing, or deleting records is very similar to the methods used with the Data control. Before a record can be written with the Update method, the AddNew or Edit methods must first be invoked. The Edit and Delete methods work on the current record.

Stored queries in a Jet database are known as QueryDef objects. To create a data set based upon the query, use the OpenRecordset method for the identified query within the QueryDefs collection. For queries that require parameters, use the queries' Parameters collection to identify parameters and assign values to them before the queries are opened with an OpenRecordset method.

Skill-Building Exercises

1. Change Exercise 25.1 so it displays the principal Data Access Objects for the sample Biblio.mdb database that comes with Visual Basic.
2. Write a program using a table-type `Recordset` that displays in a list box all film titles from the Cinema.MDB database.
3. Write a program using a snapshot-type `Recordset` that displays all film titles, running times, and ratings in a list box.
4. On your disk is a database named pop.mdb that contains population information on 209 U.S. cities as well as individual state populations. Write a program that maps the tables and fields of this simple database.
5. Based on the information learned in the preceding exercise, write a program that lists each city in a list box. When a city is clicked in the list box, have your program use the `Seek` method to retrieve the population of the state in which the city is located, and display the retrieved information.

26

Working with Objects from Microsoft Office

This chapter is focused on combining objects contained within the desktop applications of the Microsoft Office suite with a Visual Basic application. Visual Basic is much more than just a language for developing stand-alone Windows applications; it is also the scripting or *macro language* for nearly the entire suite of Microsoft desktop applications, including Word, Excel, Access, and PowerPoint. A version of Visual Basic, VBScript, is designed for use with interactive Internet applications.

Within Microsoft Office applications you can use Visual Basic for Applications to automate many program actions or to execute independent code. The newest versions of the Microsoft Office applications provide a rich design environment for writing application macros, or for developing custom applications within the framework of the applications themselves. When programming with Microsoft Office applications, you must decide whether you want to work primarily within the programming environment provided by the application and Microsoft Office, or within the Visual Basic design environment.

We focus on interacting with Microsoft applications from within stand-alone Visual Basic programs. The concepts and code presented within this chapter will be usable if you choose to work with the Visual Basic editors provided within Microsoft Office. This chapter emphasizes using Visual Basic with two of the Microsoft Office products: Excel and Access. Excel is one of the most frequently used Office products that serves to enhance Visual Basic projects. Excel's built-in functions and charting capabilities make it an extremely useful *add-on* to all kinds of programming tasks.

You have already used Data Access Objects, a library of functions encapsulating the Jet engine of Access. You were able to use DAO objects provided by an *in-process* server because the Jet engine is encapsulated inside a set of DLLs. However, the desktop program Access itself, including its user interface, can also be manipulated using Visual Basic. In this case, Access acts as an *out-of-process* server. You will use Access to print a report from a database, illustrating by example the distinction between in-process and out-of-process servers.

The newest version of Word has now migrated to Visual Basic for Applications. Visual Basic for Applications is gradually replacing Word Basic, a subset of earlier versions of Basic. Given the enormous complexity of Word and its object model, the migration of Word to Visual Basic for Applications will likely prove a relatively slow one. Word is the flagship application of the Microsoft Office suite, and there is an enormous installed base of Word Basic macros and programs. We concentrate, therefore, on Excel and Access. Needless to say, this chapter assumes you have the Microsoft Office suite, including Access, installed on your computer.

When you complete this chapter, you will be able to

- use the `CreateObject()` function to create running instances of Excel and Access;
- use the following Excel objects and collections: `Application`, `Workbook(s)`, `Worksheet(s)`, and `Range`;
- use the Excel `Cells` and `Range` methods;
- use the Excel `Average()` function in a formula;
- use the `Shell()` function to launch non-ActiveX executables;
- use the `SendKeys` statement to send keystrokes to other applications;
- use the `AppActivate` statement to shift focus to another running application;
- use the `OpenCurrentDatabase` method of the `Access.Application` objects;
- use the methods of the Access `DoCmd` object;
- use the DAO `Container` and `Document` objects;
- print a stored report from an Access database.

Chapter Objectives

The objectives of this chapter center on interacting with Microsoft Office applications from Visual Basic programs, as follows:

1. Understanding Microsoft Office applications as ActiveX servers
2. Using the `Shell()` function and `SendKeys` statement
3. Working with objects from the Excel type library
4. Working with objects from the Access type library

These objectives introduce enhancing Visual Basic applications with objects exposed by Microsoft Office applications. Given the size and complexity of these applications, this chapter cannot provide comprehensive coverage of the object hierarchy of Microsoft Office applications; rather, it gives you a flavor of how Microsoft Office objects can be combined with a Visual Basic application.

Objective 26.1 Understanding Microsoft Office Applications as ActiveX Servers

In Chapter 23, "Creating an ActiveX DLL," you constructed an ActiveX DLL containing string functions. This DLL code component operated within the same process as the client application using it. The string function DLL thus acted as an in-process server. The libraries containing Data Access Objects also act as in-process servers to your application. ActiveX DLL servers contain a type library that allows client applications to know the properties and methods exposed by the DLL.

Using ActiveX technology, free standing executables operating within their own process can also contain type libraries describing objects for use by other applications. These executables, using the underlying *component object model* technology, can become servers to client applications. The major difference between an ActiveX DLL server and an ActiveX EXE server is that the latter operates within its own process. Client applications that use the services of ActiveX EXE applications employ *cross-process* communication.

In-process communication is faster than cross-process communication; however, cross-process communication adds important flexibility to an application, especially when numerous client machines must access a single-server machine. An ActiveX EXE program on the server running within its own process can provide services to multiple client applications. The ActiveX EXE need not be stored on the client machine, unlike an ActiveX DLL. Most important, it need not be tied to a single client application: It can service multiple applications. Two or three different database reporting applications, for instance, can all avail themselves of the services of an ActiveX EXE running on the server.

Because the Microsoft Office applications are ActiveX EXE servers, they can be manipulated by other applications. This makes possible the rapid development of *document processing* applications using the constituent applications of Microsoft Office. Suppose you work in an enterprise in which elements of a spreadsheet, data in a database, and text from a word processing document must all be integrated regularly into a single report. The content varies from day to day, but the same elements are always used. Using the ActiveX capability of the Microsoft Office suite, you could construct a program to generate this report automatically.

26.1.1 Creating Instances of ActiveX EXE Objects

When working with ActiveX DLL objects, first make a reference to the type library of the DLL in the References dialog box (after choosing Project, References). With a reference established in the project, you can then declare objects defined within the type library in one of two ways:

- You can declare the object variable and then bind it with the `Set` statement, as in the following example:

```
Dim objectvar as myClass.myobject
Set objectvar = New myClass.myobject
```

The new class instance in this code occurs when the `Set` statement is executed.

- You can make the declaration with the following statement:

  ```
  Dim objectvar as New myClass.myobject
  ```

  In this code, the object is instantiated the first time a reference is made to an object property or method, as in the following example:

  ```
  objectvar.Sort = True
  ```

Visual Basic provides a special function for creating objects when working with ActiveX EXE projects: the `CreateObject()` function. This function returns an instance of an object of a specific class. The syntax for the function is as follows:

```
CreateObject(class)
```

The class argument is a string argument that contains a *fully qualified* reference to the object, including the name of the ActiveX application and the object name. This string reference uses an `appname.objecttype` syntax. Suppose, for instance, that you wanted to reference an Excel application. After making a reference to the Excel type library in the References dialog box (after choosing Project, References), you could make the following declaration:

```
Dim xlApp As Excel.Application
```

You could then create an instance of an Excel application with a `Set` statement and the `CreateObject()` function:

```
Set xlApp = CreateObject("Excel.Application")
```

The execution of this `Set` statement starts Excel running on your machine. By setting the `Visible` property of the application to `True`, a running instance of Excel will appear on the user's desktop. Exercise 26.1 demonstrates working with the `CreateObject()` function.

Exercise 26.1 Using the *CreateObject()* Function to Create a Running Instance of Excel

In this short exercise, you create a running instance of Excel, using the `CreateObject()` function, and then terminate it.

1. Create a form with two buttons (**`cmdStart`** and **`cmdStop`**), as shown in Figure 26.1. Save the form to disk as **f26-1.frm** and the project as **p26-1.vbp**. Change the `Name` property of the project to **p26_1**.
2. Choose Project, References, and make a reference to the Excel library on your computer (see Figure 26.2).

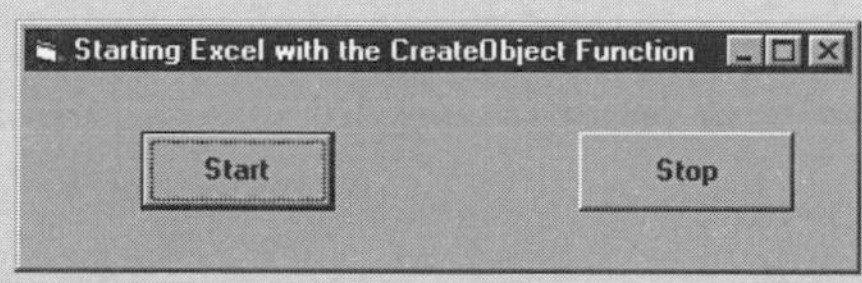

Figure 26.1

This application uses the `CreateObject()` function to start and stop Microsoft Excel.

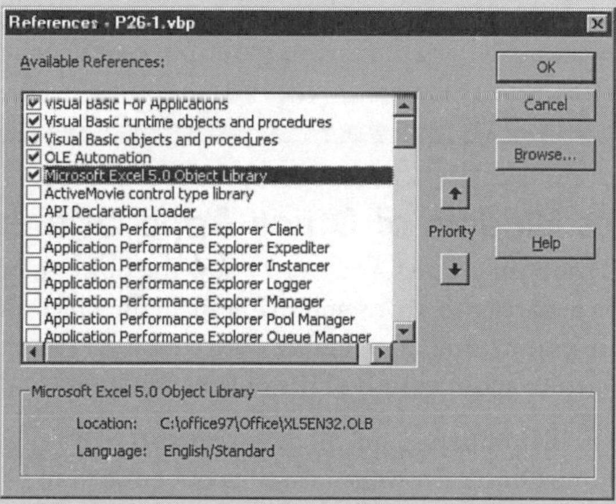

Figure 26.2

The References dialog box allows you to select another application's objects that you want available in your code.

3 At the module level of the form, make the following declaration:

```
Dim xlapp As Excel.Application
```

4 In the `cmdStart_Click()` event, add the following code:

```
Private Sub cmdStart_Click()
   Set xlapp = CreateObject("Excel.Application")
   xlapp.Visible = True
End Sub
```

5 In the `cmdStop_Click()` event, add the following code:

```
Private Sub cmdStop_Click()
   xlapp.Quit
   Set xlapp = Nothing
End Sub
```

In this code, you invoke the `Excel.Application` object's `Quit` method, and set the object variable to `Nothing`.

continues

6 Run and test the program. The Start button should start the Excel application running (with no workbook open), and the Stop button should terminate the application.

Using the Object Browser, examine the `WindowsState` property of the `Excel.Application` object. Experiment with launching the `Excel.Application` objects with the `WindowState` property set to its three possible settings (`xlNormal`, `xlMaximized`, `xlMinimized`).

Experiment further by commenting out the instruction `xlapp.Visible = True` in the `cmdStart_Click()` event. After clicking the Start button, use the Windows Task Manager utility to determine whether or not Excel is running.

26.1.2 Object Models of Office Applications

Once an ActiveX EXE has been created, you must know the object model of the application in order to navigate the hierarchy to access and manipulate the objects the application exposes. Although there is some similarity among object models, you must know the model and then learn the properties and methods of the objects contained within the application.

The Object Browser lists the properties, methods, and events of objects defined in ActiveX EXE type libraries, but the Object Browser by itself is not enough to work with these objects. You almost always need the programming reference books that accompany the developer versions of Office products to fully exploit their resources. Each application contains object methods with some syntactic idiosyncrasies.

In this chapter, we limit our attention to examining some of the major objects contained within Excel and Access. Next, however, we examine intrinsic Visual Basic commands that permit the developer to launch applications that are not ActiveX-compliant from within a Visual Basic application.

Objective 26.2 Using the *Shell()* Function and *SendKeys* Statement

Not all inter-application communication in Visual Basic involves the use of ActiveX or object linking and embedding (OLE) technology. Before the advent of OLE, Windows included simplified inter-application communication using the Windows Clipboard and a technology known as *dynamic data exchange* (DDE). You use the Windows Clipboard (a `System` object in Visual Basic) every time you copy and paste. You can easily add Clipboard functions to your Visual Basic applications, including the copying and pasting of graphics files. Visual Basic's documentation and online Help demonstrates how to add this operating system feature to your applications.

DDE, an early attempt to add programmable inter-application communication within Windows, is still available in Visual Basic. It features very fast passing of string data between applications. Because DDE is a receding technology, it is not included in this text. You should know, however, that DDE exists, and you should investigate it seriously if the inter-application communication you require is relatively simple. This technology frequently is used today for handling data transmitted from a serial port, often in applications that monitor and control automated manufacturing processes.

Two other Visual Basic instructions are important when working with inter-application communication: the `Shell()` function and the `SendKeys` statement.

26.2.1 The *Shell()* Function

A special Visual Basic function designed specifically to launch another application is the `Shell()` function. The syntax for this function is as follows:

```
Shell(pathname [,windowstyle])
```

The component `pathname` refers to the path and the name of the program to execute and any required arguments or command-line switches. It can include the directory and drive names and also the name of a document that has been associated with an executable program.

The component `windowstyle` refers to the style of the window in which the program is to be executed upon its opening. If `windowstyle` is omitted, the program is started minimized with focus. Table 26.1 explains the various styles.

Table 26.1 The *windowstyle* Constants

Constant	*Value*	*Description*
`vbHide`	`0`	The window is hidden and focus is passed to the hidden window.
`vbNormalFocus`	`1`	The window has focus and is restored to its original size and position.
`vbMinimizedFocus`	`2`	The window is displayed as an icon with focus.
`vbMaximizedFocus`	`3`	The window is maximized with focus.
`vbNormalNoFocus`	`4`	The window is restored to its most recent size and position. The currently active window remains active.
`vbMinimizedNoFocus`	`6`	The window is displayed as an icon. The currently active window remains active.

As an example of the `Shell()` function, consider the following statement:

```
X = Shell("C:\windows\Calc.exe", 1)
```

This statement uses `Shell()` to leave the Visual Basic application and run the Calculator program, an .EXE program included with Microsoft Windows. The value returned (to `X`) is the task identification number of the started program.

Exercise 26.2 Using *Shell()* to Start Two Windows Applications

In this short exercise, you use the `Shell()` function to start the Windows Calculator and the Windows Notepad programs. Figure 26.3 shows the simple interface.

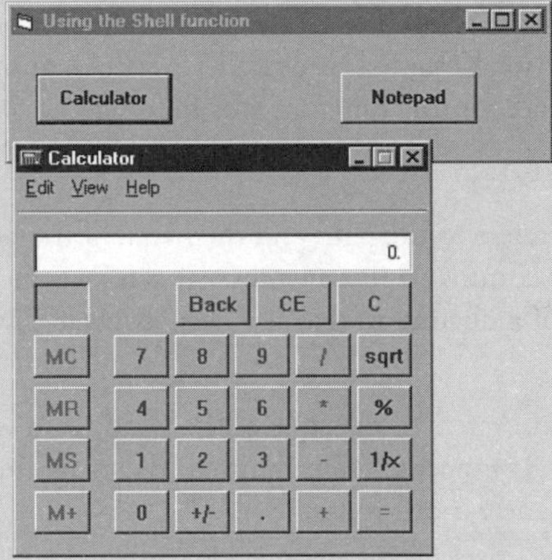

Figure 26.3

This application launches the Calculator and the Notepad .EXE files using the `Shell()` *function.*

1 Create the interface shown adding the **`cmdCalculator`** and **`cmdNotepad`** commands. Add the form caption. Save the form to disk as **f26-2.frm** and the project as **p26-2.vbp**. Change the `Name` property of the Project to **p26_2**.

2 Write the `cmdCalculator_Click()` procedure:

```
Private Sub cmdCalculator_Click()
   Dim X As Long
   X = Shell("c:\windows\calc.exe", 1) ' adjust the path statement as
                                        ' necessary
End Sub
```

3 Write the `cmdNotepad_Click()` procedure:

```
Private Sub cmdNotepad_Click()
   Dim Y As Long
   Y = Shell("c:\windows\notepad.exe", 1) ' adjust path statement as
                                           ' necessary
End Sub
```

4 Test your design.

26.2.2 The *SendKeys* Statement

A most useful Visual Basic statement is the `SendKeys` statement, which, as the name suggests, is a statement used to send one or more keystrokes from a Visual Basic application to another application. The syntax for this statement is

```
SendKeys string [,wait]
```

The `string` component is the string expression to be sent to the active window. The `wait` component is `True` or `False`. If `True`, keystrokes must be processed before control is returned to the procedure from which the keystrokes were sent. If `False`, control is returned to the calling procedure immediately after keys are sent. `False` is the default.

26.2.3 Dealing with Special Characters

Sending special characters with `SendKeys` requires knowledge of the `SendKeys` codes. The entire list of codes is found in the Help section of Visual Basic under the `SendKeys` statement. For our purposes, codes of special importance are those for the Shift, Ctrl, and Alt keys (see Table 26.2).

Table 26.2 *SendKeys* Codes for Special Characters

Key	*Code*
Alt	`%`
Backspace	`{BACKSPACE}`, `{BS}`, or `{BKSP}`
Break	`{BREAK}`
Caps Lock	`{CAPSLOCK}`
Ctrl	`^`
Del	`{DELETE}` or `{DEL}`
Down Arrow	`{DOWN}`
End	`{END}`
Enter	`{ENTER} or ~`
Esc	`{ESC}`
F1	`{F1}`
Help	`{HELP}`
Home	`{HOME}`
Ins	`{INSERT}`
Left Arrow	`{LEFT}`
Num Lock	`{NUMLOCK}`
Page Down	`{PGDN}`

continues

Table 26.2 continued

Key	*Code*
Page Up	`{PGUP}`
Print Screen	`{PRTSC}`
Right Arrow	`{RIGHT}`
Scroll Lock	`{SCROLLLOCK}`
Shift	`+`
Tab	`{TAB}`
Up Arrow	`{UP}`

As an example, suppose the `Close` window command on the close box is Alt+F4. The following statement closes the active window (meaning the window must have the focus to be closed):

```
SendKeys "%{F4}", True
```

26.2.4 The *AppActivate* Statement

The `AppActivate` statement is used to activate an application window and is closely wedded to the `SendKeys` statement. The syntax for this statement is

```
AppActivate title [, wait]
```

The `title` component is the string expression that appears in the title bar of the application window to activate. Observe that this is *not* the name of the application, but the title caption shown in the title bar. The `wait` component is `True` or `False`, as with `SendKeys`.

Consider the following example:

```
AppActivate "Calculator"
```

This statement gives the focus to the Calculator application, provided it has been opened.

Exercise 26.3 Using *SendKeys* to Close Windows Applications

Create the interface shown in Figure 26.4 to use the `SendKeys` and `AppActivate` statements to close Windows applications from within Visual Basic. From this design, you can start and close the Windows Calculator and the Windows Notepad.

1 Design the interface as shown, add the command buttons (**`cmdClose`**, **`cmdNotepad`**, and **`cmdCalculator`**). Add the form caption (exactly as shown). Save the form to disk as **f26-3.frm** and the project as **p26-3.vbp**. Change the `Name` property of the project to **p26_3**.

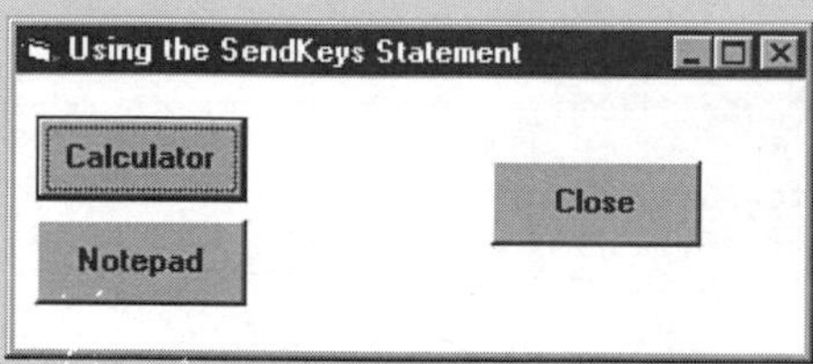

Figure 26.4

The Calculator and the Notepad applications can be stopped with the `SendKeys` *statement.*

2. Declare a flag variable at the module level.

```
Dim Message As String
```

3. Write the `cmdCalculator_Click()` procedure:

```
Private Sub cmdCalculator_Click()
    Dim X As Long

   If Message = "Notepad" Or Message = "Calculator" Then
       MsgBox "Close the Open Windows Application", , "Please"
   Else
       X = Shell("c:\Windows\calc.exe", 1) ' adjust path as necessary
       Message = "Calculator"
   End If
End Sub
```

In this procedure, error checking determines whether either the Notepad or the Calculator is already open. If the Notepad is open, the Calculator cannot be used until the Notepad is closed. If the Calculator is open, there is no need to open it again.

4. Write the `cmdNotepad_Click()` procedure:

```
Private Sub cmdNotepad_Click()
    Dim Y As Long
    If Message = "Notepad" Or Message = "Calculator" Then
       MsgBox "Close the Open Windows Application", , "Please"
    Else
        Y = Shell("c:\Windows\notepad.exe", 1) ' adjust path as necessary
        Message = "Notepad"
    End If
End Sub
```

This procedure is identical in form to the Calculator procedure.

continues

5 Write the `cmdClose_Click()` procedure:

```
Private Sub cmdClose_Click()
   If Message = "Calculator" Then
      AppActivate "Calculator"
      SendKeys "%{F4}", True
      AppActivate "Using the SendKeys Statement"
      Message = ""
   ElseIf Message = "Notepad" Then
      AppActivate "Untitled - Notepad"
      SendKeys "%{F4}", True
      AppActivate "Using the SendKeys Statement"
      Message = ""
   Else
      MsgBox "There is no Windows Application to Close", , "Sorry"
   End If
End Sub
```

The procedure first checks to determine what the `Message` is. If there is no `Message`, a message box is displayed. If the Calculator `Message` is assigned, the Calculator application is activated. Once activated, the keys (Alt+F4) are passed to close the application.

A second `AppActivate` instruction returns the focus to the Visual Basic program. Notice that the string expression in the `AppActivate` statement must match the title bar of the application to which you want to shift focus, or an error occurs.

6 Run and test your design.

Objective 26.3 Working with Objects from the Excel Type Library

In Exercise 26.1, you launched Excel using the `CreateObject()` function. In order to work with Excel, you need to review the Excel object model. Although this section cannot provide complete coverage of the Excel object model, nor cover the different ways you can manipulate these objects using Visual Basic for Applications, it does give you a flavor of manipulating Excel with ActiveX technology.

26.3.1 The Excel Object Model

The Excel object model, like the application itself, is large and complex. In Excel 95 there are 128 objects! The newer versions of Excel are even more complex. Like the hierarchy of objects with which you become familiar when working with Data Access Objects, the Excel object model is organized in a hierarchical manner. However, the Excel object model has some features that make working with it a little different from working with the DAO object model.

Consider the objects in Figure 26.5, which represent some of the most frequently used objects in Excel. At the top of the model is a single object: `Application`. This represents a running instance of Excel with no .XLS files open. Below this single object is a collection of workbooks, with each workbook representing an open .XLS file. Each workbook, in turn, contains a collection of worksheets. Finally, each worksheet contains a grouping of cells (the intersection of a column and a row), which is represented by the `Range` object.

A range can represent a single Excel cell, such as B3, or a grouping of cells, such as B3:J3. As you probably are aware from working with Excel, the letter represents the column and the number represents the row. The range B3:J3 would represent a row of nine cells on the third line of the worksheet.

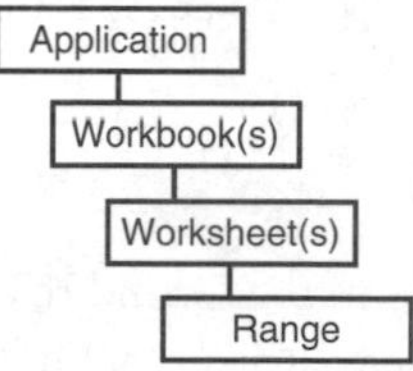

Figure 26.5

The Excel object model consists of an application that contains a collection of workbooks; each workbook contains a collection of worksheets; and each worksheet contains a range.

Examine Figure 26.6. You can observe that in this single instance of Excel, two workbooks—Funds.xls and Mf1.xls—are open. Each workbook contains a collection of worksheets identified by tabs at the bottom of the workbook. The range of cells B3:J3 in Funds.xls displays values for the Income Fund; the range of cells E2:E13 displays values for 1993.

	A	B	C	D	E	F	G	H	I	J
1		1996	1995	1994	1993	1992	1991	1990	1989	1988
2	Variable Fund	6.74	6.7	26.41	3.31	29.06	14.63	5.4	18.97	31.28
3	Income Fund	9.68	7.45	19.43	9.11	14.57	7.63	4.52	14.07	22.34
4	Encore Fund	3.19	3.67	6.53	8.44	9.39	7.5	6.81	6.88	8.46
5	Advisers Fund	9.9	6.39	18.38	5.72	5.33				
6										
7	Alger Fund	13.3	3.55	57.54	8.71	64.48	-3.35			
8	Calvert Fund	8	7.61	16.4	4.18	20.69	11.75	6.78	3.31	
9	Franklin Fund	7.58	7.66	15.87	10.2	8.9				
10	Lexington Fund	10.9	3.22	-4.95	-15					
11	N & B Fund	6.79	9.54	29.73	-8.2	29.47	25.97	-4.9	14.94	30.3
12	Scudder Fund	37.8	-3.1	11.45	-7.7	37.79	16.73			
13	TCI Fund	10.3	-1.3	41.86	-1.2	28.7	-2.26	7.2		
14										

Figure 26.6

The Funds.xls worksheet shows the asset value for a number of mutual funds across several years.

26.3.2 The *Cells()* Function

A `Range` object refers to an individual cell or a group of cells. The `Application`, `Worksheet`, and `Range` objects also have a `Cells()` function that can be used to refer to an individual cell. In this case, the row index value comes first and the column index value comes last. Consider the following code:

```
Dim xlapp As Excel.Application
Dim xlbook As Excel.Workbook
Dim xlsheet As Excel.Worksheet


Set xlapp = CreateObject("Excel.Application")
xlapp.Workbooks.Add
Set xlbook = xlapp.Workbooks(1)
Set xlsheet = xlbook.Worksheets(1)
xlsheet.Cells(2, 1).Value = 45
```

This code places the value `45` in the second row and the first column, namely, the cell that would be referred to as `xlsheet.Range("A2")`. The `Cells()` function uses only numeric arguments in order to specify a cell, unlike the `Range` function of a `Worksheet` or `Application` object, which uses letters of the alphabet to specify columns.

26.3.3 The *ActiveWorkBook* and *ActiveSheet* Methods

As you observed when working with DAOs, it is possible to navigate down the entire object model path to specify a value. Usually, however, object variables are set that allow you to easily reference the subordinate object with which you would like to work. Just as you worked primarily with the `DAO.Recordset` object in Chapter 25, "Using Data Access Objects," you work mainly with the `Worksheet` and `Range` objects in Excel. It is possible to write code like the following:

```
Application.Workbooks(1).Worksheets(1).Range("B2").Value = 56.75
```

However, you more typically bind object variables as you work down the object hierarchy, and work with code like the following:

```
Dim xlsheet as Excel.Worksheet
Set xlsheet = xlapp.Worksheets(1)
xlsheet.Range("B2").Value = 56.75
```

Any subsequent reference to the `Worksheet` object can use the object variable `xlsheet`.

Alternatively, you can use the `ActiveWorkbook` and `ActiveSheet` methods to refer to the current `Workbook` and `Worksheet` objects, depending upon the context in which you are working, You can write such code as the following:

```
Dim xlapp As Excel.Application
Dim xlbook As Excel.Workbook
Dim xlsheet As Excel.Worksheet

Set xlapp = CreateObject("Excel.Application")
xlapp.Workbooks.Add
Set xlbook = xlapp.Workbooks(1)
Set xlsheet = xlbook.Worksheets(1)
```

```
xlsheet.Cells(2, 1).Value = 45
ActiveSheet.Cells(2, 2).Value = 46
```

26.3.4 Working with the *Range* Object

If you examine the Excel library with the Object Browser, you will discover that most functions that perform calculations on cell or range values are methods of the `Application` object. These functions are the heart of Excel. By specifying formulas that act on `Range` objects of values, you make a spreadsheet truly useful. The main method of employing these functions is by assigning a new value to the `Formula` property of a `Range` object.

You will find that when working with a `Worksheet` object, you typically work with multiple `Range` objects. These `Range` objects are often used as arguments in the functions identified for Excel formulas. To make working with these multiple `Range` objects easier, each `Range` object has a `Name` property that is used when specifying a `Range` object as an argument in an Excel formula function. The next exercise illustrates how `Range` objects, their `Name` properties, and Excel functions expressed in the `Range` object's `Formula` property can be utilized.

Exercise 26.4 Calculating an Average in Excel

In this exercise, you will launch a running instance of Excel, create a new workbook, display ten randomly generated integers on a worksheet, take the average of these numbers, and save the workbook to disk as an .XLS file. Figure 26.7 shows the interface you are asked to build.

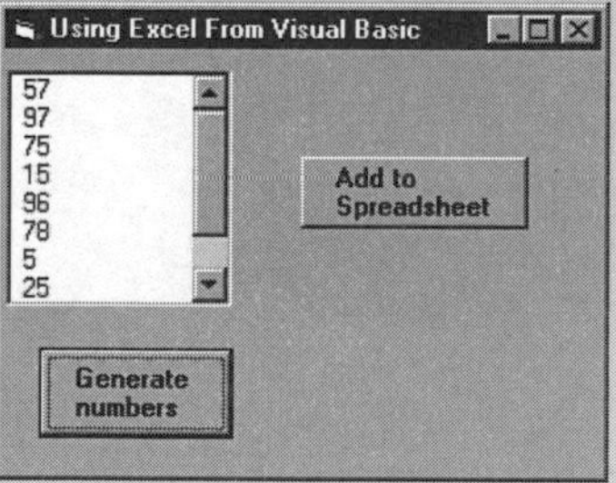

Figure 26.7

This application uses ActiveX automation to generate numbers and then add them to an Excel worksheet.

1 Add the two command buttons (**`cmdGenerate`** and **`cmdAdd`**) and a list box (**`lstNumbers`**). Change the captions of the form and the command buttons as shown. Set the `Enabled` property for `cmdAdd` to `False`. Save the form to disk as **f26-4.frm** and the projects as **p26-4.vbp**. Change the `Name` property of the project to **p26_4**.

continues

2 Set a reference to the Excel library using the References dialog box (by choosing Project, References).

3 At the form module level, make the following declaration:

```
Dim DataRange(1 To 10) As Integer
```

4 Write a procedure for the `cmdGenerate_Click()` event that will randomly generate ten integer values between 1 and 100, assign the values to the `DataRange` array, and display them in the `lstNumber` list box.

```
Private Sub cmdGenerate_Click()
   Dim i As Integer

   lstNumbers.Clear
   For i = 1 To 10
      Randomize
      DataRange(i) = Int((100 * Rnd) + 1)
      lstNumbers.AddItem Str$(DataRange(i))
   Next
   cmdAdd.Enabled = True
End Sub
```

5 Write a procedure for the `cmdAdd_Click()` event that will add these randomly generated numbers to an Excel worksheet and take their average:

```
Private Sub cmdAdd_Click()
   Dim xlapp As Excel.Application
   Dim xlbook As Excel.Workbook
   Dim xlsheet As Excel.Worksheet
   Dim xlrange As Excel.Range

   Dim i As Integer

   On Error GoTo cmdAdderr:

   Set xlapp = CreateObject("Excel.Application")
   xlapp.Visible = True
   xlapp.WindowState = xlMaximized

   xlapp.Workbooks.Add
   Set xlbook = xlapp.Workbooks(1)
   Set xlsheet = xlbook.Worksheets(1)

   For i = 1 To 10
     xlsheet.Cells(i, 2).Value = DataRange(i)
   Next

   Set xlrange = xlsheet.Range("B1:B10")
   xlrange.Name = "myRange"

   xlsheet.Range("A13:B13").Font.Bold = True
   xlsheet.Range("A13").Value = "Avg."
   xlsheet.Range("B13").Formula = "=Average(myRange)"

   xlbook.SaveAs App.Path & "\p26-4.xls"
```

```
        Set xlrange = Nothing
        Set xlsheet = Nothing
        Set xlbook = Nothing
        xlapp.Quit
        Set xlapp = Nothing
        cmdAdd.Enabled = False

        Exit Sub

    cmdAdderr:
        cmdAdd.Enabled = False
        If Not xlapp Is Nothing Then
           xlapp.Quit
           MsgBox Str$(Err) & "  " & Err.Description

        Else
           MsgBox Str$(Err) & "  " & Err.Description
           Resume Next
        End If
    End Sub
```

6 Run and test the application. The first time you run the program, you will see a worksheet that contains cells similar to those in Figure 26.8.

	A	B	C
1		9	
2		83	
3		35	
4		78	
5		2	
6		30	
7		94	
8		28	
9		100	
10		8	
11			
12			
13	**Avg.**	**46.7**	
14			
15			
16			

Figure 26.8

This worksheet shows the result of executing the ActiveX application.

In subsequent testing of the application, a dialog box will appear, asking whether you want to replace the p26-4.xls file. Select Yes to return to the Visual Basic application without error.

As you study the `cmdAdd_Click()` event, the code should become more understandable to you. In this code, you launch Excel, create an entire new workbook, and save it to disk. Like the intrinsic `Collection` object in Visual Basic, collections in Excel are 1-based. Once you have bound object variables to the first workbook in the `Workbooks`

continues

collection and the first worksheet in the `Worksheets` collection, you use the `Cells` method inside a `For...Next` loop to populate the cells with the values of the randomly generated number.

The `Range` object, `xlrange`, is bound by using the `Range` method of the `Worksheet` object `xlsheet`. The procedure then assigns the value `myRange` to the `Name` property of this `Range` object. This `Name` property is employed when you specify a formula for the individual cell `"B13"`. This formula uses the Excel `Average()` function, which expects a range for its single argument.

The workbook is then saved to disk using the `SaveAs` method.

26.3.5 Opening an Excel File

As you work more with Excel objects and consult the documentation for programming Excel, you will discover the use of a colon with named argument identifiers in Excel object methods. For instance, to open an existing Excel file, you use the `Open` method of the `Workbooks` collection. The `Open` method has a long and complex syntax depending on whether the Excel file is linked to other files and is password protected. The syntax of the Excel `Workbooks.Open` method is:

```
Workbooks.Open(fileName, [updateLinks], [readOnly], [format], [password],
[writeResPassword], [ignoreReadOnlyRecommended], [origin], [delimiter], [editable],
[notify], [converter])
```

You can consult the Excel Visual Basic for Applications Help to get a complete description of these method arguments. All arguments are optional except the first one. To assign a value to this argument, use the parameter name followed by a colon.

Consider the following code:

```
Dim xlapp As Excel.Application
Dim xlbook As Excel.Workbook
Set xlapp = CreateObject("Excel.Application")
xlapp.WindowState = xlMaximized
Set xlbook = xlapp.Workbooks.Open(filename:=App.Path & "\p26-4.xls")
```

This code would open the .XLS file you saved to disk in Exercise 26.4. You are asked to use this method of opening a workbook in the third exercise in the “Skill-Building Exercises” section at the end of this chapter.

Objective 26.4 Working with Objects from the Access Type Library

At the start of this chapter, we note the distinction between DAO objects, which serve an application from a DLL, and Access itself, which serves your application as an ActiveX executable. If you are familiar with DAO objects, there are very few times you actually need to use Access as an ActiveX EXE server. You get much speedier performance by using DAO objects because you avoid all the cross-process communication.

26.4.1 The Access Object Model

One very useful feature of Access, which is not contained within the DAO DLLs, is the Access reporting capability. Visual Basic comes with a third-party report generating utility named Crystal Reports. The Access Report editor, however, is superb and quite easy to use. If you are certain that the end user will have Access installed on the target machine of your application, you can design your program to print stored Access reports (see Exercise 26.5). Figure 26.9 displays a simplified version of the Access object model.

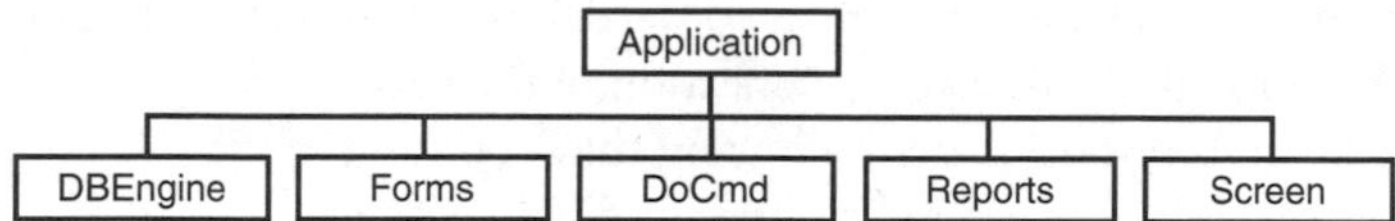

Figure 26.9

The Access object model contains all objects and collections, including the `Forms` collection, the `Reports` collection, the `DBEngine`, the `Screen` object, and the `DoCmd` object.

26.4.2 The *DoCmd* Object

The Access object model contains DAO objects within it. It also contains the `DoCmd` object as a first-tier object below the `Application` object. The `DoCmd` object permits you to execute nearly every Access menu item as well as many other built-in Access commands. The methods of the `DoCmd` object follow. Of these commands, `OpenReport` prints a report by default.

`ApplyFilter`	`FindNext`	`MoveSize`
`Beep`	`FindRecord`	`OpenForm`
`CancelEvent`	`GoToControl`	`OpenModule`
`Close`	`GoToPage`	`OpenQuery`
`CopyObject`	`GoToRecord`	`OpenReport`
`DeleteObject`	`Hourglass`	`OpenTable`
`DoMenuItem`	`Maximize`	`OutputTo`
`Echo`	`Minimize`	`PrintOut`

`Quit`	`RunMacro`	`SetWarnings`
`Rename`	`RunSQL`	`ShowAllRecords`
`RepaintObject`	`Save`	`ShowToolbar`
`Requery`	`SelectObject`	`TransferDatabase`
`Restore`	`SendObject`	`TransferSpreadsheet`
`RunCommand`	`SetMenuItem`	`TransferText`

26.4.3 The DAO *Containers* and *Documents* Collections

With DAO objects, you may have noticed the `Container` and `Document` object collections in the DAO object model. Each container in the `Containers` collection holds `Document` objects describing a particular part of the database, including stored databases, tables, table relationships, and reports. By iterating through the `Documents` collection of the `Reports` container, you can list all stored reports in an Access database. Following this, you can use the `OpenReport` method of the `DoCmd` object to select and print an individual report.

Exercise 26.5 Using DAO and Access Together

Use the DAO library to help identify stored reports for a database called Cinema2.MDB; then, select a report and print it. To work through this exercise, you must have Access on your computer (Access 95 or a later version) and a printer attached to one of your computer ports.

1 From the companion disk, copy the database named Cinema2.MDB from the data subdirectory into the root directory of your application.

2 Choose Project, References, and select both the DAO type library (DAO 3.0 or DAO 3.5) and the Access type library in the References dialog box. For Access 95, the library should be listed as MSAccess.TLB.

3 Build the interface you see displayed in Figure 26.10. The interface contains two list boxes (**`lstContainers`** on the left, and **`lstReports`** on the right) and two command buttons (**`cmdShow`** and **`cmdPrint`**). Save the form to disk as **f26-5.frm** and the project as **p26-5.vbp**. Change the `Name` property of the project to **P26_5**.

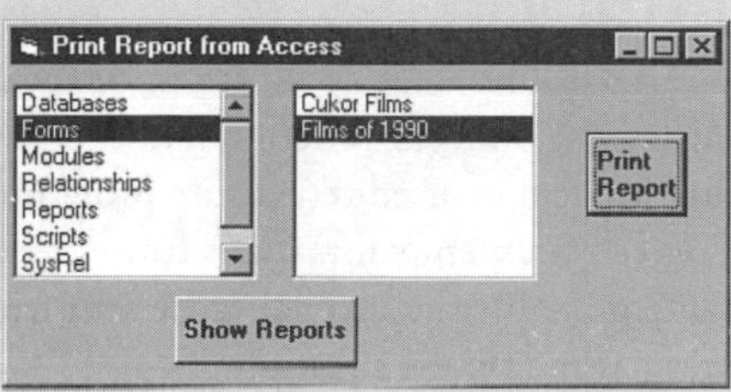

Figure 26.10

This application lists the reports prepared for the Cinema2 database, and enables you to print them using Visual Basic.

4 In the `cmdShow_Click()` event, write instructions to list every `Container` object within the Cinema2.MDB database and place them in the `lstContainers` list box. If the container holds reports, list all the reports within that container and place the report names in the `lstReports` list box.

```
Private Sub cmdShow_Click()
  Dim db As DAO.Database
  Dim doc As DAO.Document
  Dim con As DAO.Container
  Set db = DAO.Workspaces(0).OpenDatabase(App.Path & "\cinema2.mdb")
    For Each con In db.Containers
      lstContainers.AddItem con.Name
      If con.Name = "Reports" Then
        For Each doc In con.Documents
           lstReports.AddItem doc.Name
        Next
      End If
    Next
    lstReports.ListIndex = 0
    db.Close
    Set db = Nothing
End Sub
```

5 Write the `cmdPrint_Click()` event to open Access and print a specified report

```
Private Sub cmdPrint_Click()
   Dim acApp As Access.Application
   Dim rptname As String
   Set acApp = CreateObject("Access.application")
   acApp.OpenCurrentDatabase (App.Path & "\cinema2.mdb")
   rptname = lstReports.Text
   acApp.DoCmd.OpenReport rptname, acNormal
   acApp.Quit
End Sub
```

This code uses the `OpenCurrentDatabase` method of the `Access.Application` object and the `DoCmd` object to open the report. The `acNormal` constant indicates that you want to have the report printed. Other options include `acPreview` to see the report in preview mode within Access, or `acDesign` to enter the report design editor within Access.

6 Run and test your design.

Chapter Summary

This chapter introduced you to working with applications of the Microsoft Office suite as ActiveX servers. It also discussed the `Shell()` function and the `SendKeys` and `AppActivate` statements. The `Shell()` function can be used to execute programs from within a Visual Basic program. The `AppActivate` statement shifts focus from one executing program to another. The `SendKeys` statement enables keystrokes to be sent from a running Visual Basic application to another running Windows application.

Applications of the Microsoft Office suite are out-of-process ActiveX EXE servers. Through hierarchical object models, they expose their functionality for use by Visual Basic applications.

To create a running instance of an ActiveX EXE, you use the `CreateObject()` function. The object created is the singular object called `Application`. For Excel, the fully qualified name is `Excel.Application`. For Access, the fully qualified name to use with the `CreateObject()` function is `Access.Application`.

Key objects and collections when working with Excel are the `Workbook(s)`, `Worksheet(s)`, and `Range` objects. The latter can contain one or more cells. To reference a `Range` object, you can use the `Range` method of the Excel `Application` or `Worksheet` objects. `Range` objects have a `Name` property that is commonly used when the `Range` is referred to inside an Excel formula function.

The Access object model allows features of Access that are not included in the DAO library to be used by a Visual Basic application. One of these features is the Access reporting capability.

The DAO `Database` object includes a `Containers` collection. Each `Container` object in `Containers` holds a collection of `Document` objects. A `Document` object describes a single instance of part of the database, including tables and reports. Using the `Container` and `Document` objects, it is possible to list all stored reports in an Access database.

Skill-Building Exercises

1. Use the `CreateObject()` function to create a running instance of Excel. Change the caption of the running application to **Using ActiveX Automation** with the `Caption` property of the `Excel.Application` object.
2. Revise Exercise 26.4 so that the Visual Basic application places three rows of randomly generated numbers on a worksheet. Each row should contain ten randomly generated numbers between 1 and 5,000. At the end of the row, display the sum of the randomly generated numbers in one cell and the average of the randomly generated numbers in another cell.
3. Alter the preceding exercise so that the Visual Basic application opens an existing .XLS file rather than creating a new one. Refer to section 26.3.5, "Opening an Excel File," earlier in the chapter.
4. Create an application that launches the Windows Paintbrush utility using the `Shell()` function.
5. Using the Northwind.mdb file that is shipped as a sample .MDB file with Access, write an application that lists the stored reports in the database and prints a report the user selects.

27

Using the OLE Container Control

Visual Basic application development is really about software components. These components—whether they are ActiveX controls, ActiveX DLLs, or ActiveX EXEs—can be employed to rapidly construct applications. In Chapter 23, "Creating an ActiveX DLL," and Chapter 26, "Working with Objects from Microsoft Office," we explored ActiveX components in some detail. You built an ActiveX DLL in Chapter 23 and used the services of ActiveX executables in Chapter 26.

ActiveX is a new name applied to a technology formerly called *object linking and embedding* (OLE). The name, ActiveX, was applied when Microsoft resolved to pursue Internet development. What followed as a result was the definition of Internet standards and technology for cross-platform distributed computing within an enterprise (*intranets*). ActiveX makes minor modifications to pre-existing OLE technology in pursuit of smaller, more quickly downloadable components in Internet and intranet applications.

ActiveX components are at heart, however, OLE compliant components using the underlying architecture of the *component object model* (COM). Another name for the manipulation of objects exposed by ActiveX EXEs, for instance, is *OLE automation.*

Much of the functionality of OLE technology in earlier versions of Visual Basic was encapsulated in the OLE Container control. This control is retained in Visual Basic 5.0 and provides a flexible means of using OLE technology to make software components accessible to your application. The OLE Container control presents the software component's visual interface to the end user. Depending on how you employ the OLE Container control, the data produced by the visual software component is stored by your application (the client) or the application exposing the software component (the server).

Examples of OLE server applications include Microsoft Excel, Microsoft Word, Shapeware's VISIO, and several small applications included with Windows and Microsoft Office, such as Microsoft WordPad and WordArt. Although OLE is called object linking *and* embedding, a better description is object linking *or* object embedding. With object linking, a Visual Basic application is linked, by path name, to an OLE server application. The server application stores and maintains the linked data. With object embedding, the source data from the server application is actually stored in the Visual Basic client. Besides storing the source data, object embedding stores the path to the server that created the data. Thus, remember the following definitions:

- **Object linking**—To link the client application to the server application, with the server storing the data
- **Object embedding**—To link the client application to the server application, with the client storing the data

In this chapter, you will design some applications that require objecting linking and others that require object embedding. When you complete this chapter, you will be able to

- use the OLE custom control,
- link a spreadsheet object to a Visual Basic application,
- embed a spreadsheet object into a Visual Basic application,
- set OLE properties during design time,
- set OLE properties during runtime,
- save and reopen data created by an embedded server.

Chapter Objectives

The three objectives of this chapter first describe the OLE Container control and then differentiate between object linking and embedding in design time as well as in runtime. The objectives are as follows:

1. Introducing the OLE Container control
2. Working with object linking and embedding at design time
3. Working with object linking and embedding at runtime

These objectives serve to illustrate how the visual interfaces of OLE-compliant software components can be used in Visual Basic applications.

Objective 27.1 Introducing the OLE Container Control

Inter-application communication on the Windows platform has evolved rapidly with every new version release of the operating system. In the early stages of Windows, the only way to share data between applications was through the use of the Clipboard. Typically, data was copied to the Clipboard and then pasted to a second Windows application. This limitation was changed with the release of *dynamic data exchange* (DDE) technology in Windows. With DDE, it is possible to link applications and obtain raw data from the source application (or the server). The receiving application (the destination or client), however, is passive. It waits for messages from the source application.

With the advent of OLE, software components of the server application could be linked to client applications, or software components of server applications could be embedded in a client application. In early versions of OLE, the server components were often entire applications. In more recent years, server applications have been divided into a collection of objects, making them more manageable in size. You can, for instance, embed an Excel chart in a Visual Basic client rather than the entire Excel application.

With OLE, the development of *compound documents* is possible. A Word document, for instance, can include an embedded Excel worksheet or chart and store the document with the worksheet on a disk. Creating a compound document is also possible in Visual Basic using the OLE Container control.

In object linking, the client receives data from the server component, and the server component manages the persistent storage of the data on a disk. When the client application is launched, so is the linked server application—when the linked software component is activated inside the client.

With object embedding, the client maintains the data: All data managed by the server component can be contained in the OLE Container control and stored to disk by the client in a binary file. When you save the file to disk, the binary file contains a reference to the server application, the object's data, and a Metafile image of the object. This binary file can be opened in the OLE Container control at a later time.

27.1.1 Design-Time and Runtime Uses of the OLE Container Control

The inner workings of OLE technology (and the component object model) is enormously complex. Visual Basic simplifies its use with the OLE Container control. Given the complexity of OLE, it is not surprising that the OLE Container control takes some time to learn. This chapter provides an introduction to this control; however, we cannot completely examine its features in a single chapter. The OLE Container control can be used in nearly countless ways as more applications become OLE compliant.

When working with the OLE Container control, remember the following:

- Only one object can be contained within the control at a time.
- Objects can be linked or embedded at design time or during runtime, which presents four possibilities:
    1. A linked object can be created during design time.
    2. A linked object can be created during runtime.
    3. An embedded object can be created during design time.
    4. An embedded object can be created during runtime.

  This chapter demonstrates all four possibilities.

Let's examine several of the key properties of the OLE Container control. These include the `AutoActivate`, `Class`, `OLEType`, `OLETypeAllowed`, `SizeMode`, `SourceDoc`, and `SourceItem` properties.

27.1.2 Important OLE Container Control Properties

OLE properties work in the same manner as properties for other types of Visual Basic controls, with one important difference: With OLE, approximately half of the over 40 properties are only available during runtime. This next section examines only a handful of these many properties.

The *AutoActivate* Property

`AutoActivate` determines the behavior of an OLE object at runtime. The syntax is as follows:

```
object.AutoActivate [= value]
```

Table 27.1 shows the settings and their descriptions.

Table 27.1 Settings of the *AutoActivate* Property

Setting	*Value*	*Description*
`vbOLEActivateManual`	0	The OLE object cannot be activated by the user.
`vbOLEActivateGetFocus`	1	The OLE object is activated whenever the OLE object has the focus.
`vbOLEActivateDoubleclick`	2	The OLE object is activated whenever the OLE object has the focus and the user double-clicks.
`vbOLEActivateAuto`	3	If the OLE Container control contains an object, the application that provides the object is activated based on the object's normal method of activation (either when the control receives the focus or when the user double-clicks).

The *Class* Property

The `Class` property specifies the class of an embedded OLE object. The list of classes available for the OLE Container control can be examined during design time by placing an OLE Container control and examining the drop-down combo box in the Properties windows under the `Class` property. Certain classes have unique version names. If you are using a Chart from Excel 95, for instance, the `Class` property for Excel Chart objects is `Excel.Chart.5`. The syntax for this property when using OLE or OLE automation is as follows:

```
object.Class [ = string]
```

The `object` component is an OLE Container control, while `string` is a string expression specifying the class name. For example, `Word.Document.6` stands for a Microsoft Word 6.0 document.

The *OLETypeAllowed* Property

The `OLETypeAllowed` property specifies the status of an OLE control, whether it can be a linked object, an embedded object, or either. The syntax is as follows:

```
object.OLETypeAllowed [ = value]
```

Table 27.2 shows the settings.

Table 27.2 Settings for the *OLETypeAllowed* Property

Setting	*Value*	*Description*
`vbOLELinked`	0	Linked. The OLE control contains a linked (but not embedded) object.
`vbOLEEmbedded`	1	Embedded. The OLE control contains an embedded (but not linked) object.
`vbOLEEither`	2	Either (the default). The OLE control can contain a linked or an embedded object.

The *SizeMode* Property

There are four `SizeMode` settings: `Clip`, `Stretch`, `AutoSize`, and `Zoom`. `SizeMode` settings determine how an image is displayed when it contains an object. The syntax is as follows:

```
object.SizeMode [ = value]
```

Table 27.3 shows the settings and their descriptions.

Table 27.3 The *SizeMode* Property Settings

Setting	*Value*	*Description*
`vbOLESizeClip`	0	The object is displayed in actual size, but the image is clipped if the borders of the OLE Control are smaller than the image (the default).

continues

Table 27.3 continued

Setting	*Value*	*Description*
`vbOLESizeStretch`	1	The object is sized to fit the size of the OLE Control.
`vbOLESizeAutoSize`	2	The OLE Control is sized to fit the size of the image.
`vbOLESizeZoom`	3	The object is resized to fill the OLE Container control as much as possible while still maintaining the original proportions of the object.

The *SourceDoc* Property

This property determines the file name to use when an OLE object is created, or the file name to use when a file is to be utilized. The syntax is:

```
object.SourceDoc [ = name]
```

The component `object` is an OLE Container control, while `name` is a string expression specifying a file name. For example, the location of an Excel worksheet might be specified by the `SourceDoc` property, as follows:

```
OLE1.SourceDoc = "C:\Fund1.xls"
```

The *SourceItem* Property

`SourceItem` is related to `SourceDoc` by specifying the data within the `SourceDoc` (or file) to be linked when the object is created. An example would be the specific cell range within an Excel worksheet to be linked into the Visual Basic application.

The syntax for `SourceItem` follows the same syntax as `SourceDoc`, namely:

```
object.SourceItem [ = string]
```

The `object` component is an OLE Container control, while `string` is a string expression specifying the data to be linked.

Objective 27.2 **Working with Object Linking and Embedding at Design Time**

The OLE Container control is used to display an OLE object obtained from a server application. The Visual Basic developer can create the object either at design time or at runtime. When creating the object during design time, a standard OLE dialog box simplifies the decision to link or embed. The default is to embed. To link, you must click the Link check box. When creating the object at runtime, programmed instructions are used to define the server application and to set object properties.

27.2.1 Object Linking at Design Time

We begin with an example showing how object linking (not embedding) is accomplished at design time. In the first exercise, you are asked to link an Excel 5.0 worksheet, prepared for you and stored on disk as Fund1.XLS.

Exercise 27.1 Creating a Link Between Visual Basic and Excel

A common use of spreadsheets is to compare different types of stocks and mutual funds. In this exercise, the stock information was entered into an Excel 5.0 worksheet. Tie this to Visual Basic to simplify the analysis of the data by creating a link between Visual Basic and Excel.

1 Open a new project and place an OLE Container control on the form. Retain its default `Name` property of `OLE1`.

2 The Insert Object dialog box should appear (see Figure 27.1). If it does not, right-click the OLE Container control and select Insert Object from the pop-up menu. Highlight Microsoft Excel Worksheet. Select the Create from File option button.

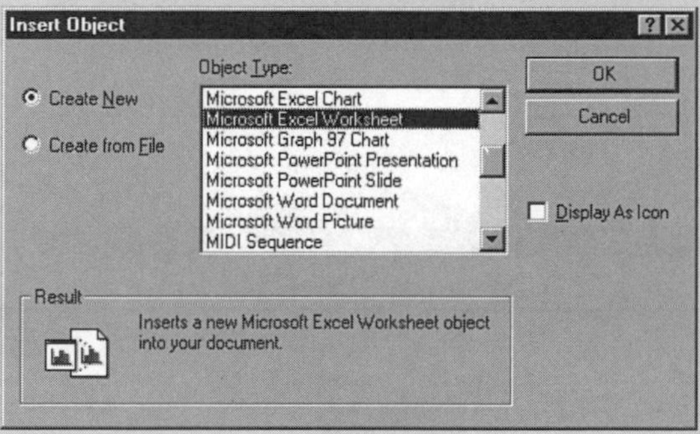

Figure 27.1

The Insert Object dialog box allows registered objects to be placed inside the OLE Container control.

3 Figure 27.2 shows the second part of theInsert Object dialog box. Set the path for the Microsoft Excel 5.0 file by performing the following steps:

  1 Make sure that the Create from File option button is checked.

  2 Type the directory in which the file is located and then click Browse.

  3 Find the file and click Insert. This inserts the file and the path name into the File location.

continues

4 Select Link. This step is important! You must tell Visual Basic that you want to link the file.

5 Click OK.

4 You will discover that the stored worksheet will not fit within the OLE space. To enlarge the space, open the Property window of the OLE1 control, and set `SizeMode` to `2 (AutoSize)`.

5 Add the caption and label as shown in Figure 27.3.

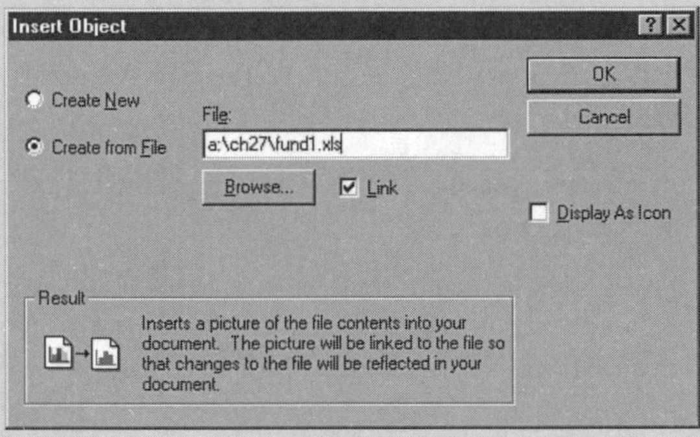

Figure 27.2

This dialog box appears when inserting an object into the OLE Container control from an existing file.

Linking a Worksheet

Performance of Mutual Funds

	1996	1995	1994	1993	1992	1991	1990	1989	1988	1984	Ave
Variable Fund	6.74	6.7	26.41	3.31	29.06	14.63	5.4	18.97	31.28	7.84	15.034
Income Fund	9.68	7.45	19.43	9.11	14.57	7.63	4.52	14.07	22.34	14.1	12.289
Encore Fund	3.19	3.67	6.53	8.44	9.39	7.5	6.81	6.88	8.46	10.8	14.34
Advisers Fund	9.9	6.39	18.38	5.72	5.33						9.144
Alger Fund	13.3	3.55	57.54	8.71	64.48	-3.35					24.035
Calvert Fund	8	7.61	16.4	4.18	20.69	11.75	6.78	3.31			9.84
Franklin Fund	7.58	7.66	15.87	10.2	8.9						10.048
Lexington Fund	10.9	3.22	-4.95	-15							-1.42
N & B Fund	6.79	9.54	29.73	-8.2	29.47	25.97	-4.9	14.94	30.3	-0.1	13.356
Scudder Fund	37.8	-3.1	11.45	-7.7	37.79	16.73					15.51
TCI Fund	10.3	-1.3	41.86	-1.2	28.7	-2.26	7.2				11.89

Figure 27.3

This display shows the OLE Container control with the linked Excel worksheet.

6 Run the application and do the following: Double-click the OLE control. This opens Microsoft Excel, allowing you to change values while working within the server application. Close Excel to return to the Visual Basic application.

7 Test the application. Make a change in the contents of the worksheet and save the revision. Close the application and run it again. Did the change you made remain?

27.2.2 Object Embedding at Design Time

Object embedding follows a similar procedure as object linking, except that the Link check box shown in Figure 27.2 is not checked. With object embedding, the link path is set; however, the source data is bound to the Visual Basic application. When the project file is saved, no other application has access to the data in the embedded object. Even so, it is possible to open the server in order to edit the object data.

Exercise 27.2 Editing Excel Data Within Visual Basic

A common use of spreadsheets is to chart the performance of different stocks and mutual funds. In the previous exercise, you entered the stock information into an Excel 5.0 worksheet. Within Visual Basic, you can edit the data in order to chart the performance of selected stocks. (Use the same Excel worksheet.) This sheet is stored on disk as Fund2.XLS.

1 Open a new project and add an OLE Container control to the form.

2 Highlight Microsoft Excel Worksheet and then Click Create from File.

3 Set the path where the Fund2.XLS file is located, but do not click Link.

4 Set the `SizeMode` of the OLE Container control to `AutoSize`.

5 Right-click to open the pop-up menu shown in Figure 27.4. This menu enables you to insert another object, delete the embedded object, edit the embedded object, or open the server application. Next, you will edit the worksheet to create a chart (see Figure 27.5).

6 Choose Edit to open the Excel server.

7 Choose Insert, Chart from the Excel server application and place the chart on a new sheet.

8 Follow the Chart Wizard to do the following:

   1 Click Chart Type 2: Line.

   2 Click Chart Sub-Type 2 for the Line Chart.

   3 Click Next.

   4 Edit the Range to read **=Sheet1!$A$1:$K$5**.

   5 Click Next.

continues

6 Click Next again.

7 Click As New Sheet, and click Finish.

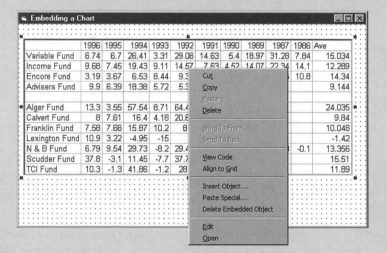

Embedding a Chart

	1996	1995	1994	1993	1992	1991	1990	1989	1987	1986	Ave
Variable Fund	6.74	6.7	26.41	3.31	29.06	14.63	5.4	18.97	31.28	7.84	15.034
Income Fund	9.68	7.45	19.43	9.11	14.57	7.63	4.52	14.07	22.34	14.1	12.289
Encore Fund	3.19	3.67	6.53	8.44	9.3					10.8	14.34
Advisers Fund	9.9	6.39	18.38	5.72	5.3						9.144
Alger Fund	13.3	3.55	57.54	8.71	64.4						24.035
Calvert Fund	8	7.61	16.4	4.18	20.6						9.84
Franklin Fund	7.58	7.66	15.87	10.2	8						10.048
Lexington Fund	10.9	3.22	-4.95	-15							-1.42
N & B Fund	6.79	9.54	29.73	-8.2	29.4					-0.1	13.356
Scudder Fund	37.8	-3.1	11.45	-7.7	37.7						15.51
TCI Fund	10.3	-1.3	41.86	-1.2	28						11.89

Cut
Copy
Paste
Delete
Bring To Front
Send To Back
View Code
Align to Grid
Insert Object...
Paste Special...
Delete Embedded Object
Edit
Open

Figure 27.4

Right-click over the OLE Container control to reveal this pop-up menu.

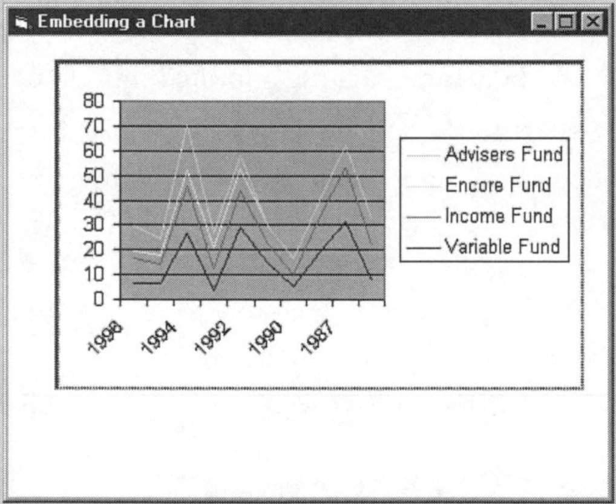

Figure 27.5

This is how the embedded chart will appear after editing.

9 Run the application. When running, click the right mouse button and choose Edit. Observe what happens. For example, click the chart area to open the Format Chart Area dialog box and then change the color.

10 Quit the application and run it again. Did the chart you made remain?

As an additional test, exit to the File Manager or DOS and check the size of the .FRX binary file for this exercise and the .FRX binary file for the last exercise. What conclusions can you draw?

Objective 27.3 Working with Object Linking and Embedding at Runtime

Runtime manipulation of the OLE Container control provides maximum flexibility, but it requires getting to know OLE Container control methods (see Table 27.4).

Table 27.4 OLE Container Control Methods

Method	*Description*
`CreateEmbed`	Creates an embedded object
`CreateLink`	Creates a linked object from the contents of a file
`Copy`	Copies the object to the system Clipboard
`Paste`	Copies data from the system Clipboard to an OLE Container control
`Update`	Retrieves the current data from the application that supplied the object, and displays that data as a picture in the OLE Container control
`DoVerb`	Opens an object for an operation, such as editing
`Close`	Closes an object and terminates the connection to the application that provided the object
`Delete`	Deletes the specified object and frees the memory associated with it
`SaveToFile`	Saves an object to a data file
`ReadFromFile`	Loads an object that was saved to a data file
`InsertObjDlg`	Displays the Insert Object dialog box
`PasteSpecialDlg`	Displays the Paste Special dialog box
`FetchVerbs`	Updates the list of verbs that an object supports
`SaveToOle1File`	Saves an object to the OLE file format

You can consult Visual Basic's online Help for the required and optional arguments for these methods, but several of the OLE Container control methods merit greater examination.

27.3.1 The *CreateLink* and *CreateEmbed* Methods

The `CreateLink` and `CreateEmbed` methods can be used to link or embed objects in an OLE Container control. For a linked object, a file (or `SourceDoc`) is a required argument. For an embedded object, the file can be an empty string if the class of the object is specified.

The *CreateLink* Method

The syntax for the `CreateLink` method is as follows:

```
object.CreateLink sourcedoc, sourceitem
```

Table 27.5 explains the syntax.

Table 27.5 The *CreateLink* Method Syntax

Component	*Description*
`object`	An OLE Container control
`sourcedoc`	A required argument specifying the file from which the object is created
`sourceitem`	An optional argument specifying the data within the file to be linked in the object

Be warned: the `CreateLink` method arguments will override any previous assignment to the OLE Container controls `SourceDoc` and `SourceItem` properties.

The *CreateEmbed* Method

The syntax for the `CreateEmbed` method is as follows:

```
object.CreateEmbed sourcedoc, class
```

Table 27.6 explains this syntax.

Table 27.6 The *CreateEmbed* Method Syntax

Component	*Description*
`object`	An OLE Container control.
`sourcedoc`	A required file name of a document used as a template for the embedded object. It must contain a zero-length string (`""`) if a source document is not specified.
`class`	An optional argument specifying the name of the class of the embedded object. It is ignored if you specify a file name for `sourcedoc`.

27.3.2 The *DoVerb()* Method and *ObjectVerbs* Property

Software components exposed by applications can specify *verbs* or operations that are specified by client applications. The `FetchVerbs` method, for example, changes the value of the `ObjectVerbs` property, a zero-based string array indicating the verbs (operations) the object in the OLE Container control currently supports. Common OLE Container control verbs are `Edit` and `Open`.

The *DoVerb()* Method

The `DoVerb()` method must specify the index value of the verb in the `ObjectVerbs` array you want activated. Otherwise, the default verb is executed. The syntax for this method is as follows:

```
Object.DoVerb[(verb)]
```

The `object` component is an object expression that evaluates to an object in the `Applies To` list, while `Verb`, which is optional, executes the `OLEObject` object within the RichTextBox control. If not specified, the default verb is executed.

The value of the `verb` argument can be one of the standard verbs supported by all objects or an index of the `ObjectVerbs` property array. Standard verbs for OLE objects, their values, and descriptions are listed in Table 27.7.

Table 27.7 Standard Verbs for OLE Objects

Constant	*Value*	*Description*
`vbOLEPrimary`	0	The default action for the object.
`vbOLEShow`	-1	Activates the object for editing. If the application that created the object supports in-place activation, the object is activated within the OLE Container control.
`vbOLEOpen`	-2	Opens the object in a separate application window. If the application that created the object supports in-place activation, the object is activated in its own window.
`vbOLEHide`	-3	For embedded objects, hides the application that created the object.
`vbOLEUIActivate`	-4	If the object supports in-place activation, `vbOLEUIActivate` activates the object for in-place activation and shows any user interface tools. If the object doesn't support in-place activation, the object doesn't activate, and an error occurs.

continues

Table 27.7 continued

Constant	*Value*	*Description*
`vbOLEInPlaceActivate`	-5	If the user moves the focus to the OLE Container control, `vbOLEInPlaceActivate` creates a window for the object and prepares the object to be edited. An error occurs if the object doesn't support activation on a single mouse click.
`vbOLEDiscardUndoState`	-6	Used when the object is activated for editing to discard all record of changes that the object's application can undo.

The following instruction would activate the object for editing:

```
OLE1.DoVerb vbOLEShow
```

This is often the default action for objects.

The *ObjectVerbs* Property

The `ObjectVerbs` property returns the list of verbs that an object supports. Its syntax is similar to the `DoVerb()` method syntax, and is written as follows:

```
object.ObjectVerbs(value)
```

Table 27.8 explains this syntax.

Table 27.8 The *ObjectVerbs* Property Syntax

Component	*Description*
`object`	An object expression that evaluates to an object in the `Applies To` list
`value`	A numeric expression indicating the element in the array
`ObjectVerbs`	A zero-based string array

Use the `ObjectVerbs` property along with the `ObjectVerbsCount` property to get the verbs supported by an object. These verbs are used to determine an action to perform when an object is activated with the `DoVerb()` method. The list of verbs in the array varies from object to object, and depends on the current conditions.

27.3.3 Object Linking at Runtime

Links to OLE objects can be created at runtime using the `CreateLink` method. Often runtime linking is executed as a form is loading. Exercise 27.3 demonstrates how to create a runtime link.

Exercise 27.3 Opening a Select Range in an Excel Worksheet During Runtime

Use the same worksheet as before, but with a new name: Fund3.XLS. Instead of placing the entire worksheet on the form, you are asked to limit your display to the cells R1C1:R5C11. Figure 27.6 shows the application you will create.

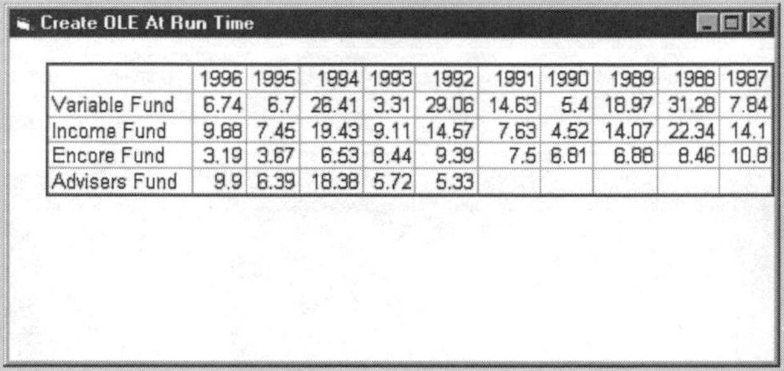

Create OLE At Run Time

	1996	1995	1994	1993	1992	1991	1990	1989	1988	1987
Variable Fund	6.74	6.7	26.41	3.31	29.06	14.63	5.4	18.97	31.28	7.84
Income Fund	9.68	7.45	19.43	9.11	14.57	7.63	4.52	14.07	22.34	14.1
Encore Fund	3.19	3.67	6.53	8.44	9.39	7.5	6.81	6.88	8.46	10.8
Advisers Fund	9.9	6.39	18.38	5.72	5.33					

Figure 27.6

This Excel `Worksheet` *object was linked to the OLE Container control in runtime.*

1 Begin a new project and add an OLE Container control to the form.

2 Click Cancel when the Insert Object dialog box appears.

3 Write the following `Form_Load()` procedure (setting the appropriate path):

```
Private Sub Form_Load()
   ChDrive App.Path
   ChDir App.Path

   OLE1.Class = "ExcelWorksheet"
   OLE1.SizeMode = vbOLESizeAutoSize
   OLE1.CreateLink "fund3.xls", "R1C1:R5C11"
End Sub
```

4 Run and test the application.

27.3.4 Object Embedding at Runtime

When an object is embedded by a client application, the data held by the Visual Basic application can be saved to disk in a binary file and then retrieved by using the `SaveToFile` and `ReadFromFile` methods. Visual Basic binary file input and output methods and statements must be used, however, in conjunction with these methods to read or write to the stored OLE file.

Exercise 27.4 Embedding Objects at Runtime

In this somewhat lengthy exercise, you will create an application that allows the user to work with an Excel chart or a WordPad document. In addition, the user can save the results of the work to disk and then reopen the file and work on it later.

1 Create the form you see in Figure 27.7. It includes a Common Dialog control (**CDlg1**), three buttons (**cmdSave**, **cmdOpen**, and **CmdClear**), two option buttons in a control array (**optObject**(0) and **optObject(1)**) inside a frame (**Frame1**), two menu objects (**mnuFile** and **mnuExit**), and an OLE Container control (**OLE1**).

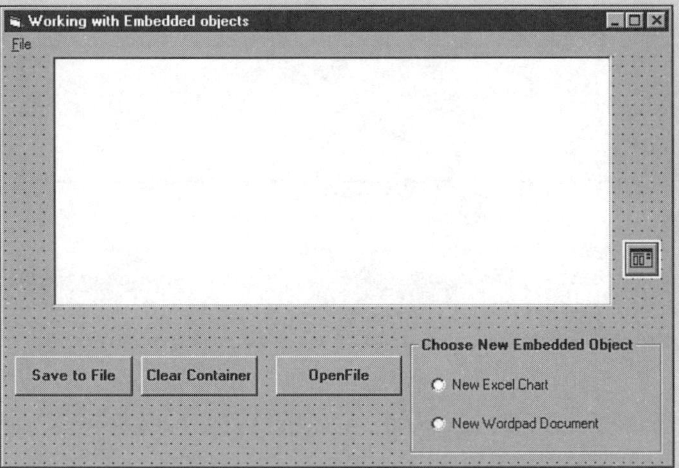

Figure 27.7

The interface for creating embedded objects at runtime includes an OLE Container control.

2 When adding the OLE Container control, select cancel when the Insert Object dialog box appears. Set the `OLETypeAllowed` property to `1` (`Embedded`), and set the `SizeMode` property to `2` (`AutoSize`).

3 Write the `Form_Load()` event, which sizes the form and positions controls on the screen. (You might have to adjust some of these settings if your screen is too small.)

```
Private Sub Form_Load()
  Dim xpos As Single, ypos As Single
  xpos = Screen.TwipsPerPixelX
  ypos = Screen.TwipsPerPixelY
  Me.Width = xpos * 640
  Me.Height = ypos * 480
  Me.Left = (Screen.Width - Height) / 2
  Me.Top = 1000
  cmdClear.Top = 5600
```

```
    cmdSave.Top = 5600
    cmdOpen.Top = 5600
    Frame1.Top = 5000
  End Sub
```

4 Write the `mnuExit_Click()` event:

```
Private Sub mnuExit_Click()
  End
End Sub
```

5 Write the `cmdClear_Click()` event, which clears the OLE Container control and sizes it:

```
Private Sub cmdClear_Click()
  OLE1.Delete
  OLE1.SizeMode = vbOLESizeAutoSize
  OLE1.Width = 6800
  OLE1.Height = 3000
End Sub
```

6 Write the `optObject_Click()` event to embed either an Excel chart or a WordPad document inside the Container control:

```
Private Sub optObject_Click(Index As Integer)
   Call cmdClear_Click  ' clear the contents
   Select Case Index
    Case 0
      OLE1.CreateEmbed "", "Excel.Chart.5"
      OLE1.SizeMode = vbOLESizeAutoSize
      OLE1.DoVerb vbOLEPrimary
    Case 1
      OLE1.CreateEmbed "", "wordpad.document.1"
      OLE1.SizeMode = vbOLESizeAutoSize
      OLE1.DoVerb vbOLEOpen
   End Select
End Sub
```

7 Write the `cmdSave_Click()` event to save the contents of the OLE Container control into a binary file:

```
Private Sub cmdSave_Click()
  Dim fnum As Integer
  Dim fname As String
  fnum = FreeFile
  CDlg1.DialogTitle = "Save Stored OLE Object"
  CDlg1.Filter = "OLE data (*.dat)|*.dat|All Files (*.*)|*.*"
  CDlg1.filename = "mydata.dat"
  CDlg1.ShowSave
  fname = CDlg1.filename
  Open fname For Binary As #fnum
  OLE1.SaveToFile fnum
  Close #fnum
End Sub
```

Observe how the `SaveToFile` method uses the file number of an open binary file as its argument.

continues

8 Write the `cmdOpen_Click()` event to open an existing binary file and place it inside the OLE Container control.

```
Private Sub cmdOpen_Click()
  Dim fname As String
  Dim fnum As Integer
  fnum = FreeFile
  ChDir App.Path
  CDlg1.DialogTitle = "Open Stored OLE Object"
  CDlg1.Filter = "OLE data (*.dat)|*.dat|All Files (*.*)|*.*"
  CDlg1.ShowOpen
  fname = CDlg1.filename
  Open fname For Binary As #fnum
  OLE1.ReadFromFile fnum
  Close #fnum
End Sub
```

9 Test the program by inserting an Excel `Chart` object (see Figure 27.8). Using the commands of the Excel server application, change the default chart to a three dimensional chart (choosing Chart, Chart Type). Perform the following steps to test the design:

1 Save the data to disk by clicking the Save to File button (`cmdSave`).

2 Clear the contents of the OLE Container control by clicking the Clear Container button (`cmdClear`).

3 Reopen the data you saved (mydata.dat) by clicking the OpenFile button (`cmdOpen`).

4 Clear the contents of the OLE Container control again and insert a new WordPad document object. Observe how the contents of the OLE Container control change as you type text in the WordPad document.

5 Click File, Update inside WordPad.

6 Exit WordPad, and save the contents to disk (as **mydata2.dat**).

7 Clear the contents of the OLE container again, and reopen the saved WordPad document.

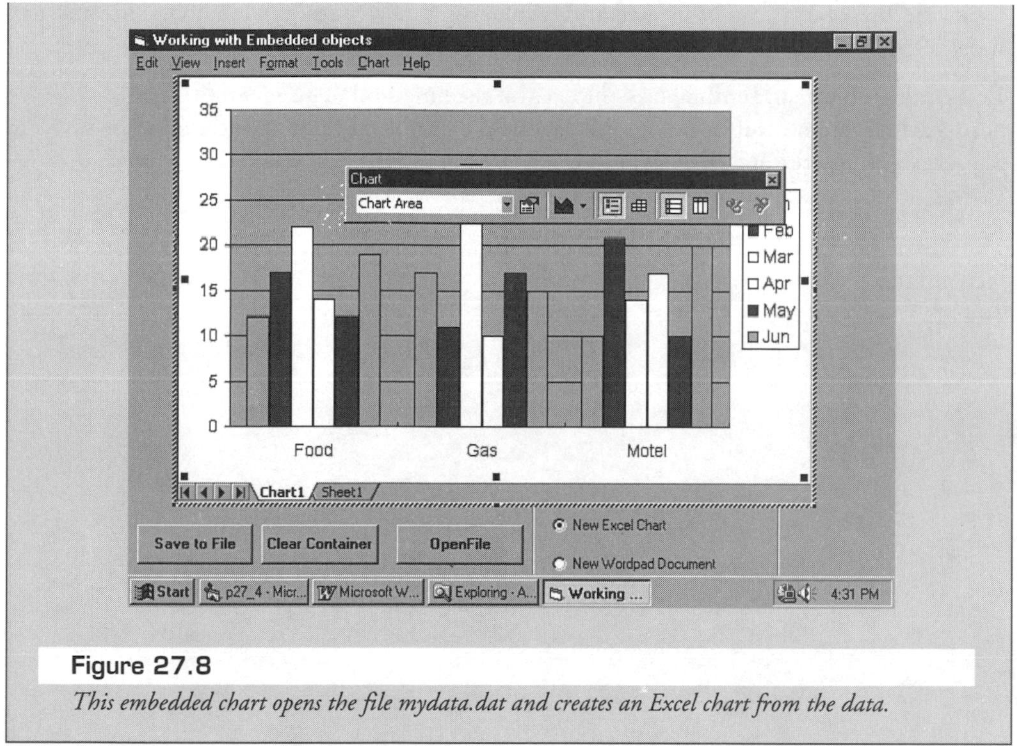

Figure 27.8

This embedded chart opens the file mydata.dat and creates an Excel chart from the data.

Chapter Summary

OLE is the abbreviation for *object linking and embedding,* but actually means object linking *or* object embedding. Object linking means to link the client application to the server application, with the server storing the data; object embedding means to link the client application to the server application, with the client storing the data. With object embedding, the link path is set; however, the source data is bound to the Visual Basic application.

Important OLE Container control properties include `AutoActivate`, `Class`, `OLETypeAllowed`, `SizeMode`, `SourceDoc`, and `SourceItem`. `AutoActivate` determines the behavior of an OLE object at runtime. `Class` specifies the class of the embedded object. `OLETypeAllowed` specifies the status of the OLE control. `SizeMode` determines how an image will be displayed. `SourceDoc` and `SourceItem` determine the file name to use when an OLE object is created and the data within the document to be linked.

The OLE Container control uses a number of methods to operate on the contents of the control. These properties are used at runtime to specify OLE behavior. These methods include `CreateEmbed`, `CreateLink`, and `DoVerb()`. The `CreateEmbed` and `CreateLink` methods can be used to embed or link objects inside an OLE Container control. The `DoVerb()` method is used to specify the OLE action you want to activate.

Links to OLE objects can be created at runtime using the `CreateLink` method. This linking most often occurs when the form is loading.

Embedding objects in runtime uses the `CreateEmbed` method. The `SavetoFile` and `ReadFromFile` methods then permit the data held by a Visual Basic application to be saved to a disk in a binary file and then retrieved.

Skill-Building Exercises

1. Add the OLE Container control to a form and link (not embed) the OLE control to Windows Paintbrush in design time. This will add a `PBrush` class to your project. Move to the Windows directory and load a bitmap image. Run your project. Double-click the bitmap to link to the Paintbrush application. Edit the bitmap, and return to the Visual Basic application.

2. Add the OLE Container control to a form and embed the fund2.XLS worksheet to the control in design time. Edit the worksheet to create a bar chart consisting of the names of the last seven funds shown on the worksheet (Alger to the TCI funds) and their average (24.035 to 11.89). Select these ranges, press the Ctrl key, and drag. Use Excel Help if you have problems. Run the application to view the bar chart comparisons of the averages of the seven funds.

3. Add two OLE Container controls to a form. Using code in the `Form_Load()` event, set the left properties of the controls to `200` and `2000`; set the top properties to `800`. Also use code in the `Form_Load()` event to create a runtime link between these two OLE Container controls and the Fund3.XLS worksheet. Size and fill the first OLE control with the names of the stocks stored in rows 7 through 13 of column 1. Size and fill the second OLE control with the stock performance averages also stored in rows 7 through 13 of column 12.

4. Using the `CreateEmbed` method, create an embedded instance of the Fund3.xls worksheet. Place the code in the `Form_Load()` event.

5. Extend the preceding exercise by allowing the user to save changes made to the worksheet to disk and then retrieve the stored OLE container data from disk. Also permit the user to remove data from the OLE Container control. Test the capabilities of the application by adding a chart to fund3.xls and saving it to disk as mysheet.dat. Then, open the saved data and display it inside the OLE Container control.

28

Building Internet Applications with ActiveX Documents

This final chapter introduces the newest innovation in ActiveX technology: *ActiveX documents*. ActiveX documents have some of the attributes of a Visual Basic form, but they are more akin to an application. ActiveX documents consist of a `UserDocuments` interface that holds controls and code. This interface can be read or activated by an executable opened by an ActiveX container. ActiveX containers include Visual Basic, the Microsoft Office Binder, and, importantly, Internet Explorer.

ActiveX documents differ from regular applications in that they cannot be run independently. They can be read or activated only by an executable (either an .EXE or .DDL), which in turn must be run inside an ActiveX container. ActiveX documents have a .VBD extension. During design, you place ActiveX documents within an ActiveX document project. When you compile your project, both the executable and all associated ActiveX document files are created for you by Visual Basic.

ActiveX documents resemble other documents with which you are familiar. A Microsoft Word document, for instance, is stored to disk with a .DOC extension. It can be read or edited only by the Microsoft Word executable (Winword.exe). An ActiveX document has a .VBD extension and can be read only by the executable created by your ActiveX document project. The chief difference between the executable you create in Visual Basic and a standard Windows executable is that the executable created by your application requires an ActiveX container, such as Internet Explorer, in order to run.

The capability of Internet Explorer to activate ActiveX documents has important implications for the emerging world of Internet applications and *intranets* (networks within organizations), which use the Internet as the basis for communication. Because the technology is new and the market response to the technology is still unknown, this chapter introduces ActiveX documents and addresses how they are activated using the Internet Explorer.

The chapter illustrates how an ActiveX document is constructed and introduces concepts for study. We cannot cover all the topics necessary for creating robust ActiveX document projects. However, you will gain an understanding of what ActiveX documents are built, how to transform a Visual Basic application into an ActiveX document, and how to retrieve data from a database using an ActiveX document. When you complete this chapter, you will be able to

- understand key factors affecting client/server design,
- create an ActiveX document,
- know the files that make up an ActiveX document project,
- observe ActiveX document classes in the Windows Registry,
- use the Hyperlink object to navigate among ActiveX documents,
- use a hyperlink to link to a Web page,
- add a form to an ActiveX document,
- convert Visual Basic applications into ActiveX document applications.

Chapter Objectives

The objectives of this chapter include an introduction to acquaint you with some of the design issues surrounding the use of ActiveX technology and the Internet. Following this, you complete a single document design, a multiple-document design, and combined user document-form design. You then convert two Visual Basic applications developed for earlier chapters, including a database application. The final exercise simulates Internet application database processing. The five objectives are as follows:

1. Understanding basic Internet application concepts
2. Creating an ActiveX document
3. Navigating among ActiveX documents
4. Adding a form to an ActiveX `UserDocument`
5. Using the ActiveX Document Migration Wizard

To test the exercises in this chapter, you will need a copy of Microsoft Explorer. This software can be downloaded from the Microsoft Web site at **http://www.microsoft.com**. An alternative is to download the Microsoft Personal Server, which allows you to run Internet Explorer on your personal computer without linking to a network provider. Personal Server is supplied with such products as Microsoft FrontPage.

Objective 28.1 Understanding Basic Internet Application Concepts

Visual Basic is a superb language for rapid development of Windows programs. Its market success, however, is attributed to its use as a development language for stand-alone applications, and more recently for client/server database applications.

We use the term *client/server* in the sense of hardware and architecture, that is, a client machine on a desk running a client application and receiving services from a network server. The services in a network environment include file and print sharing as well as application services. In the past, the primary network service accessed by a Visual Basic client application has been receiving the processing results from large-scale database management systems. Well-known DBMSs include Microsoft SQL Server, Oracle, Sybase SQL Server, Interbase, and Informix. However, you use Microsoft Access as the DBMS for this chapter.

We also use the term *client/server* with respect to software components, that is, such software components as an ActiveX DLL and ActiveX EXE written to service client applications. This mixture of the terms can be somewhat confusing. All of the software component servers used for this book are located on the local (client) machine. The software component servers we have worked with include in-process ActiveX DLLs and out-of-process ActiveX EXEs. These software components also can be placed on the network server, adding to the complexity (and flexibility) of client/server applications.

The explosion of the Internet and the use of Internet standards within enterprises in the design of intranets is rapidly changing how Visual Basic is used for client/server application development. What follows is a brief description of how client/server application design has evolved in recent years. This will provide some context by which to understand ActiveX documents.

28.1.1 Visual Basic and Two-Tier Client/Server Applications

A typical Visual Basic application in an enterprise is a database application. Queries, whether update queries (`INSERT`, `DELETE`, `UPDATE`, and so on) or read-only `SELECT` queries, are typically submitted by a Visual Basic application on the client computer to a DBMS hosted on a network server. If data is returned by the query, it is often presented to the user on-screen or processed and formatted for printed output.

This *two-tier model* describes most client/server applications in use today. In this environment, the application designer has important choices to make about how computing is to be distributed. For instance, will calculations on retrieved data be performed by the client application or by a stored procedure (stored code) on the server? How will new rows of data be added to the database? Will they be added one by one as the client application finishes entering the information required to complete a row of data, or will the data be stored temporarily on the client machine and multiple rows updated in a batch operation? Where will validation of field values take place? Should validation occur on the client side or should a stored network procedure check input and return an error for invalid field (column) values?

28.1.2 Software Components and the Internet

Simultaneous with the evolution of client/server designs, the Internet and the World Wide Web (WWW) have exploded. The Internet and its standard protocols hold an enormous appeal to information managers struggling with multiple platforms in their enterprises (for example, Windows, UNIX, and Macintosh). Just a short while ago, debates raged about the best combination of operating system platform, back-end DBMS, and front-end development tools to rapidly deploy client/server applications. These debates have essentially been silenced. There appears very little doubt that delivery of distributed information in the enterprise and among enterprises will be dominated by Internet compliant applications. New debates, however, have taken their place. These debates center on which software component model best suits Internet-compliant client/server design.

Using Internet protocols and World Wide Web technology for client/server applications holds additional appeal because it suggests a way of lowering the maintenance costs of complex client machines. The typical desktop machine hosting a suite of office productivity applications can prove very costly to maintain. By moving more data storage and application program storage to the server, client machines can become “thinner.” For performance reasons, software components can be stored on the hard disk of the client machine, or loaded into the local machine’s memory from the server when needed.

Transforming the Web into the main platform for client/server applications poses some significant challenges. Originally, Web pages were composed of static text and images that could be read by a Web browser. HTML (Hypertext Markup Language), the standard defining Web documents, evolved to permit limited interactivity between the reader of the Web document and the document itself. Form fields describing such objects as buttons, text boxes, and list boxes were added to the HTML standard. These built-in HTML objects allow the end user to submit information to the Web server and have data displayed in a fairly primitive interface.

Standard HTML, however, cannot provide the rich user interface of a program written specifically for a native platform, such as Windows or a graphical UNIX environment. To overcome these limits, the HTML standard allows objects and applets to be referenced in the HTML page. If the browser can read the code, the standard allows the code to be executed on the local machine and displayed within the browser. Such code includes ActiveX components and Java *applets.*

28.1.3 ActiveX Documents

ActiveX documents are really something entirely different from HTML or software components embedded within HTML pages. ActiveX documents contain no HTML whatsoever. Web browsers that have ActiveX capability can read ActiveX documents. The only Web browser at the time of this writing that also serves as an ActiveX container is Microsoft Internet Explorer. You can navigate from an HTML document to an ActiveX document using Internet Explorer and back again, but when you load the ActiveX document, you are no longer really Web browsing: You are executing a software executable within an ActiveX container.

The *UserDocument* Object

The base object of an ActiveX document is the `UserDocument`. As an object, it is supported by properties, methods, and events. The `UserDocument` object resembles a standard Visual Basic `Form` object with several exceptions. For example, the `UserDocument` object has most, but not all, of the events that are found on a `Form` object. The events present on a form that are not found on the `UserDocument` include `Activate()`, `Deactivate()`, `LinkClose()`, `LinkError()`, `LinkExecute()`, `LinkOpen()`, `Load()`, `QueryUnload()`, and `Unload()`.

Events present on the `UserDocument`, but not found on a `Form` object include `AsycReadComplete()`, `EnterFocus()`, `ExitFocus()`, `Hide()`, `InitProperties()`, `ReadProperties()`, `Scroll()`, `Show()`, and `WriteProperties()`. Observe that you cannot load or unload a `UserDocument`, although you can show and hide one. To load a `UserDocument`, you must use the `Initialize()` event (which also is used for a `Form` object). The syntax for this event is

```
object_Initialize()
```

The following is an example:

```
UserDocument_Initialize()
```

Another difference between a `Form` object and a `UserDocument` is that you cannot place embedded objects (such as an Excel or Word document) or an OLE Container control on a `UserDocument`.

UserDocument Objects and the Internet

When Internet Explorer finds a `UserDocument`, it checks to see if the document is currently registered on the client machine. If the `UserDocument` document is not registered on the client machine, or if the `UserDocument` document on the server is a newer version, the `UserDocument` document and any components the `UserDocument` document requires must be downloaded to the client machine and registered in the client machine's Registry. The `UserDocument` document then executes within Internet Explorer.

For software development and deployment, this is almost a revolutionary capability. In one sense, using Internet Explorer, you have a *universal client application* or at least a universal client application container. All new client-side applications can be controlled from one source: ActiveX documents located on the server. Upgrades to client-side applications can now be automatic.

Consider the implications for a far-flung, worldwide enterprise. The distribution costs for software plummet and upgrades and improvements to the software are automatic. If an ActiveX document upgrades one of its components, such as an improved Grid control for displaying result sets from queries, the upgraded component is automatically installed whenever the user next browses to the ActiveX document.

Should ActiveX documents become widespread, any piece of software can be distributed anywhere on the Web at the click of a button. This is a powerful vision for distributed computing.

28.1.4 ActiveX Documents and Security

The popularity of ActiveX documents will depend in large measure on whether security concerns about ActiveX components can be resolved. At present, ActiveX technology is primarily used within corporate intranets on secured networks behind firewalls, and not on the public Internet. Because ActiveX components are executable code and are downloaded onto user machines, they, and the ActiveX components they contain, have the capability of creating mischief.

Microsoft has developed a digital signature for ActiveX controls that uses a third-party certification authority to vouch for the component's code and to ensure it has not been tampered with. The market acceptance of digitally signed ActiveX controls, however, is still uncertain.

Objective 28.2 Creating an ActiveX Document

An ActiveX document is a component. You create the component by completing an ActiveX document .EXE project within Visual Basic. The project executable, an .EXE, defines the type library for the documents that can be read by the executable. When the ActiveX project is compiled, so are the associated ActiveX documents included in the project.

Each ActiveX document is given its own class ID and registered in the Registry of your computer. When the ActiveX document is placed on a server, through a program created by the Setup Wizard, a .CAB (*cabinet*) file is created. The .CAB file instructs Internet Explorer to download the document and its components and register them on the end user's machine.

With this brief background for ActiveX document technology, let's build such a document and view it.

Exercise 28.1 Building an ActiveX Document

In this exercise, you will build an ActiveX document, compile it to create an executable, then read the document created by the executable from within Internet Explorer. As you proceed with the exercise, pay close attention to the file extensions that make up an ActiveX project. You need Internet Explorer installed on your machine to complete this exercise. Figure 28.1 shows the design interface you are asked to create.

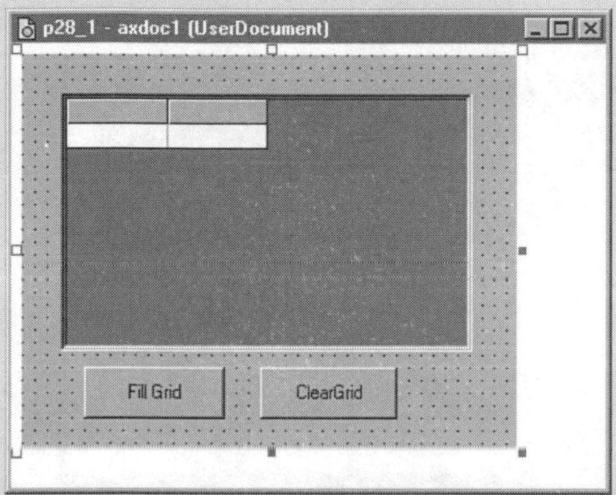

Figure 28.1

This ActiveX EXE interface contains the MSFlexGrid control and two command buttons.

1 Choose File, New Project. When the New Project dialog box appears, select ActiveX Document EXE.

2 In the Project Explorer, you will observe a User Documents folder with one `UserDocument` included in the folder with a default name of UserDocument1 (UserDocument1), as shown in Figure 28.2. Double-click the folder, UserDocument, if you cannot see UserDocument1(UserDocument1).

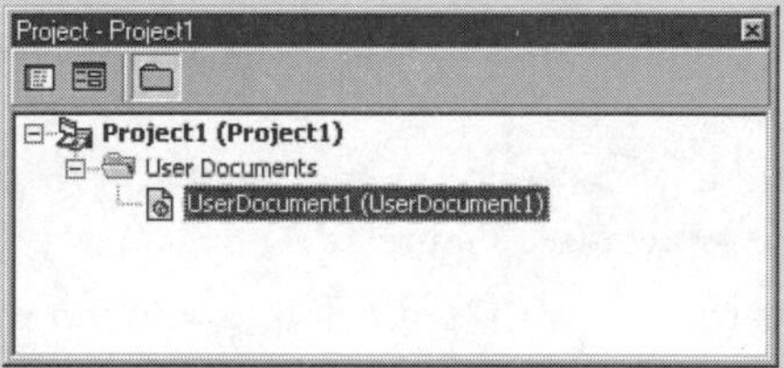

Figure 28.2

Click the Project1 icon if the UserDocument1 icon is not visible.

3 Choose View Object (click the icon) in the Project Explorer to examine the Project1-UserDocument1 interface. Press F4 and change its `Name` property to **axdoc1** (see Figure 28.3).

4 Save this file to disk as axdoc1.dob. `UserDocuments` in design time have a .DOB extension. This is similar to Visual Basic forms. A .DOB file can be read in a text editor similar to the way a form's .FRM file is read.

continues

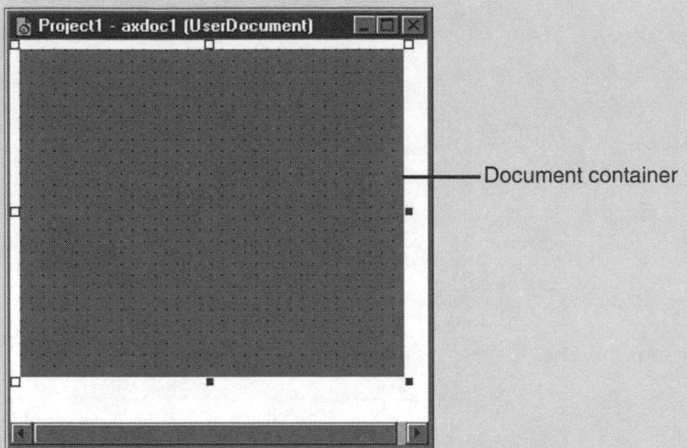

Figure 28.3

The shaded area of the interface is the document container; it can be sized and moved within the project window.

5 Change the `Name` property of the project to **p28_1** and save the project to disk as **p28_1.vbp**. The project `Name` property will now become the name of the type library in which the ActiveX documents will be placed and defined.

6 Choose Project, Components and click the MSFlexGrid control. This will add the control to the toolbox.

7 Add a MSFlexGrid to the `UserDocument` and size it so it has a width of about 4,000 twips and a height of 2,500 twips. Change the `Name` property of the Grid control to **`grid1`**.

8 Add two command buttons (**`cmdFill`** and **`cmdClear`**) to the `UserDocument`, and add the captions as shown (see Figure 28.1).

9 Add code that fills the grid with four columns and ten rows of numbers and also clears the grid.

```
Private Sub cmdClear_Click()
  grid1.Clear
End Sub

Private Sub cmdFill_Click()
  Dim i As Integer, j As Integer

  grid1.Cols = 4
  grid1.Rows = 10
  grid1.FixedCols = 0
  grid1.FixedRows = 0
  For i = 0 To 3
    For j = 0 To 9
       grid1.Col = i
       grid1.Row = j
       grid1.Text = (i + 1) * (j + 1)
```

```
    Next
  Next
End Sub
```

10 *Important:* Choose File, Make p28_1.exe to compile the project. *Do not run the project at this time.*

11 Examine the files that have been created in the root directory of the application (see Figure 28.4). You should see the following files:

- **axdoc1.dob**—The design-time file for editing the `UserDocument`
- **axdoc.vbd**—The compiled ActiveX document, which can be read by an ActiveX container application, such as Internet Explorer
- **p28_1.vbp**—The project file to open when you want to change the project
- **p28_1.exe**—The executable file that is registered on your computer

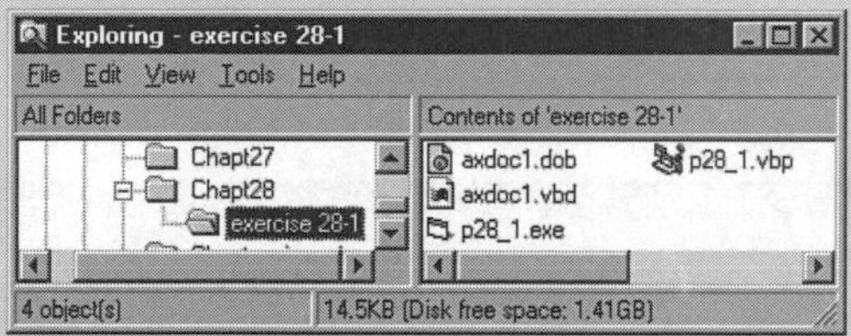

Figure 28.4

Compiling an ActiveX document project leads to the creation of four files.

12 Run the Regedit.exe program. (Use Microsoft Explorer and choose Tools, Find to locate this program.) Under the HKEY_CLASSES_ROOT folder you should see an entry p28_1.axdoc1 specifying that the ActiveX document is now registered for your computer (see Figure 28.5).

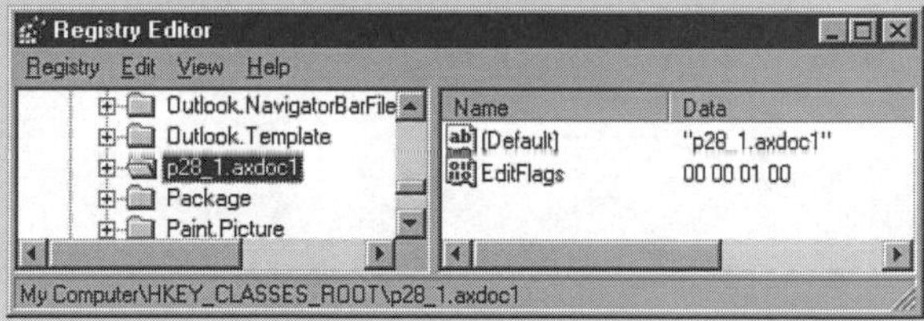

Figure 28.5

ActiveX document applications must be registered for your computer.

continues

13 Run Internet Explorer. You do not need to connect to an Internet server provider if you have installed Personal Web Server. Rather, click to display your home page. If you are connected to an Internet service provider, press the Stop button on Internet Explorer because you do not need the service provider to build this application.

14 Choose File, Open in Internet Explorer. When the Open dialog box appears, select Browse. *Important:* When the Browse dialog box appears, be sure to select All Files under the combo box named Files of Type. Move to the folder that contains the saved project.

15 Open the file axdoc1.vbd. If you have trouble opening the file, make sure Visual Basic is not in Run mode. You might need to exit Visual Basic entirely.

16 The ActiveX document should be displayed within the browser, as shown in Figure 28.6. Test the program by clicking both buttons. If you encounter an error, you must return to Visual Basic, make the correction, and *recompile* the executable before testing it again from within Internet Explorer.

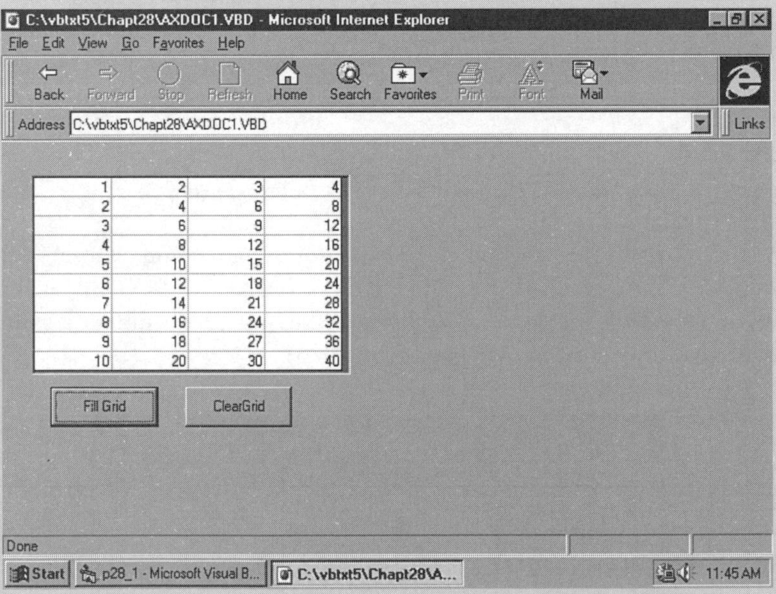

Figure 28.6

Internet Explorer allows you to load an ActiveX .VBD file once it knows the address.

With Internet Explorer running, choose File, Close if you plan to quit at this time. In addition, before you shut down your computer, press Shift+Alt+Del to examine the Close dialog box. If an ActiveX document program is still running, click it and click End Task.

Objective 28.3 Navigating Among ActiveX Documents

Every ActiveX document is a separate software component. As such, it must be registered on the user's computer when it is downloaded. A project can contain multiple `UserDocuments` defining ActiveX documents. These documents can be navigated similarly to the way one navigates among Web documents using hyperlinks.

Each `UserDocument` has a `Hyperlink` property that returns a `Hyperlink` object. The `Hyperlink` object has three navigational methods: `GoBack`, `GoForward`, and `NavigateTo`. By specifying the path (URL) of the ActiveX document, you can navigate among ActiveX documents with the `NavigatoTo` method.

28.3.1 The *GoBack* Method

The `GoBack` method executes a hyperlink jump back in the history list, for which the syntax is simply

```
object.Hyperlink.GoBack
```

The `object` in this syntax is a container that supports OLE hyperlinking. `GoBack` signifies that the container will jump to the location that is back in the history list. However, if the object is in a container that does not support OLE hyperlinking, then this method will raise an error.

28.3.2 The *GoForward* Method

The `GoForward` method executes a hyperlink jump forward in the history list, provided there is one. Like the `GoBack` method, the syntax is limited to

```
object.Hyperlink.GoForward
```

The `object` must support OLE hyperlinking.

As a general rule, use error checking when using `GoBack`, `GoForward`, or `NavigateTo`. As an example, the following code will move to the next `UserDocument` in the history list, if there is one:

```
Private Sub cmdGoForward_Click()
     On Error GoTo noDocInHistory
     UserDocument.Hyperlink.GoForward
     Exit Sub
noDocInHistory:
     Resume Next
End Sub
```

28.3.3 The *NavigateTo* Method

The `NavigateTo` method is used to jump to a URL. For example, the following code presumes an ActiveX document named axdMyDoc exists:

```
UserDocument.Hyperlink.NavigateTo "c:\mydocs\axdmydoc.vbd"
```

The syntax for this method is as follows:

```
object.Hyperlink.NavigateTo Target [, Location [, FrameName]]
```

Table 28.1 explains this syntax.

Table 28.1 The *NavigateTo* Method Syntax

Component	*Description*
`object`	An application that is registered as supporting hyperlinking is started to handle the request.
`Target`	A string expression specifying the location to jump to. This can be a document, such as axdmydoc.vdb, or an URL.
`Location`	A string expression specifying the location within the URL specified in `Target` to jump to. If `Location` is not specified, the default document will be jumped to.
`FrameName`	A string expression specifying the frame within the URL specified in `Target` to jump to. If `FrameName` is not specified, the default frame will be used.

To link to an URL, either type in or code the URL. For example, the following code will jump to the URL placed in the `txtURL` text box:

```
Private Sub cmdNavigateTo_Click()
  Hyperlink.NavigateTo txtURL.Text
End Sub
```

To jump to the computer science Web site at the University of Oregon, type in

```
http://cs.uoregon.edu
```

The hyperlink will navigate to this Web location.

Exercise 28.2 Using the *NavigateTo* Method of the *Hyperlink* Object

This exercise asks you to use the `NavigateTo` method to jump move one `UserDocument` (one .vbd file) to a second `UserDocument` (a second .vbd file). The names of the two `UserDocument` files are axdoc1a.vbd and axdoc2.vbd, respectively.

1 With Exercise 28.1 open, save axdoc1.dob to disk as **axdoc1A.dob**. Save p28_1.vbp to disk as **p28_2.vbp**.

2 Open project p28_2 and change the `Name` property of the project to **p28_2**. Change the `Name` property of the user document to **axdoc1A**.

3 Add a label to axdoc1A, and name the label `lblJump`. Change the caption of the label to **Go there**. Make the label look like a hypertext link on a Web page by changing the label `ForeColor` to royal blue and add an underline attribute to the font (see Figure 28.7).

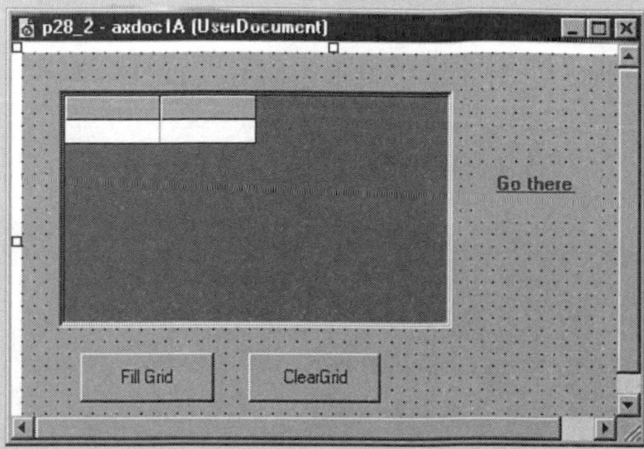

Figure 28.7

Adding a hypertext link to an ActiveX document requires a label and an associated label `Click()` event.

4 In the `lblJump_Click()` event, add the following code:

```
Private Sub lblJump_Click()
   On Error GoTo noDocFound
   UserDocument.Hyperlink.NavigateTo App.Path & "\axdoc2.vbd"
   Exit Sub
noDocFound:
   Resume Next
End Sub
```

5 Insert a new `UserDocument` into the project by choosing Project, Add User Document. Save the new `UserDocument` to disk as **axdoc2.dob**. Change the `Name` property of the document to `axdoc2`. Save the entire project.

6 Add two labels to axdoc2 so that it appears similar to Figure 28.8. The upper label (`lblName`) has its alignment centered and its `BorderStyle` property set to `FixedSingle`. The caption to this label is set in code. The lower label (`lblJumpBack`), has the caption **Return**, has `ForeColor` set to royal blue, and `Font` set to underlined.

7 In the `UserDocument_Initialize()` event add the following code:

```
Private Sub UserDocument_Initialize()
   lblName.Caption = UserDocument.Name
End Sub
```

continues

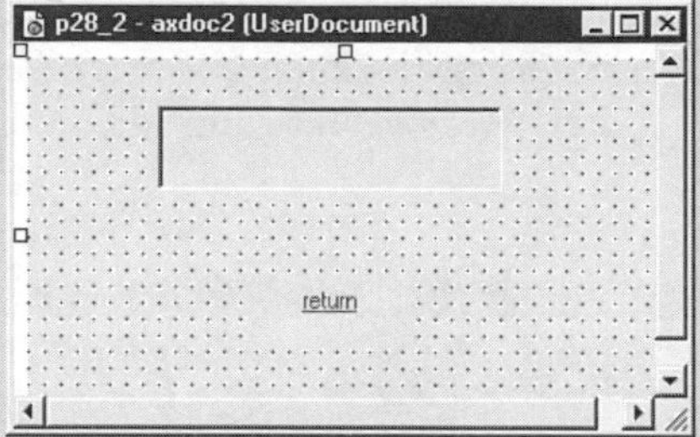

Figure 28.8

This second ActiveX document contains a return link back to the first ActiveX document.

Remember that although the `UserDocument` appears similar to a form, it is a component and has an `Initialize()` event rather that a `Load()` event.

8 In the `lblJumpBack_Click()` event, add the following code:

```
Private Sub lblJumpBack_Click()
   On Error GoTo noDocFound
   UserDocument.Hyperlink.NavigateTo App.Path & "\axdoc1A.vbd"
   Exit Sub
noDocFound:
   Resume Next
End Sub
```

9 Compile the executable, but do not run it. Observe the files created in the application directory and the new entries made to the Registry.

10 Test the application by launching Internet Explorer and opening the file axdoc1A.vbd. Test the design by navigating between the two ActiveX documents. Once this is done, click the back and forward arrows located on the Microsoft Explorer menu bar.

With Internet Explorer running, choose File, Close if you plan to quit at this time. Otherwise, continue with the next exercise.

Exercise 28.3 Linking to an URL

This brief exercise is similar to the last, except that it asks you to link to an URL rather than to another ActiveX user document. Figure 28.9 shows the Internet Explorer interface you are asked to create.

1 Open a new ActiveX document .EXE project, and open the UserDocument1 interface.

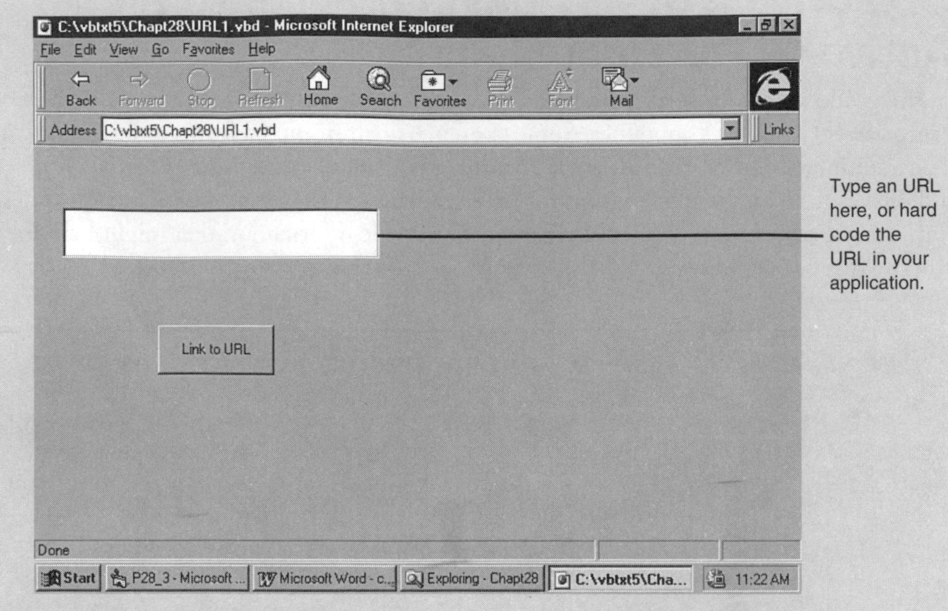

Figure 28.9

This interface allows you to type in an URL and to navigate to this location.

2 Change the name of the project to **p28_3**. Change the name of the `UserDocument` to **URL1**.

3 Add a command button (**`cmdNagivateTo`**) and a text box (**`txtURL`**) to the document. Change the caption on the command button to **Link to URL**, and remove the text from the text box.

4 Add the `cmdNagivateTo_Click()` event:

```
Private Sub cmdNavigateTo_Click()
  txtURL.Text = "C:\WEBSHARE\WWWROOT\DEFAULT.HTM"
  Hyperlink.NavigateTo txtURL.Text
End Sub
```

In this code, substitute the `txtURL.Text` statement to link to an URL address of your choosing. We decided to link to the default home page on our computer.

5 Save the `UserDocument` to disk as **axURL1.dob** and save the project as **P28_3**.

6 Make the P28_3 executable (again do not run the program).

7 Run Internet Explorer and open the file URL1.vdb. Click the command button to navigate to the URL address indicated by your code.

With Internet Explorer running, choose File, Close if you plan to quit at this time. Otherwise, continue with the next section.

Objective 28.4 Adding a Form to an ActiveX *UserDocument*

In this section you are asked to add a modal form to an ActiveX `UserDocument`. Remember, a `UserDocument` requires a container. Some containers permit modal forms to be displayed. The addition of a form to an ActiveX document is a simple procedure. What is more difficult is deciding on good reasons for adding forms. As the next exercise demonstrates, a form can be added to show special effects or to provide information that might help the user work with a `UserDocument`.

Exercise 28.4 Calling a Form from a *UserDocument*

Create a `UserDocument` and add a single form to it, adding command buttons to both the `UserDocument` (to call the form) and the form (to execute a form-designed event procedure). Figure 28.10 shows the Internet Explorer interface you are asked to create.

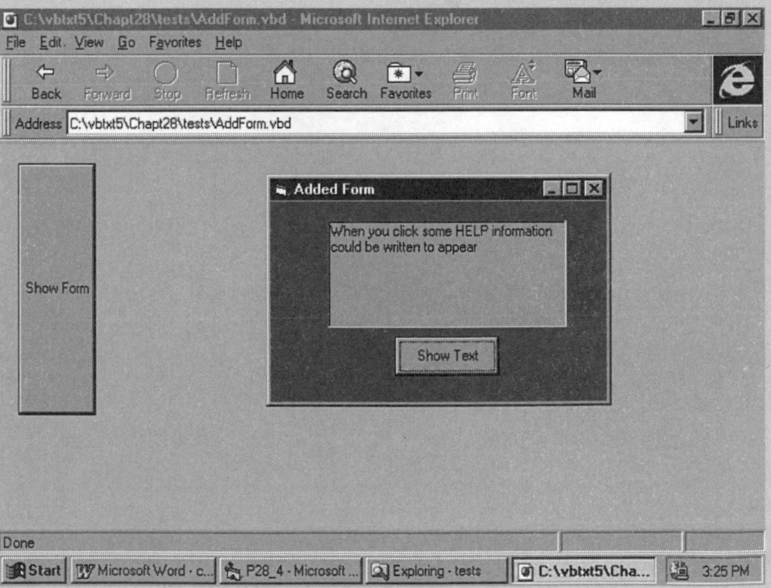

Figure 28.10

This application opens with only the `UserDocument` visible.

1 Open a new ActiveX document .EXE project and open the UserDocument1 interface.

2 Change the name of the project to **p28_4**. Change the name of the `UserDocument` to **AddForm**.

3 Choose Project, Add Form, and select Form. This will add a new form to the project.

4 Change the name of the form to **frmAdd**.

5 On the form, place a label (`lblText`), set `WordWrap` to `True`, and remove the label caption.

6 Add a command button to the form (**`cmdShow`**) and change the caption to **Show Text**.

7 Write the following `cmdShow_Click()` procedure:

```
Private Sub cmdShow_Click()
  lblText.Caption = "When you click some HELP information could be written
  ➥to appear"
End Sub
```

8 Return to the `UserDocument` (`AddForm`), and add a command button to this form (**`cmdShowForm`**) and change the button caption to **Show Form**.

9 Write the following `cmdShowForm_Click()` procedure:

```
Private Sub cmdShowForm_Click()
  frmAdd.Show vbModal

End Sub
```

10 Save the `Form` as **frmAdd.frm**, and save the `UserDocument` as **AddForm.dob**.

11 Make the P28_4 executable (again do not run the program).

12 Run Internet Explorer and open the file AddFile.vdb. Click Show Form to open the form. With the form open, click Show Text. As a further test, close the form and click Show Form again.

With Internet Explorer running, choose File, Close if you plan to quit at this time. Otherwise, continue with the next section.

Objective 28.5 Using the ActiveX Document Migration Wizard

There will be times when you want to convert a Visual Basic application you have written to an ActiveX `UserDocument`. Rather than starting from scratch, it would be most helpful if there were a tool for performing the conversion. Fortunately, there is such a tool, known as the ActiveX Document Migration Wizard. This wizard is available as an add-in tool. The ActiveX Document Migration Wizard performs the following tasks:

- Copies form properties to a new `UserDocument`.
- Copies all controls from the form to the `UserDocument`, and retains their names.

- Copies all code written for the form to the `UserDocument`.
- Copies all event handlers into ActiveX document objects, as the code is being copied. For example, it replaces the word `Form` with `UserDocument`. (`Form_Click()` becomes `UserDocument_Click()`.)
- Comments out illegal code, such as `Me.Left` and `End`.
- Changes the project type to an ActiveX EXE or DLL.

Exercise 28.5 Using the ActiveX Document Migration Wizard

Refer to Chapter 16, "Lists and Arrays," and convert Exercise 16.8, "Creating an Application with Control and Variable Arrays," to an ActiveX `UserDocument` application. Figure 28.11 shows the converted application running within Microsoft Explorer.

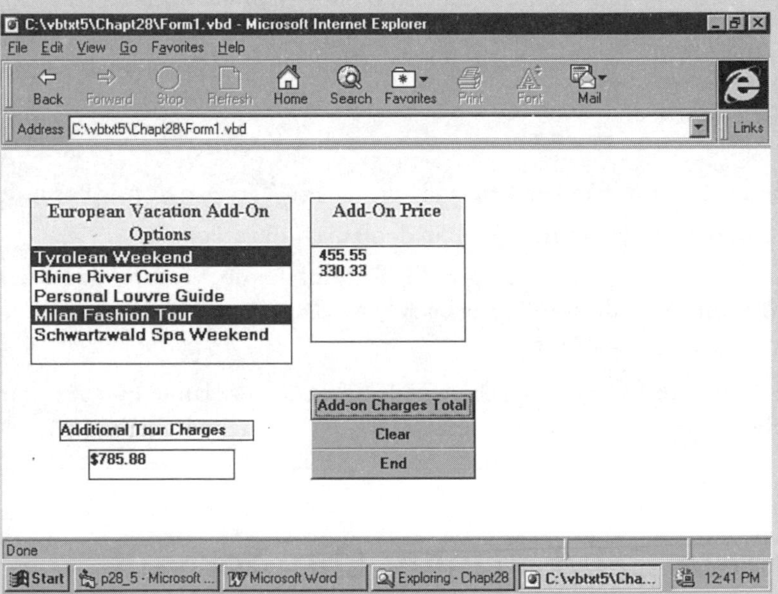

Figure 28.11

The form used in designing this application has been converted into a `UserDocument`.

1. Copy Exercise 16.8 and copy and paste files F16_8.FRM, P16_8.VBP, P16_8.VBW in a new folder.
2. Open the project file and change the name of the form to **Vacation**. Change the project name to **p28_5**.
3. Save the project as **P28_5**.

4 With the project still open, select Add-Ins, Add-In Manager.

5 Click the VB ActiveX Document Migration Wizard check box, and click OK. This places the ActiveX Document Migration Wizard on the Add-Ins menu (see Figure 28.12).

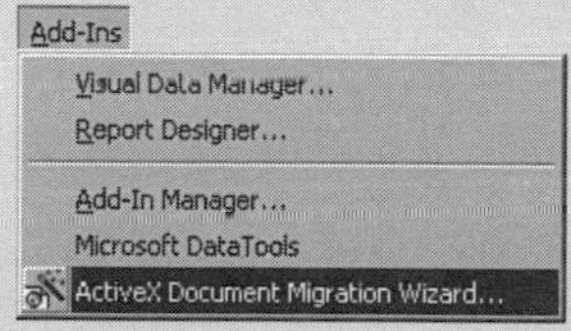

Figure 28.12

Select the ActiveX Document Migration Wizard to start the conversion of a Visual Basic application to an ActiveX document application.

6 Select the Document Migration Wizard, and perform the following steps:

   1 Click Next.
   2 Click Vacation to insert a check mark when asked to select a form, and click Next.
   3 Click Comment out invalid code?
   4 Click Remove original forms after conversion?
   5 Click Convert to ActiveX EXE?
   6 Click Next.
   7 Click Yes to review the summary report and then click Finish. During the conversion process, you will receive a message stating that the wizard has found some invalid code. Because not all Visual Basic code works with Internet Explorer, the wizard will comment out invalid code.

7 Examine the converted application. A `UserDocument` with the name Vacation should have replaced the original form.

8 Save the project.

9 Make a P28_5 executable file. If you are asked to select a start-up file, select (none).

10 Run Internet Explorer and open the file Vacation.vbd.

11 Test your application. While in Internet Explorer, click the End button. Observe what happens. Return to the project and examine the revised code. What changes have been made?

Exercise 28.6 Retrieving Data from a Database Using an ActiveX Document

For this final, but somewhat complex, exercise, refer to Chapter 24, "Accessing and Manipulating Databases with Visual Basic," and copy Exercise 24.4, "Using a Query to Retrieve a Set of Records," into a new folder and convert this Visual Basic database application into an Internet ActiveX application. Figure 28.13 shows the converted application running within Microsoft Explorer.

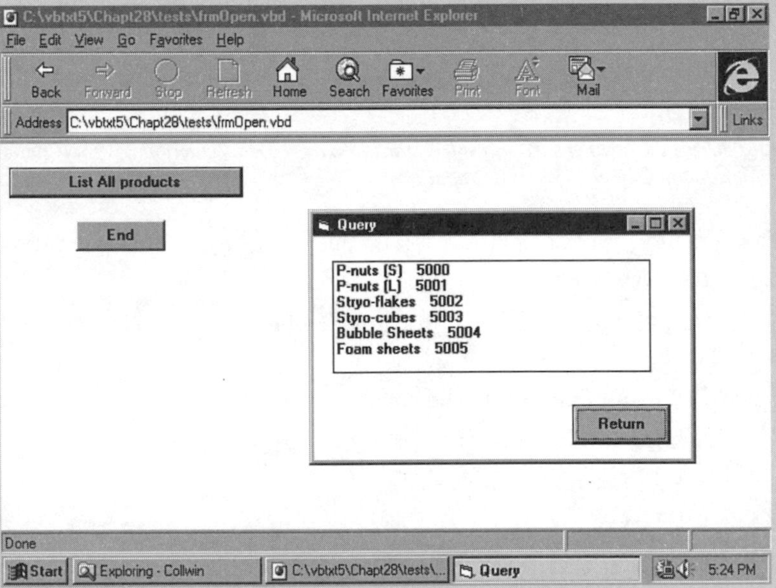

Figure 28.13

This interface contains a `UserDocument`*, which in turn contains the command button List All Products that opens a form named Query with the query results.*

1 Copy the files to a new folder. These include the two form files (F24-4a.frm and F24-4b.frm), the project file (p24-4.vbp), and the database file (Design4.mdb).

2 Rename the form files **F28-6a.frm** and **F28_6b.frm**, respectively.

3 Run the project to determine if it works as before. Review Figures 24.7 and 24.8 if you need to review the original interface and the results.

4 Rename the project, and save it as **p28_6.vbp**.

5 With the project still open, select the ActiveX Document Migration Wizard from the Add-Ins menu.

6 With the wizard open, perform the following steps:

1 Click Next.
2 Click `frmOpen` to insert a check mark when asked to select a form, but to not insert a check mark for `frmQuery`. Click Next.
3 Click Comment out invalid code?
4 Click Remove original forms after conversion?
5 Click Convert to ActiveX EXE?
6 Click Next.
7 Click Yes to review the summary report and then click Finish.

7 Examine the converted application. A `UserDocument` with the name `frmOpen(frmOpen.dob)` and a form with the name `frmQuery(F28.6.frm)` should exist.

8 For `frmQuery`, comment out the line

```
'frmOpen.Show
```

9 Click the Data control and check the path for Design4.mdb. If needed, change this path.

10 Place the form inside the `UserDocument`.

11 Save the project.

12 Make p28_6 an executable file.

13 Run Internet Explorer, and open the file frmOpen.vdb.

Test your design. Does it work? If it does, you have just simulated a client/server environment in which the database resides on the server and the query instructions reside on the client.

14 Close Internet Explorer.

15 As a final test, run RegEdit.EXE again, and check the Registry on your computer. The programs for all six programs should now be registered on your machine.

Chapter Summary

ActiveX documents consist of a user document interface that holds controls, including forms and code. This interface can be read by an ActiveX container, such as Internet Explorer. ActiveX documents have a .VBD extension. This file, with its extension, is created for you when you create an ActiveX executable; it is the .VBD file that is opened by Internet Explorer.

In this chapter, we strove to simulate a client/server Internet architecture, in the sense that a client application running on the desktop can receive services over the Internet from a server application, which involved access to a stored Access database.

ActiveX documents contain no HTML whatsoever and are therefore essentially unlike HTML-based Internet documents. The base object of an ActiveX document is the `UserDocument` object—an object supported by properties, methods, and events. An event used for a Form object, but not for a `UserDocument` is `Load()`. To load a user document, use the `Initialize()` event.

Before Internet Explorer can gain access to an ActiveX `UserDocument`, the `UserDocument` must be registered on the client machine. Run Regedit.exe to determine if a `UserDocument` has been registered.

Building an ActiveX document requires several new steps, including selecting the ActiveX Document Exe when asked to begin a new project. Following this, a `UserDocument` object is displayed, rather than a form. Controls and forms can be added to this new type of object. When the `UserDocument` is saved, it requires a .DOB extension. A `UserDocument` is not run. Rather is it compiled into an executable, which leads to an .EXE and .VBD files. Because of these requirements, many Visual Basic developers build an application outside of ActiveX and then use the ActiveX Document Migration Wizard to transform forms into ActiveX `UserDocuments`.

It is possible to navigate between ActiveX documents using hyperlinks. Each `UserDocument` has a `Hyperlink` property that returns `Hyperlink` objects. Three navigational methods are `GoBack`, `GoFoward`, and `NavigateTo`. The syntax for `NavigateTo` is the most complex and the most powerful:

```
object.Hyperlink.NavigateTo Target [, Location [, FrameName]]
```

With `NavigateTo`, it is possible to link to a second `UserDocument`, as in the following example:

```
UserDocument.Hyperlink.NavigateTo "c:\mydocs\axdmydoc.vbd"
```

Jumping to a Web page is also possible, given the URL. For example:

```
Hyperlink.NavigateTo txtURL.Text
```

Adding a form to a `UserDocument` is straightforward: Choose Project, Add Form. To show the form, use the `Show` method and make the form modal, as in the following example:

```
FrmAdd.Show vbModal
```

The ActiveX Document Migration Wizard is a powerful tool. It allows you to convert a Visual Basic application into an ActiveX application. In conversion, it copies all form properties, controls and their names, code, and event handlers. It even comments out code that is illegal and replaces `Form` with `UserDocument`, when appropriate.

With the ActiveX Migration Wizard, selected forms can be transformed into `UserDocuments`, leaving others as forms. You used this technique in Exercise 28.6, in which the `UserDocument` served to execute the query and the form served to display the query results.

Skill-Building Exercises

1. Modify Exercise 28.3 to navigate to a different Web page, such as the Web page for your college or university, or to a friend's personal Web page. To run this exercise, you will need the services of a Web server. In your design, add error-trapping code. Do not add quotation marks if you type in the Web address.
2. Modify Exercise 28.3 by adding a combo box that will allow you to select one URL from several. Use the back arrow to demonstrate that your design is capable of linking to more than one Web site.
3. Convert the third exercise in the "Skill-Building Exercises" section at the end of Chapter 24 to run on the Internet. Use the Cinema2.mdb database. Write a program that lists the year of the film's release and its title from the Film Titles table, and place the results in a list box.
4. Modify Exercise 28.6 to allow it to contain two possible query selections: one to list all products and one to list all customers. Use Design4.mdb once again. Place the results of the all-products query on a form named **All Products**; place the results of the all-customers query on a form named **All Customers**.
5. Refer to Chapter 22, "Modular Design," and convert Exercise 22.7, "Building Classes to Enhance the Ski Rental Shop Application," to run on the Internet. Place the order form on a `UserDocument` (see Figure 22.9); leave the invoice form as a form (see Figure 22.10).

Appendices

ANSI Code	*Character*
38	&
39	'
40	(
41	)
42	*
43	+
44	,
45	-
46	.
47	/
48	0
49	1
50	2
51	3
52	4
53	5
54	6
55	7
56	8
57	9
58	:
59	;
60	<
61	=
62	>
63	?
64	@
65	A
66	B
67	C
68	D
69	E
70	F

ANSI Code	*Character*
71	G
72	H
73	I
74	J
75	K
76	L
77	M
78	N
79	O
80	P
81	Q
82	R
83	S
84	T
85	U
86	V
87	W
88	X
89	Y
90	Z
91	[
92	\
93	]
94	^
95	_
96	`
97	a
98	b
99	c
100	d
101	e
102	f
103	g

The ANSI Character Set

ANSI values 8, 9, 10, and 13 have no graphical representation, but—depending upon the application—can affect the visual display of text.

The abbreviation *n/a* indicates characters that are not supported by Microsoft Windows.

ANSI Code	*Character*	*ANSI Code*	*Character*
0	*n/a*	19	*n/a*
1	*n/a*	20	*n/a*
2	*n/a*	21	*n/a*
3	*n/a*	22	*n/a*
4	*n/a*	23	*n/a*
5	*n/a*	24	*n/a*
6	*n/a*	25	*n/a*
7	*n/a*	26	*n/a*
8	[backspace]	27	*n/a*
9	[tab]	28	*n/a*
10	[linefeed]	29	*n/a*
11	*n/a*	30	*n/a*
12	*n/a*	31	*n/a*
13	[carriage return]	32	[space]
14	*n/a*	33	!
15	*n/a*	34	"
16	*n/a*	35	#
17	*n/a*	36	$
18	*n/a*	37	%

ANSI Code	*Character*
104	h
105	i
106	j
107	k
108	l
109	m
110	n
111	o
112	p
113	q
114	r
115	s
116	t
117	u
118	v
119	w
120	x
121	y
122	z
123	{
124	|
125	}
126	~
127	*n/a*
128	*n/a*
129	*n/a*
130	*n/a*
131	*n/a*
132	*n/a*
133	*n/a*
134	*n/a*
135	*n/a*
136	*n/a*

ANSI Code	*Character*
137	*n/a*
138	*n/a*
139	*n/a*
140	*n/a*
141	*n/a*
142	*n/a*
143	*n/a*
144	*n/a*
145	‘
146	’
147	*n/a*
148	*n/a*
149	*n/a*
150	*n/a*
151	*n/a*
152	*n/a*
153	*n/a*
154	*n/a*
155	*n/a*
156	*n/a*
157	*n/a*
158	*n/a*
159	*n/a*
160	[space]
161	¡
162	¢
163	£
164	¤
165	¥
166	|
167	§
168	¨
169	©

ANSI Code	*Character*
170	ª
171	«
172	¬
173	—
174	®
175	-
176	°
177	±
178	2
179	3
180	′
181	µ
182	¶
183	*n/a*
184	,
185	1
186	°
187	»
188	_
189	_
190	_
191	¿
192	À
193	Á
194	Â
195	Ã
196	Ä
197	Å
198	Æ
199	Ç
200	È
201	É
202	Ê

ANSI Code	*Character*
203	Ë
204	Ì
205	Í
206	Î
207	Ï
208	_
209	Ñ
210	Ò
211	Ó
212	Ô
213	Õ
214	Ö
215	x
216	Ø
217	Ù
218	Ú
219	Û
220	Ü
221	Y
222	_
223	ß
224	à
225	á
226	â
227	ã
228	ä
229	å
230	æ
231	ç
232	è
233	é
234	ê
235	ë

ANSI Code	*Character*
236	ì
237	í
238	î
239	ï
240	_
241	ñ
242	ò
243	ó
244	ô
245	õ
246	ö
247	÷
248	ø
249	ù
250	ú
251	û
252	ü
253	y
254	_
255	ÿ

Data Access Object Collections, Properties, and Methods

DBEngine Object

Collections	*Properties*	*Methods*
Errors	DefaultPassword	BeginTrans
Workspaces	DefaultType	CommitTrans
Properties	DefaultUser	CompactDatabase
	IniPath	CreateDatabase
	LoginTimeout	CreateWorkspace
	SystemDB	Idle
	Version	OpenConnection
		RegisterDatabase
		RepairDatabase
		Rollback
		SetOption

Workspace Object

Collections	*Properties*	*Methods*
Connections	DefaultCursorDriver	BeginTrans
Databases	IsolateODBCTrans	Close
Groups	LoginTimeout	CommitTrans
Properties	Name	CreateDatabase
Users	Type	CreateGroup
	UserName	CreateUser
		OpenConnection
		OpenDatabase
		Rollback

Database Object

Collections	*Properties*	*Methods*
Containers	CollatingOrder	Close
Properties	Connect	CreateProperty
QueryDefs	Connection	CreateQueryDef
Recordsets	DesignMasterID	CreateRelation
Relations	Name	CreateTableDef
TableDefs	QueryTimeout	Execute
	RecordsAffected	MakeReplica
	Replicable	NewPassword
	ReplicaID	OpenRecordset
	Updatable	PopulatePartial
	Version	Synchronize
	V1xNullBehavior	

TableDef Object

Collections	*Properties*	*Methods*
Fields (default)	Attributes	OpenRecordset
Indexes	ConflictTable	RefreshLink
Properties	Connect	CreateField
	DateCreated	CreateIndex
	KeepLocal	CreateProperty
	LastUpdated	
	Name	
	RecordCount	
	Replicable	
	ReplicaFilter	
	SourceTableName	
	Updatable	
	ValidationRule	
	ValidationText	

Recordset Object

Collections	*Properties*	*Methods*
Fields	AbsolutePosition	AddNew
Properties	BatchCollisionCount	Clone
	BatchCollisions	Close
	BatchSize	CopyQueryDef
	BOF	CopyQueryDef
	Bookmark	Delete
	Bookmarkable	Edit
	CacheSize	FillCache
	CacheStart	FindFirst
	Connection	FindLast
	DateCreated	FindNext
	EditMode	FindPrevious

continues

Collections	Properties	Methods
	EOF	GetRows
	Filter	Move
	LastModified	MoveFirst
	Index	MoveLast
	LockEdits	MoveNext
	Name	MovePrevious
	NoMatch	OpenRecordset
	PercentPosition	Requery
	RecordCount	Seek
	RecordStatus	Update
	Restartable	
	Sort	
	StillExecuting	
	Transactions	
	Type	
	Updatable	
	UpdateOptions	
	ValidationRule	
	ValidationText	

Field Object

Collections	Properties	Methods
Properties	AllowZeroLength	AppendChunk
	Attributes	CreateProperty
	CollatingOrder	GetChunk
	DataUpdatable	
	DefaultValue	
	FieldSize	
	ForeignName	
	Name	
	OrdinalPosition	

Collections	Properties	Methods
	OrdinalValue	
	Required	
	Size	
	SourceField	
	SourceTable	
	Type	
	ValidateOnSet	
	ValidationRule	
	ValidationText	
	Value	
	Visible	

Index Object

Collections	Properties	Methods
Fields (default)	Clustered	CreateField
Properties	DistinctCount	CreateProperty
	Foreign	
	IgnoreNulls	
	Name	
	Primary	
	Required	
	Unique	

QueryDef Object

Collections	*Properties*	*Methods*
Fields	CacheSize	Cancel
Parameters	Connect	CreateProperty
Properties	DateCreated	Execute
	KeepLocal	OpenRecordset
	LastUpdated	
	LogMessages	
	MaxRecords	
	Name	
	ODBCTimeout	
	Prepare	
	RecordsAffected	
	Replicable	
	ReturnsRecords	
	SQL	
	StillExecuting	
	Type	
	Updatable	

Index

SYMBOLS

B

C

D

G

J

K

N

O

T

X

Y

Z